DFG

Chemistry and Physics of Macromolecules

The 'Deutsche Forschungsgemeinschaft' and its 'Sonderforschungsbereiche'

The Deutsche Forschungsgemeinschaft (DFG) serves science and the humanities in all fields by financing research projects. In many scientific areas, the Deutsche Forschungsgemeinschaft has taken over the task of strengthening cooperation among researchers and coordinating basic research with state research promotion. The Deutsche Forschungsgemeinschaft advises parliaments and public authorities on scientific matters, cultivates relations between science and industry and promotes cooperation between German scientists and foreign researchers. Special attention is paid to promoting young scientists.

The DFG distinguishes between funding individual projects (normal procedure – Normalverfahren) and cooperative activities (priority programmes – Schwerpunktprogramme, research groups – Forschergruppen, special collaborative programmes – Sonderforschungsbereiche, central research facilities – Hilfseinrichtungen der Forschung).

In special collaborative programmes groups of scientists from various disciplines combine their efforts in order to pursue a large scale investigation of a major problem. They aim at strengthening research at universities by concentrating personnel and material, coordinating research within and between universities as well as between universities and research institutions outside and by providing a long-term financial basis.

In 1988 the Deutsche Forschungsgemeinschaft funded 156 Sonderforschungsbereiche with more than 335 million marks. This amounts to almost a third of the DFG's total budget of DM 1.122 billion in 1988. 75% of the funds are provided by the Federal Government via the Ministry of Education and Science (Bundesminister für Bildung und Wissenschaft), 25% come from the eleven states (Bundesländer).

Distribution:
VCH, P.O. Box 10 11 61, D-6940 Weinheim (Federal Republic of Germany)
Switzerland: VCH, P.O. Box, CH-4020 Basel (Switzerland)
United Kingdom and Ireland: VCH (UK) Ltd., 8 Wellington Court, Cambridge CB1 1HZ (England)
USA and Canada: VCH, Suite 909, 220 East 23rd Street, New York, NY 10010–4606 (USA)

ISBN 3-527-27715-3 (VCH, Weinheim)
ISBN 0-89573-982-8 (VCH, New York)
ISSN 0930-4398

DFG Deutsche Forschungsgemeinschaft

Chemistry and Physics of Macromolecules

Final Report of the Sonderforschungsbereich
"Chemie und Physik der Makromoleküle"
1969–1987

Edited by
Erhard W. Fischer, Rolf C. Schulz,
and Hans Sillescu

Sonderforschungsbereiche

Deutsche Forschungsgemeinschaft
Kennedyallee 40
D-5300 Bonn 2
Telefon: (0228) 885-1
Telefax: (0228) 885-2221

Published jointly by
VCH Verlagsgesellschaft mbH, Weinheim (Federal Republic of Germany)
VCH Publishers Inc., New York, NY (USA)
Library of Congress Card-No. 90-38615

British Library Cataloguing-in-Publication Data:

Chemistry and physics of macromolecules: final report of the Sonderforschungsbereich
"Chemie und Physik der Makromoleküle" 1969–1987.
1. Polymers
I. Fischer, Erhard W. II. Schulz, Rolf C. III. Sillescu, Hans
IV. Deutsche Forschungsgemeinschaft
547.7
ISBN 0-89573-982-8
ISBN 3-527-27715-3

Deutsche Bibliothek Cataloguing-in-Publication Data:

Chemistry and physics of macromolecules : final report of the Sonderforschungsbereich
"Chemie und Physik der Makromoleküle" 1969 – 1987 / DFG,
Deutsche Forschungsgemeinschaft. Ed. by Erhard W. Fischer ... –
Weinheim ; Basel (Switzerland) ; Cambridge ; New York, NY : VCH, 1991
(Sonderforschungsbereiche)
ISBN 3-527-27715-3 (Weinheim ...)
ISBN 0-89573-982-8 (New York)
NE: Fischer, Erhard W. [Hrsg.]; Sonderforschungsbereich Chemie und Physik
der Makromoleküle <Mainz; Darmstadt>

Production Manager: L & J Publikations-Service GmbH, D-6940 Weinheim
Composition: Hagedornsatz GmbH, D-6806 Viernheim
Printing and Bookbinding: Wilhelm Röck, Druckerei und Großbuchbinderei,
D-7012 Weinsberg
Printed in the Federal Republic of Germany

Preface

Two decades have passed since the Sonderforschungsbereich 41 "Chemie und Physik der Makromoleküle" special collaborative programme on chemistry and physics of macromolecules was founded in 1969. During this period, Polymer Science has multiplied in size, and many new branches have come into existence. Support by the Deutsche Forschungsgemeinschaft (DFG) – 54 million DM in total from 1969 to 1987 – has helped the research groups in the Sonderforschungsbereich 41 to participate in the world wide endeavour and to initiate new developments. In a way, the present book documents the achievements of the Sonderforschungsbereich 41 and is a sign of gratefulness for our sponsor, the Deutsche Forschungsgemeinschaft. However, it is not intended as a museum with all obtained results meticulously exposed. Our goal was rather a review written by authors who have gained knowledge and insight from their own research in the field.

The selection of topics, while influenced by personal preferences of the authors, reflects the coherence of a research programme characterized by intense internal cooperation within the Sonderforschungsbereich 41. The title "Chemie und Physik der Makromoleküle" signals the joint efforts of chemists and physicists exploring polymer properties on a molecular level. Though important fields are not covered (e.g., networks, biopolymers) the selection of themes presented should be of interest to many polymer scientists. It ranges from preparative polymer chemistry and physical chemistry to experimental and theoretical polymer physics. Due to its documentary character, the book concentrates on research done in the Sonderforschungsbereich 41. However, the authors have always tried to cover related work and provide guidelines to related literature. Some of the authors report on research started in the Sonderforschungsbereich 41, but continued and developed further in their present institution. In the General Introduction, we give some reference to the relation of research projects in the Sonderforschungsbereich 41 with particular chapters of the book.

This is also an opportunity to express our sincere thanks, not only to the

Deutsche Forschungsgemeinschaft in general for the financial support of the Sonderforschungsbereich 41 but, particularly, to the staff in the DFG office for all the help and advice we received over the years. We always could be sure that we are not dealing with some bureaucracy but with human beings who share our problems in reconciling financial regulations with unexpected developments in the progress of research projects.

We also thank the many referees who spent much time on our research proposals and often provided good advice along with positive criticism.

Finally, we thank for the guidelines and detailed help provided by the DFG during the preparation of this book.

Contents

1 General Introduction

1.1 Polymer Science in the Federal Republic of Germany

The Sonderforschungsbereich 41 "Chemie und Physik der Makromoleküle" was concerned with basic research about polymeric materials, which play a very important role in chemical industry. The beginning of Polymer Chemistry dates back to the decade 1920–1930 as H. Staudinger at the Universität Freiburg recognized the nature of macromolecules and developed already the theoretical base for understanding the synthesis and properties of this completely new kind of materials.

The main activities in this field were located in Germany until the beginning of World War II, but after this war the leading role was taken over by the USA. There were several attempts to improve the situation and to support basic research in Polymer Science more extensively. The DFG accomplished a "Schwerpunktprogramm" (priority programme) in this field during 1953–1958 and in 1953 the "Forschungsgesellschaft Kunststoffe e.V." was founded and built up the Deutsche Kunststoff-Institut in Darmstadt. The DFG published two memoirs about the situation: in 1960 the memorandum on "Forschung auf dem Gebiet der Kunststoffe und Synthesefasern" and in 1971 on "Physik der Polymeren". In both papers an increase of activities was strongly recommended including the foundation of new institutions concerned with Polymer Science. With regard to this point the suggestions were not realized, but in 1969 an important step was taken by the foundation of the Sonderforschungsbereich 41. The development of this organization is described in the following paragraph.

The newest development in the field of Polymer Science in the Federal Republic of Germany was initiated by the "Wissenschaftsrat", which issued in 1980 the "Empfehlungen zur Förderung der Polymerforschung in der Bundesrepublik Deutschland". On the basis of this memorandum the Max-Planck-Gesellschaft decided in 1982 the foundation of the "Max-Planck-Institut für Polymerforschung" in Mainz. One member and one former project leader of the Sonderforschungsbereich 41 were appointed as directors of the new institution: Prof. E.W. Fischer, Mainz, and Prof. G. Wegner, Freiburg. There is no doubt that the success and the reputation of the Sonderforschungsbereich 41 was one of the main reasons for the election of Mainz as location for the new institute.

1.2 Objectives and Developments of the Sonderforschungsbereich 41

In 1946 W. Kern and G.V. Schulz were appointed to professorships at the Universität Mainz, reopened after World War II. Thus, in these times, Mainz, as well as Freiburg, were the only universities in the Federal Republic of Germany in which the research and teaching of polymer science was performed. From these initial activities, several groups formed to work in different research fields of polymer chemistry at the Universität Mainz. These initial efforts in chemistry and physical chemistry of macromolecules were supplemented by H. A. Stuart, working in the field of polymer physics. In 1959, this new center became internationally well known because of a IUPAC Symposium on Macromolecules that took place in Mainz and Wiesbaden.

On the basis of these polymer research activities in Mainz, the DFG-Research Centre "Sonderforschungsbereich 41" was founded in 1969 with the topic "Chemistry, Physics and Biological Functions of Macromolecules". This action was taken in order to combine groups, which were working at the Universität Mainz, the Technische Hochschule Darmstadt and the Deutsches Kunststoffinstitut Darmstadt into one combined research programme, and to coordinate the different projects. By 1970 already 22 project managers directing 45 projects had joined the Sonderforschungsbereich 41. The essential tasks of the Sonderforschungsbereich 41 were:

- to cooperate between the fields of preparative chemistry, physical chemistry and the physics of polymers,
- to plan and to carry out long range research programmes, taking into consideration development and promotion of contemporary polymer topics.

According to these requirements the initially broad range of topics was already narrowed in the first years. In addition, the focal points changed gradually. Moreover, emeritations and further professorships generated personnel and, consequently, thematic changes, as the years went by. Thus, in 1972 for example, the projects dealing with biological functions were deleted and the remaining projects were combined within the following research fields:

- synthesis and degradation of oligomers and polymers,
- kinetic and elementary steps of polymerization reactions,
- solubility characteristics of polymers,
- structure and characteristics of solid, high molecular weight polymers.

Since 1979 the former thematic arrangement of the different project fields, was concentrated further to two final focal points:

- polymers with particular structural features and
- structure, dynamics and molecular interactions in polymeric systems

which subsequently resulted in the involvement of 27 project managers directing approximately 32 projects. The main fields are summarized in Chapter 1.3.

Contacts to foreign colleagues as well as cooperations with other institutes were particular goals requested by the Sonderforschungsbereich. From 1971 to 1985, 110 scientists coming from the USA, Japan, Latin America and several European countries including the Eastern countries stayed in Mainz as visiting scientists. They participated in projects of the Sonderforschungsbereich for periods of three months to two years.

On the other hand, many project managers have received invitations from famous foreign institutes for stays as visiting scientists. At every IUPAC Congress and Symposium, as well as Gordon Conferences and international meetings in the polymer field, members of the Sonderforschungsbereich 41 have given presentations. Between 1969 and 1984 approximately 1000 papers, reviews and book contributions have appeared, as can be evidenced from the regularly published reports of the Sonderforschungsbereich.

A further, very important result is that several young scientists have advanced to professor positions in Polymer Science as a result of their scientific work in the Sonderforschungsbereich 41. The high scientific reputation fostered by the Sonderforschungsbereiche is easily demonstrated by the fact that, between 1974 and 1981, nine project managers have accepted professorships in both German as well as foreign universities.

At the end of the Sonderforschungsbereich 41 in 1987 Mainz had developed into a world wide known leading Center of Polymer Science. At the Chemistry Department approximately 50 % of the dissertations completed per year deal with topics from the field of Macromolecular Chemistry and Physics. The various fields of Polymer Science are represented in the curriculum and according to this a wide range of lectures and seminars are continually offered. The new Max-Planck-Institut für Polymerforschung, officially opened in 1989, is jointly working with the different polymer oriented groups at the university.

1.3 Topics and Results

In this section we describe in some more detail how the interplay of chemical and physical methods has worked in the Sonderforschungsbereich 41. We have selected a number of topics from the broad research programme of the Sonderforschungsbereich which have led to important results. The origin and development of the projects is outlined and related to the following chapters of this book. We also refer to the complete project listing in the Appendix (p. 537 ff.).

1.3.1 Pathways to New Polymers

The chemically oriented research groups were primarily concerned with the synthesis of new polymers, macromolecules which either have hitherto unknown chemical composition or which have very regular structure (e.g., ordered arrangement of building blocks or monodispersity of chain length). The purpose of this work was to obtain

- polymers with new chemical and physical properties,
- well defined polymer models for clarifying general structure property relationships,
- guidelines for using synthetic polymers in simulating specific reactivities and functions of biological macromolecules.

Methods generally used for the characterization of small molecules are usually insufficient in investigations of properties resulting from the composition and structure of macromolecules. For this reason, new methods were developed in the Sonderforschungsbereich 41 in order to improve the ability to characterize polymers and to study their structures.

There are many different pathways to new polymers. In order to illustrate the strategies applied in the Sonderforschungsbereich 41, we have chosen some examples. We start with new polymers from new building blocks.

Liquid crystalline polymers have been a field of particularly active research during the past years. Though liquid crystalline phases have been known for a long time in systems of small rod-like molecules it was considered uncertain whether this property could also apply to macromolecules. Also, the effect of the polymeric backbone was unknown. A deciding aspect was the idea that the mesogenic side group had to be positioned at a certain minimum distance from the main chain in order to obtain nematic or smectic phases. With the introduction of "spacers" between the main chain and the mesogenic side groups, the first liquid crystal side chain polymers, which were able to combine the typical properties of mesophases with those of polymeric systems, had become available [1]. Cooperation with physically oriented groups of the Sonderforschungsbereich provided optimal conditions for unfolding the unique properties of this new class of polymeric materials. The combined efforts of polymer chemists, physicists, and physical chemists, well documented in joint projects of the Sonderforschungsbereich (see project list in the Appendix), have led to a rapidly growing field of interdisciplinary polymer research with a wide range of potential applications, e. g. in opto-electronics, or for LC-displays. In Chapter 9 a review of these activities is given (see also applications of ^{2}H-NMR in Chapter 12.3.2.1 and of the Kerr effect in Chapter 13.4).

Polymerization in ordered systems: The idea of transferring molecular order from monomer to polymer systems has played an important role in different projects of the Sonderforschungsbereich 41. A famous example is the solid state polymerization of diacetylenes which led to large polymer single crystals [2]. The monomers are ordered in the lattice in a particular way where the conjugated triple bonds form the steps of a ladder with distances determined by the size of the side groups. The topochemical reaction occurs through rotation of the diacetylene molecules thus forming a polymer chain with alternating triple and double bonds. This polymerization reaction results in almost perfect polymer single crystals which have stimulated much interest among solid state physicists.

Reactions in oriented systems play a decisive role in living organism, e. g., in the synthesis of nucleic acids and proteins or in bio-membranes. For a long time it was an open question whether polymer chemists could help in the efforts to elucidate the function of cell membranes. The activities in Mainz, relating function and organization in monomer and polymer systems finally led to results with far-reaching consequences. Since the rigid packing of monomers in a crystal lattice limited the potential of topochemical solid state polymerization it appeared attractive to look for more mobile ordered systems such as liquid crystals, micelles, and particularly mono-, bi- and multilayer systems. Polymerization in these mobile structures opened up a wide variety of applications. Reactions in monolayers at gas-liquid interfaces led to stable polymer films of perfect orientation. From mixtures of natural and polymerizable liquids models for reactions in biomembranes and technical membranes were obtained.

Liposomes, made from polymerizable liquids, could be polymerized for the first time resulting in stable vesicles. Meanwhile it has become possible to build polymer vesicles from mixed liposomes containing active enzymes which can interact with proteins. The possibility of designing ordered polymer membranes with high stability and a wide variability of well defined functions exhibits an enormous potential of applications in chemical technology, life sciences, and medicine. We have omitted in this book a particular chapter on polymer membrane systems since an extensive recent review is available [3].

Ring-opening polymerization: Ordered monomer systems are no necessary condition for obtaining ordered polymers. The first large single crystals of polyoxymethylene were grown from 1, 3, 5-trioxane solutions by cationic polymerization [4]. Here, crystal growth occurs by insertion of trioxane units in the chain folds at the front surface thus forming extended chain crystals of polyoxymethylene. In a way, these polymer single crystals were among the fruits of long and eventful endeavours in Mainz starting with ring-opening polymerization of trioxane [5] including the design of regular copolymers with a predetermined sequence of monomer units [6] and the formation of oxacycles described in Chapter 5 of this book where also the metathetical ring-opening polymerization of cycloolefins is reviewed among other pathways to macrocyclic compounds.

Polymer reactions and functional polymers: Methods for generating polymers from the respective monomers are limited by the fact that some monomers cannot be synthesized or because they cannot be polymerized. In such cases one starts with a polymer containing a certain preliminary stage and adds the desired side-groups onto the polymeric backbone by means of the appropriate reactions. This principle which was introduced by Staudinger as "polymer analog conversion" is extremely variable and is the basis of technologies such as the production of cellulose derivatives or polyvinyl alcohol. In the Sonderforschungsbereich 41, the synthesis of polymers with functional groups and subsequent modification was extensively used as a pathway to new polymers [7, 8], and to polymers serving as carriers in preparative chemistry (see below). We mention the synthesis of the first stable polyradicals [9], complexation of polymers with small molecules such as iodine or electron donors and acceptors [10], and the development of "polymer reagents" [11, 12] which can be compared with the well-known ion-exchangers since the reactions can be performed in column systems under mild conditions.

Polymer carriers in polypeptide synthesis: Whereas in earlier investigations in the Sonderforschungsbereich 41 solid state carriers were used for the synthesis of oligopeptides [13] the liquid-phase method [14, 15] brought many advantages and has led to the synthesis of a large number of biologically active peptides and model peptides, e. g., a breakthrough in the preparation of collagen models [16]. The liquid-phase method has also found international recognition in connection with conformational analysis of less soluble peptides. The investigations have opened up new perspectives in predicting the three-dimensional structure and physical properties of the polypeptides to be synthesized.

New polyelectrolytes: Because of the long experience with polyelectrolytes [17, 18] this subject was also introduced in the Sonderforschungsbereich from the beginning. Different polyanions, polycations and zwitterions have been studied. In the connection with studies about modification of polyamides

[19] the amide groups in polyamide 6 were reduced yielding poly(iminohexamethylene), a new main chain charged polycation [20]. More recently a polyampholyte blockcopolymer was obtained by living anionic polymerization [21]. Polymer tetraalkylammonium salts with well defined charge densities (aliphatic ionenes) provide ideal model systems for investigations of solid polyelectrolyte crystals which have started in the Sonderforschungsbereich 41 and are presently being studied in the Max-Planck-Institut für Polymerforschung [22].

1.3.2 Kinetics, Mechanisms, and Polymer Characterization

Kinetic studies play an important role on the pathway to new polymeric materials since they elucidate the reaction mechanisms by which the substances are formed. Thus, the anionic polymerization of styrene in polar solvents first appeared rather puzzling until the hypothesis of chain growth by three different active species could be proven through analysis of the kinetic data and the resulting molecular weight distributions [23]. The kinetics of anionic polymerization reactions has been investigated during the whole period of the Sonderforschungsbereich. The present state of the art is reviewed in Chapter 8 along with recent studies on the kinetics and mechanism of group transfer polymerization.

Polymers obtained by radical polymerization play an outstanding role in chemical industry. The challenge of analyzing the complexity of polymer formation has resulted in many successful kinetic studies in the Sonderforschungsbereich which are covered to some extent in Chapter 7 whereas kinetic studies of radical copolymerization is reviewed in Chapter 6.

Neither kinetic studies nor the synthesis of oligomers and polymers can be performed without the availability of appropriate techniques for polymer characterization. Research in the Sonderforschungsbereich 41 concentrated on the development of new materials for gel permeation chromatography [24], and techniques for optimizing flow properties and separation capacity of column systems [25]. In particular, the separation of oligomers (see also Chapter 4), identification, molecular weight determination of macrocycles, and the determination of molecular weight distributions were successfully treated in the Sonderforschungsbereich. Though the technical perfection of present day commercial instruments was achieved in industrial development much of the basic knowledge resulted from research in the Sonderforschungsbereich.

1.3.3 New Methods in Polymer Physics

One of the main objectives of the physicochemical and physical projects in the Sonderforschungsbereich 41 was understanding macroscopic polymer properties on a molecular scale. Here, many important results followed from the development of new physical methods or their modification for application to polymers. The potential of small angle X-ray and neutron scattering techniques proved to be particularly fruitful in structural studies of polymer systems. Thus, selective deuteration yielded the first experimental evidence that polymer chains have the same shape of a random coil in bulk polymer glasses as in dilute θ-solutions [26], and the principle of contrast variation by deuteration [27] has found wide application in the structure analysis of complex biopolymer systems. Further applications of elastic and quasielastic neutron scattering are described in Chapters 15–17.

Light scattering techniques, a classical domain of dilute polymer solution studies, have not only been refined and extended in range in the Sonderforschungsbereich, but also modified and applied to many new fields in polymer physics. Examples are the application of forced Rayleigh scattering to polymer diffusion in melts and networks (Chapter 11) and of quasielastic light scattering techniques in studies of collective polymer dynamics (Chapter 15).

A particular fruitful example of methodical developments in the Sonderforschungsbereich 41 is *solid state deuteron NMR*. The easy access to interesting deuterated polymer systems (also used in neutron scattering) by cooperation with polymer chemists was certainly helpful for early convincing demonstrations of the new method's potential which is now being utilized in many laboratories in fundamental and industrial research (see Chapter 12).

1.3.4 Conformation and Interaction of Macromolecules in Polymer Systems

The behaviour of macromolecules in solution was studied in experiments sensitive to static properties, e.g., phase diagrams in dependence of concentration, temperature, and pressure, but also in shear flow experiments where intermolecular interactions can result in chain fracture. In both cases, pressure variation has led to interesting new results which are described in Chapter 10. In amorphous bulk polymers, the early successful neutron scattering study of chain conformation in polymethylmethacrylate [26] was followed by investigations in many other amorphous polymers by neutron scattering and various other techniques [28].

Neutron scattering was particularly helpful in the analysis of partially crystalline polymers. Ever since the discovery of chain folding during the crystallization of polyethylene in dilute solution (1957), much discussion has centered on the problem of the structure of the surfaces of the lamellar crystallites and the amorphous regions located between them. Even today it is under debate whether tight chain folds are formed at the crystal surface during crystallization (adjacent reentry) or whether Flory's "random switchboard" model is more appropriate. With the aid of neutron scattering in concentrated mixtures of protonated and deuterated polyethylene and polyethylen oxide it was possible to demonstrate that crystallization from the molten state was much more adequately described by means of a "solidification model" according to which chain segments arrange themselves into the growing crystal front without extensive diffusion movement or larger alternations in the mean radius of gyration. Polyethylene crystals from dilute solution could also be shown to have an amorphous covering layer in accordance with the Flory-model. We should further mention the work on random copolymer crystallization which led to concepts in which a successive "insertion-crystallization" of lamina into the amorphous regions plays a significant role (see [29] and Chapter 16).

1.3.5 Macromolecules in Motion

One of the most obvious differences between polymeric systems and low molecular weight substances is their different molecular motion. The manner in which the rapid thermal movement of the chain segments is transferred to larger chain regions depends on the morphology of the surroundings. The partially crystallized polymers are of special interest in this respect. Here, chain segment motion in the amorphous regions is on the same time scale as in polymer melts, although the chains are spatially confined due to their fixation within the crystallites. In a certain temperature range below the melting point, the chains can also move within the crystallites. This motion was analyzed in the Sonderforschungsbereich 41 in comparison with chain motion in alkane crystals. Further advances became possible after the development of new ^{2}H-NMR methods which provide information on the type and timescale of rotational motions in a large dynamic range from microseconds to many seconds. By selective deuteration, one can observe separately, say, the main chain and side group motion in polymers, or identify the influence of low molecular weight additives on the mobility of special portions of the macromolecules. This is of particular importance in the attempts to understand the influence of plasticizers on such important mechanical properties as elasticity, hardness and toughness (for further examples see Chapter 12).

Since deuterated compounds are also required for neutron scattering, it was possible from the beginning to use different methods in order to study the same systems. Quasielastic neutron scattering only detects rapid segment movement, but has the advantage of also providing information on spatial correlations. The example of polydimethylsiloxane melts has demonstrated that chain motion within a range of up to about 5 nm can be accounted for by the Rouse model, and that accordingly, no evidence for chain entanglement is observable (Chapter 15). On the other hand, studies of translational diffusion of polymer chains in entangled as well as chemically cross-linked polystyrene melts were in perfect agreement with the Doi-Edwards tube model whereby polymer chains diffuse in a snake like fashion (reptation) along their own chain contour (Chapter 11). Both results are not in contradiction to one another since they refer to motion on different length scales smaller or larger, respectively, than the diameter of the Doi-Edwards tube.

1.3.6 Cooperative Phenomena in Polymers

In liquid crystalline polymers, the joint efforts of research groups applying different physical methods (X-ray- and neutron scattering, diverse optical techniques, dielectric relaxation, ^{2}H-NMR) have led to a detailed portrait of the cooperative ordering and dynamics in these complex systems (Chapters 9, 12, 13). A similar effort – which still persists in groups originating from the Sonderforschungsbereich 41 as well as among polymer physicists and chemists all over the world – concentrates on cooperative phenomena in polymer blends. One peculiarity of these systems is the small combinatorial contribution to free energy resulting in a strong enthalpic influence in favour of decomposition of the polymer mixtures. New theoretical developments on the dynamics of spinodal decomposition have stimulated collaboration of theoretical and experimental groups in the Sonderforschungsbereich 41 where theoretical predictions are tested by various experimental techniques (see Chapters 14–18). An important recent theoretical result is the unexpected large difference between numerical simulations and the Flory-Huggins mean field theory of intermolecular interactions. Even more puzzling are recent results from photon correlation spectroscopy indicating the existence of very slow collective motional modes in addition to the interdiffusion mode of concentration fluctuations in polymer blends. These examples demonstrate that polymer blends can be considered a field of active research in the foreseeable future.

An important cooperative phenomenon in polymers, which has been under vigorous debate for decades, is the process of glass formation from the melt. The analysis of data from X-ray scattering and quasielastic light scattering could be interpreted by assuming that density fluctuations in the molten state are determined by thermodynamic properties of the materials whereas additional fluctuations of order parameters are important for the glass state. Whereas X-ray, neutron, and light scattering are sensitive to cooperative motions, ^{2}H-NMR and dielectric relaxation can be applied to probe slow local motions of well defined molecular vectors at the glass transition. (Examples are discussed in Chapters 12 and 15.) At present, a revived world wide interest in the glass transition and the glass state has stimulated research in all areas of solid state physics where molecular disorder is involved. Among spin glasses, orientational glasses, amorphous metals, supercooled liquids, high tech ceramics, and advanced "window glasses" the family of polymer glasses is still the most versatile and open to surprising new properties.

1.4 References

[1] H. Finkelmann, H. Ringsdorf, J. H. Wendorff: Model considerations and examples of enantiotropic liquid crystalline polymers. Polyreactions in Ordered Systems 14, Makromol. Chem. 179 (1978) 273.

[2] G. Wegner: Topochemical polymerization of monomers with conjugated triple-bonds, Makromol. Chem. 154 (1972) 35.

[3] H. Ringsdorf, B. Schlarb, J. Venzmer: Molecular architecture and function of polymeric oriented systems: models for the study of organization, surface recognition and dynamics of biomembranes, Angew. Chem. Int. Ed. Engl. 27 (1988) 113–158.

[4] R. Mateva, G. Wegner, G. Lieser: Growth of polyoxymethylene crystals during cationic polymerization of Trioxane in Nitrobenzene, J. Polym. Sci. C 11 (1973) 369.

[5] W. Kern, V. Jaacks: Die Bedeutung der Polyoxymethylene für die Entwicklung der makromolekularen Chemie, Koll. Z. u. Z. Polym. 216 (1967) 286.

[6] R. C. Schulz, K. Albrecht, Q.V. Tran Thi, J. Nienburg, D. Engel: Some new copolymers by ionic polymerization, Polym. J. 12 (1980) 639 (spec. issue, 5th Internat. Symp. on cationic and other ionic polymerization).

[7] R. C. Schulz: Some new examples of polymer modification, Makromol. Chem. Suppl. 13 (1985) 123.

[8] R. C. Schulz: Reaktionen an Makromolekülen in Houben-Weyl, Methoden der Organischen Chemie, 4th ed., Vol. E 20, Part 1. Thieme, Stuttgart 1987, 608–625.

[9] D. Braun: Polyadicals, Encycl. Polym. Sci. Tech. 15 (1971) 429.

[10] R. C. Schulz: Addition compounds and complexes with polymers and models, Pure Appl. Chem. 38 (1974) 227.

[11] H. Schuttenburg, G. Klump, U. Kaczmar, S. R. Turner, R. C. Schulz: N-Chlor-Nylon as polymer reagents, J. Macromol. Sci. Chem. A 7 (1973) 1085.

[12] E. J. Günster, R. C. Schulz: N-Trifluor acetylpolyamid 66 als polymeres Reagens, Makromol. Chem. 180 (1979) 1891.

[13] M. Rothe, J. Mazanek: Nebenreaktionen bei der Festphasensynthese von Peptiden, Liebigs Ann. Chem. (1974) 439, Tetr. Letters (1972) 3795.

[14] M. Mutter, E. Bayer: The liquid-phase method for petide synthesis, in E. Gross, J. Meienhofer (eds.): The Peptides. Academic Press, New York 1980 285.

[15] K. Geckeler, V.N. R. Pillai, M. Mutter: Applications of soluble polymeric supports, Adv. Polym. Sci. 39 (1981) 65.

[16] E. Heidemann et al.: Polytripeptide durch repetitive Peptidsynthese und Verbrückung von Oligopeptiden, Makromol. Chem. 180 (1979) 905.

[17] W. Kern: Die Polyacrylsäure, 1. Mitt. Z. Physik. Chem. (A) 181 (1938) 249.

[18] W. Kern, V.V. Kale, B. Scherhag: Zur Polymerisation und Copolymerisation des Natriumvinylsulfonats, Makromol. Chem. 32 (1959) 37.
[19] R. C. Schulz: Chemische Modifizierung von Polyamiden, Plaste und Kautschuk 28 (1981) 485.
[20] R. C. Schulz, Th. Perner: Synthesis and some reactions of linear poly(iminohexamethylene), Brit. Polymer J. 19 (1987) 181.
[21] R. C. Schulz, M. Schmidt, E. Schwarzenbach, J. Zöller: Some new polyelectrolytes, Makromol. Chem. Macromol. Symp. 26 (1989) 221.
[22] L. Dominguez, W. H. Meyer, G. Wegner: Solid polyelectrolytes: Single crystalline ionenes, Makromol. Chem. Rapid Commun. 8 (1987) 151.
[23] L. L. Böhm, M. Chmelir, G. Löhr, B. J. Schmitt, G. V. Schulz: Zustände und Reaktionen von Carbanionen bei der anionischen Polymerisation von Styrol, Fortschr. Hochpol. Forsch. 9 (1972) 1.
[24] W. Heitz: Gelpermeationchromatographie, Angew. Chem. 82 (1970) 675; Z. Anal. Chem. 277 (1975) 323.
[25] G. Meyerhoff: Extension of GPC techniques, Sep. Sci. 6 (1971) 239.
[26] R. G. Kirste, W. A. Kruse, J. Schelten: Die Bestimmung des Trägheitsradius von Polymethylmethacrylat im Glaszustand durch Neutronenbeugung, Makromol. Chem. 162 (1972) 299.
[27] H. B. Stuhrmann, J. Haas, K. Ibel, M.H. J. Koch, R. R. Chrichton: Low angle neutron scattering of ferritin studied by contrast variation, J. Mol. Biol. 100 (1976) 399.
[28] E.W. Fischer, M. Dettenmaier: Structure of polymeric glasses and melts, J. Non-Cryst. Solids 31 (1978) 181.
[29] M. Stamm, E.W. Fischer, M. Dettenmaier, P. Convert: Chain conformation in the crystalline state by means of neutron scattering methods, Faraday Discussions 68 (1979) 263.

2 From Oligonuclear Phenolic Compounds to Calixarenes

Volker Böhmer and Hermann Kämmerer*

2.1 Introduction

In a series of patents starting in 1907, Leo H. Baekeland described condensation products of phenol and formaldehyde, which became the first totally synthetic polymeric products of commercial value. These polymers, known as phenolic resins, are produced and used for more and more sophisticated purposes although their total structure is still not completely understood [1]. This is mainly due to the fact, that the reaction of the difunctional formaldehyde and the trifunctional (two ortho-, one para-position) phenol finally leads to a completely insoluble cross-linked product.

Starting probably with the work of Koebner [2] a large variety of model compounds have been prepared in which a definite number of phenolic units are connected in a definite way via methylene groups. (Compounds with dimethylene ether or other links are less frequently described.) They may be regarded as either components of a complicated mixture of novolaks (or resols) or as representative segments of the cross-linked network of a hardened phenolic resin. Formerly they were the only basis for the correlation of physical properties with molecular structure, and still they are required for the calibration of modern analytical techniques. This variety of well defined, structural uniform oligomers makes them also attractive as model compounds for fundamental studies like neighboring group effects, or "long range" cooperative effects, especially via the influence of intramolecular hydrogen bonding which is the most determinant factor for all ortho-linked oligomers.

Among all these compounds cyclic oligo[(2-hydroxy-1,3-phenylene)-methylene]s which Gutsche named calixarenes [3] are of growing interest. "Host-guest-chemistry" [4], "inclusion phenomena" [5], "supramolecular

*Institut für Organische Chemie der Universität Mainz, J.-J.-Becher-Weg 18–20, D-6500 Mainz

chemistry" [6] are topics of current chemical research, and calixarenes are readily available potential host molecules, which again can be prepared in a large variety [3]. Thus, a topic once related to a special part of Macromolecular Chemistry developed into a very actual field in Macrocyclic Chemistry. It is the aim of this chapter, to follow some lines of this development. However, this can be neither a strictly historical report nor an exhaustive survey of the current state. From a structural point of view it will be restricted to molecularly uniform oligomers of monovalent phenols while derivatives of divalent phenols like pyrocatechol or resorcinol are omitted.

2.2 Synthesis of Structurally Uniform Oligomers

2.2.1 General Principles, Protective Groups

Definite oligomers may be synthesized either by the direct condensation of phenols or phenolic oligomers with formaldehyde or by the condensation of hydroxymethylated compounds with phenols or phenolic oligomers. Structures which are unsymmetric relative to the newly formed methylene bridge are available only by the latter strategy. A further advantage of this pathway is, that readily removable phenols may be taken in excess to suppress side reactions. Instead of hydroxymethylated compounds which are available by the reaction of phenols with formaldehyde under alkaline conditions [7], the use of chloromethylated (or bromomethylated) compounds is convenient or necessary in special cases [8,9].

The rational synthesis of definite oligomers usually requires suitable protective or blocking groups. In the case of phenolics, halogen (chlorine, bromine) is used to protect (or block) ortho or para positions from undergoing reaction. Dehalogenation is performed under very mild conditions (room temperature, normal pressure) with hydrogen in alkaline solution using Raney nickel as catalyst [1,9]. This method may be applied for all compounds with nonreducible substituents, such as alkyl, aryl or -COOR.

Bromine atoms may also be selectively eliminated with Zn/NaOH in the presence of chlorine atoms [10]. As an independent protecting group, the t-butyl group can be used, which is eliminated by transalkylation in toluene in the presence of $AlCl_3$, leaving halogen atoms unchanged in the same molecule [11]. Among all these possibilities only hydrogenation is sufficiently smooth not to affect hydroxymethyl groups [12].

2.2.2 Linear Oligomers

Using these different protective groups a variety of convergent strategies are now available to synthesize linear (and in principle also branched) oligomers. In addition selective coupling reactions in ortho-position to the phenolic hydroxyl group are possible using bromomagnesium salts in refluxing benzene [13]. Under dry conditions magnesium acts as a coordination site for formaldehyde *(1)*, and facilitates the formation of a quinone methide, the proposed reaction intermediate [14], which is coordinated in a similar manner *(2)*:

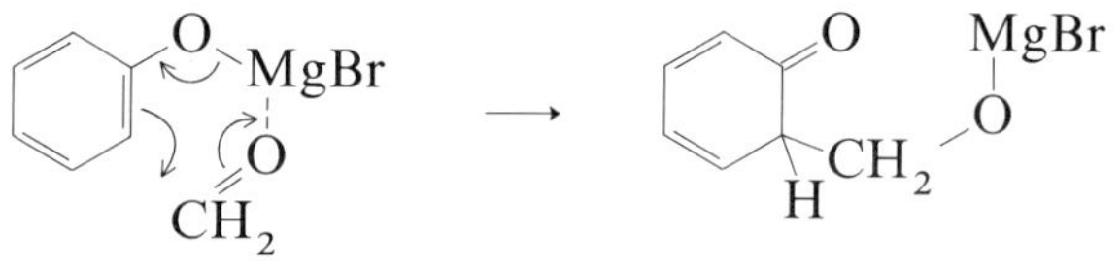

(1)

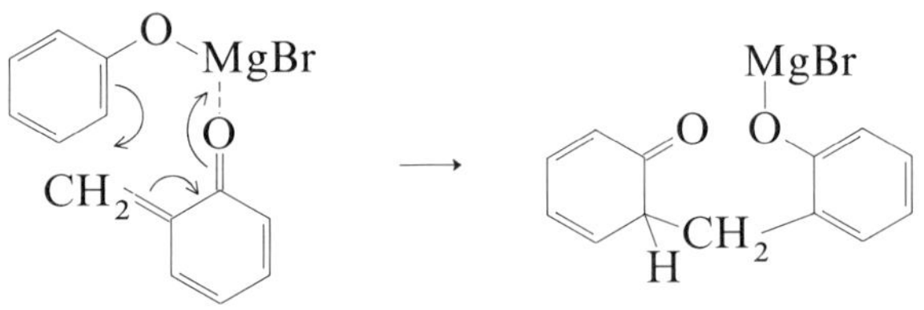

(2)

The oligomerization again may be performed as a "duplication reaction" with formaldehyde or a "stepwise extension" with salicylic alcohol (or generally an ortho-hydroxybenzyl alcohol). In this way Casiraghi et al. synthesized linear ortho-linked oligomers with up to nine phenol units without protecting the para-position [15], while the same compounds were initially obtained by Kämmerer et al. via the corresponding oligomers of p-chlorophenol [9].

Under similar conditions novolaks with alkylidene bridges, for instance from acetaldehyde [16] or isobutyraldehyde [17] can be prepared, which attract considerable theoretical interest, since stereoisomers are available as a consequence of the unsymmetrically substituted bridge (-CHR-). Indeed, if suitable chiral alkoxyaluminium chlorides are used instead of the magnesium derivatives, not only regioselective substitution is possible, but also enantioselective ortho-hydroxyalkylation [18] as well as enantiocontrolled synthesis of dinuclear compounds [19] has been realized.

Many oligomers containing further functional groups have been synthesized [1] either directly or by chemical modification (see 2.2.4). In addition, the use of modern chromatographic techniques now allows all desired structures to be separated even on a preparative scale.

2.2.3 Cyclic Oligomers, Calixarenes

One of the most surprising facts in the field of phenol-formaldehyde oligomers is, that many macrocyclic products are more readily available than the corresponding linear oligomers. A cyclic structure was already postulated by Zinke in 1944 for the high melting condensation products he obtained under alkaline conditions from p-alkylphenols and formaldehyde (see 2.3). However, the complex composition of these reaction mixtures was not elucidated before Gutsche started his investigations on these "calixarenes" [3].

He showed, that the direct condensation of p-t-butylphenol with formaldehyde leads to a mixture of cyclic (and probably linear) compounds, containing not only methylene, but also dimethylene ether bridges, and that the composition of these mixtures is greatly dependent on reaction conditions [20]:

OH; n; $+ (CH_2O)_n$; base; OH; CH_2; n; +; CH_2-O; OH; CH_2; 4 (3)

n = 4–8

The mechanism for the formation of these cyclic compounds is not yet understood in detail [21]. It has been shown for instance that under the drastic reaction conditions cleavage of methylene bridges and restructuring of phenolic units occurs, as calix(4)arenes can be obtained from calix(8)arenes. It is reasonable to assume, that calix(8)arenes are obtained as the "kinetically" controlled product, calix(4)arenes as the "thermodynamically" stable product and calix(6)arenes by a certain "template effect". It seems clear, that even-numbered oligomers are favoured for as yet unknown reasons, but small amounts of odd-numbered oligomers were also obtained [22, 23].

Synthesis conditions are now well defined to prepare calixarenes with 4, 6, or 8 p-t-butylphenol units [20] with yields up to 70 % in a preparative manner and even on an industrial scale. Calixarenes with several other alkyl substituents (like i-propyl, t-pentyl, t-octyl or n-alkyl up to octadecyl and phenyl) have been prepared in a similar way, but interestingly the calix(4)arenes are less accessible, than the calix(6)- or calix(8)arenes [3, 21], (although 16 (!) covalent bonds are newly formed by the reaction of 8 molecules of formaldehyde with 8 molecules of a phenol in the one-pot synthesis of a calix(8)arene). In fact for the most simple alkyl phenol, p-cresol (as well as for other p-substituted phenols) the one-pot synthesis fails [3].

A stepwise synthesis of calixarenes (first mentioned by Hayes and Hunter (see [3]) was systematically studied and extended by Kämmerer and Happel [24–27]. Starting with an o-halogen-p-alkyl phenol it consists of subsequent hydroxymethylation and condensation steps. The halogen, which was used to protect one o-position is eliminated in the penultimate step (the whole sequence is an excellent example of the use of halogen as protective group), and the linear monohydroxymethylated precursor is cyclized under high dilution conditions in refluxing acetic acid.

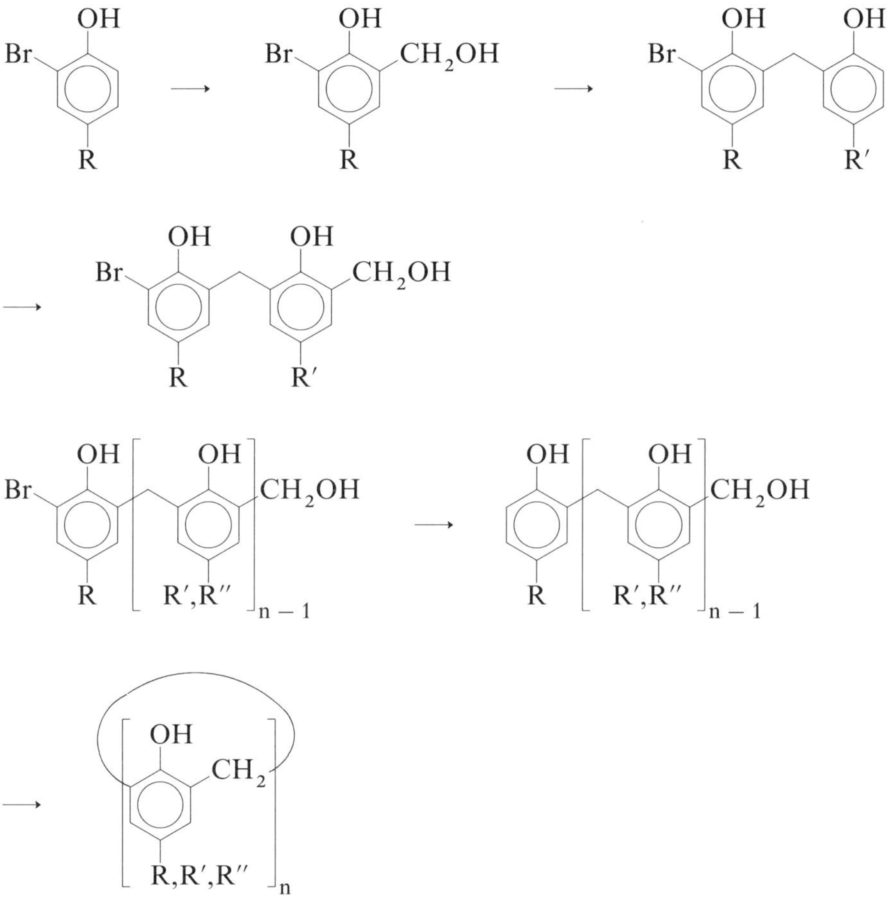

OH Br R → OH Br CH_2OH R → OH OH Br R R′

→ OH OH Br CH_2OH R R′

(4)

OH Br [OH CH_2OH R′,R″]$_{n-1}$ R → OH [OH CH_2OH R′,R″]$_{n-1}$ R

→ [OH CH_2 R,R′,R″]$_n$

By this synthetic pathway the ring size is unambiguously determined. It also makes accessible compounds with different (alkyl) substituents in p-position (R, R', R"). Thus, calixarenes with up to seven phenolic units (p-cresol, p-t-butylphenol) were obtained, however, an obvious disadvantage of this method is the long reaction-sequence leading also to a low overall yield.

A rather straightforward way to calix(4)arenes with different substituents in p-position consists of the condensation of a suitable linear trimer (or dimer) with the corresponding bis(bromomethyl)ated phenol (or dimer), which is mostly done now in dioxane, using $TiCl_4$ as a catalyst (and probably as a template) [28–30].

(5)

It allows the preparation of compounds with reducible substituents such as nitro, azophenyl or halogen. By this way also several asymmetrically substituted calix(4)arenes were obtained, consisting of three (in the order ABCC) or four different phenolic units [29, 30] or containing a single m-substituted unit [31]. They are especially interesting as the basis for chiral host molecules. Unfortunately this method fails in the attempted synthesis of calix(5)- and calix(6)arenes.

When α,ω-(4-hydroxyphenyl)alkanes are reacted with bis(bromomethyl)-phenols, calix(4)arenes are obtained in which two opposite p-positions are connected by an aliphatic chain of various length (5 to 16 C-atoms) *(6)* [32, 33]. Thus the cone conformation is fixed, and the size and shape of the cavity as well as the geometry of the OH-region can be varied. First attempts to prepare in a similar way compounds in which two calix(4)arene moieties are connected via their p-positions by one *(7)*, two or four aliphatic chains were successful too [34].

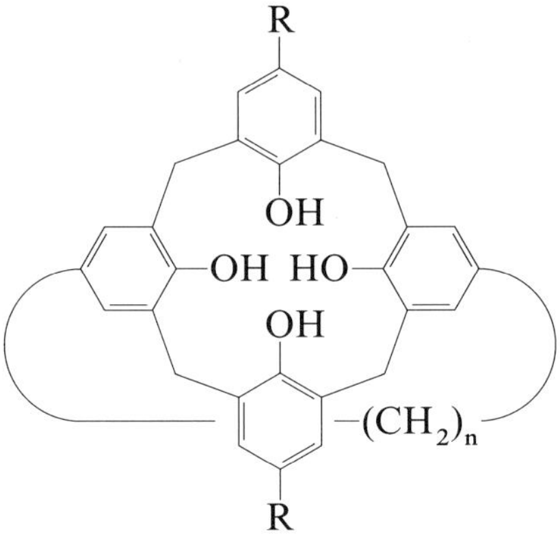

R
OH
OH HO
OH
$(CH_2)_n$
R

(*6*)

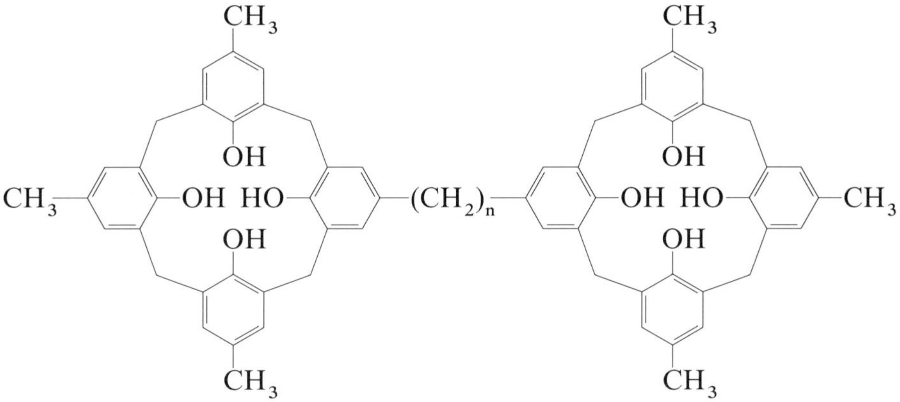

CH_3 CH_3
OH OH
CH_3 OH HO $(CH_2)_n$ OH HO CH_3
OH OH
CH_3 CH_3

(*7*)

2.2.4 Chemical Modifications

Phenolic oligomers may be modified by all those reactions which are known to occur with phenols such as:

a) reaction of the phenolic hydroxyl groups,
b) (electrophilic) substitution of the aromatic nucleus or elimination of substituents,
c) modification of substituents (or methylene bridges).

a) Initially ether or ester derivatives of oligo[(hydroxyphenylene)methylene]s were synthesized to prove the general structure of those compounds (for instance to show the presence of free, reactive OH-groups), or to have a further characterization via a derivative (e. g. an acetate). This derivatization has more importance with calixarenes.

Thus, interesting compounds with new complexing properties towards metal cations (see 2.4.4) have been obtained from calixarenes by the introduction of diethylene glycol residues *(8)* [35] and by the reaction with halogenaceticacid derivatives [36–39] or halogenketones (e. g. bromoacetone) *(9)* [40].

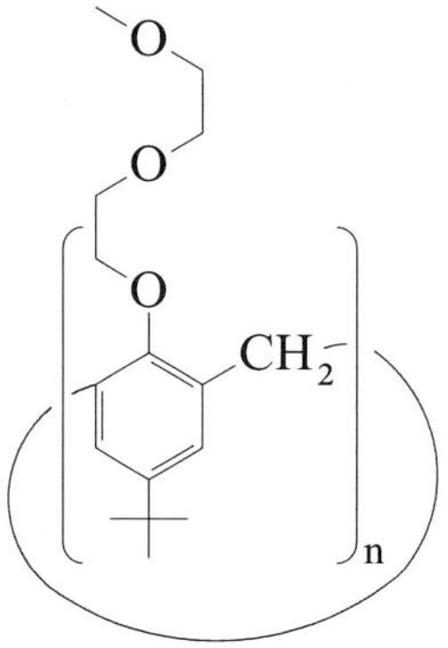

a) $R = OC_2H_5$

b) $R = N(C_2H_5)_2$

c) $R = CH_3$

(8) 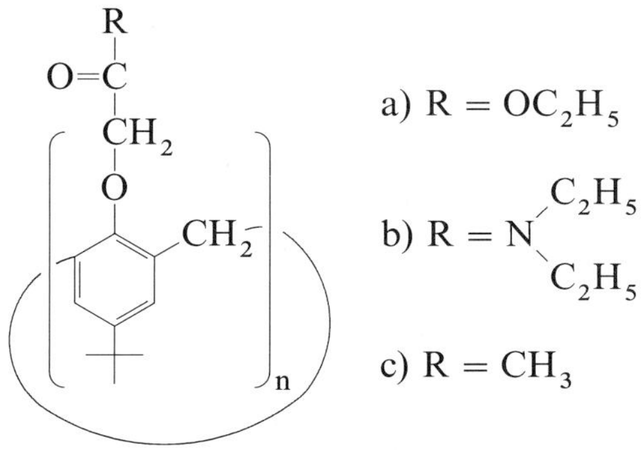 *(9)*

While with calix(6)- or calix(8)arenes all these derivatives are more or less flexible (compare 2.4.1), conformationally rigid derivatives are obtained from calix(4)arenes (with the exception of tetramethyl ethers). They are not always fixed in the so-called "cone-conformation" which means that all O-substituents are at one side of the plane through the methylene carbons. Examples for "partial-cone" or "1,3-alternate" derivatives are also known, and the reasons leading to these different conformations are not yet well understood, although some relation with the reaction rate was found in the formation of p-substituted benzoates [41].

For calix(4)arenes methods are elaborated to obtain derivatives in which only two opposite OH-groups have reacted while the remaining OH-groups are free for further reactions. If suitable bifunctional reagents are used, "macrobicyclic" derivatives may be obtained *(10)* from calix(4)arenes in which two opposite oxygens are connected by various bridges [42, 43]:

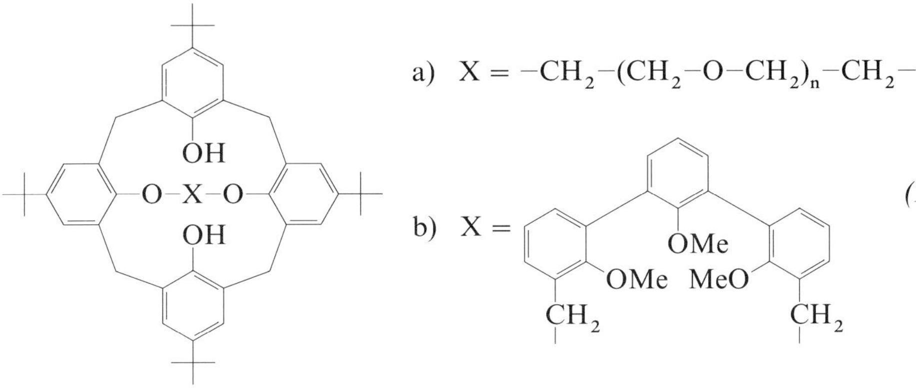

a) $X = -CH_2-(CH_2-O-CH_2)_n-CH_2-$

b) X =

(10)

b) Starting with the easily available t-butylcalixarenes many derivatives with other substituents in p-position have been prepared, after elimination of the t-butyl group by transalkylation [24, 44] (the analogous reaction is possible with t-octylcalixarenes [35]). This has been done by direct electrophilic substitution of either the free calix(4)arene or its tetramethylether. Among many derivatives with a large variety of functional groups in p-position the p-benzoylated compounds, showing a "deep cavity" should be mentioned [45].

p-Nitrocalixarenes, which are an interesting starting material for further derivatization, have been obtained via the sulfonated products [46] or by direct nitration [47]. Another interesting route to introduce new substituents in the p-position is the Claisen rearrangement of allylethers to p-allylphenols [44, 48]. It is noteworthy, that this reaction can be done selectively also with mono- or diallylethers of calix(4)arenes [44], leading finally to derivatives with different p-substituents in a definite position.

c) Especially by the variation of these p-allylcalixarenes a large variety of new functionalized compounds has been prepared, since the allyl group was for instance transformed into CH_2-CH=O, CH_2-CH_2-OH, CH_2-CH_2-Br, CH_2-CH_2-N_3, CH_2-CH_2-NH_2, CH_2-CH_2-CN in tetratosylates of calix(4)arenes, from which the tosyl group may be removed under mildly basic conditions [3, 48]. The end of this development cannot yet be foreseen.

2.2.5 Synthetic Matrix Reactions

Acrylates [49] or methacrylates [50] of oligo[(2-hydroxy-1,3-phenylene)methylene]s deserve some special remarks. They may be used in copolymerization with monomers like styrene or acrylonitrile, leading to the introduction of "blocks" of adjacent acrylic or methacrylic acid units into the polymer chain [51]. Under suitable conditions a cyclopolymerization of diesters is possible

[52] which (after cleavage of the ester bonds) leads to more or less alternating copolymers if mixed diesters of e.g. methacrylic- and acrylic acid are used [53].

If these esters of acrylic- or methacrylic acid and phenolic oligomers are treated with an excess of free radicals (e.g. from AIBN) in very dilute solution, the "polymerization" may be restricted to the single molecule. Thus, after cleavage of the ester bonds oligoacrylates or methacrylates (or suitable derivatives) are obtained, in which the degree of oligomerization should be determined by the number of phenolic units in the starting molecule. Therefore, the whole reaction sequence has been regarded as a model for a "synthetic matrix reaction" [54].

$$+\ n\ Cl{-}C({=}O){-}CH{=}CH_2 \xrightarrow{-n\ HCl} \quad \downarrow + R\cdot \qquad (11)$$

$R{-}[CH_2{-}CH(COOH)]_{n-1}{-}CH_2{-}CH(COOH){-}R$

$R = CH_3{-}\dot{C}(CH_3){-}C{\equiv}N$

(from excess AIBN)

Unfortunately in practice side reactions (e.g. intramolecular head-to-head addition, disproportionation or combination with the excess of radicals) cannot be completely suppressed [55, 56], which strongly limits the practical value, although these reactions gave some mechanistic insight. Interestingly the amount of desired product increased with increasing length of the phenolic matrix, and it is tempting to try similar reactions (initiated for instance by UV irridation) with cyclic compounds.

2.3 Phenolic Oligomers in the Solid State

2.3.1 Cyclic Oligomers

The first single-crystal X-ray diffraction studies in the field of phenolic oligomers were reported for t-butylcalix(4)arene [57] and for the octaacetate of t-butylcalix(8)arene [58], proving for the first time definitively the ring size of these two oligomers. Later several crystal structures of calix(4)arenes have been published [59–61], showing the molecule always in the so-called "cone-conformation", which is favoured by the cyclic array of intramolecular hydrogen bonds. The distances between adjacent oxygen atoms are found in the range of 2.63–2.67 Å in all cases, although slight conformational differences are reflected by the different inclination of the phenyl rings with respect to the plane of the methylene groups. A similar conformation is found for calix(5)arene [62] and a complete cycle of intramolecular hydrogen bonds is present also in calix(6)arenes (their "pinched" conformation may be visualized by the combination of two three ring fragments of a calix(4)arene [63, 64]) and t-butylcalix(8)arene [65].

X-ray structures of several bridged calix(4)arenes *(6)* show [33, 66], that the shape of the cavity is more and more distorted, when the length (n = number of carbon atoms) of the connecting aliphatic chain decreases. As an example in Fig. 2.1 the compound with n=5 is shown, in which two phenolic units (connected by the aliphatic chain) are nearly parallel. Here the distances of adjacent oxygen atoms reach an average value of 2.93 Å.

Solvent entrapment in the crystal lattice often occurs with calixarenes, although solvent-free crystals have been obtained from t-octylcalix(4)arene [59]. However, the guest molecules are not always located inside the cavity of the host molecules [59, 60]. The host/guest ratio may be different from one and different ratios have been found for the same host/guest pair [60, 61]. Nevertheless, even without knowing the exact structure of the inclusion type, calixarenes may be used to separate for instance isomeric xylenes in a preparative scale [67].

Several X-ray structures of calixarene derivatives (ethers or esters) have been published, showing the calix(4)arene derivatives mostly in the cone-conformation [36, 37, 39, 40]. An early exception is the tetraacetate in the partial cone conformation [68], while structures of other conformations were not published. Interestingly the "crowned" calix(4)arene *(10a)* changes its conformation from cone to a distorted partial cone (which is present already in the free *(10b)*) while complexing an alkali cation [43].

In addition to the alkali cation complexes with neutral ligands derived from t-butylcalix(4)arene (e. g. the tetra-acetamide [39]) the crystal structures of t-butylcalix(4)arene with titanium(IV), iron(III), and cobalt(II) [69], titanium

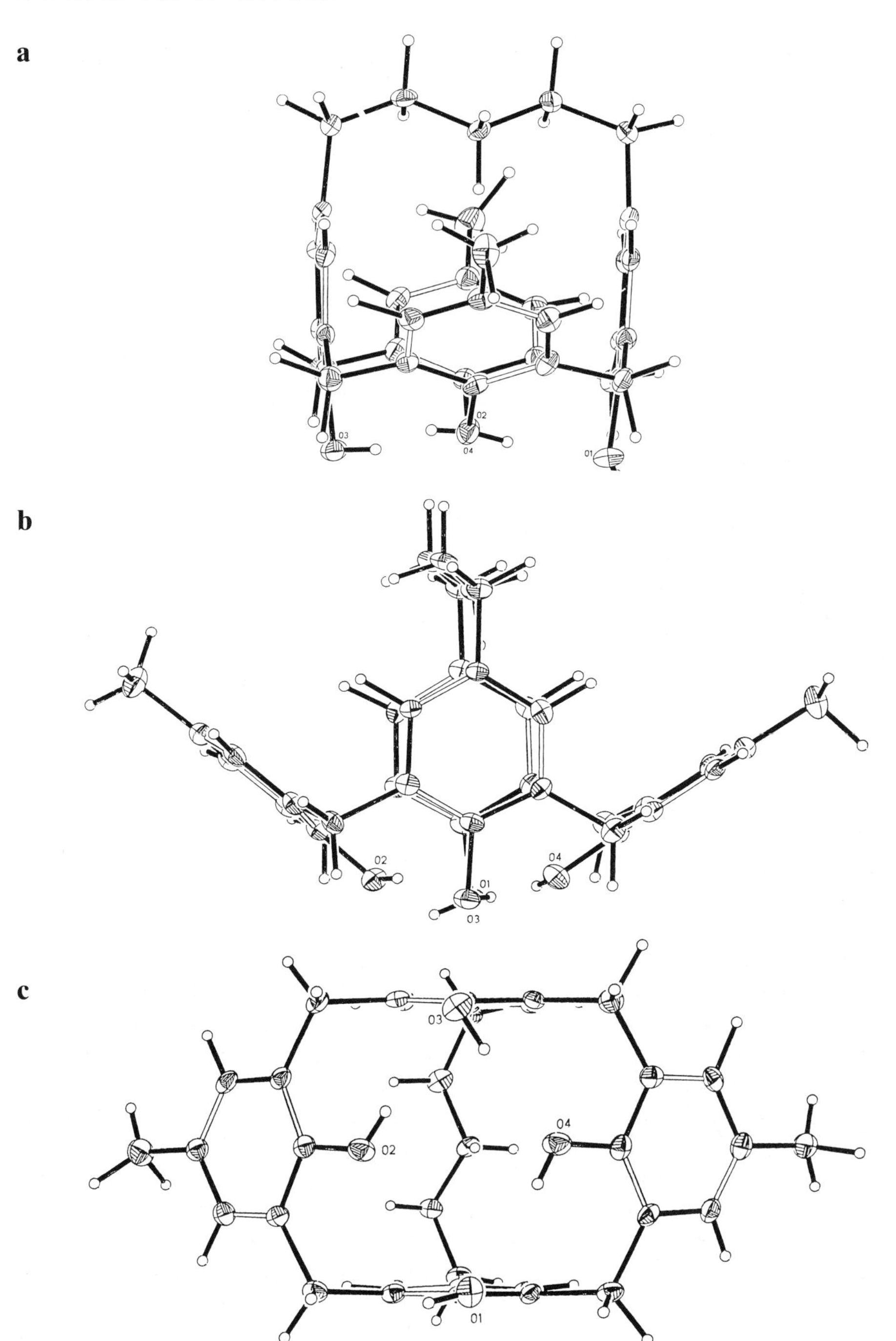

Fig. 2.1a–c: Molecular structure of a bridged calix(4)arene (*6*, R = CH_3, n = 5) with distorted cone-conformation seen from different directions turned by 90°.

complexes with t-butylcalix(6)arene [70] and the bimetallic complex of europium(III) with t-butylcalix(8)arene [71] should be mentioned.

2.3.2 Linear Compounds

Crystal structures for ortho-linked linear oligomers were first reported by Casiraghi et al. [15], somewhat later by Paulus and Böhmer [72]. Obviously the general principle of all structures again is a maximum of intramolecular hydrogen bonds between hydroxyl groups of adjacent phenolic units. In the crystal lattice the molecules are ordered by further intermolecular hydrogen bonds either to indefinite chains or to cyclic dimers. In this arrangement each hydroxyl acts simultaneously as a donor and an acceptor ("isodromic hydrogen bonds" [15]) and never as a double acceptor. While no special direction of hydrogen bonds is found for oligomers with a nonsubstituted ortho-position, a methyl or t-butyl group in one ortho-position obviously directs the hydrogen bonds in the opposite direction [72].

The intramolecular O-O-distances for all known structures are in the range of 2.62–2.71 Å for methylene bridged compounds (slightly higher values of 2.71–2.79 Å were found for ethylidene-bridged trimers [16]) and the intermolecular O-O-distances are only slightly longer. This indicates that all hydrogen bonds (inter- as well as intramolecular) have nearly the same strength which is similar to those in calixarenes.

Bond distances, bond angles, and torsion angles are practically equal in all compounds and comparable with literature values. The overall conformation of the molecules therefore is determined by the torsion angles for the rotation around the δ-bonds to the methylene (or ethylidene) bridge. They are in the range of 80–100°. The resulting dihedral angles between the aromatic planes are found between 105° and 130°.

Three subsequent phenolic units may be placed in a chairlike anti- or trans-position or in a boatlike syn- or cis-position (compare Fig. 2.2). In the case of the ethylidene bridged oligomers this sequence is determined by the relationship of the methyl substituent to the O-H · · · O-bond system which occurs at the opposite side. This means anti-conformations correspond to racemic dyads, while syn-conformations correspond to mesodyads. While two subsequent anti-conformations are possible for the next higher homologue (compare Fig.2.2), the two syn-conformations related to the cone like structure of calix(4)arenes were not found for tetranuclear compounds. Up to now only structures with ortho-ortho-linked phenolic oligomers were studied. Thus, in principle the question is open, if the conformation of oligo[(2-hydroxy-1,3-phenylene)methylene]s is determined by intramolecular hydrogen bonds, or if the usual conformation of oligo[(1,3-phenylene)methylene]s just favours intramolecular hydrogen bonds.

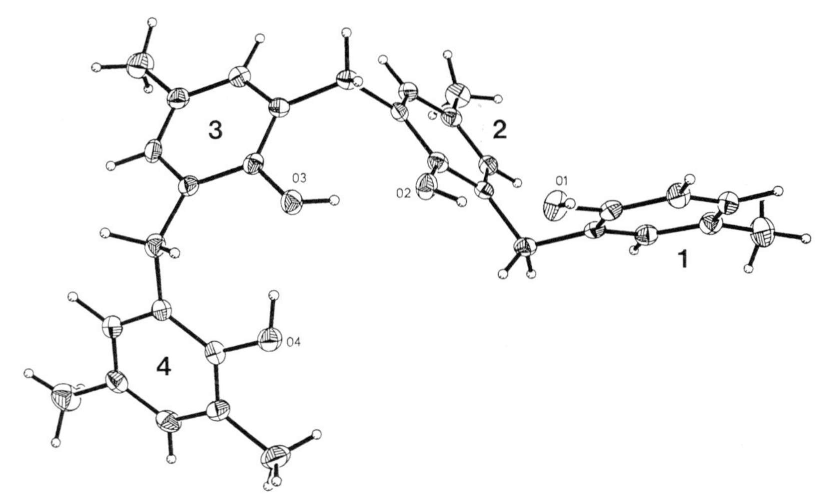

Fig. 2.2: Molecular structure of a linear tetramer (*15*, $R = CH_3$, $n = 4$), showing intramolecular hydrogen bonds between the adjacent phenolic units. Rings 1, 2 and 3 are anti or trans, rings 2, 3 and 4 are syn or cis arranged like in calixarenes.

2.4 Reactions and Properties in Solution

2.4.1 Spectroscopic Studies

Intramolecular hydrogen bonds between the hydroxy groups of adjacent phenolic units are also the most determinant factor for the conformation of oligo[(2-hydroxy-1,3-phenylene)methylene]s in solution (and, as a consequence, for their chemical reactivity). In aprotic solvents intra- as well as intermolecular O-H · · · O-hydrogen bonds (in some cases also intramolecular O-H · · · π-hydrogen bonds) were observed by IR-spectroscopy and also by ^{1}H-NMR-spectra [1]. Photo-CIDNP (chemically induced dynamic nuclear polarization) studies on linear oligomers showed, that the effect is always found for the phenolic unit forming the most stable radical. This can be explained by rapid hydrogen transfer between ortho-ortho-linked phenoxy radical and phenolic units [73]. In contrast to an earlier interpretation [74] these results provide further evidence for an intramolecular hydrogen bonded system.

Conformational analysis of calixarenes was done by dynamic ^{1}H-NMR spectra. In the region of the methylene protons, calix(4)arenes show a sharp singlet at higher temperatures which broadens with decreasing temperature to a broad doublet and finally to a pair of doublets (AB-system with a coupling constant J = 12–14 Hz, typical for geminal protons) at lower temperatures [75, 76]. These data are best interpreted in terms of the (more or less rapid) interconversion of the two opposite cone-conformations (in which 4 "axial" and 4 "equatorial" hydrogen atoms can be distinguished at the methylene groups), while other conformations cannot be detected from the NMR-spectra.

The mechanism of this interconversion where the hydroxyl groups have to pass through the annulus is not yet known. (It could proceed for instance in a concerted manner or step by step with partial cone and alternate conformations as intermediates.) However, the overall Gibbs-energy of activation can be calculated from these temperature dependent NMR-spectra. For calix(4)-arenes in apolar solvents it is in the range of $\Delta G^{\neq} = 62–67$ kJ mol^{-1} [1,76]. A similar temperature dependence is found for larger calixarenes (where at low temperature the pattern may be more complicated). As expected cyclic penta- and heptanuclear compounds are more flexible ($\Delta G^{\neq} = 51–54$ kJ mol^{-1}) than tetranuclear compounds. Surprisingly calix(8)arenes (which should be even more flexible) show exactly the same temperature dependence of the ^{1}H-NMR spectrum as calix(4)arenes [76]. Obviously this is caused by special conformations stabilized by intramolecular hydrogen bonds, because with pyridine as solvent the cyclic octamer shows the expected higher flexibility ($\Delta G^{\neq}$ = 36 kJ mol^{-1}) in comparison to the tetramer ($\Delta G^{\neq}$ = 56 kJ mol^{-1}).

In bridged calix(4)arenes *(6)* the cone conformation is fixed which follows from the stable, temperature independent AB-system. However, an increasing mobility of the calixarene part with increasing length of the connecting chain is shown by photo-CIDNP-spectra [77]. For shorter chains the calixarene part becomes more and more distorted and the intramolecular hydrogen bonds become weaker, which can be seen by IR- as well as by ^{1}H-NMR-spectra [33]. The cone-conformation of calix(4)arenes (or other conformations) may be fixed if large residues are bond to the phenolic oxygen. Obviously this is not possible for calix(6)- or calix(8)arenes which may interconvert by passing the p-substituent through the annulus [78].

2.4.2 Acidity, pK-Values

The "hyperacidity" of 2,2'-dihydroxydiphenylmethanes and o-o-linked oligomers is long known. The lower pK_{a1} values in comparison with isomers or closely analogous compounds are easily explained by a stabilization of the monoanion by strong intramolecular hydrogen bonds, an effect which was also reported for similar compounds [1].

A detailed study became possible using "well designed" oligomers, consisting of alkyl phenol units and one single nitrophenol unit [79]. By its higher acidity this nitrophenol unit clearly is dissociating in the first step which can be easily monitored by the spectral changes. Thus, structural influences on this dissociation can be studied.

For compounds with a p-nitrophenol unit at one end of the molecule *(12)* pK_{a1} decreases with increasing chain length, indicating an increased stabilization of the monoanion in comparison to the undissociated compound [80].

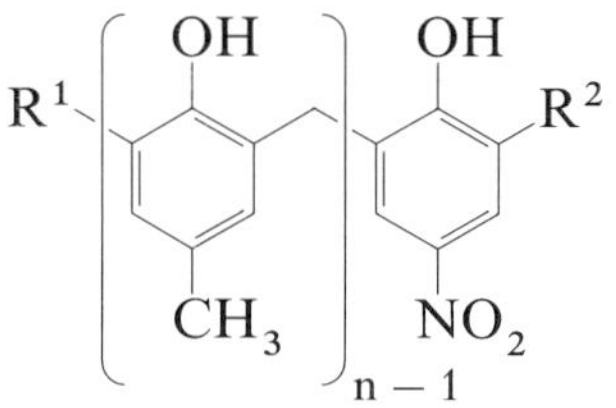

$R^1, R^2 = H, CH_3, C(CH_3)_3$

$n = 1-4$

(12)

This can also be deduced from the UV-spectra of the monoanions, where the absorption coefficient as well as the wavelength of the absorption maximum related to the p-nitrophenol unit decrease, if the strength of the intramolecular hydrogen bond increases. However, a chain of intramolecular hydrogen bonds must be assumed also for the undissociated molecules, since even for tetranuclear oligomers a bulky substituent R^1 in o-position to the hydroxyl group at one end of the molecule (which disturbes these hydrogen bonds) causes a higher acidity at the other end of the molecule.
The lowest pK_{a1}-values were found for trimers with the p-nitrophenol unit in the middle where the anionic position can be stabilized from both sides *(13)*.

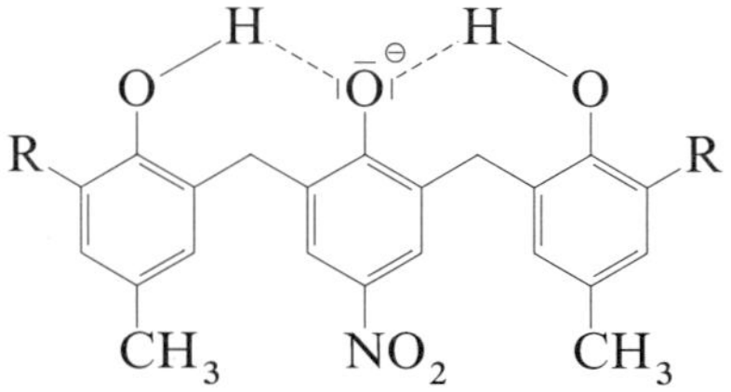

$R = H, C(CH_3)_3$

(13)

Calix(4)arenes containing the same trimeric element *(14)* show pK_{a1}-values depending strongly on the p-substituent in the opposite phenolic unit [81]. Differences up to 3 pK-units are observed if R^2 is varied as indicated, keeping R^1 constant. Obviously the cyclic array of intramolecular hydrogen bonds is very sensitive to small conformational changes which may be caused already by these different substituents.

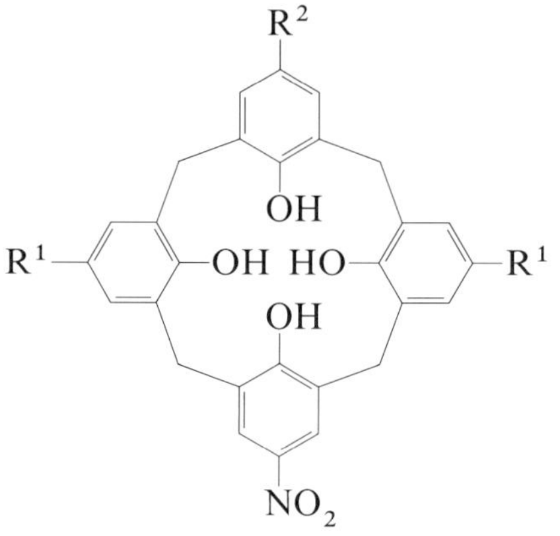

$R^1 = CH_3, C(CH_3)_3, C_6H_{11}$

$R^2 = CH_3, C(CH_3)_3, C_{12}H_{25}, C_6H_{11}$

(*14*)

The special situation in calix(4)arenes is also demonstrated by the very low pK_{a1} found for p-nitrocalix(4)arene (*14* with $R^1=R^2=NO_2$) which cannot be explained by a statistical factor [82].

2.4.3 Kinetic Studies

First attempts to correlate the reactivity of oligo[(2-hydroxy-1,3-phenylene)-methylene]s with the chain length were reported by Imoto et al. for the acid catalyzed reaction with formaldehyde (see 2.1). More detailed studies are reported for the bromination (molecular bromine in acetic acid) as an example of an electrophilic substitution. For several isomeric or similar dinuclear compounds containing the same reacting p-cresol (or o-cresol) unit, a lower reactivity was found if an intramolecular hydrogen bond is possible between the phenolic hydroxyl groups, especially if this hydrogen bond is directed by bulky substituents from the adjacent towards the reacting phenolic unit [83]. This can be explained by a smaller electron donating effect of a hydroxyl group accepting an intramolecular hydrogen bond in comparison to a "free" hydroxyl group. The directing effect of an o-substituent (methyl, t-butyl) is transmitted from one end of the molecule to the other end in linear, ortho-linked oligomers *(15)* even up to the hexamer [84].

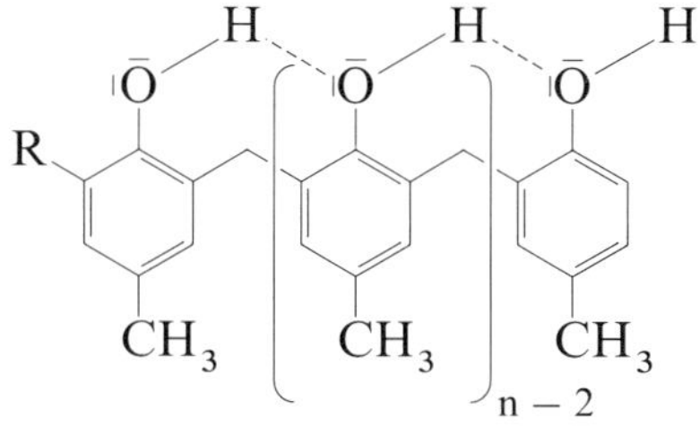

$R = H, CH_3, C(CH_3)_3;\ n = 2-6$ (*15*)

And like for the pK-values, the strongest effect is observed, when the interior phenolic unit is influenced by hydrogen bonds from both adjacent units [85].

The reactivity of chloromethyl groups in oligonuclear phenolic compounds was studied in methanolysis and aminolysis (p-nitroaniline in DMSO) reactions [86].

CH_3 [...] $CH_2Cl + RH \longrightarrow CH_3$ [...] $CH_2{-}R + HCl$ (16)

n = 2–4;

R = OCH_3, NH–C_6H_4–NO_2

The methanolysis, leading to hydroxybenzyl methyl ethers, is a S_N1-reaction for the nondissociated chloromethylated phenols [87]. The reaction rate is nearly independent of the structure in isomeric dimers [86]. The aminolysis, leading to hydroxybenzyl amines, proceeds via the reversible elimination of HC1 and the subsequent addition of p-nitroaniline to the intermediate quinone methide [88]. This first elimination step is intramolecularly assisted by the adjacent hydroxyl group in chloromethylated 2,2-methylene-diphenols. This leads to rate accelerations by factors greater than 10^3 in comparison to the isomeric 2,4- or 4,4-methylene-diphenols [86]. This accelerating effect is more pronounced for o-linked trimers and tetramers, which suggests again a chain of intramolecular hydrogen bonds.

2.4.4 Complexation of Cations, Host-Guest Interactions

Calixarenes are able to carry metal cations from an alkaline source-phase through hydrophobic liquid membranes into a neutral receiving phase [89, 90].

The effect is most pronounced for Cs^+ ions, where the highest transport rate is found for the cyclic octamer, while the cyclic tetramer shows the highest selectivity. The structure of the complex is still unclear, the cation may be surrounded by the oxygen atoms or it may be situated in the cavity after loss of its hydration shell. A pronounced maximum of the transport ability in bridged calixarenes *(6)* for $n = 8$ is better understandable with the latter assumption [33].

Strong complexation of uranylium cations (UO_2^{2+}) has been observed for sulfonated calix(5)- and calix(6)arenes *(17)* (and also their derivatives with $R = CH_2COOH$), while the corresponding calix(4)arenes form only weak complexes [91].

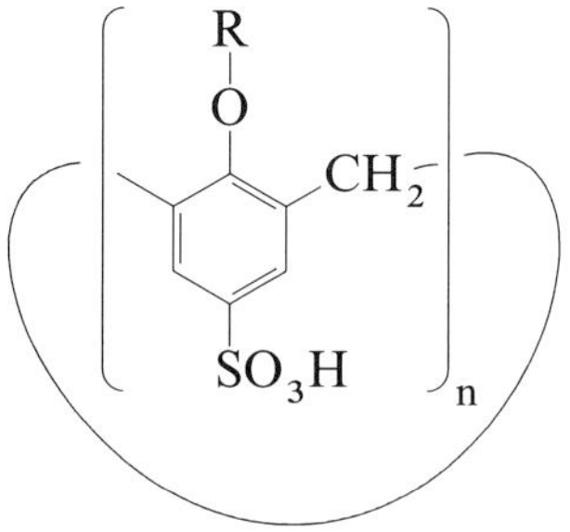

(17)

Selectivity factors of 10^{12} to 10^{17} in comparison with other divalent cations like Zn^{2+}, Ni^{2+} or Cu^{2+} are explained by "coordination-geometry selectivity" since the UO_2^{2+} ion demands a planar penta- or hexacoordination sphere (in contrast to the other cations).

An easily available class of neutral ligands for (mainly alkali) cations are calixarene ethers with ester, amide or ketone functions as donor groups (see 2.2.4). In many cases they show much higher stability constants than the classical crown ethers and remarkable selectivities for instance in the complexation of sodium [37] have been reported. Since slight variations of the residues fixed to the phenolic oxygen lead to appreciable differences in the stability constants, this group of compounds offers a vast field of different neutral ligands. The remarkable kinetic stability which may be obtained with calixspherands *(10b)* [43] makes these ligands attractive for medical purposes.

Gutsche has reported, that tert-butylamine is complexed within the cavity of p-allylcalix(4)arene, probably through ion-pairing, where the ammonium cation is situated in the cavity of the calixarene anion, while for other combinations of amine/calixarene obviously the "hole-size" does not fit sufficiently [92, 93]. Indeed, Shinkai has demonstrated such a "hole-size selectivity" in the complexation of sulfonated calix(4,6,8)arene ethers *(17)* with suitable dyes (phenol blue respectively anthrol blue [94]). He also reported the first (but certainly not the last) example of an enzyme like catalysis with calixarenes. The acid-catalyzed hydration of 1-benzyl-1,4-dihydronicotinamide was accelerated by sulfonated calix(6)arenes by factors of up to 1220 in comparison with non-cyclic analogues and the reaction proceeds according to the Michaelis-Menten scheme [95].

2.5 Concluding Remarks

The above reported results did not include patents. But the rapid development of calixarene chemistry also in the field of applied research is best demonstrated by the increasing number of patents. According to Chemical Abstracts where only the first patent of a family is reported, from 1975 to 1979 one patent, from 1980 to 1984 eight patents and from 1984 to the beginning of 1988 seventeen patents involved calixarenes. The applications reach from ion sequestering compounds (recovery of uranium from sea water, ionsensitive electrodes) to additives in glues or other polymers (antioxidants). For these purposes calixarenes were added to polymers or polymer mixtures, incorporated in polymer chains or fixed on a polymer backbone. For all these applications the easy access to the calixarene system is an important factor, and surely the number of sophisticated molecules or structures on the basis of calixarenes as building blocks will further increase in the future.

During the preparation of this book the development especially in the field of calixarenes became more and more rapid. For a detailed information two books [96, 97] are recommended which recently appeared.

One of the tools which were not discussed in this chapter but will gain increasing importance to understand complex mixtures like phenolic resins and their precursors are various computational calculations. A recent and representative example taken again from calixarene chemistry may be mentioned [98].

2.6 References

[1] A. Knop, L. A. Pilato: Phenolic Resins. Springer, Berlin-Heidelberg-New-York-Tokyo 1985.

[2] W. Koebner: Der Aufbau der Phenoplaste, Angew. Chem. 46 (1933) 251–262.

[3] C. D. Gutsche: The calixarenes, Topics in Current Chemistry 123 (1984) 1–188.

[4] See for instance: Host-Guest Complex Chemistry I, II, III, Topics in Current Chemistry 98 (1981); 101 (1982); 121 (1984) (eds.: F. Vögtle, E. Weber).

[5] The Journal of Inclusion Phenomena appears 1990 in its eighth year.

[6] J.-M. Lehn: Supramolekulare Chemie – Moleküle, Übermoleküle und molekulare Funktionseinheiten, Angew. Chem. 100 (1988) 91–116.
[7] H. Kämmerer, G. Happel: Eine präparative Monohydroxymethylierung höhermolekularer, molekulareinheitlicher Oligo[(5-alkyl-2-hydroxy-1,3-phenylen)-methylen]e mit Formaldehyd, Makromol. Chem., Rapid Commun. 1 (1980) 461–466.
[8] V. Böhmer, J. Deveaux, H. Kämmerer: Die Synthese reiner Oligo[(hydroxy-5-nitro-1,3-phenylen)methylen]e, Verbindungen mit mehreren ortho- bzw. para-Nitrophenolbausteinen im Molekül, Makromol. Chem. 177 (1976) 1745–1770.
[9] H. Kämmerer, H. Lenz: Über molekulareinheitliche Phenol-Formaldehyd-Kondensate mit einheitlicher Konstitution, Makromol. Chem. 27 (1958) 162–165.
[10] A. A. Mosfegh, B. Mazandarani, A. Nahid, G. H. Hakimelahi: The synthesis of hetero-halogenated derivatives of phloroglucide analogues, Helv. Chim. Acta 65 (1982) 1229–1232.
[11] V. Böhmer, D. Rathay, H. Kämmerer: The tert-butyl group as a possible protective group in the synthesis of oligo(hydroxy-1,3-phenylene)methylenes, Org. Prep. Proced. Int. 10 (1978) 113–121.
[12] H. Kämmerer, G. Happel, V. Böhmer: Halogen as a readily cleavable protective group for reactive positions in phenols and phenolic compounds, Org. Prep. Proceed. Int. 8 (1976) 245–248.
[13] G. Casnati, G. Casiraghi, A. Pochini, G. Sartori, R. Ungaro: Template catalysis via non-transition metal complexes, new highly selective syntheses on phenol systems, Pure & Appl. Chem. 55 (1983) 1677–1688.
[14] A. Arduini, A. Pochini, R. Ungaro, P. Domiano: o-Quinone methides, Part 3. X-ray crystal structure and reactivity of a stable o-quinone methide in the E-configuration, J. Chem. Soc., Perkin Trans. I (1986) 1391–1395.
[15] G. Casiraghi, M. Cornia, G. Sartori, G. Casnati, V. Bocchi, G. D. Andreetti: Selective step-growth phenol-aldehyde polymerization, 1. Synthesis, characterization and X-ray analysis of linear all-ortho-oligonuclear phenolic compounds, Makromol. Chem. 183 (1982) 2611–2634.
[16] G. Casiraghi, M. Cornia, G. Ricci, G. Casnati, G. D. Andreetti, L. Zetta: Selective step-growth phenol-aldehyde polymerization, 3. Synthesis, characterization, and X-ray analysis of regular all-ortho ethylidene-linked oligonuclear phenolic compounds, Macromolecules 17 (1984) 19–28.
[17] G. Casiraghi, M. Cornia, G. Balduzzi, G. Casnati: Non-transition metal assisted regio and stereoselective synthesis of alkylidene-bridged oligomeric phenolic compounds, Ind. Eng. Chem. Prod. Res. and Develop. 23 (1984) 366.
[18] F. Bigi, G. Casiraghi, G. Casnati, G. Sartori, L. Zetta: Enantioselective ortho-hydroxyalkylation of phenols promoted by chiral alkoxyaluminium chlorides, J. Chem. Soc., Chem. Commun. (1983) 1210–1211.
[19] G. Casiraghi, M. Cornia, G. Casnati, L. Zetta: Chiral novolacs. Enantiocontrolled synthesis of alkylidene-linked binuclear phenolic compounds, Macromolecules 17 (1984) 2933–2934.
[20] C. D. Gutsche, B. Dhawan, K. H. No, R. Muthukrishnan: Calixarenes, 4. The synthesis, characterization, and properties of the calixarenes from p-tert-butylphenol, J. Am. Chem. Soc. 103 (1981) 3782–3792.
[21] B. Dhawan, S. Chen, C. D. Gutsche: Calixarenes, 19. Studies of the formation of calixarenes via condensation of p-alkylphenols and formaldehyde, Makromol. Chem. 188 (1987) 921–950.

[22] A. Ninagawa, H. Matsuda: Formaldehyde polymers, 29. Isolation and characterization of calix(5)arene from the condensation product of 4-tert-butylphenol with formaldehyde, Makromol. Chem., Rapid Commun. 3 (1982) 65–67.
[23] Y. Nakamoto, S. Ishida: Calix(7)arene from 4-tert-butylphenol and formaldehyde, Makromol. Chem., Rapid Commun. 3 (1982) 705–707.
[24] H. Kämmerer, G. Happel, V. Böhmer, D. Rathay: Die stufenweise Synthese von 4,11,18,25-Tetra-tert-butyl(1.1.1.1)metacyclophan-7,14,21,28-tetraol und 4,11-Dimethyl(1.1.1.1)metacyclophan-7,14,21,28-tetraol, Monatsh. Chem. 109 (1978) 767–773.
[25] H. Kämmerer, G. Happel: Stufenweise Darstellung eines Cycloheptamers aus p-Kresol, 4-tert-Butylphenol und Formaldehyd. Vergleich mit einem phenolischen, heptanuklearen Kettenoligomer, Makromol. Chem. 181 (1980) 2049–2062.
[26] H. Kämmerer, G. Happel, B. Mathiasch: Schrittweise Synthesen und Eigenschaften einiger Cyclopentamerer aus methylenverbrückten (5-Alkyl-2-hydroxy-1,3-phenylen)-Bausteinen, Makromol. Chem. 182 (1981) 1685–1694.
[27] H. Kämmerer, G. Happel: Die stufenweise Synthese o,o'-methylenverbrückter Cyclohexamerer mit p-Kresol- oder p-Kresol- und 4-tert-Butylphenol-Bausteinen, Vergleich mit ähnlich strukturierten Kettenoligomeren, Monatsh. Chem. 112 (1981) 759–768.
[28] V. Böhmer, P. Chhim, H. Kämmerer: A new sythetic access to cyclic oligonuclear phenolic compounds, Makromol. Chem. 180 (1979) 2503–2506.
[29] V. Böhmer, F. Marschollek, L. Zetta: Calix(4)arenes with four differently substituted phenolic units, J. Org. Chem. 52 (1987) 3200–3205.
[30] V. Böhmer, L. Merkel, U. Kunz: Asymmetrically Substituted Calix(4)arenes, J. Chem. Soc., Chem. Commun. (1987) 896–897.
[31] H. Casablanca, J. Royer, A. Satrallah, A. Taty-C, J. Vicens: Synthesis of calix-(4)arenes presenting no plane of symmetry, Tetrahedron Lett. 28 (1987) 6595–6596.
[32] V. Böhmer, H. Goldmann, W. Vogt: The first example of a new class of bridged calixarene, J. Chem. Soc., Chem. Commun. (1985) 667–668.
[33] H. Goldmann, W. Vogt, E. Paulus, V. Böhmer: A series of calix(4)arenes, having two opposite p-positions connected by an aliphatic chain, J. Am. Chem. Soc. 110 (1988) 6811–6817.
[34] V. Böhmer, H. Goldmann, W. Vogt, J. Vicens, Z. Asfari: The synthesis of double-calixarenes, Tetrahedron Lett. 30 (1989) 1391–1394.
[35] V. Bocchi, D. Foina, A. Pochini, R. Ungaro, G. D. Andreetti: Synthesis, ^{1}H NMR, ^{13}C NMR spectra and conformational preference of open chain ligands on lipophilic macrocycles, Tetrahedron 38 (1982) 373-378.
[36] M. A. McKervey, E. M. Seward, G. Ferguson, B. Ruhl, S. J. Harris: Synthesis, X-ray crystal structures, and cation transfer properties of alkyl calixaryl acetates, a new series of molecular receptors, J. Chem. Soc., Chem. Commun. (1985) 388–390.
[37] A. Arduini, A. Pochini, S. Reverberi, R. Ungaro, G. D. Andreetti, F. Ugozzoli: The preparation and properties of a new lipophilic sodium selective ether ester ligand derived from p-t-butylcalix(4)arene, Tetrahedron 42 (1986) 2089–2100.
[38] S.-K. Chang, I. Cho: New metal cation-selective ionophores derived from calixarenes: Their syntheses and ion-binding properties, J. Chem. Soc., Perkin Trans. I (1986) 211–214.

[39] A. Arduini, E. Ghidini, A. Pochini, R. Ungaro, G. D. Andreetti, G. Calestani, F. Ugozzoli: p-t-Butylcalix(4)arene tetraacetamide: A new strong receptor for alkali cations, J. Incl. Phenom. 6 (1988) 119–134.
[40] G. Ferguson, B. Kaitner, M. A. McKervey, E. Seward: Synthesis, X-ray crystal structure, and cation transfer properties of a calix(4)arene tetraketone, a new versatile molecular receptor, J. Chem. Soc., Chem. Commun. (1987) 584–585.
[41] M. Iqbal, Th. Mangiafico, C. D. Gutsche: Calixarenes 21. The conformations and structures of the products of aroylation of the calix(4)arenes, Tetrahedron 43 (1987) 4917–4930.
[42] C. Alfieri, E. Dradi, A. Pochini, R. Ungaro, G. D. Andreetti: Synthesis, and X-ray crystal and molecular structure of a novel macrobicyclic ligand: Crowned p-t-butyl-calix(4)arene, J. Chem. Soc., Chem Commun. (1983) 1075–1077.
[43] D. N. Reinhoudt, P. J. Dijkstra, P. J. A. In't Veld, K. E. Bugge, S. Harkema, R. Ungaro, E. Ghidini: Kinetically stable complexes of alkali cations with rigidified calix(4)arenes. X-ray structure of a calixspherand sodium picrate complex, J. Am. Chem. Soc. 109 (1987) 4761–4762.
[44] See for instance C. D. Gutsche, L.-G. Lin: Calixarenes, 12. The synthesis of functionalized calixarenes, Tetrahedron 42 (1986) 1633–1640.
[45] C. D. Gutsche, P. F. Pagoria: Calixarenes, 16. Functionalized calixarenes: The direct substitution route, J. Org. Chem. 50 (1985) 5795–5802.
[46] S. Shinkai, K. Areki, T. Tsubaki, T. Arimura, O. Manabe: New synthesis of calixarene-p-sulfonates and p-nitrocalixarenes, J. Chem. Soc., Perkin Trans. I (1987) 2297–2299.
[47] K. No, Y. Noh, The synthesis of p-nitrocalix(4)arene, Bull. Korean Chem. Soc. 7 (1986) 314–316.
[48] C. D. Gutsche, J. A. Levine, P. K. Sujeeth: Calixarenes, 17. Functionalized calixarenes: The Claisen rearrangement route, J. Org. Chem. 50 (1985) 5802–5806.
[49] H. Kämmerer, G. Hegemann, N. Önder: Die Veresterung von p-Kresol und Oligo[(hydroxyphenylen)methylen]-verbindungen mit Acrylsäure- und Propionsäurechlorid, Makromol. Chem. 183 (1982) 1435–1444.
[50] H. Kämmerer, J. Pachta, J. Ritz: Spektroskopische Analyse mehrfacher Methacrylsäureester von Oligo[(2-hydroxy-1,3-phenylen)methylen]en mittels UV-, IR-, ^{1}H NMR- und Massenspektren, Makromol. Chem. 178 (1977) 1229–1247.
[51] S. Polowinski: Copolymerization on matrices, Eur. Polym. J. 14 (1978) 563–566.
[52] H. Guéniffey, H. Kämmerer, C. Pinazzi: Cyclopolymérisation du diméthacrylate de dihydroxy-2,2'-diméthyl-5,5'-diphénylméthane, Makromol. Chem. 165 (1973) 73–81.
[53] V. Böhmer, R. Funk, J. Kielkiewicz, W. Vogt: Die Darstellung und Polymerisation von Diestern aus 2,2'-Methylendiphenolen und Acryl-, Methacryl- bzw. Crotonsäure, Makromol. Chem. 185 (1984) 1905–1913.
[54] H. Kämmerer: Was sind Matrizenreaktionen?, Chemiker-Zeitung 96 (1972) 7–15.
[55] H. Kämmerer, J. Pachta: Nebenreaktionen bei der radikalischen, intramolekularen Cycloaddition von 2,6-Bis(2-methacryloyl-oxy-5-methylbenzyl)-4-methylphenylmethacrylat. Beitrag zu einer Matrizenreaktion, Colloid. Polym. Sci. 255 (1977) 656–663.

[56] H. Kämmerer, J. Pachta: Nebenreaktionen der radikalisch ausgelösten Cyclisierung von 2,2'-Methylen-bis(4-methyl-1,2-phenylen)-dimethacrylat, Makromol. Chem. 178 (1977) 1659–1670.
[57] G. D. Andreetti, R. Ungaro, A. Pochini: Crystal and molecular structure of cyclo{quater[(5-t-butyl-2-hydroxy-1,3-phenylene)methylene]} toluene (1:1) clathrate, J. Chem. Soc., Chem. Commun. (1979) 1005–1007.
[58] G. D. Andreetti, R. Ungaro, A. Pochini: X-ray crystal and molecular structure of the p-t-butylphenol-formaldehyde cyclic octamer cyclo{octa[(5-t-butyl-2-acetoxy-1,3-phenylene)methylene]}, J. Chem. Soc., Chem. Commun. (1981) 533–534.
[59] G. D. Andreetti, A. Pochini, R. Ungaro: Molecular inclusion in functionalized macrocycles, Part. 6. The crystal and molecular structures of the calix(4)arene from p-(1,1,3,3-tetramethybutyl)phenol and its 1:1 complex with toluene, J. Chem. Soc., Perkin Trans. II (1983) 1773–1779.
[60] R. Ungaro, A. Pochini, G. D. Andreetti, V. Sangermano: Molecular inclusion in functionalized macrocycles, Part 8. The crystal and molecular structure of calix(4)arene from phenol and its (1:1) and (3:1) acetone clathrates, J. Chem. Soc., Perkin Trans. II (1984) 1979–1985.
[61] R. Ungaro, A. Pochini, G. D. Andreetti, P. Domiano: Molecular inclusion in functionalized macrocycles, Part 9. The crystal and molecular structure of p-t-butylcalix(4)arene-anisole (2:1) complex: a new type of cage inclusion compound, J. Chem. Soc., Perkin Trans. II (1985) 197–201.
[62] M. Coruzzi, G. D. Andreetti, V. Bocchi, A. Pochini, R. Ungaro: Molecular inclusion in functionalized macrocycles, Part 5. The crystal and molecular structures of 25,26,27,28,29-pentahydroxycalix(5)arene-acetone (1:2) clathrate, J. Chem. Soc., Perkin Trans. II (1982) 1133–1138.
[63] R. Perrin, R. Lamartine: Etude de la production de resines phenoliques bien definies a partir de precurseurs purs, Makromol. Chem., Makromol. Symp. 9 (1987) 69–78.
[64] G. D. Andreetti, F. Ugozzoli, A. Casnati, E. Ghidini, A. Pochini, R. Ungaro: Crystal and molecular structure of p-tert-butylcalix(6)arene 1:1 tetrachloroethylene clathrate, Gaz. Chim. Ital. 119 (1989) 47–50.
[65] C. D. Gutsche, A. E. Gutsche, A. I. Karaulov: Calixarenes, 11. Crystal and molecular structure of p-tert-butylcalix(8)arene, J. Incl. Phen. 3 (1985) 447–451.
[66] E. Paulus, V. Böhmer, H. Goldmann, W. Vogt: The crystal and molecular structure of two calix(4)arenes bridged at opposite para positions, J. Chem. Soc., Perkin Trans. II (1987) 1609–1615.
[67] Y. Nakamoto, S. Ishida: Inclusion and Separation of Organic Compounds Using Calixarenes, Abstracts of the 12th International Symposium on Macrocyclic Chemistry, Hiroshima 1987, 92.
[68] C. Rizzoli, G. D. Andreetti, R. Ungaro, A. Pochini: Molecular inclusion in functionalized macrocycles. 4. The crystal and molecular structure of the cyclo{tetrakis[(5-t-butyl-2-acetoxy-1,3-phenylene)methylene]}-acetic acid (1:1) clathrate, J. Mol. Struct. 82 (1982) 133–141.
[69] M. M. Olmstead, G. Sigel, H. Hope, X. Xu, P. Power: Metallocalixarenes: syntheses and X-ray crystal structures of titanium(IV), iron(III), and cobalt(II) complexes of p-tert-butylcalix(4)arene, J. Am. Chem. Soc. 107 (1985) 8087–8091.

[70] G. D. Andreetti, G. Calestani, F. Ugozzoli, A. Arduini, E. Ghidini, A. Pochini, R. Ungaro: Solid state studies on p-t-butylcalix(6)arene derivatives, J. Incl. Phen. 5 (1987) 123–126.
[71] B. M. Furphy, J. M. Harrowfield, D. L. Kepert, B.W. Skelton, A. H. White, F. R. Wilner: Bimetallic lanthanide complexes of the calixarenes: Europium(III) and tert-butylcalix(8)arene, Inorg. Chem. 26 (1987) 4231–4236.
[72] E. Paulus, V. Böhmer: Die Kristallstruktur von Oligo[(2-hydroxy-1,3-phenylen)-methylen]en, Makromol. Chem. 185 (1984) 1921–1935.
[73] L. Zetta, V. Böhmer, R. Kaptein: Rapid hydrogen atom transfer in oligophenols. A photo-CIDNP study, J. Mag. Res. 76 (1988) 587–591.
[74] L. Zetta, A. DeMarco, G. Casiraghi, M. Cornia, R. Kaptein: Exposure of hydroxyl groups in phenol-acetaldehyde oligomers, as investigated by photo-CIDNP ^{1}H NMR and infrared spectroscopy, Macromolecules 18 (1985) 1095–1100.
[75] G. Happel, B. Mathiasch, H. Kämmerer: Darstellung einiger oligomerer Cyclo{oligo[(2-hydroxy-1,3-phenylen)methylen]}e. Spektroskopische Untersuchung ihrer Pseudorotation, Makromol. Chem. 176 (1975) 3317–3334.
[76] C. D. Gutsche, L. J. Bauer, Calixarenes 13. The conformational properties of calix(4)arenes, calix(6)arenes, calix(8)arenes, and oxacalixarenes, J. Am. Chem. Soc. 107 (1985) 6052–6059.
[77] V. Böhmer, H. Goldmann, R. Kaptein, L. Zetta: Photo-CIDNP studies on calixarenes and bridged calixarenes, J. Chem. Soc., Chem. Commun. (1987) 1358–1360.
[78] C. D. Gutsche, L. J. Bauer: Calixarenes, 14. The conformational properties of the ethers and esters of the calix(6)arenes and the calix(8)arenes, J. Am. Chem. Soc. 107 (1985) 6059–6063.
[79] V. Böhmer, W. Lotz, J. Pachta, S. Tütüncü: Die Darstellung von Oligo[(hydroxy-1,3-phenylen)methylen]en mit Nitrophenol- und Alkylphenolbausteinen, Makromol. Chem. 182 (1981) 2671–2686.
[80] V. Böhmer, E. Schade, C. Antes, J. Pachta, W. Vogt, H. Kämmerer: A chain of intramolecular hydrogen bonds, an important factor for the acidity of oligo-[(2-hydroxy-1,3-phenylene)methylene]s, Makromol. Chem. 184 (1983) 2361–2376.
[81] V. Böhmer, E. Schade, W. Vogt: The first dissociation constant of calix(4)arenes, Makromol. Chem., Rapid Commun. 5 (1984) 221–224.
[82] S. Shinkai, K. Araki, H. Koreishi, T. Tsubaki, O. Manabe: On the acidity of the hydroxyl groups in calix(4)arenes and the dissociation-dependent conformational change, Chem. Lett. (1986) 1351–1354.
[83] V. Böhmer, D. Stotz, K. Beismann, W. Niemann: Kinetik der Bromierung von Phenolen und phenolischen Mehrkernverbindungen, 4. Der Einfluß des Nachbarbausteins auf die Reaktivität von Dihydroxydiphenylmethanen, Monatsh. Chem. 114 (1983) 411–423.
[84] V. Böhmer, K. Beismann, D. Stotz, W. Niemann, W. Vogt: Kinetics of the bromination of phenols and oligonuclear phenolic compounds, 6. Far reaching effects via chains of intramolecular hydrogen bonds, Makromol. Chem. 184 (1983) 1793–1806.
[85] V. Böhmer, D. Stotz, K. Beismann, W. Vogt: Kinetik der Bromierung von Phenolen und phenolischen Mehrkernverbindungen, 5. Dreikernverbindungen mit dem reaktiven Baustein in der Mitte, Monatsh. Chem. 115 (1984) 65–77.

[86] V. Böhmer, G. Stein: Kinetische Untersuchungen zur Reaktionsfähigkeit von Chlormethylgruppen in Methylendiphenolen, Makromol. Chem. 185 (1984) 263–279.
[87] G. Stein, V. Böhmer, W. Lotz, H. Kämmerer: Kinetische Untersuchung der Methanolyse von chlormethylierten Phenolen, Z. Naturforsch. 36b (1981) 231–241.
[88] G. Stein, H. Kämmerer, V. Böhmer: Kinetics of the reaction of chloromethylated phenols with aniline and substituted anilines, J. Chem. Soc., Perkin Trans. II (1984) 1285–1291.
[89] R. M. Izatt, J. D. Lamb, R.T. Hawkins, P. R. Brown, S. R. Izatt, J. J. Christensen: Selective M^+-H^+ coupled transport of cations through a liquid membrane by macrocyclic calixarene ligands, J. Am. Chem. Soc. 105 (1983) 1782–1785.
[90] S. R. Izatt, R.T. Hawkins, J. J. Christensen, R. M. Izatt: Cation transport from multiple alkali cation mixtures using a liquid membrane system containing a series of calixarene carriers, J. Am. Chem. Soc. 107 (1985) 63–66.
[91] S. Shinkai, H. Koreishi, K. Ueda, T. Arimura, O. Manabe: Molecular design of calixarene-based uranophiles which exhibit remarkably high stability and selectivity, J. Am. Chem. Soc. 109 (1987) 6371–6376.
[92] L. J. Bauer, C. D. Gutsche: Calixarenes, 15. The formation of complexes of calixarenes with neutral organic molecules in solution, J. Am. Chem. Soc. 107 (1985) 6063–6069.
[93] C. D. Gutsche, M. Iqbal, I. Alam: Calixarenes, 20. The interaction of calixarenes and amines, J. Am. Chem. Soc. 109 (1987) 4314–4320.
[94] S. Shinkai, K. Araki, O. Manabe: Does the calixarene cavity recognise the size of guest molecules? On the 'hole-size selectivity' in water-soluble calixarenes, J. Chem. Soc., Chem. Commun. (1988) 187–189.
[95] S. Shinkai, S. Mori, H. Koreishi, T. Tsubaki, O. Manabe: Hexasulfonated calix-(6)arene derivatives: A new class of catalysts, surfactants, and host molecules, J. Am. Chem. Soc. 108 (1986) 2409–2416.
[96] C. D. Gutsche: Calixarenes, in: F. J. Stoddart (ed.): Monographs in Supramolecular Chemistry, vol. 1, The Royal Society of Chemistry, Cambridge 1989.
[97] J. Vicens, V. Böhmer (eds.): Calixarenes, a Versatile Class of Macrocyclic Compounds, Kluwer, Dordrecht 1990.
[98] P. D. J. Grootenhuis, P. A. Kollman, L. C. Groenen, D. N. Reinhoudt, G. J. van Hummel, F. Ugozzoli, G. D. Andreetti: Computational study of the structural, energetical, and acid-base properties of calix[4]arenes, J. Am. Chem. Soc. 112 (1990) 4165–4176.

3 Synthesis of Monodisperse Oligomers

Manfred Rothe*

3.1 Introduction: Importance of Oligomers

Oligomers are defined as the low members of the homologous series of chain or ring molecules, which can be still distinguished from the consecutive members by different physical properties. Depending on the chemical structure of the chains this applies to compounds of molecular weights up to 1000–2000 approximately. Therefore, oligomers can be removed from the corresponding polymers by physical methods and separated into the individual monodisperse homologs, into dimers, trimers, tetramers etc. According to their structure they are subdivided into linear, branched, and cyclic homo- and cooligomers.

Monodisperse oligomers represent ideal model compounds for the corresponding polymers as they are low molecular weight compounds of the same chemical structure and of exactly defined degrees of polymerization. Physical studies of complete series of monodisperse oligomers furnish exact data concerning the relation between chain length and physical properties, e. g. the formation of stable conformations beginning with an exactly determined chain length. By means of defined oligomers certain physical properties and spectral data of polymers can be assigned, e. g. to certain end groups, to ring structures, occasionally even to certain conformations. Tables listing all known monodisperse oligomers and their physical properties have been compiled in the course of the studies performed in the Sonderforschungsbereich 41 [1].

As to their chemical behaviour, oligomers must on principle have the same properties as the corresponding polymers; they are, however, much easier accessible to all investigations owing to their low molecular weight and relatively high solubility. For this reason they can be used for studying the structure of the polymers and – in close relation – for the elucidation of the mechanisms of polymerization. In the Sonderforschungsbereich 41 the mechanism of the

* Lehrstuhl Organische Chemie II, Universität Ulm, D-7900 Ulm

cationic caprolactam polymerization and the structure of the resulting polyamides were elucidated in this way by our group [2,3] using synthetic N-(oligo-ε-aminocaproyl)caprolactam hydrochlorides for comparison.

These investigations are based on the occurrence of oligomers as intermediates in all polyreactions. In many polymers they have been found in more or less significant amounts due to equilibria between different chains and between chains and rings. Important conclusions on the structure of the corresponding polymers can be drawn from isolation and structure determination of oligomers. In particular, the type of linkage between the monomer units in the polymer and the structure of unknown end groups can be exactly detected. Finally, it is to be expected that the behaviour of oligomers under the conditions of the polyreaction will furnish unequivocal evidence for the mechanism of polymer formation and also of side reactions. If required, such reactions are performed with the addition of equimolar amounts of monomers in order to investigate the first propagation step separately. New functional groups resulting from side reactions can be detected much easier in the oligomer range, especially if they occur only in small amounts in the polymers.

Today, polydisperse oligomers are of particular interest as telechelics, which are used e. g. to produce polymers with hard and soft segments. They are dealt with in a separate chapter by W. Heitz.

On the other hand, the strict monodispersity of the molecules in the field of oligomers is the first and indispensable prerequisite for their use as model compounds for investigations of physical properties as a function of chain length and ring size. Studies of this kind were one of the main subjects of the Sonderforschungsbereich 41 in connection with work on conformational problems. Otherwise, average values are obtained for the physical properties which entail a considerable misinterpretation of the results, in particular in view of the high chain length dependence observed in the low molecular weight range. A remarkable example found during the studies in our laboratory [4] is the exact determination of the critical chain length for onset of helix formation of poly-L-prolines. Using monodisperse oligoprolines we obtained a completely different result compared with the preceding investigation of polydisperse low molecular polymers [5].

Monodisperse polymeric homologs are also indispensable as reference compounds for the detection and structure determination of long chains in the polymers, which at best have been obtained so far in form of chromatograms of the polymers showing successive peaks. This also applies in particular to the detection of large, possibly even polymeric rings in the polymerization products.

At the same time, oligomers obtained in monodisperse form can serve to show the molecular homogeneity and purity of the compounds isolated from the polymers. In view of only slight differences in all properties, the presence of small amounts of the lower or higher homologs often cannot be detected without the data given by authentic reference compounds. Therefore, methods to synthesize monodisperse oligomers have been the object of systematic studies in the Sonderforschungsbereich 41. The results obtained shall be dealt with in general and in detail in this paper.

3.2 Synthesis of Monodisperse Oligomers

3.2.1 Oligomer Formation

Monodisperse oligomers can be obtained by two basically different approaches. First, the polymers can be subjected to extraction or to thermal or solvolytic degradation to obtain mixtures of oligomers which subsequently have to be fractionated by chromatography. A similar procedure can be applied to mixtures of low molecular weight polymers which are formed by oligomerization of monomers under suitable, i. e. particularly mild conditions: short reaction times, temperatures as low as possible and high initiator concentrations. In this way we succeeded to elucidate the initial stages of the cationic lactam polymerization [2, 3, 6] in the course of the work of the Sonderforschungsbereich 41.

However, it is generally not possible to obtain in this way monodisperse linear oligomers in sufficiently large amounts for physical investigations. In fact, only cyclic oligomers of rather small ring size are present in sufficient concentration in the polymers to be isolated in pure form. On the other hand, the physical properties of short chain oligomers depend to a large extent on their end groups. Hence, linear homologs from oligomer mixtures differ far less distinctly than the corresponding rings and therefore have not yet been separated by chromatography on a preparative scale.

For this reason, the second and basically different route to oligomers, vic. the stepwise synthesis, is strongly preferred for these purposes. It starts from the monomers which are coupled in a stepwise manner under mild conditions according to the methods applied in low molecular weight organic chemistry. Furthermore, oligomers of different but defined chain length can be coupled as well to give homologs of a considerable degree of polymerization (segment synthesis). For a controlled stepwise or segment synthesis it is essential to use appropriate monofunctional monomer derivatives blocked temporarily at one of the two end groups. The coupling reaction leads to oligomeric derivatives protected at both ends which cannot polymerize any further even under harder conditions. They can be only contaminated by the shorter oligomeric educts protected at one end. However, in most cases the latter as well as possible by-products can be easily separated in the low molecular weight range due to different structures of the end groups. At the end of the synthesis the protecting groups are removed under mild conditions.

Stepwise and segment synthesis involving isolation and purification of the oligomeric intermediates has led so far to the formation of monodisperse chains containing 10 or more units for polycondensation and polyaddition products. In the case of vinyl polymers, however, the synthesis of monodisperse oligomers encounters difficulties during the first steps already, mainly due to the formation of stereoisomers at each coupling step which can hardly be fractionated. In this connection, a new remarkable method using a controlled polymerization of vinyl monomers bound to an oligomeric matrix was suggested by Kämmerer in Mainz [7].

3.2.2 Principles of Oligomer Synthesis

The stepwise and the segment syntheses of oligomers involving monofunctional coupling of the monomer units are the only routes known so far to obtain sufficiently large amounts of strictly monodisperse chain molecules. This also applies to very large rings with more than 50 ring atoms which have properties very similar to those of the polymer chains.

Previous efforts were directed to the synthesis of rather high molecular weight monodisperse oligomers, at least up to chain lengths beyond which all physical properties completely correspond to those of the polymers. Surprisingly often this was observed already from about 6–10 monomer residues onwards. First examples of such syntheses, in which the monodispersity of the products was checked chromatographically, were reported in the fifties and sixties by Kern and Wirth on oligophenylenes [8], by Kern and Heitz on oligourethanes [9], by Zahn on oligoesters and oligoamides [10], by Kämmerer on oligomeric phenol-formaldehyde condensates [11], as well as by Goodman on oligopeptides [12], and by Rothe on oligoamides [13] and oligopeptides [4].

To achieve stepwise or segment coupling reactions, common procedures of organic chemistry can be applied. As briefly indicated above, the principle of the synthesis is as follows: one of the two reactive functional groups is blocked temporarily by a reversibly cleavable protecting group under mild conditions, whereas the other end group is activated by conversion into a highly reactive derivative. The methods applied for this purpose have been developed during the last 30 years in the field of peptide, macrolide, and oligonucleotide synthesis. Coupling with another monomer molecule which is also monofunctionally protected – though at the other chain end – yields the dimer derivative blocked at both chain ends which is purified by recrystallization, distillation, or chromatography. Unreacted monomer derivatives can be easily removed due to their higher solubility or the different chemical structure of the end groups. Subsequently, one of the two protecting groups of the dimer is selectively cleaved and the activation and coupling steps can be repeated with the monoprotected oligomer. If higher oligomers shall be coupled to each other by segment synthesis, it is of crucial importance to consider the appropriate chain lengths of the oligomers involved in order to facilitate as far as possible the separation of unreacted educts from the end products by recrystallization or chromatography.

In case of the conventional stepwise or segment synthesis in solution, the maximum polymerization degree which can be achieved is met at molecular weights between 1000 and 1500, because the solubility of the coupling components drastically decreases in all solvents appropriate for synthesis. More important is that at higher chain lengths the properties of the polymeric homologs obtained do not differ sufficiently to allow a quantitative separation of educts and products or of possible by-products and to furnish the proof of this separation. This is the case even if longer chains are coupled to each other.

To overcome these solution problems the solid-phase method developed by Merrifield [14] in the peptide and oligonucleotide field can be also applied to synthetic polymers [15]. The basic idea is to bind the monomer unit, which is to be lengthened in a stepwise manner, covalently at one end to an insoluble polymeric support serving as a protecting group. The coupling reactions proceed in heterogeneous phase, allowing easy removal of excess educts as well as of by-products of the synthesis – all of which are soluble – by filtration and washing. The oligomeric chain, on the contrary, remains bound to the polymeric support during all steps of the synthesis which are performed in an analogous way and is only removed from the support in the end.

The decisive prerequisite for a successful synthesis of monodisperse oligomers according to this procedure is the strictly complete reaction of the resin-bound chain molecules at each step. Otherwise, mixtures of oligomers of all chain lengths formed during the synthesis will inevitably occur in the end product after the final removal from the support, because there is no possibility of purifying the polymer-bound intermediates. As indicated above, such mixtures can by no means be fractionated to yield molecularly homogeneous compounds.

Cyclic oligomers are frequently formed from polycondensates by thermal degradation at high temperatures. Generally, only the small monomeric and dimeric rings are volatile and can be easily obtained in pure state by vacuum sublimation. A controlled synthesis of cyclic oligomers consists of three stages: 1. the synthesis of monodisperse open-chain oligomers with free end groups as has been described above, 2. the activation of one of these groups, and 3. the intramolecular coupling of the reactive chain ends of these derivatives to form a ring under reaction conditions which prevent the polyreaction, that means applying the high dilution principle.

3.2.3 Synthesis of Monodisperse Oligoamides, Homo-Oligopeptides, and Sequential Oligopeptides

The controlled stepwise synthesis of oligomers can be elaborated best in the field of oligocondensates, in particular of oligoamides and oligopeptides, because in these cases the coupling reactions proceed in clearly distinguished steps. Therefore, the synthesis of monodisperse linear and cyclic oligoamides of the nylon types, especially of nylon-6 [16], as well as the preparations of collagen models, i. e. of homooligopeptides of proline [17,18] and sequential oligopeptides consisting of glycine and proline [19], were studied in the Sonderforschungsbereich 41 by the group of the present author. This work finally led to the synthesis of the most comprehensive series of monodisperse oligomers known so far. Linear oligoamides and peptides containing up to 175 chain atoms and cyclic oligoamides with up to 70 ring atoms could be obtained. They were used to study the dependence of the physical properties and the conformation on chain length and ring size, resp.

3.2.3.1 Linear Oligoamides of the Nylon Type and Collagen Models

Solution Synthesis

The coupling of several ω-amino acid molecules to form linear nylon oligomers can be performed according to methods common in peptide chemistry.

Previous studies of the synthesis of oligoamides had used activated derivatives (azides, mixed carbonic acid anhydrides) for the coupling reaction, which can lead to hardly separable by-products, as recent results have shown. For this reason, we decided to link the ω-amino acid residues by means of phosphorous acid diester chlorides or pyrophosphites (Equ. 1). These reagents attack the carboxylic group of N-protected amino acids with the intermediate formation of mixed phosphorous acid anhydrides which subsequently react with amino acid esters to form the amide bond. Activation and amide bond formation can be achieved in a one-pot reaction giving high yields and pure products [20].

$$(CH_3)_3COCO{-}[NH(CH_2)_5CO]_m{-}OH + Cl{-}P(OR)_2$$

$$\downarrow + NR_3$$

$$(CH_3)_3COCO{-}[NH(CH_2)_5CO]_m{-}O{-}P(OR)_2 \quad (1)$$

$$\downarrow + H[NH(CH_2)_5CO]_n{-}OC(CH_3)_3$$

$$(CH_3)_3COCO{-}[NH(CH_2)_5CO]_{m+n}{-}OC(CH_3)_3$$

In this way linear oligomers of the ε-aminocaproic acid could be synthesized in monodisperse form up to the decamer. The carbobenzoxy residue served as amino protecting group, the tert.butyl ester was used to block the carboxylic group. The coupling reaction was performed by means of phosphorous acid catechol ester chloride (o-phenylene chlorophosphite), the solvent used was diethyl phosphite, which possesses a high dissolving power for oligoamides of considerable chain length in contrast to all other solvents used so far for the synthesis of oligoamides. Consequently, even higher oligomers could be coupled to each other. Compared with the mixed carbonic acid anhydride method usually applied so far, this coupling method is considerably easier to handle and works also at rather high temperatures (100 °C). In this way, yields can be increased and a wide variety of coupling compounds of different chain length can be used. The above-mentioned combination of protecting groups offers the advantage to be selectively removable. The carbobenzoxy group could be removed by catalytic hydrogenation, whereas the tert.butyl ester was cleaved by anhydrous trifluoroacetic acid, each without attack on the other protecting group. Consequently, amide bond formation at either the amino or the carboxylic group can be performed with the same protected educt.

The above-mentioned limit of the coupling of polyamide units in solution is a consequence of their restricted solubility, due to cross-linking by strong intermolecular hydrogen bonds. In contrast, oligomers containing N-substituted amide groups can be expected to possess a drastically increased solubility, because these molecules cannot form hydrogen bonds to each other, but only to polar protic solvents.

For this reason, Rothe and coworkers [17] in the Sonderforschungsbereich 41 elaborated the synthesis of homo-oligopeptides of L-proline which are of interest as model compounds for the fibrous protein collagen and also as molecules with the shape of stiff rods. Till then the synthesis of free oligoprolines had encountered considerable difficulties and could be achieved only up to the dipeptide. These difficulties result, on the one hand, from steric hindrance of the peptide formation between several proline residues as secondary amine derivatives, on the other hand, from the special sensitivity of the Pro-X-bond (X = optional amino acid) to intramolecular nucleophilic attack from the

N-terminal end of the peptide chain, due to the high formation tendency of the 6-membered proline diketopiperazine for conformational reasons.

By means of the active ester method N-protected and free monodisperse oligo-L-prolines were synthesized up to the pentadecapeptide. The purity of all these compounds was shown by thin layer electrophoresis. The tert.butyloxycarbonyl residue served as amine protecting group. Peptide coupling was performed by reaction of N-protected oligoproline p-nitrophenyl esters with the alkali salts of free oligoprolines in aqueous pyridine at constant pH 9.0. Again, the choice of the appropriate chain length of the educts for peptide coupling was decisive for the formation of chromatographically homogeneous products. Subsequently, cleavage of the amino protecting group by trifluoroacetic acid and treatment with a weak basic anion exchanger furnished the free oligoprolines which are characterized by high solubility in water and high specific rotation values up to $-500°$.

In the same way, oligo-tripeptides of the sequence Gly-Pro-Pro protected at both ends and containing up to 7 tripeptide residues were synthesized and studied as collagen models in cooperation with the group of J. Engel, Biocenter, Basle [21].

Solid-Phase Synthesis

As outlined above, the main difficulties arising during the synthesis of series of monodisperse oligomers result from solubility problems and decreasing differences in physical properties between consecutive members with increasing chain lengths. Thus, the unambiguous chromatographical proof of their molecular homogeneity becomes impossible.

The work performed in the Sonderforschungsbereich 41, however, has shown the possibility of overcoming these difficulties by means of the solid-phase method developed by Merrifield and by application of electrophoretic techniques.

For the solid-phase synthesis a cross-linked chloromethyl polystyrene (2 % divinylbenzene) was used as insoluble polymeric support, to which the first α- or ω-amino acid residue of the oligomer chain was bound as the benzyl ester. As is well known, deprotection and stepwise coupling with the successive units take place in heterogeneous medium, and excess of educts and by-products of the synthesis – all of which are soluble – can be easily removed by filtration. In contrast, the oligomer chain remains fixed to the support during all following steps and will be cleaved only at the end of the synthesis.

The crucial prerequisite for the preparation of monodisperse oligomers is the really quantitative conversion of the polymer-bound oligoamide at each step. Otherwise, mixtures of oligomers of all intermediate chains will be obtained. Therefore, in the ω-amino acid series the usual solid-phase techniques were decisively improved by two modifications. After the acidic cleavage of the tert.butyloxycarbonyl protecting group the neutralization step was performed by means of strong bases, such as tetramethylguanidine. This turned out to be necessary, because the basicity of the standard tert.amines (e.g. triethylamine) was not sufficient to release the strongly basic ω-amino groups from their protonated forms. Moreover, the extension of the oligoamide chain was achieved [15,22] by the phosphite method using o-phenylene chlorophosphite which had proved well suited for the oligoamide synthesis in homogeneous solution. Amide coupling was carried out by a short reaction at 100°C in diethylphosphite as solvent. This allowed to use even the higher, sparingly soluble protected oligoamides for chain extension by several units at every step. Thus the number of possibly incomplete coupling steps is drastically reduced and the chromatographic separability of the oligomers obtained in successive steps is considerably improved. Hence, truncated sequences can be detected more easily.

Using this procedure we prepared monodisperse oligoamides with degrees of polymerization up to $n = 25$. In the initial steps the chain was extended by two monomer units at a time, and in subsequent steps by four units. The monodispersity of the oligomers obtained was proved by thin-layer and high-voltage paper electrophoresis in acetic acid/formic acid/water in the pH range of 1.8–0. A complete separation was attained up to a degree of polymerization of $n = 17$. Moreover, oligomers with $n = 21$ and 25 could be still distinguished by small, but defined differences in their electrophoretic behaviour. This is where the procedure reaches its limits. The purity of the oligomers was quantitatively determined in a stepwise synthesis up to penta-ε-aminocaproic acid (99.6%) (see Table 3.1). Pentacosa-ε-aminocaproic acid, $H[NH(CH_2)_5CO]_{25}OH$, (mol.wt. 2847, ≃ 218 A in the extended state) is one of the longest monodisperse chain molecules of identical monomer units synthesized so far.

In an analogous synthesis series of γ-aminobutyric acid oligomers up to a polymerization degree of 20 and of ω-aminoundecanoic acid up to $n = 5$ were prepared [25].

Similar synthetic problems also arise during the solid-phase synthesis of oligo-L-prolines, which can serve as model compounds for the polyproline helices and collagen.

The usual stepwise procedure was employed for the synthesis of homologous series of oligoprolines with up to 20 L-proline residues. A segment condensation on the resin using a protected triproline for chain extension permitted the preparation of oligoprolines with chain lengths as high as 40 residues (mol. wt. 3903) in analytically pure form (Table 3.2).

Difficulties were encountered at the dipeptide stage of the stepwise approach [26] which are attributed to the formation of the cyclodipeptide c-(Pro-Pro) during the neutralization step with tert.amine by intramolecular aminolysis of the proline benzyl ester group linked to the resin. Consequently, hydroxymethyl groups are formed on the resin and may be esterified in the following coupling steps with the protected proline. New and shorter oligo-

Table 3.1: Linear Oligoamides of ε-Aminocaproic Acid $H[NH(CH_2)_5CO]_nOH$.

n	Mol.Wt.	m.p. (°C) (calcd.*)	m.p. (°C) (found)
3	357.5	203.5	202–203
5	583.8	208.1	205–206
7	810.1	210.1	208–210
9	1036.5	211.3	209–210
11	1262.8	211.9	209–212
13	1489.1	212.4	210–211
15	1715.3	212.8	207–208
17	1941.5	213.1	209–211
21	2394.5	213.4	212–213
25	2847.0	213.5	212–213.5

*According to the equation of van der Wyk [23] using the constants a = 2.048×10^{-3} and b = 15.0×10^{-5}; see [24].

mers are then incorporated into the resin resulting in a contamination of the final product.

This new side reaction found by us [26] and subsequently by Merrifield [27] and others [28] is of general importance in solid-phase peptide synthesis, if proline- or N-methylamino acid-containing dipeptide-resins are involved. In the present case it can be easily overcome by coupling directly a protected tripeptide to the chloromethyl resin, because the dipeptide stage is left out.

The high purity of the oligoprolines obtained was established by electrophoresis, ion exchange, and gel permeation chromatography. For all oligomers with n up to 30, no contamination with shorter homologs could be detected even in the crude products cleaved from the resin by paper electrophoresis. Oligoprolines with n = 1–7 were completely separated by ion exchange chromatography on Aminex A 6; 0.1 % truncated sequences could be detected. High-voltage paper electrophoresis enabled us to separate even the highest members of the series (n = 19 and 20 in the stepwise synthesis, n = 37 and 40 in the segment condensation).

The purity of the synthesized oligoprolines in the crude state even after more than 20 coupling steps is surprisingly high. We attribute this result to the helix formation starting already at the tripeptide stage, as was found during our conformational studies in the course of the work of the Sonderforschungsbereich 41 described below (see 3.3.1; [17,18]). Accordingly, the rigid helical

Table 3.2: Oligo-L-prolines $\mathrm{H[N{-}CHCO]_nOH}$.

n	Mol.Wt.	m.p. (°C)	$[\alpha]_D^{22*}$	E_F-Values**			
1	115.2	221	− 86.5	1.00			
2	212.3	144	−171	0.84			
3	309.4	122	−220	0.72			
4	406.5	170	−291	0.60			
5	503.6	189	−338	0.52	1.00		
6	600.7	209	−374		0.91		
7	697.8	228	−394		0.77		
8	794.9	>280	−412		0.72		
9	892.0	>280	−435		0.66		
10	989.1	>300	−456		0.60	1.00	
11	1086.2	>300	−464			0.92	
12	1183.3	>300	−472			0.83	
13	1280.4	>300	−480			0.75	1.00
14	1377.5	>300	−491			0.70	0.94
15	1474.6	>300	−496			0.64	0.90
16	1571.7	>300	−499				0.86
17	1668.8	>300	−505		1.00		0.81
18	1765.9	>300	−509		0.95		
19	1863.0	>300	−513		0.90		
20	1960.1	>300	−520		0.84		
21	2057.3	>300	−521				
22	2154.5	>300	−523	1.00			
25	2445.7	>300	−526	0.93	1.00		
28	2737.0	>300	−528		0.92	1.00	
30	2931.3	>300	−530			0.95	
31	3028.5	>300	−532			0.93	1.00
34	3320.1	>300	−530	1.00			0.93
37	3611.4	>300	−528	0.93	1.00		
40	3902.7	>300	−532		0.94		

* c = 1, in water.
** Paper electrophoresis; buffer: formic acid/acetic acid/water (1:1:3), 1000–5000 V, 40–115 mA, 40–240 min.; paper 2043b, Schleicher-Schüll.

structure of the polymer-bound oligoprolines is expected to prevent the formation of randomly coiled peptide chains. Thus, the access of the coupling components to the terminal amino group is considerably improved.

Tripeptides in which one or two proline residues are replaced by other amino acids are more flexible and therefore show less complete segment coupling reactions on the polymeric support than expected. They must be purified thoroughly, e.g. by GPC in aqueous solution. Monodisperse oligotripeptides of the structures H[Pro-Gly-Gly]$_n$OH, n = 1–8, H[Pro-Pro-Gly]$_n$OH, n = 1–3, and H[Gly-Pro-(OAc)Hyp]$_n$OH, n = 1–4, were synthesized in electrophoretically pure form as defined collagen models [19]. Heidemann in Darmstadt obtained the oligotripeptides H[Pro-Ala-Gly]$_n$OH, n = 1–4.

3.2.3.2 Cyclic Oligoamides of the Nylon Type

Solution Synthesis

As briefly described in a preceeding Chapter (3.2.2) cyclic oligomers are generally obtained by ring-closure of linear oligomers activated at the carboxylic group at high dilution. The activation can be performed in two different ways applying an activated oligoamide derivative or the free oligoamide which is activated temporarily by condensation agents.

In the first case a reactive but stable derivative, preferably an activated ester, is prepared. The activated group must be introduced at an early step of the oligomer synthesis because higher oligomers cannot be activated quantitatively nor separated from the activated derivative for solubility reasons.

The second approach consists in the direct conversion of linear oligoamides into the corresponding rings by condensation agents with elimination of water. In our investigations unprotected oligoamides of several ω-amino acids were reacted with chlorophosphites or pyrophosphites to form mixed anhydrides as active intermediates which interact spontaneously with the free amino group at the other chain end to yield cyclic oligoamides. The advantage of this method is due to the use of easily accessible and stable educts and diethylphosphite as a good solvent for the polar oligoamides, the very simple handling techniques, and the efficiency of the ring-closure leading to higher yields than other methods. Too high a dilution must be avoided in this case because the activation of the oligomers by chlorophosphite is bimolecular and thus strongly concentration dependent. Hence, cyclo-oligomerization is favoured.

Obviously, the successful ring-closure largely depends on the selection of the suitable dilution. Concentrations as low as 0.001 M are required to suppress not only polycondensation but also cyclo-oligomerization leading to very large homologous rings. Starting with the N-protected activated oligoamide derivative, the protecting group must be removed in an acidic medium to yield the amine salt. In this way it is ensured that cyclization cannot occur before the solution has been suitably diluted.

Ring-closure is achieved by means of an excess of tert.amine leading to the "monomeric" rings in high yields of 60–80 %. At higher concentrations up to 0.1 M the intermolecular condensation and the cyclization of each chain molecule take place at the same time. Thus, homologous series of ring oligomers are obtained which can be efficiently fractionated by GPC after removal of linear oligomers by ion exchangers. They can be identified by mass spectrometry.

As an example, we condensed the amino acid β-alanine by the phosphite method under these conditions and obtained a series of cyclic oligo-β-alanines containing 3–11 β-alanine residues [29]. In the nylon-6 series, ring-closure of the individual linear oligomers yielded the complete series of monodisperse rings beginning with the dimer up to the decamer of caprolactam containing 14–70 ring atoms [16]. This series is of technical interest, too, because it can help to identify the ring compounds occurring in the polyamide in relatively large amounts which are extracted by water in the industrial process. With the aid of the phosphite cyclization (Equ. 2), the cyclic decamer of caprolactam was synthesized for the first time. In addition, the cyclic octamer and nonamer which had been obtained earlier only in impure state could be prepared in monodisperse form.

$$\begin{array}{c} H_2N\text{———————}COOH \\ \downarrow + X-P(OR)_2 \\ H_2N\text{———————}CO-O-P(OR)_2 \\ \downarrow - HO-P(OR)_2 \\ \underbrace{HN\text{———————}CO}_{} \end{array} \qquad (2)$$

$$X = Cl,\ O-P(OR)_2$$

Solid-Phase Synthesis of Cyclic Collagen Models

The cyclization of linear oligoamides rapidly reaches the limits of solubility with increasing chain lengths as well, in spite of the use of large quantities of solvent. In particular, it can be supposed that in case of higher concentrations of the linear educts, cyclo-oligomerizations will lead to extraordinarily large macrocycles the properties of which can be hardly distinguished from those of long oligoamide and polyamide chains. This accounts for the difficulty of detecting such large rings in the polyamide.

As a contribution to solve this problem we applied the method of ring-closure on polymeric supports developed by Fridkin et al. [30]. The basic idea is to bind the peptide chains by an active ester bond to the cross-linked support. Due to its rigid structure they should be sterically isolated from each other. At low degrees of substitution the peptide chains should then behave as at high dilution. In this way, intermolecular reactions of the linear activated peptides leading to polycondensation on the resin should be suppressed, but ring-closure by intramolecular aminolysis is possible. All the rings formed are removed from the support during the cyclization reaction and must be found in solution, whereas all oligomeric chains formed due to insufficient steric isolation remain fixed to the support. Thus, the separation problem mentioned above should be solved.

We therefore took up the problem again and studied the ring-closure of the linear oligopeptide Pro-Pro-Gly in order to obtain cyclic collagen models. The protected peptide was attached to a cross-linked (4-hydroxy)phenyl-sulfonyl polystyrene with the aid of carbodiimide by an active ester bond. After deprotection and neutralization of the protonated terminal amino group with tert.amine, the free polymeric peptide active ester was allowed to cyclize (see Equ. 3).

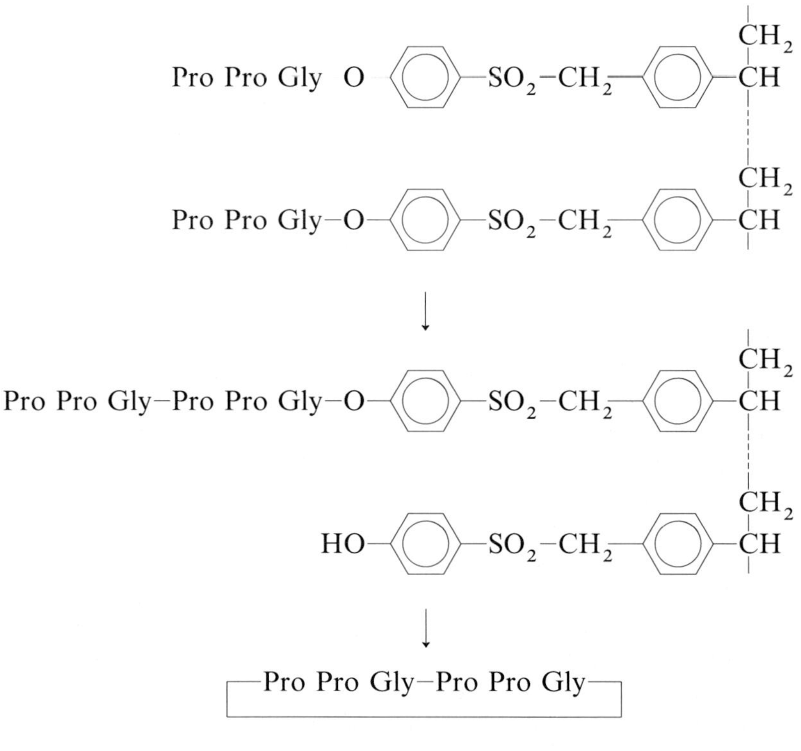

Pro Pro Gly O–C_6H_4–SO_2–CH_2–C_6H_4–CH(CH_2)–

Pro Pro Gly–O–C_6H_4–SO_2–CH_2–C_6H_4–CH(CH_2)–

↓ (3)

Pro Pro Gly–Pro Pro Gly–O–C_6H_4–SO_2–CH_2–C_6H_4–CH(CH_2)–

HO–C_6H_4–SO_2–CH_2–C_6H_4–CH(CH_2)–

↓

cyclo(–Pro Pro Gly–Pro Pro Gly–)

Remarkably, in all the cyclizations studied so far, large rings with up to 36 ring atoms were obtained, which were obviously formed during intraresin reactions by cyclo-oligomerization. They could be separated and identified by GPC and MS as described above. In this way, a series of ring oligomers of the structure cyclo-(Pro-Pro-Gly)$_n$ with n = 2–4 was obtained in moderate yields [31].

It can be concluded from this work that the cross-linked polystyrene support is not rigid enough to ensure a steric isolation of the grafted peptide chains. Therefore, the application of this principle will open a route to prepare pure cyclic oligoamides with very large ring sizes by cyclo-oligomerization of short, easily accessible chains.

3.3 Physical Properties of Synthesized Monodisperse Oligoamides and Oligopeptides

The final chapter of this report deals with the physical properties of homologous linear and cyclic oligoamides and oligopeptides. A few examples synthesized during the work in the Sonderforschungsbereich 41 will be presented to point out the importance of complete series of monodisperse oligomers.

3.3.1 Chain-Length Dependence of the Conformation of Linear Oligo-L-Prolines

Poly-L-proline exists in two unique helical conformations which are stabilized mainly by steric constraints due to the five-membered pyrrolidine ring and by polymer-solvent interactions. Polyproline I represents a compact right-handed helix containing only cis peptide bonds, whereas polyproline II is a highly extended left-handed helix with all-trans peptide bonds.

In our group monodisperse oligo-L-prolines were used as defined models to study helix stabilization with increasing chain-lengths and helix-helix interconversion by changes in solvent. Table 3.2 shows the characteristically high optical rotation of oligoprolines up to n = 40 in aqueous solution corresponding to polyproline II. In poor solvents such as n-propanol the peptides show a slow mutarotation leading to less negative values. This is due to a change in conformation analogous with the conversion of polyproline II into polyproline I which is caused by different solvation of the carbonyl groups. This conformational change is reversible in good solvents such as water or acetic acid. The differences in the optical rotation strongly increase with increasing chain-lengths and amount to as high as 500° with the decapeptide.

Hexaproline is the first member of the oligoprolines which can be isolated in the solid state from the suitable solvent in form of helix I and helix II which can be distinguished by IR-spectroscopy. In solution either helix can be identified by its characteristic CD-spectrum.

The CD-spectra of the oligoprolines up to n = 40 in aqueous and n-propanol solutions can be seen in Figs. 3.1 and 3.2. They show the features typical of helix I and helix II, resp. [4, 18]. Oligoprolines I are characterized by 3 Cotton

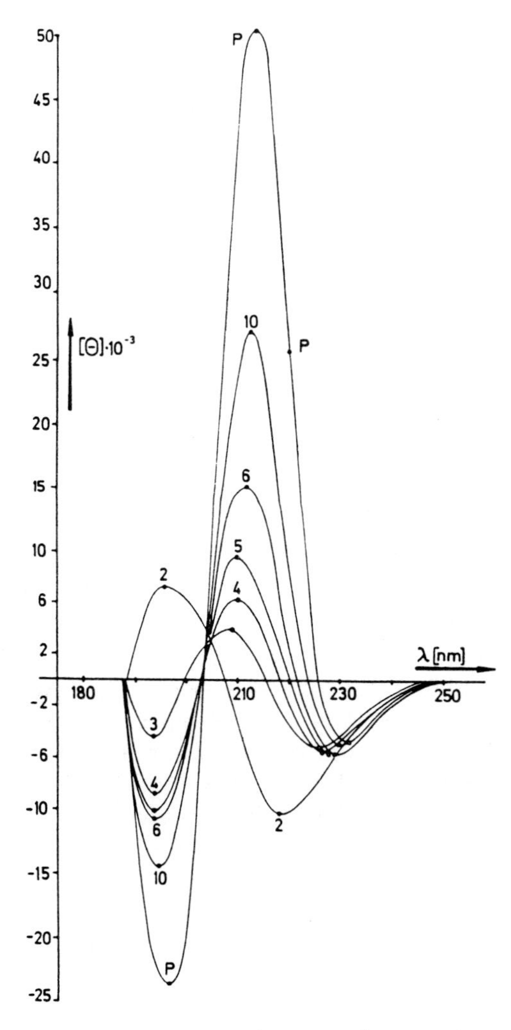

Fig. 3.1: CD spectra of Oligo-L-prolines (n = 2–10) and of Poly-L-proline I (P) in Water/n-Propanol (1:9).

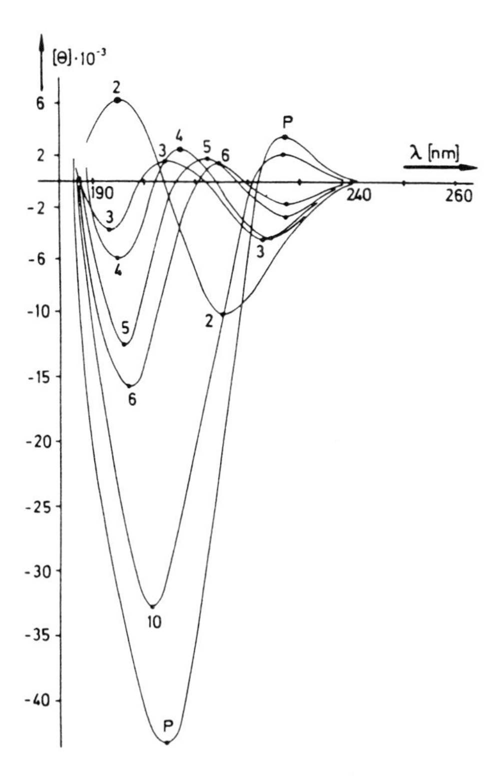

Fig. 3.2: CD spectra of Oligo-L-prolines (n = 2–10) and of Poly-L-proline II (P) in Trifluoroethanol.

effects in n-propanol: a weak negative absorption maximum at 225–232 nm, a strong positive maximum at 208–213 nm, and a strong negative absorption at 193–198 nm. Oligoprolines II show only 2 Cotton effects in water: a weak positive absorption at 227–228 nm, and a strong negative band at 197–205 nm. The Cotton bands are shifted to higher wavelengths with increasing chain lengths; the intensities increase as well reaching the characteristic ellipticities of the polymers at $n \simeq 20$. Helical conformations are formed already with triproline as indicated both by mutarotation and by CD measurements. In fact, the polyproline I helix contains 3 1/3 proline residues per turn; thus triproline can form a helix turn stabilized by electrostatic interaction of the charged end groups [4, 18].

3.3.2 Ring-Size Dependence of Physical Properties

Cyclic oligomers or at least their low members differ basically from linear oligomers in their physical properties. In contrast to the short chains they usually crystallize excellently and are far better soluble in water or organic solvents. Generally, they possess markedly higher, defined melting points and can be sublimed or distilled in vacuo due to their compact structure.

On the other hand, X-ray studies show that cyclic oligomers with large ring sizes are arranged as parallel double chains linked at their ends. Such rings possess properties very similar to those of the polymeric chains. In these cases they can also serve as models for polymers free of end groups and in particular for turns in polymers.

From the beginnings of oligomer chemistry the cyclic oligomers of caprolactam have aroused special interest. In 1958, caprolactam oligomers up to the nonamer were the first examples of a homologous series of ring molecules which was found in polymers [13]. They are formed in relatively large quantities (3–4 %) in the hydrolytic caprolactam polymerization. For technical application of nylon-6, they have to be extracted from the polyamide by hot water. The identification of the extracted compounds is considerably facilitated with the aid of the synthesized rings. Their properties and chromatographical separability are listed in Table 3.3.

Smaller rings are subjected to particularly strong steric restrictions so that only few conformations are allowed. Therefore they can serve as models for

Table 3.3: Cyclic Oligomers of Caprolactam $\overline{[NH(CH_2)_5CO]_n}$.

n	Mol.Wt.	m.p. (°C)	R_F (BEW*)
2	226.3	348	0.52
3	339.5	244	0.45
4	452.6	250	0.40
5	565.8	254	0.34
6	678.9	258	0.30
7	792.1	243	0.28
8	905.3	246	0.27
9	1018.4	240	0.25
10	1131.8	248	0.19

* TLC: n-butanol/acetic acid/water (4 : 1 : 1).

these particular conformations occurring in polymers. An example taken from our work is the cyclic triproline, a nearly rigid ring consisting of 3 L-proline residues with 3 cis peptide bonds. It can be regarded as model for the poly-proline I helix which also contains cis peptide bonds and 3 1/3 proline residues per turn, as can be seen from a top view down the helix in which every fourth residue takes an almost identical position. Hence small rings can be considered as projections of helices onto the paper plane.

In the case of cyclotriprolyl and the polyproline I helix, this high conformational coincidence of rings and helices is convincingly demonstrated by means of the CD-spectra, because in both conformations the steric orientation of the peptide chromophores is much alike [4] (see Fig. 3.3).

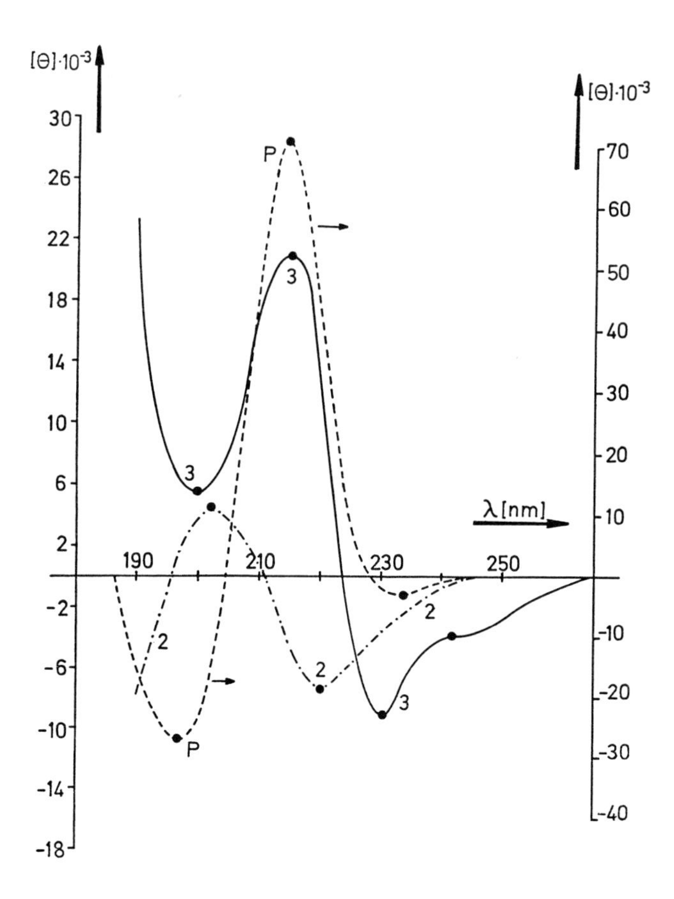

Fig. 3.3: CD spectra of Cyclo-di-L-prolyl (2), Cyclo-tri-L-prolyl (3), and of Poly-L-proline I (P) in Trifluoroethanol.

3.4 Outlook

After our synthetic and conformational work in the Sonderforschungsbereich 41 was terminated in 1975, more than 100 further papers on monodisperse oligomers have appeared till now, which emphasize the increasing importance of oligomers as polymer models, especially in the field of vinyl polymers.

Forthcoming studies will have to consider the substantial progress in chromatographic separation made during the last 15 years and to improve the synthetic methods applied so far accordingly.

Then the goal of synthesizing markedly longer chains and larger rings in monodisperse form should be achieved by employing HPLC and solid-phase synthesis, thus progressing towards the range of polymers and contributing to higher exactness of results in this field.

3.5 References

*[1] M. Rothe: Physical data of oligomers, in J. Brandrup, E. H. Immergut (eds.): Polymer Handbook, 2nd ed. Wiley, New York 1975, VI-1–VI-48.

*[2] M. Rothe, G. Bertalan, J. Mazánek: Die kationische Caprolactam-Polymerisation, Chimia 28 (1974) 527–533.

*[3] M. Rothe, G. Bertalan: Mechanism of the cationic polymerization of lactams, in T. Saegusa, E. Goethals (eds.): Ring-Opening Polymerization, ACS Symposium Series, No. 59. Amer.Chem.Soc., Washington 1977, 129–144.

*[4] M. Rothe, R. Theysohn, K.-D. Steffen, H.-J. Schneider, M. Zamani, M. Kostrzewa: Helixbildung bei Prolinpeptiden, in E. Scoffone (ed.): Peptides 1969, Proc. 10th Europ. Peptide Symposium, North-Holland Publ. Co., Amsterdam 1971, 179–188.

[5] A. Yaron, A. Berger: Helix formation in short poly-L-proline chains, Bull. Res. Counc. Israel 10A (1961) 46–47.

*[6] G. Bertalan, M. Rothe: Amidin-Endgruppen bei der mit Caprolactam-hydrochlorid initiierten Caprolactam-Polymerisation, Makromol.Chem. 172 (1973) 249–254, 1015.

[7] H. Kämmerer, J. S. Shukla, G. Scheuermann: Die Herstellung einer molekulareinheitlichen Tetramethylacrylsäure mittels verschiedener Matrizen, Makromol. Chem. 116 (1968) 72–77.

[8] H. O. Wirth, F. U. Herrmann, W. Kern: Synthese und Eigenschaften von p-Oligophenylenen mit p-Xylol und mit Durol als Grundbausteinen, Makromol. Chem. 80 (1964) 120–140 and preceding papers.

* Contributions of the Sonderforschungsbereich 41, 1969–1975, are marked by *.

[9] W. Heitz, H. Höcker, W. Kern, H. Ullner: Darstellung und Eigenschaften von linearen Oligourethanen mit Phenylendgruppen und von cyclischen Oligourethanen aus Diethylenglycol und Hexamethylendiisocyanat, Makromol.Chem. 150 (1971) 73–94 and preceding papers.
[10] H. Zahn, G.B. Gleitsmann: Oligomere und Pleionomere von synthetischen faserbildenden Polymeren, Angew.Chem. 75 (1973) 772–783 and preceding papers.
[11] H. Kämmerer: Über Phenol-Formaldehyd-Kondensate definierter Konstitution und einheitlicher Molekülgröße, Angew.Chem. 70 (1958) 390–398.
[12] M. Goodman, R.P. Saltman: NMR studies on linear homo-oligopeptides: a perspective view, Biopolymers 20 (1981) 1929–1948.
[13] M. Rothe: Polymerhomologe Ringamide in Polycaprolactam, J. Polymer Sci. 30 (1958) 227–238.
M. Rothe: Lineare und cyclische Oligomere, Makromol.Chem. 35 (1960) 183–199.
M. Rothe: Lineare und cyclische Oligomere. Habilitationsschrift Univ. Halle 1960.
[14] R.B. Merrifield: Solid phase peptide synthesis, I. The synthesis of a tetrapeptide, J.Amer.Chem.Soc. 85 (1963) 2149–2154.
G. Barany, R.B. Merrifield: Solid-phase peptide synthesis, in E. Gross, J. Meienhofer (eds.): The Peptides, Vol. 2A. Academic Press, New York 1980, 1–284.
[15] M. Rothe, H.-J. Schneider, W. Dunkel, Makromol.Chem. 96 (1966) 290–294.
M. Rothe, W. Dunkel: Synthesis of monodisperse oligomers of ε-aminocaproic acid up to a degree of polymerization of 25 by the Merrifield method, J.Polymer Sci., Polymer Lett. 5 (1967) 589–593.
*[16] U. Kress, Diss. Univ. Mainz 1974.
*[17] M. Rothe, R. Theysohn, K.-D. Steffen: Synthese von Oligo-L-prolinen bis zum Pentadecapeptid, Tetrahedron Lett. (1970) 4063–4066.
*[18] M. Rothe, J. Mazánek: Synthese von Oligo-L-prolinen bis zum Eikosameren nach der Festphasenmethode, Tetrahedron Lett. (1972) 3795–3798.
M. Rothe, H. Rott, J. Mazánek: Solid-phase synthesis and conformation of monodisperse high molecular weight oligo-L-prolines, in A. Loffet: Peptides 1976, Proc. 14th Europ. Peptide Symposium, Edit. l'Univ. Bruxelles 1976, 309–318.
*[19] Th. Doll: Diss. Univ. Mainz 1975.
[20] R.W. Young, K.H. Wood, R.J. Joyce, G.W. Anderson: The use of phosphorous acid chlorides in peptide synthesis, J.Amer.Chem.Soc. 78 (1956) 2126–2131.
*[21] P. Bruckner, B. Rutschmann, J. Engel, M. Rothe: A chemical synthesis of collagen-like peptides with the sequence Z(Gly-Pro-Pro)$_n$OBut and their characterization with circular dichroism and ultracentrifugation, Helv.Chim.Acta 58 (1975) 1276–1287.
*[22] E. Kiss: Diss. Univ. Mainz 1974.
[23] K.H. Meyer, A. van der Wyk: Solubilité des séries homologues et polymères-homologues, Helv.Chim.Acta 20 (1937) 1313–1320.
[24a] H. Zahn, D. Hildebrand: Zur Kenntnis der linearen Oligoamide der ε-Aminocapronsäure, Chem.Ber. 90 (1957) 320.

* Contributions of the Sonderforschungsbereich 41, 1969–1975, are marked by *.

[24b] H. Zahn, D. Hildebrand: Nonakis-, Decakis-, Undecakis- und Dodecakis-ε-aminocapronsäure, Chem.Ber. 92 (1959) 1963.
*[25] E. Bigdeli: Diss. Univ. Mainz 1970.
*[26] M. Rothe, J. Mazánek: Nebenreaktionen bei der Festphasen-Peptidsynthese als Folge der Bildung von Cyclopeptiden, Angew.Chem. 84 (1972) 290–291; Angew.Chem.Int.Ed.Engl. 11 (1972) 293–294.
M. Rothe, J. Mazánek: Intrachenare und interchenare Aminolyse der Benzylesterbindung zum polymeren Träger und ihre Auswirkungen, Liebigs Ann.-Chem. (1974) 439–459.
[27] B.F. Gisin, R.B. Merrifield: Carboxyl-catalyzed intramolecular aminolysis. A side reaction in solid-phase peptide synthesis, J.Amer.Chem.Soc. 94 (1972) 3102–3106.
[28] M. C. Khosla, R. R. Smeby, F. M. Bumpus: Failure sequences in solid-phase peptide synthesis due to the presence of an N-alkylamino acid, J.Amer.Chem.Soc. 94 (1972) 4721–4724.
*[29] M. Rothe, D. Mühlhausen: Cyclooligocondensation of amino acids: macrocyclic amides of β-alanine, Angew.Chem. 91 (1979) 79–80; Angew.Chem.Int.Ed.Engl. 18 (1979) 74–75.
[30] M. Fridkin, A. Patchornik, E. Katchalski: A synthesis of cyclic peptides utilizing high molecular weight carriers, J.Amer.Chem.Soc. 87 (1965) 4646–4648.
*[31] M. Rothe, A. Sander, W. Fischer, W. Mästle, B. Nelson: Interchain reactions (cyclo-oligomerizations) during the cyclization of resin-bound peptides, in M. Goodman, J. Meienhofer (eds.): Peptides. Proc. 5th Amer. Peptide Symposium. Wiley, New York 1977, 506–509.

* Contributions of the Sonderforschungsbereich 41, 1969–1975, are marked by *.

4 Telechelic Oligomers

Walter Heitz*

Abstract

Telechelics are strictly bifunctional oligomers or low molecular weight polymers. According to their glass transition temperature and melting temperature resp. as well as their field of application soft and hard segments of telechelics can be distinguished. Telechelic soft segments are obtained by cationic, anionic, radical, and metal catalyzed reactions. Telechelic hard segments are preferably synthesized by polycondensation.

4.1 Introduction

The synthesis of block structures is a means to adjust properties of polymers. Condensation polymers with block structures are stable only if one of the constituent blocks is chemically inert against reactions like transesterification. Preferably this stable block consists of a pure CC-chain. It must have exactly two functional end groups in order to be no factor limiting molecular weight in the subsequent polycondensation process. These two functional groups can be either located at both chain ends (telechelics) [1] or one chain end is bifunctional (macromonomers) [2]. The use of telechelics in polycondensation results in blockcopolymers, macromonomers will give graft copolymers (Fig. 4.1).

The characterization of telechelics needs the determination of the functionality. Any end group analysis will give only average values. An efficient chromatographic method is necessary to guarantee that bifunctionality is caused by one polymer homologous series.

* Zentrum für Materialwissenschaft, Fachbereich Physikalische Chemie, Polymere, Philipps-Universität, D-3550 Marburg

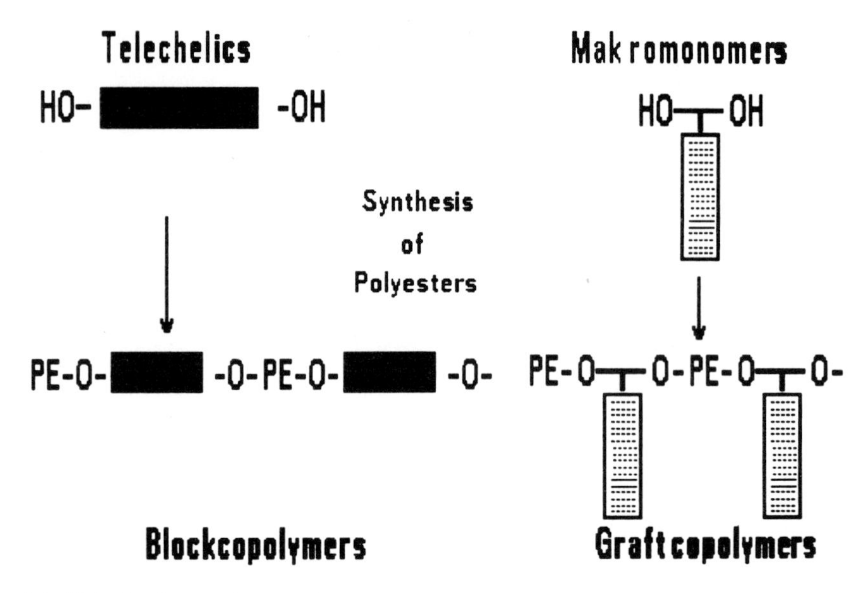

Fig. 4.1: Telechelics and macromonomers as precursors.

Using telechelics an essential aim is the modification of polycondensation products by soft segments. Soft segments should have a glass transition temperature well below the application temperature of the final block copolymer. These telechelic soft segments can be obtained by cationic, anionic, or radical polymerization.

Telechelic hard segments have a high glass transition temperature or melting point resp. In principal they can be obtained by polymerization of vinyl monomers with appropriate substituents. But the resulting main chain is constituted of sp^3 CC-bonds with a limited thermal stability. Aromatic units in the main chain will give enhanced thermal stability. Therefore polycondensation is the proper way to synthesize telechelic hard segments.

The inhomogeneity (M_w/M_n) is smaller than given by classical theories of polymerization (Table 4.1). The main reason is that the long chain approximation cannot be applied in this case.

Table 4.1: M_w/M_n values of mono- and bifunctional oligomers.

Structure	Polym.	M_n	M_w/M_n exp.	M_w/M_n theor.
$Ac(OCH_2CH_2CH_2CH_2)_nOAc$	cationic	390 535	1.27[a)] 1.46[a)]	1.25 1.43
$NC-C(CH_3)_2-(CH_2-CH=CH-CH_2)_n-C(CH_3)_2-CN$	radical	275 330	1.23 1.27	
$H-(C_6H_2(CH_3)_2-O)_n-H$	oxydativ	530	1.41[b)]	

a) corrected for end groups, b) excluding monomer

4.2 Telechelics by Cationic Polymerization

The cationic polymerization is usually initiated by protons. This will result in dead end groups in the polymerization of vinyl monomers. Kennedy has offered a solution to this problem using the inifer polymerization of i-butene [3]. Telechelic poly(vinyl ether) have been prepared by Higashimura using HI/I_2 as the initiating system [4].

A number of commercial telechelic polytetrahydrofurans are strictly bifunctional as shown by GPC [5]. Two functional end groups can be introduced either by bifunctional initiation and killing the living end groups by a proper nucleophile. The disadvantage of this method is that one mole of an expensive initiator is needed to produce one mole of polymer. The second method to introduce functional end groups uses a proper transfer agent. Only catalytic amounts of initiator are necessary in this case.

Carboxylic acid anhydrides are such transfer agents. Using acetic anhydride a transfer reaction will produce an ester end group and regenerates the initiating species, an acylated THF cation. In fact, polymerization of THF in presence of acetic anhydride produces a polytetrahydrofuran with strictly two acetate end groups (Fig. 4.2, [6]).

The determination of the transfer constant at conversions lower than 5 % results in a value of $C_{Ac_2O} = 0.059$. The low transfer constant expresses the

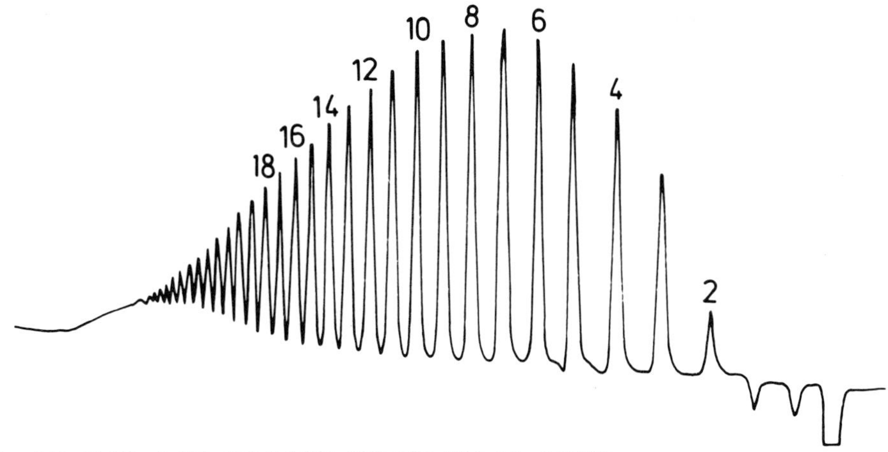

Fig. 4.2: GPC of $CH_3COO(CH_2CH_2CH_2CH_2O)_nOCCH_3$.

that THF is a much better nucleophile than acetic anhydride. Due to this low transfer constant THF is consumed much faster than acetic anhydride in a kinetically controlled reaction. But approaching the equilibrium concentration of THF, the net conversion of THF is approaching zero and now the thermodynamically controlled equilibration reaction is dominating.

This is reflected in the dependence of molecular weight from the reaction time. In case of short reaction times, when the conversion is low, the molecular weight increases with time and goes through a maximum. The THF conversion is approaching zero, simultaneously acetic anhydride is consumed in a slow reaction; as a consequence, the molecular weight is going down. The central point of this reaction is not the polymerization but the equilibration, as this determines the total reaction time. At long reaction times a limited P_n is obtained which is given by the ratio of consumed THF to consumed end group forming species.

$$P_n = \frac{\Delta\,[THF]}{[Ac_2O] + [I]} \tag{1}$$

Acetic anhydride is completely consumed at the end of the reaction. Any desired molecular weight can be adjusted by the ratio of the reactants (Fig. 4.3).

The polymerization of THF is an equilibrium polymerization (Fig. 4.4).

$$P_n^* + THF \rightleftarrows P_{n+1^*} \tag{2}$$

$$K_n = \frac{[P^*_{n+1}]\,[THF]_e}{[P^*_n]} \tag{3}$$

At polymerization degrees < 20 the equilibrium concentration is dependent on the average degree of polymerization, but of course within one experiment identical for all degrees of polymerization.

To calculate the equilibrium constant we make use of the long chain approximation, i. e. the difference in concentration of two neighbouring species is neglected. The mole ratio of two neighbouring active species is equal to the probability of monomer addition. At high molecular weights this probability is practically one and thus the equilibrium constant is the reciprocal of the equilibrium concentration. But at low molecular weights this is not admissible and we observe a dependence of the equilibrium concentration from the degree of polymerization. In general the calculation of the equilibrium constant is only possible if we know the concentration of two neighbouring active species. Being at the equilibrium the distribution of active species should be the same as in the final product. Therefore the mole ratio of two neighbouring species is obtained from the intensity ratio of two neighbouring peaks in GPC. A peculiarity of this distribution is that P = 1 is present in very small amounts, i. e. K_2 must be larger than the other K_n.

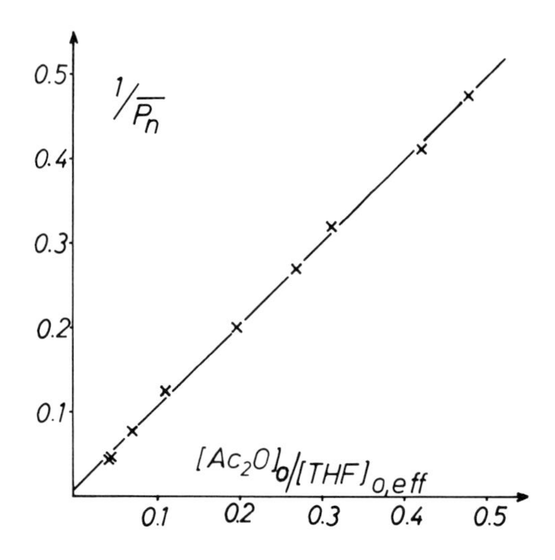

Fig. 4.3: Dependence of equilibrium P_n in THF polymerization in presence of Ac_2O ($HSbF_6$, 10 °C, CH_2Cl_2).

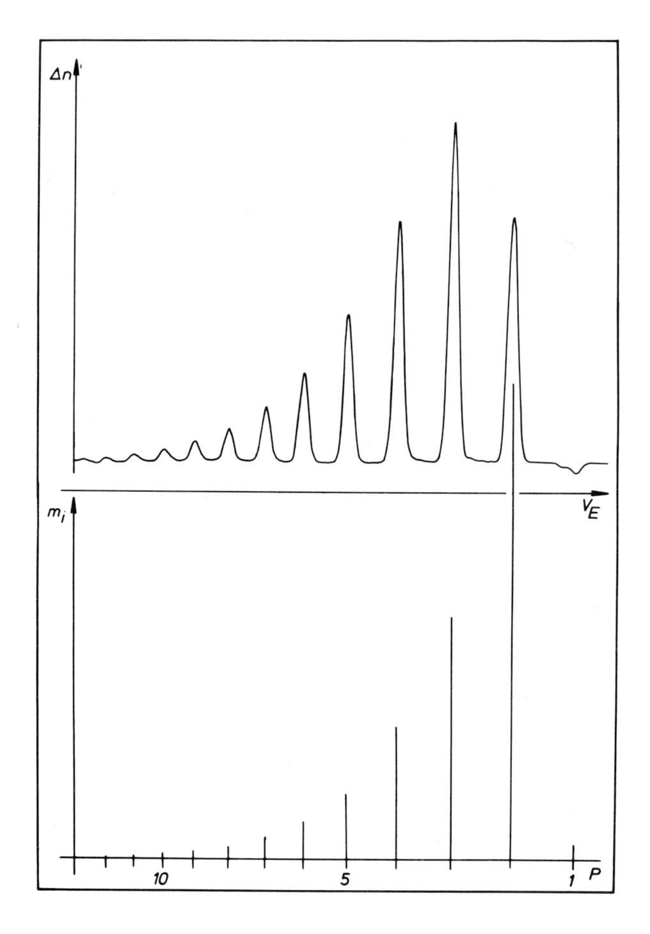

Fig. 4.4: GPC of $CH_3COO(CH_2CH_2CH_2CH_2O)_nOCCH_3$ and the corresponding mass distribution in equilibrium (48 h), $THF/Ac_2O = 3.8$, $P_n = 2.73$.

As shown by Table 4.2 the equilibrium constants are equal within the limits of error with the exception of K_2.

Assuming that the equilibrium constant K_2 is larger than the other K_n it can be rigorously shown [7] that

$$\frac{P_n - 2}{P_n - 1} = b - \frac{1}{ab} - \frac{b}{a} + \frac{2}{a} \tag{4}$$

$$K_n\,[M]_e = b;\; b < 1;\; K_2 = aK_n$$

Table 4.2: Equilibrium constants of the THF polymerization determined by GPC.

		n	K_n
$P_1^* + M \rightleftharpoons P_2^*$	$K_2 = [P_2^*]/[P_1^*][M]_e$	2	(1.86)
$P_2^* + M \rightleftharpoons P_3^*$	$K_3 = [P_3^*]/[P_2^*][M]_e$	3	0.24
$P_3^* + M \rightleftharpoons P_4^*$	$K_4 = [P_4^*]/[P_3^*][M]_e$	4	0.25
|	|	5	0.24
|	|	6	0.24
|	|	7	0.25
|	|	8	0.25
$\overline{P}_n = 5.03$		9	0.23
$[M]_e = 3.27$ mole l^{-1}		10	0.24
$\overline{K}_n = 0.24$			

If K_2 is much larger than K_n, which is equivalent to P_1 not being formed or a »1, then

$$\frac{P_n - 2}{P_n - 1} = K_n\,[M]_e \tag{5}$$

The molecular weight distribution at equilibrium conditions is a Schulz-Flory distribution. Nevertheless M_w/M_n gives values of about 1.3. The main reason is that the long chain approximation cannot be applied in this case (Table 4.1).

Other acid anhydrides can also be used as transfer agents. Of special interest are acrylic, methacrylic and maleic anhydride [7].

$$\text{THF} + (CH_2{=}CH\overset{\displaystyle O}{\overset{\|}{C}})_2O \longrightarrow CH_2{=}CH\overset{\displaystyle O}{\overset{\|}{C}}{-}O[(CH_2)_4{-}O]_n\overset{\displaystyle O}{\overset{\|}{C}}CH{=}CH_2 \tag{6}$$

$$+ \left(CH_2{=}\overset{CH_3}{\overset{|}{C}}{-}\underset{O}{\underset{\|}{C}}\right)_2 O \longrightarrow CH_2{=}\overset{CH_3}{\overset{|}{C}}{-}\underset{O}{\underset{\|}{C}}{-}O[(CH_2)_4{-}O]_n\overset{O}{\overset{\|}{C}}{-}\underset{CH_3}{\underset{|}{C}}{=}CH_2$$

$$+ \text{maleic anhydride } \left(O{<}\begin{matrix}\overset{O}{\overset{\|}{C}}{-}CH\\ \| \\ \underset{O}{\underset{\|}{C}}{-}CH\end{matrix}\right) \longrightarrow \left([(CH_2)_4O]_n{-}\overset{O}{\overset{\|}{C}}{-}CH{=}CH{-}\overset{O}{\overset{\|}{C}}{-}O\right)_x$$

The corresponding acids are stronger acids and these anhydrides are weaker nucleophiles. So we can expect that the transfer constants are smaller. The experimental consequence is that the molecular weight obtained at the beginning of the reaction is higher and the total reaction time, necessary to come to equilibration, is getting longer. Using acrylic and methacrylic anhydride a THF with two polymerizable end groups was obtained. With maleic anhydride this reaction is so slow that no useful amounts are introduced into the PTHF chain in reasonable times. The necessary reaction time increases in the expected order acetic <acrylic<methacrylic anhydride.

$$\text{THF} \xrightarrow{H^{\oplus}} \sim\!\sim\!\sim\overset{\oplus}{O}\!\!\left\langle\text{(ring)}\right. \quad \text{I} \tag{7}$$

$$\text{I} + AcOCH_2CH_2CH_2CH_2OAc \not\rightarrow AcO[(CH_2)_4O]_nAc$$

$$\text{I} + CH_2{=}CHCOO(CH_2)_4OOCCH{=}CH_2 \not\rightarrow CH_2{=}CH\overset{O}{\overset{\|}{C}}O[(CH_2)_4O]_n\overset{O}{\overset{\|}{C}}CH{=}CH_2$$

$$\text{I} + CH_2{=}\overset{CH_3}{\overset{|}{C}}{-}COO(CH_2)_4OOC\overset{CH_3}{\overset{|}{C}}{=}CH_2 \not\rightarrow CH_2{=}\overset{CH_3}{\overset{|}{C}}{-}CO[(CH_2)_4O]_n\overset{O}{\overset{\|}{C}}{-}\underset{CH_3}{\underset{|}{C}}{=}CH_2$$

Butanediol acrylate is better available and more easy to handle than acrylic acid anhydride. According to the given reaction scheme the anhydride reacts as a nucleophile with the oxonium ion. Esters are better nucleophiles than anhydrides. But the expected reaction failed to take place. Of course polymerization is observed but within one week there is no consumption of these esters within the limits of error. It is the equilibration process which does not take place. With the ester of the dimer equilibration is observed.

The experiments described up to now were done with protonic acids like $HSbF_6$. By use of none-protic initiators like acylium or trialkyloxonium ions the course of the reaction is dramatically changed. The system is getting highly viscous and it is not stirable after a short reaction time. Using acetic anhydride the viscosity is going down slowly. The consumption of the anhydride is much slower than by initiation with protonic acids. Acrylic anhydride and methacrylic anhydride do not result in a reduction of the viscosity. Even after one week of reaction time their concentration is not changed. The equilibration process is cut off in the absence of protons. The results are summarized in Equ. (8).

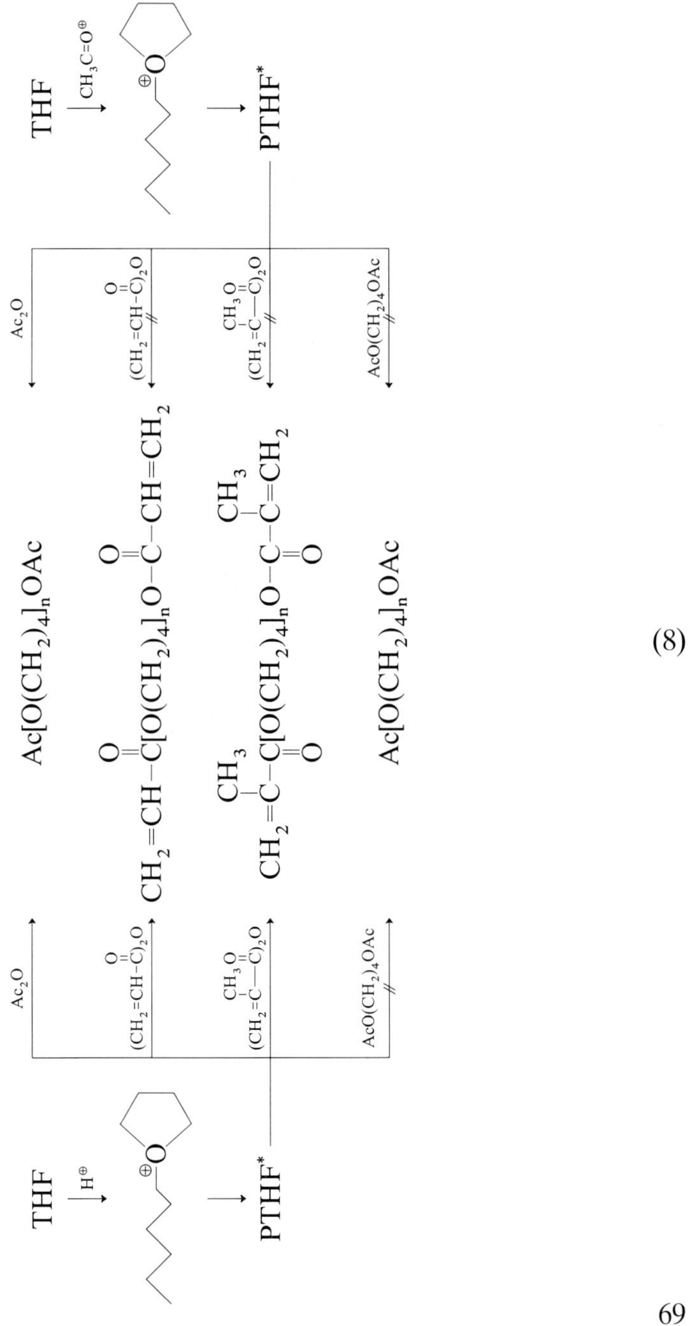

(8)

Polymerizing THF in presence of the given transfer agents polymerization is obtained by initiation either with protons or acylium cations. With the exception of butanediol diacetate the polymerization initiated by protons is followed by an equilibration resulting in PTHF with ester end groups. The anhydrides are completely consumed. With none-protic initiators a very slow consumption of acetic anhydride is observed. The other anhydrides and esters do not react in this case.

The polymerization-depolymerization equilibrium necessitates oxonium ions. Ester end groups are dead in the sense of this equilibrium. We can either assume that protonated esters can split off carboxylic acids and form oxonium ions or an activated monomer mechanism taking place. This explains why the addition of acetic acid results in an enhancement of the equilibration. Du Pont claims in a patent the polymerization of THF in presence of a mixture of acetic acid/acetic anhydride. The reason for this process is primarily not an increase of polymerization rate but an increase in the depolymerization process and so the anhydride is consumed faster.

4.3 Telechelics by Anionic Polymerization

Anionic polymerization is supposed to be the safest method for synthesizing telechelics. It was extensively used to prepare telechelic polystyrenes [8]. Of special interest is the synthesis of telechelic soft segments of polybutadiene. The necessary high fraction of 1.4-structure is obtained usually by initiation using lithium organic compounds in hydrocarbons as solvent. Butyllithium is soluble in hexane, simple dilithium organic compounds are only soluble in polar solvents. This caused an intensive search for well soluble dilithium organic compounds. These compounds are extremely interesting for the synthesis of block copolymers. With telechelics they result in the incorporation of an initiator fragment of considerable size resulting in a reduction of the chain flexibility.

Another possibility to synthesize telechelics by anionic polymerization was demonstrated by D. N. Schulz [9]. The tetrahydropyranyl unit was used protecting the OH function. A similar result was obtained by the vinylether group [10].

$$HO-(CH_2)_6-Cl \longrightarrow {/\!\!/}\!-O(CH_2)_6-Cl \tag{9}$$

$$\xrightarrow[-\ ButCl]{+\ ButLi} {/\!\!/}\!-O-(CH_2)_6-Li \xrightarrow{{/\!\!/}\!-\!{/\!\!/}} {/\!\!/}\!-O(CH_2)_6\!\!-\!\!(CH_2-CH=CH-CH_2)_n Li$$

Table 4.3: Polybutadiene initiated by Li-compounds.

Microstructure	1.2	1.4 cis	1.4 trans
ButLi	13	35	52
$= -O(CH_2)_6Li$	22	40	38

Vinylethers are stable against bases and are therefore suitable for the protection of hydroxy groups. Vinyl ethers introduce a polar group to the initiating system. Vinyl and benzylethers have only a minor influence on the micro structure of polybutadiene (see Table 4.3).

4.4 Telechelics by Radical Polymerization

By radical polymerization of vinylmonomers telechelics can be obtained if the two reaction steps causing the formation of the end groups of the polymer chain result in functional groups. In a radical polymerization an initiator fragment is fixed as an end group of the chain during the initiation step. This has been shown with different monomers and initiators [11, 12]. Investigations with labelled initiators confirm this conclusion even at high molecular weights [13–16]. Some initiators however will initiate by hydrogen transfer [17].

Besides the initiator controlled synthesis of telechelics the telomerization may also result in α,ω-bifunctional segments. Telomerization is a polymerization in presence of an effective transfer agent. The resulting products, the telomers describe this procedure of synthesis and make no statement about the functionality.

4.4.1 Mechanistic Aspects

4.4.1.1 Elementary Steps

The classical scheme of radical polymerization is valid also in the synthesis of telechelics.

Initiation	$I \longrightarrow 2\,R\cdot$	(10)
	$R\cdot + M \longrightarrow RM\cdot$	(11)
Propagation	$RM_n\cdot + M \longrightarrow RM_{n+1}\cdot$	(12)
	$(RM_n\cdot + XH \longrightarrow RM_nH + X\cdot)$	(13)
Termination	$2\,RM_n\cdot \longrightarrow RM_{2n}R$	(14)
	$\longrightarrow RM_nH + RM_n$ (minus H)	(15)
	$RM_n\cdot + R\cdot \longrightarrow RM_nR$	(16)
	$\longrightarrow RM_nH + R$ (minus H)	(17)

The decomposition of the initiator (Equ.10) must occur into one definite primary radical which will not give secondary reactions before initiating a chain. Reactions like the partial decarboxylation of oxybenzoyl radicals will give rise to different initiating radicals [18–20], making the analysis of the reaction product nearly impossible [8]. The initiator is the most expensive chemical of the reaction. The economical value of the synthesis is dependent upon the fact that the addition of the first monomer unit occurs with a high efficiency. The value of the efficiency is dependent upon the differences in the chemical nature of initiator and monomer [21–22].

On this basis it is to be expected that the efficiency of AIBN in the polymerization of butadiene is high whereas its initiation of the polymerization of ethylene is extremely ineffective. The efficiency is also dependent upon the monomer concentration and the viscosity. But in the synthesis of telechelics the viscosity of the reaction mixture is low. Primary radicals are consumed by the initiation (Equ.11), the primary radical termination (Equ.16/17) and by Equ. (18).

$$2\,R\cdot \xrightarrow{k_c} RR \tag{18}$$

The primary radical termination can be neglected at higher molecular weights, thus the efficiency defined as the fraction of initiating radicals is

$$f = \frac{k_1[R\cdot][M]}{k_1[R\cdot][M] + k_c[R\cdot]^2} \tag{19}$$

At constant rate of formation of primary radicals the efficiency is relatively insensitive to monomer concentrations at >5 mol/l. Fig. 4.5 gives an estimate of the dependence of the efficiency from monomer concentration.

In the synthesis of telechelics by bulk polymerization of butadiene ([M] = 11 mol/l) using AIBN the efficiency found was 0.9. Using Equ.(19) it results $k_c[R\cdot]/k_i = 1$. This value is used to construct the curve of Fig. 4.5. Data of Moad [23] confirm the dependence given in Equ. (19).

High monomer concentrations are of importance in the synthesis of telechelics. Fragments of the initiator can be incorporated also by primary radical termination which becomes of increasing importance with decreasing molecular weight. The effectiveness of the incorporation of the initiator (apparent efficiency f_a) is easily available with telechelics.

$$f_a = \frac{\Delta\,[M]}{e/2\ [I]_o\ P_n} \tag{20}$$

Δ [M] – monomer consumption
e – functionality
$[I]_o$ – initiator concentration

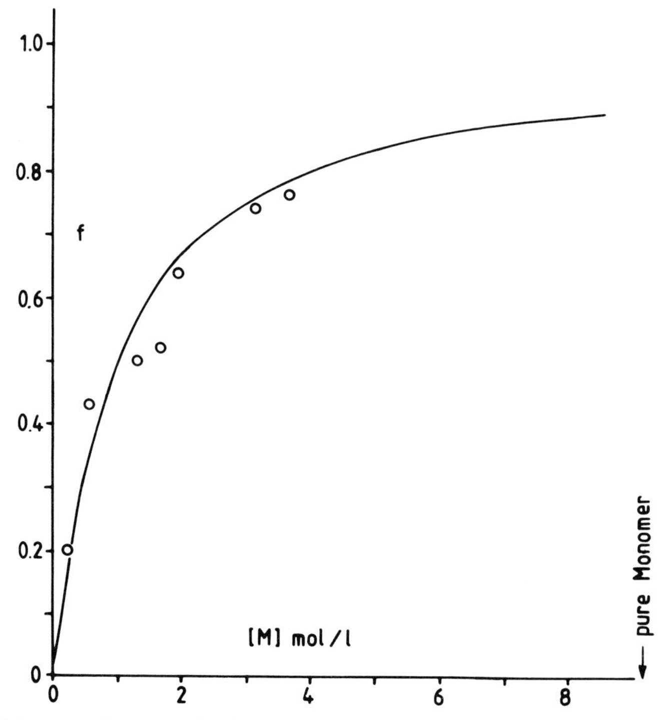

Fig. 4.5: Efficiency of AIBN in the polymerization of styrene; values of Moad [23], curve calculated with Equ. (19) using $k_c[R\cdot]/k_i = 1$.

The initiator is completely consumed during the synthesis ($[I]_o = \Delta[I]$. Apparent efficiencies of 0.8–0.95 are found. The rate of the propagation reaction is of importance for the selection of the experimental conditions. But even with rapidly polymerizing monomers carrying off the heat of polymerization is no problem when using flow reactors [24]. The propagation reaction can be disturbed by transfer reactions. If solvents have to be used they must be as carefully selected for low transfer constants as in normal polymerizations. Telechelics can be obtained if the termination reaction is a combination. A reaction between two macroradicals (Equ. 14/15) is the common mode of termination. Termination by primary radicals can also cause combination (Equ. 16) and disproportionation (Equ. 17). The ratio of combination to disproportionation can be different in these cases. The cross reaction is in many cases a combination. This makes the synthesis of telechelics sometimes possible at $M < 10^3$ in cases where macroradicals react by disproportionation. Combination of macroradicals is not equate with a safe synthesis of telechelics. The ratio of the rate constants of combination to disproportionation is 8 for alkyl radicals [25], but the high reactivity of alkyl radicals causes a hydrogen transfer and results in monofunctional polyethylene (see 4.4.1.4).

4.4.1.2 Dead End Polymerization

There is one decisive difference between a usual radical polymerization and the synthesis of telechelics by radical polymerization: a normal radical polymerization is taken to high conversions of the monomer, i.e., the initiator must be present at the very end of the reaction to convert residual monomer to polymer; in the synthesis of telechelics the initiator is the most expensive reagent. Primary radicals formed at the final stage of the reaction must still find sufficient monomer present. The decomposition of the initiator has a higher activation energy and in turn a stronger temperature dependence than the propagation reaction. The reaction temperature must be chosen in that way that only a fraction of the monomer is consumed within 10 half lives of initiator decomposition. These are the conditions of the "dead end" polymerization derived theoretically by Tobolsky in 1958 [26].

The monomer consumption in the radical polymerization is given by Equ. (21).

$$\ln[M]_o/[M] = k_1[I]_o \{1-([I]/[I]_o)\} \tag{21}$$

with

$$k_1 = \frac{k_p}{k_t^{0.5}} \frac{1}{k_d^{0.5}} 2.85 f \tag{22}$$

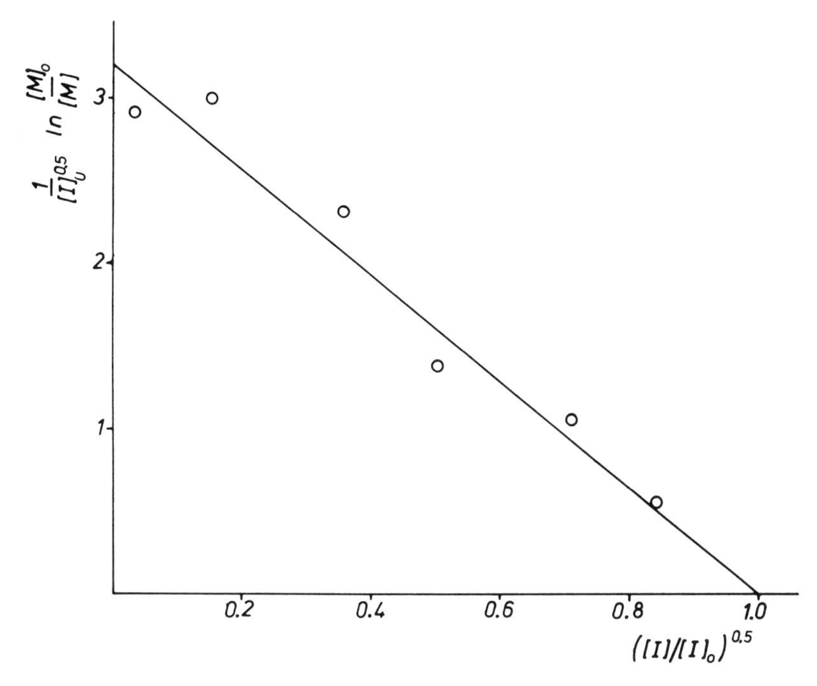

Fig. 4.6: Polymerization of styrene with azobis(methyl-i-butyrate) in toluene at 98 °C (Equ. 21).

A plot of data obtained in the polymerization of styrene with an azo initiator according to Equ. (21) shows that complete comsumption of the initiator end up in a limited conversion of styrene [27] (Fig. 4.6). With $[I]_\infty = 0$ Tobolsky derived

$$\ln([M]_0/[M]_\infty) = k_1[I]_0^{0.5} \tag{23}$$

$[M]_\infty$ – monomer concentration at the end of the reaction

With a given initiator concentration the polymerization reaction stops at an adjustable conversion (Fig. 4.7). From Fig. 4.6 k = 3.2 is obtained, this value is used for the slope of the straight line in Fig. 4.7. Equ. (23) does not take into account the primary radical termination and the monomer consumption in the initiation step. An induced decomposition of the initiator is probably the reason for the deviation from Equ. (23) found in the polymerization of butadiene using diethyl peroxydicarbonate (Fig. 4.8).

According to Equ. (23) the conversion of the monomer increases with the initiator concentration. Primary radicals formed at high conversion of monomer have barely a chance to react with monomer properly. The ratio of monomer to initiator is decisive for a clean reaction. Thus the yield per volume

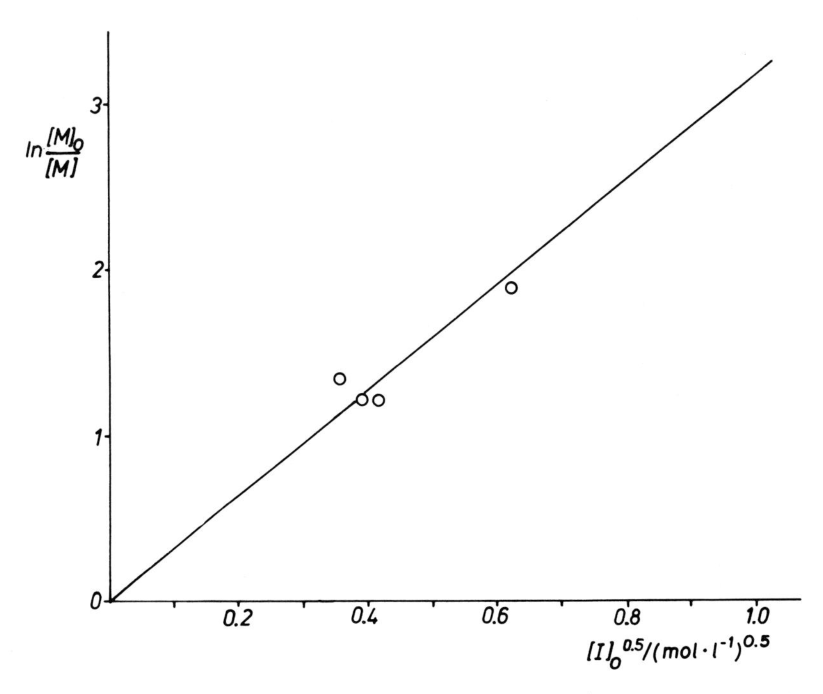

Fig. 4.7: "Dead end" conversion of styrene with azobis(methyl-i-butyrate) at 98 °C (Equ. 23).

Table 4.4: Percentages of combination for different monomers (25 °C) [28].

styrene	100
p-chlorostyrene	100
p-methoxystyrene	81
methyl/ethyl acrylate	100
methylmethacrylate	33
ethyl methacrylate	32
n-butyl methacrylate	25
acrylonitrile	100
methacrylonitrile	35

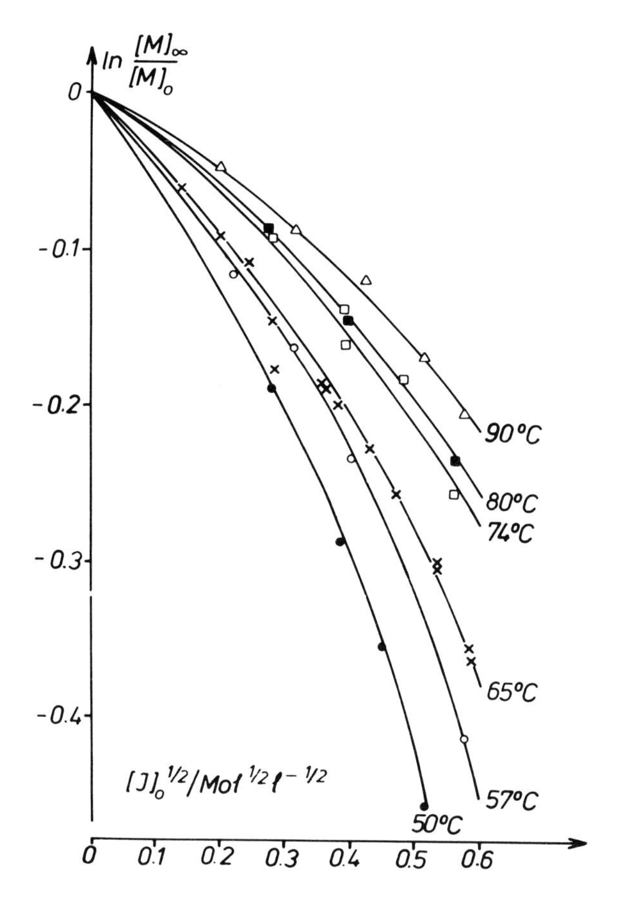

Fig. 4.8: "Dead end" conversion in the bulk polymerization of butadiene initiated by diethylperoxy dicarbonate.

has to be reduced if a solution polymerization is used. Bulk polymerization is the most economic way so synthesize telechelics. By limiting the conversion an oligomer solution with the monomer as solvent is formed. Because of the low viscosity of an oligomer solution there are no heat transfer problems in contrast to polymer synthesis.
The termination in the initiator controlled synthesis of telechelics must be a combination. Chain termination occurs almost exclusively by combination in the case of butadiene.

Percentages of combination for different monomers are given by Bamford (Table 4.4) [28]. Recently it has been shown that the termination reaction in the styrene polymerization is about 80 % combination and 20 % disproportionation [29–32]. But the ratio of rate constants for Equ. (16) to Equ. (14) is about sixty [28] for styrene/AIBN, thus primary radical termination plays an important role if low molecular weights are produced with high amounts of initiator.

4.4.1.3 The Initiator

Azo Compounds

The azo initiator most extensively studied is AIBN. No induced decomposition is observed. The cyano-i-propyl radical has a moderate reactivity. The ransfer constant to polymer is low. In the absence of monomers about 90 % of the primary radicals react by combination to form tetramethyl succinonitrile [33]. By disproportionation methacrylonitrile *(5)* is formed besides i-butyronitrile *(4)*. Most of the metacrylonitrile *(5)* is found incorporated in the trimer *(6)*. If the monomer conversion is too high in the synthesis of telechelics the formation of methacrylonitrile *(5)* is a possible side reaction. The incorporation of *(5)* will give a functionality > 2. The cyano-i-propyl radicals can react to some extent to form the ketene imin *(7)*. But this is a reversible reaction.

$$\underset{(1)}{\text{AIBN}} \xrightarrow{-N_2} \underset{(2)}{NC-\overset{CH_3}{\underset{CH_3}{C}}\cdot} \longrightarrow \underset{(3)}{NC-\overset{CH_3}{\underset{CH_3}{C}}-\overset{CH_3}{\underset{CH_3}{C}}-CN}$$

$$(2) \rightleftharpoons \underset{(7)}{NC-\overset{CH_3}{\underset{CH_3}{C}}-N=C=\overset{CH_3}{\underset{CH_3}{C}}}$$

$$(2) \longrightarrow \underset{(4)}{NC-\overset{CH_3}{\underset{CH_3}{C}}-H} + \underset{(5)}{NC-\overset{CH_2}{\overset{\|}{\underset{CH_3}{C}}}} \qquad (24)$$

$$(5) \longrightarrow \underset{(6)}{NC-\overset{CH_3}{\underset{CH_3}{C}}-CH_2-\overset{CH_3}{\underset{CN}{C}}-\overset{CH_3}{\underset{CH_3}{C}}-CN}$$

The efficiencies found with AIBN for the polymerization of butadiene and styrene in bulk or highly concentrated solutions are $f_a > 0.9$.

The use of other azo initiators follows two strategies: the modification of the methyl and the nitrile group resp.

The functional azo initiators most often used are derived from 3-keto-valeric acid and 4-keto-pentanol.

$$\begin{matrix} CH_3 \\ | \\ C{=}O \\ | \\ CH_2 \\ | \\ CH_2R \end{matrix} \longrightarrow \begin{matrix} CH_3 & & CH_3 \\ | & & | \\ C{=}N{-}N{=}C & & \\ | & & | \\ CH_2 & & CH_2 \\ | & & | \\ CH_2R & & CH_2R \end{matrix} \longrightarrow \begin{matrix} & CH_3 & & CH_3 & \\ & | & & | & \\ NC{-}&C&{-}N{=}N{-}&C&{-}CN \\ & | & & | & \\ & CH_2 & & CH_2 & \\ & | & & | & \\ & CH_2R & & CH_2R & \end{matrix} \qquad (25)$$

$(8a)$ R = COOH
$(8b)$ R = CH_2OH

The rate constants of decomposition of *(8a), (8b)* is similar to AIBN. This route of synthesis could be extended by use of other substituted ketones to prepare telechelics with isocyanate groups [34]. AIBN can be quantitatively converted to esters.

$$AIBN \longrightarrow \begin{matrix} O & CH_3 & & CH_3 & O \\ \backslash\backslash & | & & | & // \\ C{-}&C{-}N{=}N{-}&&C{-}&C \\ / & | & & | & \backslash \\ RO & CH_3 & & CH_3 & OR \end{matrix} \qquad (26)$$

$(9a)$ R = CH_3
$(9b)$ R = $(CH_2)_nOH$

The use of *(9a)* results in a telechelic with methylester end groups. They are relativley stable towards hydrolysis and need hard conditions to be hydrolyzed. If the conversion to the ester is made with an excess of diol an azo initiator with hydroxylic functionality *(9b)* is obtaines [35]. They have the advantage that their solubility is higher than *(8b)*.

Diacyl Peroxides

Primary radicals derived from diacyl peroxides (*10*) can initiate polymerization, but they can also lose carbon dioxide before initiation.

$$R{-}C(=O){-}O{-}O{-}C(=O){-}R \ (10) \longrightarrow 2\,R{-}C(=O){-}O\cdot \xrightarrow{M} Pol. \qquad (27a)$$

$$\downarrow -CO_2$$

$$2\,R\cdot \xrightarrow{M} Pol. \qquad (27b)$$

$(10a)$ R = Alkyl
$(10b)$ R = Aryl

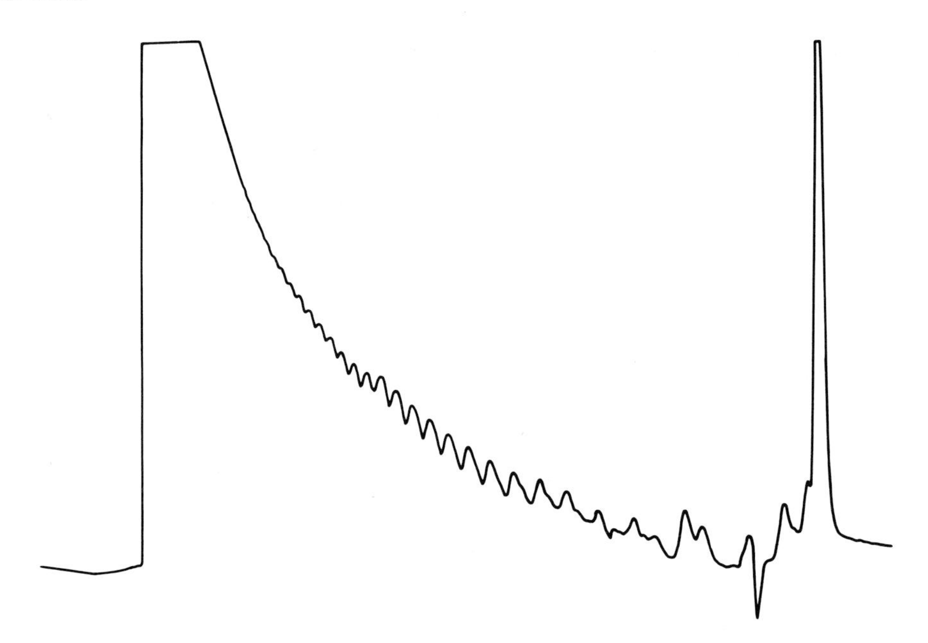

Fig. 4.9: GPC of oligostyrene from styrene and dibenzoyl peroxide.

With aliphatic diacyl peroxides *(10a)* there is quantitative decomposition to alkyl radicals before initiation occurs. Aromatic diacyl peroxides *(11b)* will initiate in both ways, thus the initiator fragment is fixed via CC-bonds or ester groups at the chain end [18, 19]. This complicated situation is clearly shown by GPC (Fig. 4.9). The chromatogram indicates several polymer homologous series. Phenyl and alkyl radicals have a high reacitivity which increases the tendency of transfer reactions.

Dialkylperoxy Dicarbonate

Dialkylperoxy dicarbonate *(11)* will form primary radicals which will

$$\underset{(11)}{\mathrm{RO\overset{}{\underset{\|}{\underset{O}{C}}}OO\underset{\|}{\underset{O}{C}}OR}} \longrightarrow 2\,\mathrm{RO\underset{\|}{\underset{O}{C}}O\cdot} \xrightarrow{\mathrm{M}} \mathrm{Pol.} \tag{28}$$

initiate without losing carbon dioxide; an exception is R= tert.-butyl. Razuvaev et al. [36] have reported using ^{14}C labelled dicyclohexylperoxy dicarbonate and employing a tracer technique for end group analysis, that the primary radical ROCOO· formed by the decomposition of the initiator reacts with monomers such as styrene and MMA much faster than it undergoes decarboxylation. They conclude that the RO· radical plays a very small role in initiation. End groups of carbonate but no fragments from the incorporation of RO· could be detected with R = methyl, ethyl in the polymerization of ethylene. With R = tert. butyl, the loss of carbon dioxide results in the formation of tert. butoxy radicals which are cleaved to methyl radicals before initiation [37].

Hydrogen Peroxide

Hydrogen peroxide can be cleaved to hydroxy radicals thermally, in redox systems and by strong acids. The most wellknown redox initiator is the ferrous ion/hydrogen peroxide system [38].

$$Fe^{2+} + H_2O_2 \longrightarrow Fe^{3+} + OH^- + \cdot OH \qquad (29)$$

The ferric ions may be reduced by hydrogen peroxide.

$$Fe^{3+} + H_2O_2 \longrightarrow Fe^{2+} + H^+ + HOO\cdot \qquad (30)$$

This redox system is often used in the presence of alcohols like methanol or i-propanol. Due to the high reactivity of OH radicals an H-abstraction from the solvent has to be considered.

$$HO\cdot + CH_3OH \longrightarrow H_2O + \cdot CH_2OH \qquad (31)$$

The reactivity of the hydroxyl radical will give rise to other transfer reactions if the conditions are favourable.

4.4.1.4 Monomers

Ethylene

The termination reaction in the polymerization of ethylene is predominantly combination. From the behaviour of alkyl radicals [37] we can estimate for the polymer $k_{td}/k_{tc} = 0.13$. On this basis one should expect a functionality of 1.8 if two monofunctional macroradicals of polyethylene react together. But the high reactivity of alkyl radicals forces the reaction into another direction. The macroradicals of ethylene abstract hydrogen from any source in the system. This is normally the solvent. Taking a solvent with a very small transfer constant (e.g. tert. butanol) the macroradical reacts with the initiator.

By reaction of ethylene with AIBN in tert. butanol a product with a functionality of 1.7 is obtained [37]. In a typical run (60 bar ethylene at room temperature, 80 °C reaction temp., 0.1 mol AIBN, 250 ml tert. butanol, 1 l steel autoclave) 0.54 mol of oligomer are formed corresponding to an apparent efficiency $f_a = 0.46$. From GC and MS different polymer-homologous series can be identified.

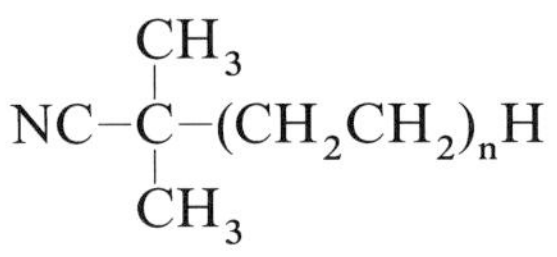

$$NC-C(CH_3)_2-(CH_2CH_2)_nH \qquad (12a)$$

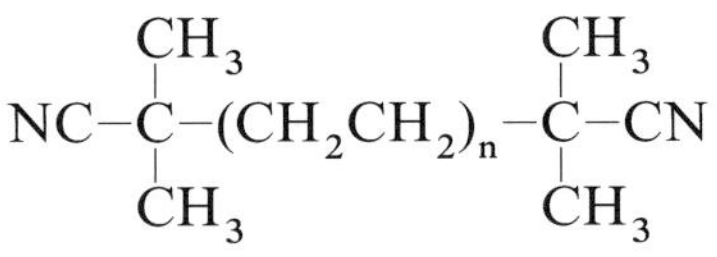

$$NC-C(CH_3)_2-(CH_2CH_2)_n-C(CH_3)_2-CN \qquad (12b)$$

$$NC-C(CH_3)_2-(CH_2CH_2)_nCH_2-C(CH_3)(CN)-C(CH_3)_2-CN \qquad (12c)$$

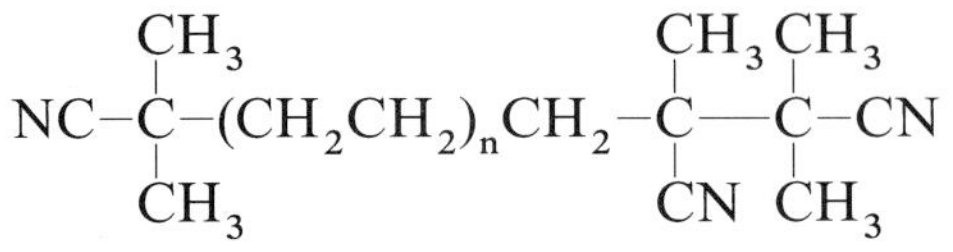

(12c) can become the major peak series in GC if the ethylene pressure during the polymerization is lowered to 10 bars. The average functionality of this product is higher than two. *(12c)* can be formed by incorporation of methacrylonitrile, the resulting macroradical is very inefficient towards the addition of ethylene and reacts with primary radicals. Another route of formation would originate from an H-abstraction from tetramethyl succinonitrile *(3)*, to form the initiating radical.

With tert. butanol as the solvent no polymer-homologous series derived from the solvent could be detected. Using methanol about 10 % of the product is $HOCH_2(CH_2CH_2)_nH$. With benzene about 20 % of the product is formed according to

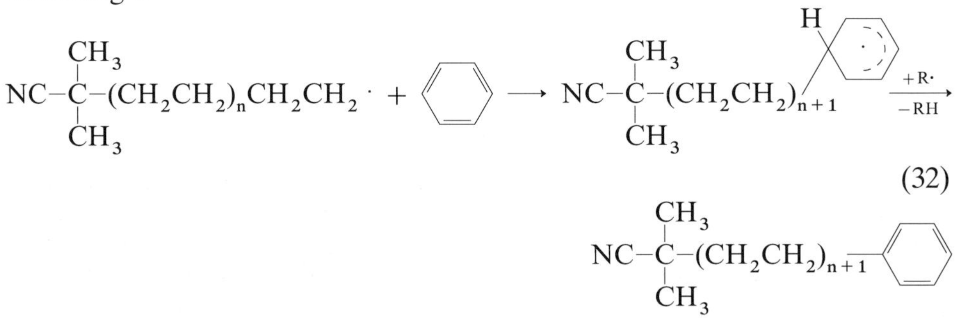

$$NC-C(CH_3)_2-(CH_2CH_2)_nCH_2CH_2^{\cdot} + C_6H_6 \longrightarrow NC-C(CH_3)_2-(CH_2CH_2)_{n+1}-C_6H_6^{\cdot}(H) \xrightarrow[-RH]{+R\cdot} NC-C(CH_3)_2-(CH_2CH_2)_{n+1}-C_6H_5 \qquad (32)$$

Thermal decomposition of dialkylperoxy dicarbonates results in primary radicals initiating polymerization without loss of carbon dioxide.

$$\mathrm{ROC(=O)OOC(=O)R} \longrightarrow \mathrm{ROC(=O)O}\cdot \longrightarrow \mathrm{ROC(=O)O(CH_2CH_2)_nCH_2CH_2}\cdot \xrightarrow{\mathrm{CH_3CH(OC(=O)OOC(=O)OR)}} \tag{33}$$

$$\mathrm{ROC(=O)O(CH_2CH_2)_{n+1}H + CH_3CHO + ROC(=O)O\cdot + CO_2}$$

GC shows that mainly a monofunctional series is obtained. Bifunctional product is formed in small amounts. The average functionality is 1.1–1.2. Substituting the α-hydrogen of the initiator by a methyl group should avoid the transfer to the initiator and give higher functionalities. But with R=tert. butyl the primary radicals are not stable and decompose partially to methyl radicals. The alkanes with odd number of carbon atoms are observed as a prominent peak series in GC. The mean functionality is about 0.7.

The use of bis(3-methoxycarbonyl)propionyl-peroxide *(13)* in the polymerization of ethylene results in oligomers with the functionality 1.1.

$$\mathrm{CH_3O{-}C(=O){-}CH_2CH_2C(=O){-}O{-}O{-}C(=O)CH_2CH_2{-}C(=O){-}OCH_3} \qquad (13)$$

Model reactions with dihexanoyl peroxide showed that this functionality does not result from a disproportionation.

$$\mathrm{C_5H_{11}C(=O){-}O{-}O{-}C(=O)C_5H_{11}} \xrightarrow{-\mathrm{CO_2}} \mathrm{C_5H_{11}}\cdot \longrightarrow \underset{(14a)}{\mathrm{C_5H_{11}(CH_2CH_2)_nCH_2CH_2}\cdot} \longrightarrow \tag{34}$$

$$\underset{(14b)}{\mathrm{C_5H_{11}(CH_2CH_2)_{2n+2}C_5H_{11}}} + \underset{(14c)}{\mathrm{C_5H_{11}(CH_2CH_2)_{n+1}H}}$$

$$+ \underset{(14d)}{\mathrm{C_5H_{11}(CH_2CH_2)_nCH{=}CH_2}}$$

(14c) is formed to 90 % at 95 °C, but the olefin *(14d)*, to be expected from disproportionation cannot be detected. The major product is formed by an induced decomposition of the initiator.

$$(14a) + C_3H_7-\underset{\displaystyle H}{\underset{|}{C}}H-CH_2\overset{O}{\overset{\|}{C}}-O-O-\overset{O}{\overset{\|}{C}}-C_5H_{11} \xrightarrow{-CO_2} \tag{35}$$

$$(14c) + C_3H_7-CH{=}CH_2 + C_5H_{11}\cdot$$

Substituting the β-hydrogen of the diacyl peroxide by methyl the induced decomposition is avoided. Using bis(3.3-dimethylbutyryl) peroxide *(15)* in the polymerization of ethylene the product contains 5.6 CH_3-groups/mol.

$$CH_3-\underset{CH_3}{\overset{CH_3}{C}}-CH_2-\overset{O}{\overset{\|}{C}}-O-O-\overset{O}{\overset{\|}{C}}-CH_2-\underset{CH_3}{\overset{CH_3}{C}}-CH_3 \quad (15)$$

Oligoethylenes which are clearly telechelic in character are obtained by telomerization with carbon tetrachloride.

GC (Fig. 4.10) shows one polymer-homologous series in a purity of 95%. As shown by MS, some of the side product results from a transfer with the solvent (hexane).

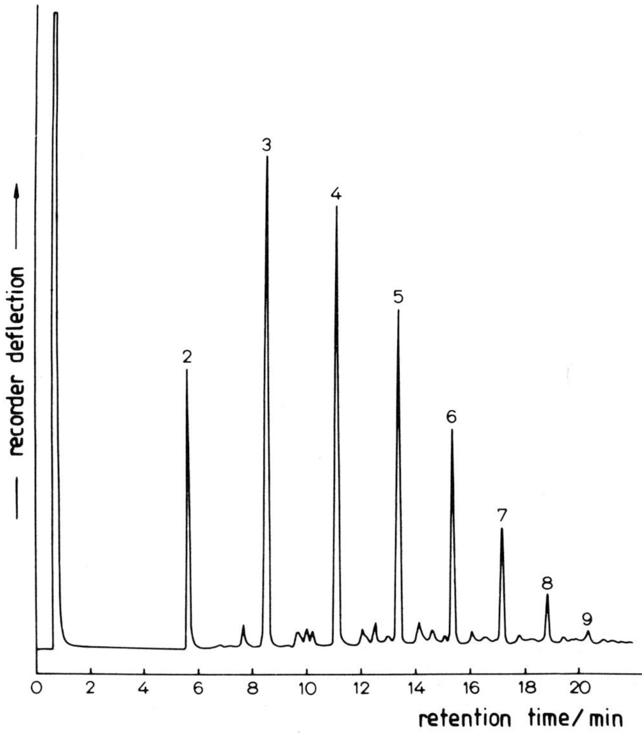

Fig. 4.10: GC of $CCl_3(CH_2CH_2)_nCl$ obtained from ethylene and carbon tetrachloride; figures at the peaks = degree of polymerization.

Styrene

To the present understanding the termination in the polymerization of styrene is about 80 % combination. Therefore in the initiator-controlled synthesis of telechelics 80 % of the product should be bifunctional and 20 %, with half the molecular weight, should be monofunctional. The average functionality should be 1.8 if this is the only reaction controlling the structure of the product.

GPC of the product obtained from AIBN and styrene shows one peak series [27]. From N-analysis together with M_n (vapour pressure osmometry) the functionality is two. No gelation occurred when these oligomers where coupled after hydrolysis of the nitrile groups and the molecular weight increase was more than 30 fold.

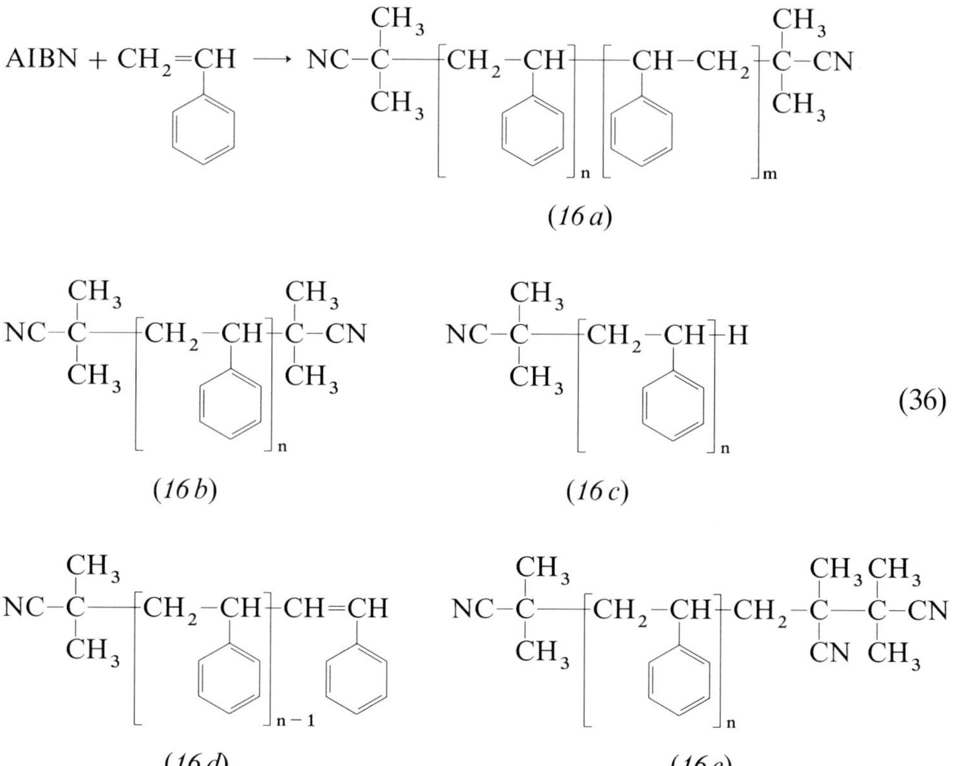

$$\text{AIBN} + CH_2{=}CH(C_6H_5) \longrightarrow NC{-}C(CH_3)_2{-}[CH_2{-}CH(C_6H_5)]_n{-}[CH(C_6H_5){-}CH_2]_m{-}C(CH_3)_2{-}CN$$

(*16a*)

$$NC{-}C(CH_3)_2{-}[CH_2{-}CH(C_6H_5)]_n{-}C(CH_3)_2{-}CN$$

(*16b*)

$$NC{-}C(CH_3)_2{-}[CH_2{-}CH(C_6H_5)]_n{-}H$$

(*16c*)

$$NC{-}C(CH_3)_2{-}[CH_2{-}CH(C_6H_5)]_{n-1}{-}CH{=}CH(C_6H_5)$$

(*16d*)

$$NC{-}C(CH_3)_2{-}[CH_2{-}CH(C_6H_5)]_n{-}CH_2{-}C(CH_3)(CN){-}C(CH_3)_2{-}CN$$

(*16e*)

(36)

Therefore it can be ruled out that the structures (*16c*), (*16d*), and (*16e*) are present in a significant amount. According to Moad [23, 39] the primary radical termination on the basis of NMR-studies is unimportant in the formation of oligomers. But the analytical data point out the significance of primary radical termination.

The end groups of dinitrile telechelic oligostyrene *(16a)*, *(16b)* can be converted to other functional groups [40]. By hydrolysis carboxylic end groups are formed. Catalytic hydrogenation results in amino end groups which can be converted into isocyanate end groups.

Methyl-2,2-azoisobutyrate (MAIB) can initiate the polymerization of styrene to form α, ω-bis(2-methoxycarbonylmethylethyl) oligostyrene *(17)*.

$$CH_3O-C(=O)-C(CH_3)_2-N=N-C(CH_3)_2-C(=O)-OCH_3 \;(17) \xrightarrow{\text{styrene}} CH_3O-C(=O)-C(CH_3)_2-\left[CH_2-CH(C_6H_5)\right]_n-C(CH_3)_2-C(=O)-OCH_3 \quad (37)$$

At molecular weights < 1000 the functionality is two and one peak series is observed in GPC (Fig. 4.11). With increasing molecular weight the functionality is reduced from 2.0 to 1.8, but if the conversion of monomer is too high, side reactions occur and the product has an increased functionality.

Two azo initiators allow to introduce functionality without the necessity of a reaction step with the final polymer: 3,3-azo-bis(3-cyanovaleric acid) *(8a)* and 4,4-azobis(cyanopentanol) *(8b)*.

Bamford and Jenkins [41] described in a series of papers the preparation and use of telechelic oligostyrene with carboxylic end groups.

$$(8a) \longrightarrow HOOC(CH_2)_2-C(CH_3)(CN)-\left[CH_2-CH(C_6H_5)\right]_n-C(CH_3)(CN)-(CH_2)_2-COOH \quad (38)$$

The product was chain extended with diols and an eleven fold increase in molecular weight was observed; when reacted with a telechelic oligoacrylonitrile with OH end groups a block copolymer consisting of 7.4 blocks resulted.

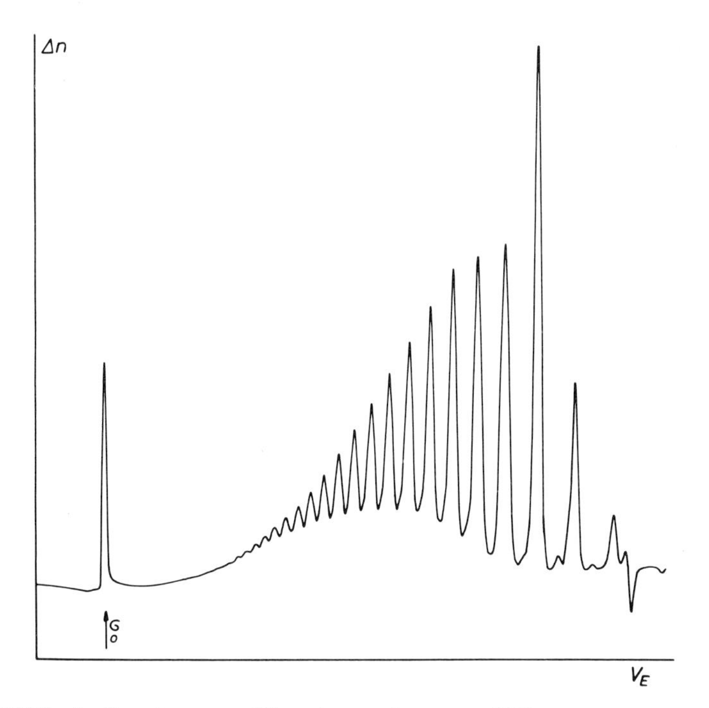

Fig. 4.11: GPC of oligostyrene with ester end groups *(17)*.

Butadiene

Functional oligodienes are of commercial importance. Their preparation was reviewed by French [42] and by Schnecko [43].

It is of primary importance in the synthesis of telechelic oligobutadienes that the conversion is limited to 20–40%. The low rate constant of polymerization makes bulk polymerization to an easy to handle process. Using AIBN, a polybutadiene with two nitrile end groups is formed [24].

$$\text{AIBN} \longrightarrow \text{NC}-\underset{\text{CH}_3}{\overset{\text{CH}_3}{\underset{|}{\overset{|}{\text{C}}}}}-(\text{CH}_2-\text{CH}{=}\text{CH}-\text{CH}_2)_n-\underset{\text{CH}_3}{\overset{\text{CH}_3}{\underset{|}{\overset{|}{\text{C}}}}}-\text{CN} \qquad (39)$$

(*18*)

The average functionality of two has been confirmed up to M_n of 2000.

In Fig. 4.12 a GPC of *(18)* is given together with the mass distribution. The polymerization degree one is present only in very small amounts. Initiated by AIBN the polymerization of butadiene shows a high efficiency. Solution polymerization results in values of $f_a = 0.7–0.8$, whereas in bulk polymerization efficiencies $f_a = 0.9–0.95$ are found. This points to the fact that butadiene is a good radical scavenger. The cyanoisopropyl radicals are immediately scavenged forming allyl radicals. Combination of the allyl radicals results in the polymerization degree two.

The microstructure corresponds to values typical for radical polymerization: 15% 1.4-cis, 60% 1.4-trans and 25% 1.2 structure. The fraction of 1.2 structure shows some dependence from the reaction temperature. The 1.2 fraction changes from 21 to 27% in the temperature range between 50 and 90 °C. Azobisisobutyrates can be used to prepare telechelic butadienes with ester end groups. The methyl ester has the advantage that it is easier to purify than other esters and is prepared in quantitative yields from AIBN. By

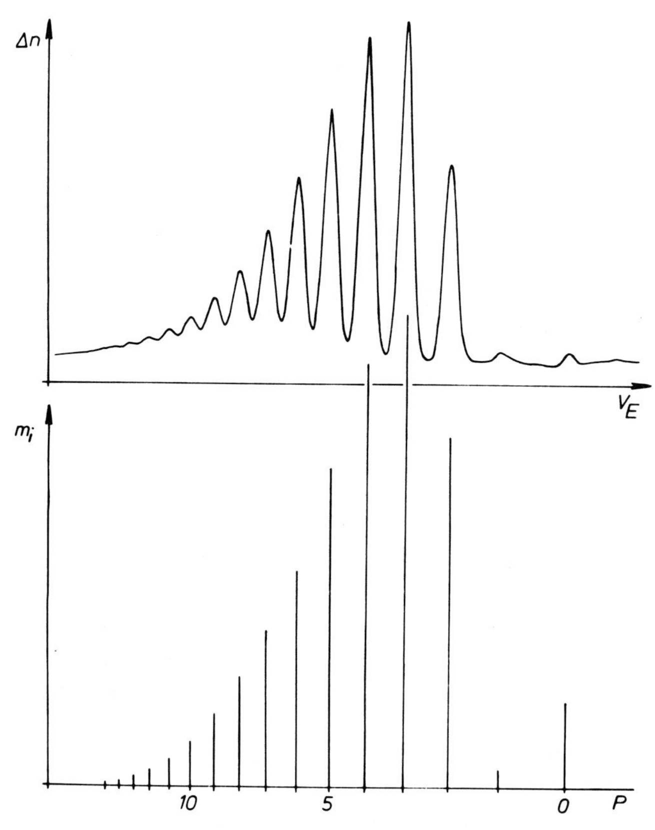

Fig. 4.12: GPC of oligobutadiene *(18)* and the mass distribution.

bulk polymerization of butadiene with azobis(methylisobutyrate) α, ω-bis(2-methoxycarbonyl-2-methyl ethyl) oligobutadiene *(19)* is formed with an ester functionality of two and an efficiency $f_a > 0.9$.

$$CH_3O(O{=})C{-}C(CH_3)_2{-}N{=}N{-}C(CH_3)_2{-}C({=}O)OCH_3 \longrightarrow \tag{40}$$

$$CH_3O(O{=})C{-}C(CH_3)_2{-}(CH_2{-}CH{=}CH{-}CH_2)_n{-}C(CH_3)_2{-}C({=}O)OCH_3$$

(*19*)

Telechelics of butadiene, isoprene and chloroprene with hydroxy end groups have been prepared with 4.4-azobis(4-cyano-n-pentanol).

$$HO(CH_2)_3{-}C(CH_3)(CN){-}N{=}N{-}C(CH_3)(CN){-}(CH_2)_3OH \longrightarrow \tag{41}$$

$$HO(CH_2)_3{-}C(CH_3)(CN){-}(CH_2{-}CH{=}CH{-}CH_2)_n{-}C(CH_3)(CN){-}(CH_2)_3OH$$

The functionality was usually greater than two. Reed [44] studied in detail the effect of initiator, monomer concentration and temperature. The results are in agreement with a "dead end" polymerization. Diene telechelics with carboxyl end groups were prepared from 4.4-azobis(4-cyanovaleric acid) [45]. The functionalities are

$$HOOC(CH_2)_2{-}C(CH_3)(CN){-}N{=}N{-}C(CH_3)(CN){-}(CH_2)_2COOH \longrightarrow \tag{42}$$

$$HOOC(CH_2)_2{-}C(CH_3)(CN){-}(CH_2{-}CH{=}CH{-}CH_2)_n{-}C(CH_3)(CN){-}(CH_2)_2COOH$$

somewhat greater than two and the M_w/M_n values are in a similar range as in Table 4.1.

Dialkylperoxy dicarbonates allow to prepare oligobutadienes with carbonate end groups [24]. The efficiency of the reaction is 0.95. The functionality is two. The carbonate end groups are easy to hydrolyze and the chain can be hydrogenated.

Hydrogen peroxide is frequently used in the preparation of functional oligobutadienes [46]. Hydroxy radicals are very reactive giving rise to side reactions. The functionality is 2.2–2.3. The product contains mono-, bi- and trifunctional molecules [46]. The fraction of bifunctional molecules is the highest for low conversions.

$$H_2O_2 \xrightarrow{} HO(CH_2-CH=CH-CH_2)_nOH \qquad (43)$$

4.5 Telechelics by Metal Catalyzed Reactions

Polymerization of butadiene using different Rh-compounds, specially $Rh(NO_3)_3$, in allyl alcohol results in oligobutadienes with a carbonyl functionality of 1 and a hydroxyl functionality between 0–1 [47, 48]. Under proper conditions telechelics are formed. The catalysis by $Rh(NO_3)_3$ can be accelerated by the addition of alcoholate or phosphine. The oligomers have exclusively 1,4-transstructure. The combined use of alcoholate and triphenylphosphine results in oligobutadienes with 40 % 1,4-cis-, 35 % 1,4-trans- and 25 % 1,2 structure. Determination of the structure of isolated oligomers shows that both functional groups are incorporated by insertion of a complete allyl alcohol unit. The polymer has the following structure:

$$CH_2{=}C(CH_2OH){-}[CH_2{-}CH{=}CH{-}CH_2]_n{-}CH_2{-}CH_2{-}CHO \qquad (44)$$

The carbonyl functionality is formed in the initiation step by hydride transfer from allyl alcohol to Rh. In the termination step allyl alcohol is incorporated and the molecule is split off by β-hydride transfer. The analogous use of allyl chloride or allyl bromide results in telechelic oligobutadienes containing two halogens per molecule, but the structure of isolated oligomers shows that only one allyl halogenide unit is incorporated. The second terminal halogen is fixed to a butadiene unit. Again the characteristic group $CH_2 = C(CH_2Cl)$-. . .is found, suggesting a termination by β-hydride transfer.

4.6 Telechelic Hard Segments

Besides the synthesis of telechelic soft segments telechelic hard segments are of increasing importance. Besides a high glass transition temperature and cristallinity they should have a good thermal stability. This is achieved by the incorporation of aromatic units in the main chain.

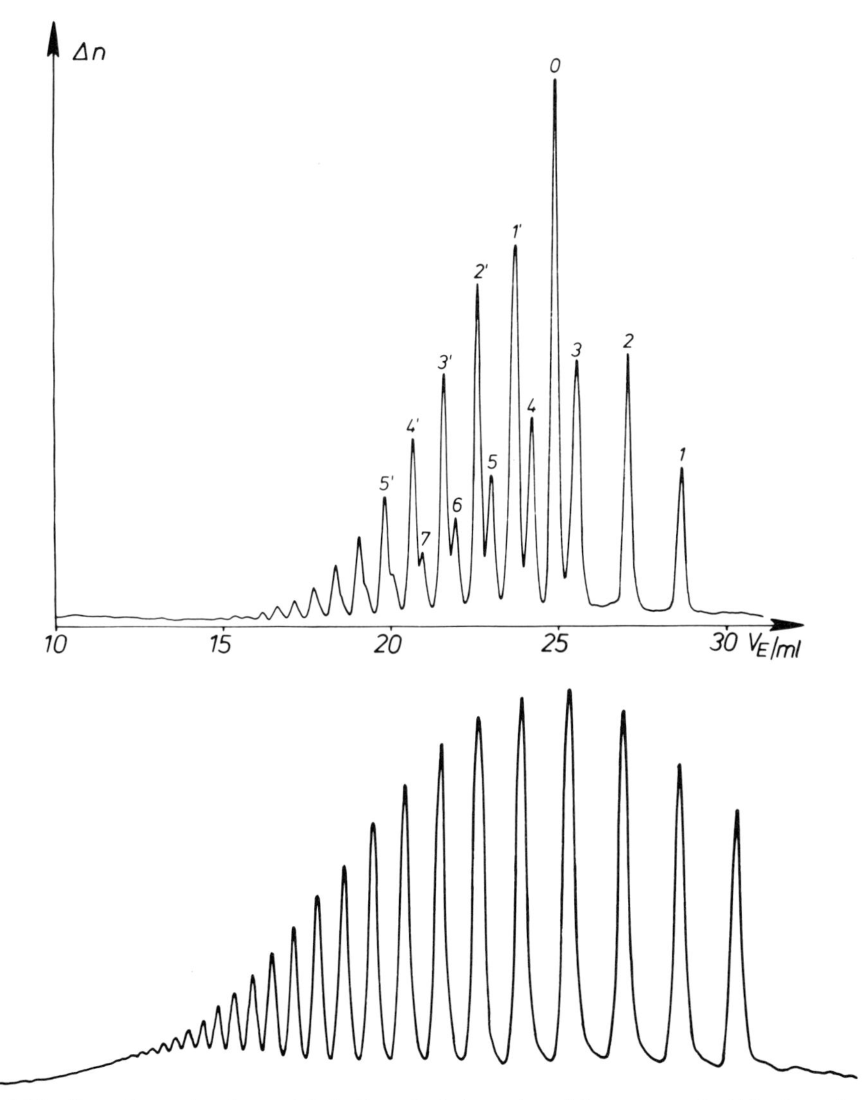

Fig. 4.13: Copolymerization of 2.6-dimethylphenole with tetramethylbisphenole A; (a) short reaction time; (b) long reaction time.

The oxydative polymerization of 2.6-dimethylphenol with oxygen in presence of copper/amine catalysts results in polymers which have exactly the functionality 1. By cooxydation of 2.6-dimethylphenol with tetramethylbisphenol A bifunctional polymers are obtained [49]. At the beginning of the reaction mono- and bifunctional molecules are produced, at the end of the reaction only the bifunctional polymer homologous series is present as shown by GPC (Fig. 4.13).

The synthesis of unsubstituted poly(oxyphenylene) starts from 4-bromophenolate in a copper catalyzed reaction at 150–200 °C. Telechelic poly(oxyphenylene) is obtained by cooxydation with a bisphenol like 4,4'-dihydroxybiphenyl. The main problem of this reaction is the dehalogenation. Under proper conditions the amount of dehalogenation is lower than 0.5 %.

Telechelics of poly(thiophenylene) are obtained by reacting 1.4-dichlorobenzene with an excess of sodium sulfide. The resulting block with thiolate end groups is reacted with 4-chloro benzoic acid in a one pot reaction [50].

Telechelics of the PPO type are amorphous materials. Telechelics of the structure of poly(oxyphenylene) and poly(thiophenylene) are crystalline materials with melting points above 200 °C (see Table 4.5).

Table 4.5: Properties of telechelic hard segments.

		T_g/°C	T_m/°C
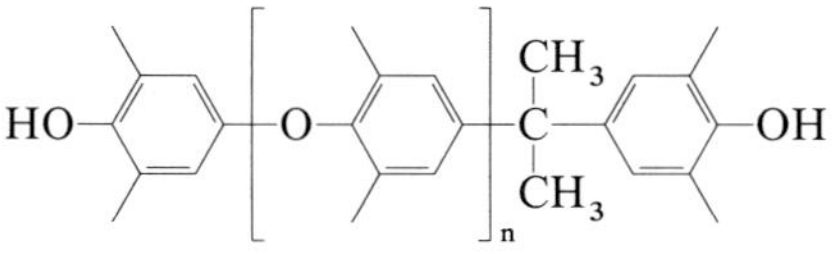 HO–[–O–]n–$C(CH_3)_2$–OH	amorphous	160–215	–
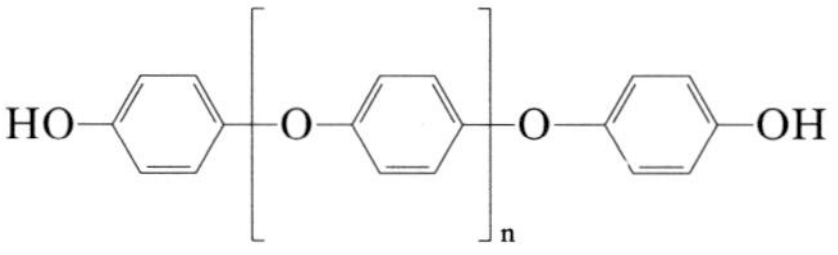 HO–[–O–]n–O–OH	crystalline	~ 85	210–270
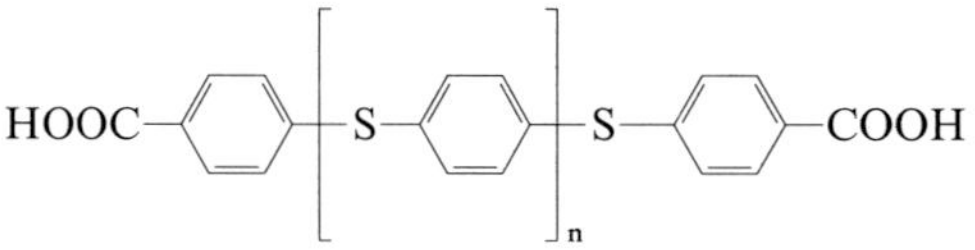 HOOC–[–S–]n–S–COOH	crystalline	~ 85	260–270

4.7 References

[1] C. A. Uraneck, H. L. Hsieh, O. G. Buck: Telechelic polymers, J. Polym. Sci. 46 (1960) 535.
[2] P. F. Rempp, E. Franta: Macromonomers: synthesis, characterization and applications, Adv. Polym. Sci. 58 (1984) 1.
[3] J. P. Kennedy: Synthesis of telechelic polymers by cationic techniques and application of the products, J. Macromol. Sci., Chem., A 21 (1984) 929.
[4] M. Miyamoto, M. Sawamoto, T. Higashimura: Synthesis of telechelic living poly-(vinyl ethers), Macromolecules 18 (1985) 123.
[5] W. Stix, W. Heitz: Cationic telomerization of THF by acetic anhydride, Makromol. Chem. 180 (1979) 1367.
[6] H.-J. Kress, W. Stix, W. Heitz: Telechelics, 7. Polytetrahydrofuran with acetate end groups, Makromol. Chem. 185 (1984) 173.
[7] H.-J. Kreß, W. Heitz: Polytetrahydrofuran with acrylate and methacrylate end groups, Makromol. Chem., Rapid Commun. 2 (1981) 427.
[8] W. Heitz: Telechele als Bausteine für neuartige Polymere, Angew. Makromol. Chem. 145/146 (1986) 37.
[9] D. N. Schulz, A. F. Halasa, A. E. Oberster: Anionic polymerization initiators containing protected functional groups and functionally terminated diene polymers, J. Polymer. Sci. 12 (1974) 153.
[10] Ch. Walter: Versuche zur Darstellung von Butadientelechelen durch anionische Polymerisation, Diss. Marburg 1981.
[11] W. Kern, H. Kämmerer: Die chemische Molekulargewichtsbestimmung von Polystyrolen, II. J. Prakt. Chem. N. F. 161 (1943) 289.
[12] W. Kern, H. Kämmerer: Über den Einbau peroxydischer Katalysatoren in die Makromoleküle von Polymerisaten bei der Polymerisation von Metharylsäuremethylestern, Vinylacetat und Methacrylnitril, Makromol. Chem. 2 (1948) 127.
[13] H. Kämmerer, G. Sextro: Die Herstellung einiger gekennzeichneter, aliphatischer Azoverbindungen und ihre Anwendung als Initiatoren der Polymerisation, Makromol. Chem. 137 (1970) 183.
[14] J. C. Bevington, H. G. Troth: Further tracer studies of azoisobutyronitrile as an initiator for radical polymerization, Trans. Faraday Soc. 58 (1962) 186.
[15] K. Berger, G. Meyerhoff: Disproportionierung und Kombination als Abbruchmechanismen bei der radikalischen Polymerisation von Styrol 1, Makromol. Chem. 176 (1975) 1923.
[16] G. Moad, E. Rizzardo, D. H. Solomon, S. R. Johns, R. I. Willing: Application of ^{13}C-labelled initiators and ^{13}C NMR to the study of the kinetics and efficiency of initiation of styrene polymerization, Makromol. Chem. Rapid Commun. 5 (1984) 785.
[17] D. Braun, K. Becker: Kinetik der Polymerisationsauslösung mit aromatischen Pinakolen, Makromol. Chem. 147 (1971) 91.
[18] J. C. Bevington: The nature of the initiation reaction in the polymerization of styrene sensitized by benzoyl peroxide, Proc. Roy. Soc. London, Ser. A 239 (1957) 420.
[19] J. C. Bevington, J. Toole: Further study of the decomposition of benzoyl peroxide in the presence of styrene, J. Polym. Sci. 28 (1958) 413.

[20] C. A. Barson, J. C. Bevington: A tracer study of the benzoyloxy radical, Tetrahedron 4 (1958) 147.
[21] J. C. Bevington: End groups in polymer molecules, Makromol. Chem. 34 (1959) 152.
[22] J. C. Bevington, J. Toole, L. Trosarelli: A study of two bromine-containing peroxides as initiators for the radical polymerization of styrene, Makromol. Chem. 32 (1959) 57.
[23] G. Moad, E. Rizzardo, D. Solomon, S. Johns, R. Willing: Application of ^{13}C-labelled initiators and ^{13}C NMR to the study of the kinetics and efficiency of initiation of styrene polymerization, Makromol. Chem., Rapid Commun. 5 (1984) 793.
[24] W. Heitz, P. Ball, M. Lattekamp: Oligobutadiene mit Ester- und Carbonatendgruppen, Kautschuk, Gummi, Kunststoffe 34 (1981) 459.
[25] R. A. Sheldon, J. Kochi: Pair production and cage reaction of alkyl radicals, J. Am. Chem. Soc. 92 (1970) 4393.
[26] V. V. Tobolsky: Dead-end radical polymerization, J. Am. Chem. Soc. 80 (1958) 5927.
[27] W. Konter, B. Bömer, K. H. Köhler, W. Heitz: Oligostyrenes by radical polymerization, Makromol. Chem. 182 (1981) 2619.
[28] C. H. Bamford, C. F. Tipper: Comprehensive Chemical Kinetics, Vol. 14a, Free Radical Polymerization, Elsevier, Amsterdam 1976.
[29] O. F. Olaj, J.W. Breitenbach, B. Wolf: Untersuchungen über Molekulargewichtsverteilungen von Hochpolymeren, 6. Mitt.: Zum Problem des Disproportionierungsabbruchs bei der gestarteten Polymerisation des Styrols, Monatsh. Chem. 95 (1964) 1646.
[30] K. C. Berger, G. Meyerhoff: Disproportionierung und Kombination als Abbruchmechanismen bei der radikalischen Polymerisation von Styrol 1, Makromol. Chem. 176 (1975) 1983.
[31] K. C. Berger: Disproportionierung und Kombination als Abbruchmechanismen bei der radikalischen Polymerisation von Styrol 2, Makromol. Chem. 176 (1975) 3575.
[32] G. Gleixner, O. F. Olaj, J.W. Breitenbach: Kombination und Disproportionierung von Modellradikalen für die wachsende Polystyrolkette, Makromol. Chem. 180 (1979) 2581.
[33] A. F. Bickel, W. A. Waters: The decomposition of aliphatic azocompounds: II. Rec. Trav. Chim. Pay-Bas 69 (1950) 1490.
[34] D. Ghatge, S. P. Vernekar, P. P. Wadgaonkar: A new free-radical initiator for the syntheses of polymers with isocyanato end groups, Makromol. Chem., Rapid Commun. 4 (1983) 307.
[35] R. Walz, B. Bömer, W. Heitz: Monomeric and polymeric azoinitiators, Makromol. Chem. 178 (1977) 2527.
[36] G. A. Razuvaev, L. M. Terman, D. M. Yanovskii: The nature of the radicals in initiation of polymerization by organic peroxydicarbonats, Dokl. Akad. Nauk. SSSR 161 (1965) 614; C. A. 63 (1965) 1869 h.
[37] W. Guth, W. Heitz: Telechele Oligomere 3. Oligoäthylene durch radikalische Polymerisation mit Diacylperoxiden als Initiator, Makromol. Chem. 177 (1976) 3159.
[38] N. Uri: Inorganic free radicals in solution, Chem. Rev. 50 (1951) 375.

[39] G. Moad, D. Solomon, S. Johns, R. Willing: Fate of the initiator in the azobis-(isobutyronitrile)-initiated polymerization of styrene, Macromolecules 17 (1984) 1094.
[40] W. Heitz, W. Konter, W. Guth, B. Bömer: Preparation of oligomers with functional end groups by polymerization reactions, Pure Appl. Chem., Macromol. Chem 8 (1973) 65.
[41] C. Bramford, A. D. Jenkins, R. P. Wayne: The coupling of polymers, part 2. Vinyl polymers and block copolymers, trans. Faraday Soc. 56 (1960) 932.
[42] D. M. French: Functionally terminated butadiene polymers, Rubber Chem. Techn. 42 (1969) 71.
[43] H. Schnecko, G. Degler, H. Dongowski, R. Caspari, G. Angerer, E. S. Ng: Synthesis and characterization of functional diene oligomers in view of their practical applications, Angew. Makromol. Chem. 70 (1978) 9.
[44] S. F. Reed: Telechelic diene prepolymers, I. Hydroxyl-terminated polydienes, J. Polym. Sci. A1, 9 (1971) 2029.
[45] S. F. Reed: Telechelic diene prepolymers, II. Carboxyl-terminated polydienes, J. Polym. Sci. A1, 9 (1971) 2147.
[46] O. S. Falkowa, V. I. Valuev, R. A. Sklyakhter, K. M. Avanesova, M. D. Korolkova, Y. L. Spirin: Chain transfer to a polymer during a hydrogen-peroxide-induced radical polymerization of 1,3-butadiene, Sint. Fiz.-Khim. Polim. 15 (1975) 16.
[47] W. Heitz, K. Arlt, W. Mehnert: Telechele Oligomere 1. Die Rhodium-katalysierte Synthese von telechelen Oligobutadienen, Makromol. Chem. 177 (1976) 1625.
[48] K. Arlt, W. Heitz: Telechelic Oligomers, 4[a]. Synthesis of telechelic oligobutadienes by Rh-catalysis, Makromol. Chem. 180 (1979) 41.
[49] W. Risse, W. Heitz, D. Freitag, L. Bottenbruch: Preparation and characterization of poly[oxy(2,6-dimethyl-1,4-phenylene)] with functional end groups, Makromol. Chem. 186 (1985) 1835.
[50] L. Freund, W. Heitz: Telechelic poly(thio-1.4-phenylene)s and poly(thio-1.4-phenylene)-block-polyamides, Makromol. Chem. 191 (1990) 815.

5 Cyclic and Macrocyclic Compounds

Hartwig Höcker* and Rolf C. Schulz**

5.1 Introduction

Cyclic molecules generally are prepared following high dilution principle techniques [1]. The starting materials are bifunctional compounds for which under the chosen dilution conditions the probability of the intramolecular reaction is significantly larger than that of the intermolecular reaction. The dilution principle was applied to the preparation of macrocyclic polystyrene. On the other hand, any ring-opening polymerization reaction is capable of forming cyclic oligomers and polymers by back-biting and end-biting reaction mechanisms. Two prominent examples will be considered here, i.e., the cationic ring-opening polymerization of oxacycles and the metathetical ring-opening polymerization of cycloolefins.

The distribution of cyclics in the kinetically controlled regime may differ significantly from that under thermodynamical control. In certain cases such as the electron transfer induced polymerization of divinylidene compounds the individual reaction steps are irreversible. Thus, the cyclics concentration formed under kinetical control is frozen in.

* Lehrstuhl für Textilchemie und Makromolekulare Chemie der RWTH Aachen, Veltmanplatz 8, D-5100 Aachen

** Institut für Organische Chemie der Universität Mainz, J.-J.-Becher-Weg 18–20, D-6500 Mainz

5.2 Ring-Chain Equilibria

Systems which result in ring-chain equilibria have been reviewed in detail by Semlyen [2, 3]. The equilibrium concentration of the cyclic oligomers yields valuable experimental information on the conformation of the respective low molecular weight chain molecules. For Gaussian chains, the Jacobson-Stockmayer cyclization theory [4] may be applied to systems of this kind. According to this theory the equilibrium constant K_x of a macrocycle with degree of polymerization x is proportional to $x^{-5/2}$.

It should be noted that K_x (which to a first approximation is equal to the inverse of the respective ring concentration) is independent of the initial monomer concentration. Consequently, for each system, a maximum initial monomer concentration is formed up to which only cyclics are formed.

The cyclic oligomer distribution in the kinetically controlled regime may differ significantly from that under thermodynamical control. Under certain conditions, the discrimination between stepwise and chain growth mechanism may vanish as the following example indicates. From the product distribution it may be deduced that the metathesis polymerization of unstrained cycloolefins such as cyclooctene and cyclododecene follows a stepwise growth mechanism. Oligomers are formed primarily, their concentration being higher than in equilibrium. The rate constant of the back-biting reaction has to be larger than that of the insertion reaction. In the contrast, for strained cycloolefins such as cyclopentene and norbornene a chain growth mechanism is followed. The polymer formation is kinetically favoured. The rate constant of the insertion reaction has to be larger than that of the back-biting reaction. Catalysts with reduced activity (partially poisoned catalysts) will result in a reduced yield of polymers in the first case (unstrained cycloolefins) and will not afford oligomers at all in the second case (strained cycloolefins).

5.3 Cycloolefins and Cycloparaffins Obtained by Metathesis Polymerization of Cyclomonoolefins

The metathesis reaction of cycloolefins is a polymerization reaction which allows the establishment of a ring-chain equilibrium and hence the preparation of cyclic oligomers. A suitable catalyst is WCl_6 in conjunction with $EtAlCl_2$ or $(CH_3)_4Sn$. The active species is assumed to be a tungsten carbene complex *(1)* [5]

$$\left(\begin{matrix}CH{=}CH\\(CH_2)_n\end{matrix}\right) + L_xW{=}CR_2 \rightleftarrows \left(\begin{matrix}L_xW-CR_2\\ | \quad\quad | \\HC-CH\\(CH_2)_n\end{matrix}\right) \rightleftarrows \left(\begin{matrix}L_xW \quad CR_2\\ \| \quad\quad \| \\CH \quad CH\\(CH_2)_n\end{matrix}\right) \qquad (1)$$

(1) *(2)* *(3)*

which forms a metallacyclobutane intermediate *(2)* with the olefin and eventually a new carbene complex *(3)*. The new carbene complex may be regarded formally as the insertion product of a cycloalkene into the metal-carbene bond. The cyclic oligomers are then formed by the back-biting reaction of a long chain carbene complex, i. e., by the backward reaction of Equ. (1).

Fig.5.1 shows a gel permeation chromatogram of the reaction products of cyclododecene. As seen from the figure, no oligomer is particularly favoured although the tetramer of cyclododecene was chosen as the starting olefin. Because of the relatively large difference in molecular weight (166 units) between the single oligomers, a preparative separation by gel permeation chromatography may be achieved.

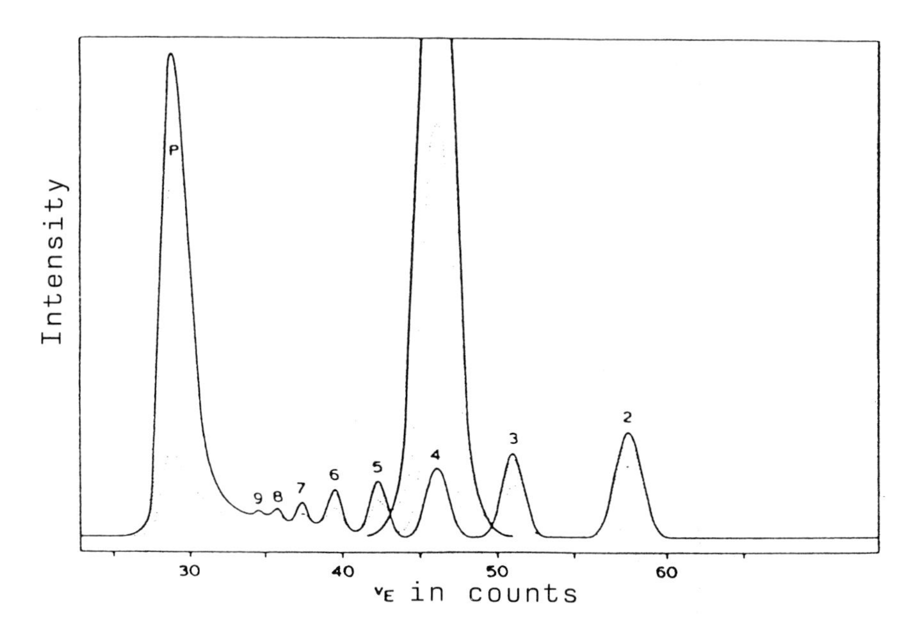

Fig. 5.1: GPC of a homologous series of oligomers and a polymer (at the exclusion limit) as generated from the tetramer of cyclododecene, $C_{48}H_{88}$; the broken lines show distributions before equilibrium is achieved.

As published in detail elsewhere [6], the following characteristics are observed:

- The oligomers exhibit cyclic structure, as proven by means of spectroscopic methods, in particular by mass spectroscopy.
- At low initial monomer concentration up to a critical concentration, the so-called cut-off point, only cyclic oligomers are observed and practically no polymer is detected.
- The cut-off point depends on the chemical structure of the monomer. For norbornene the critical initial monomer concentration was found to be 0.125 mol/l and for cyclooctene 0.2 mol/l (Fig.5.2).

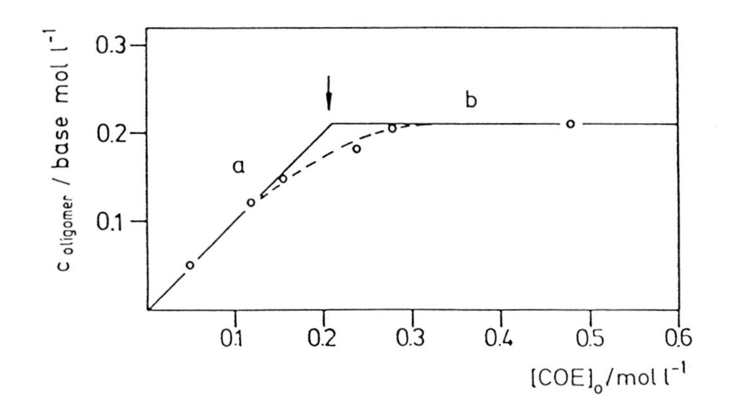

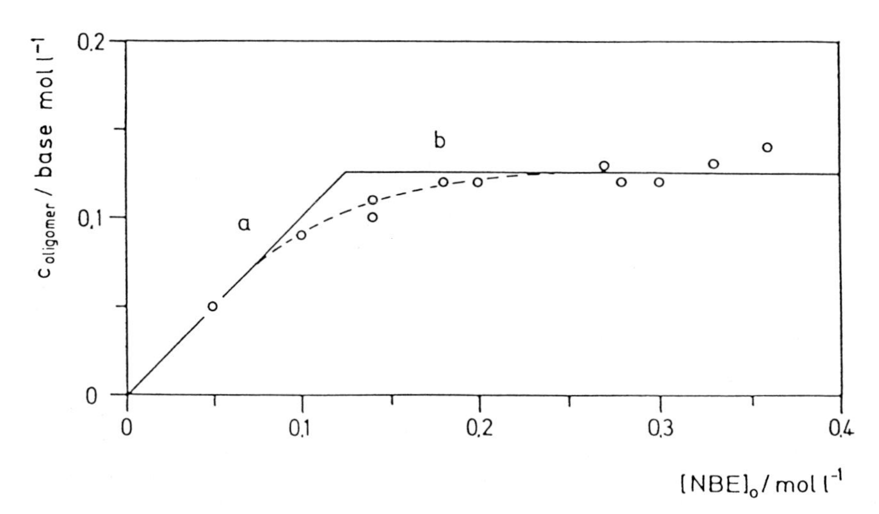

Fig. 5.2: 'Cut-off' point found in the metathesis reaction of cyclooctene (COE) and norbornene (NBE).

- At initial monomer concentrations higher than those corresponding to the cut-off point, the overall oligomer concentration in the solution of the reaction product remains constant; the excess monomer is converted into high polymer.
- In the kinetically controlled regime, the cyclic oligomer concentration is given by

$$[M_x] = A_t \cdot x^{-3/2}\alpha^x$$

with x being the degree of polymerization,

α = 1-1/P_n and
A_t = constant,

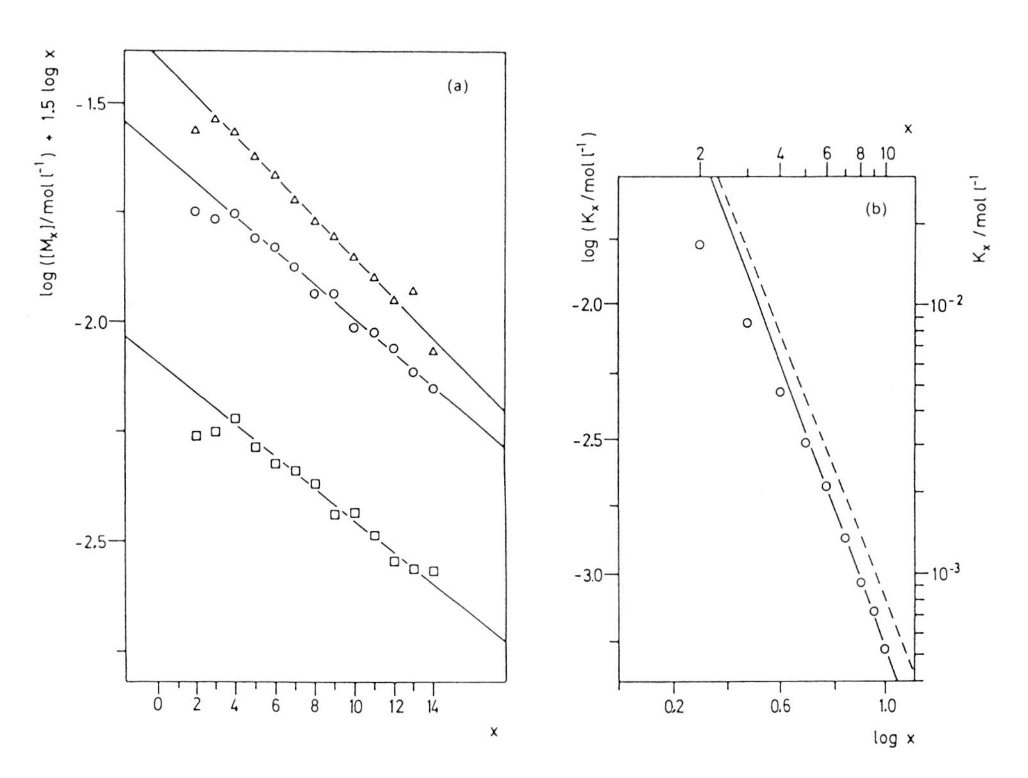

Fig. 5.3: Distribution of COE oligomers (a) in the kinetically controlled regime: □, 6.5%; ○ 21.0%; △ 35.0% conversion; (b) in the thermodynamically controlled regime. The slope of the broken line is −2.5. The deviation of the experimental points from this slope is partially attributed to the non-Θ character of the equilibration solvent (chlorobenzene).

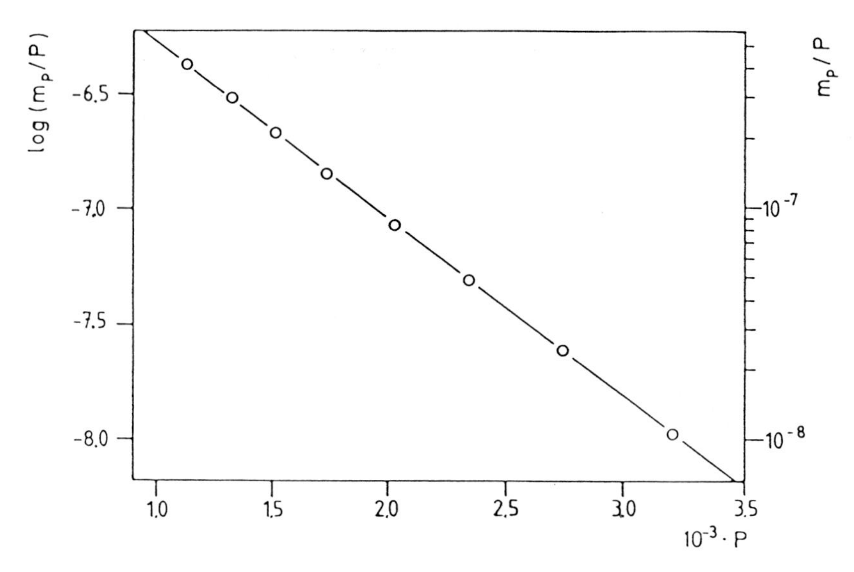

Fig. 5.4: Most probable distribution of the polymer population generated by the metathesis reaction of norbornene.

P_n being the number-average degree of polymerization (Fig.5.3a); in the thermodynamically controlled regime the distribution is given by the Jacobson-Stockmayer theory (Fig.5.3b)

$$K_x \sim C_x^{-3/2} \cdot x^{-5/2}$$

C_x being the characteristic ratio.

- The distribution of the (open chain) polymer may be represented by a most probable distribution function (Fig. 5.4).

5.4 Properties of Cycloolefins and Cycloparaffins

The cyclic oligomers of cycloalkenes, obtained by the metathesis reaction [7], show an increasing portion of trans-configurated double bonds with increasing degree of oligomerization. Thus, the dimer of cyclooctene exhibits a trans-

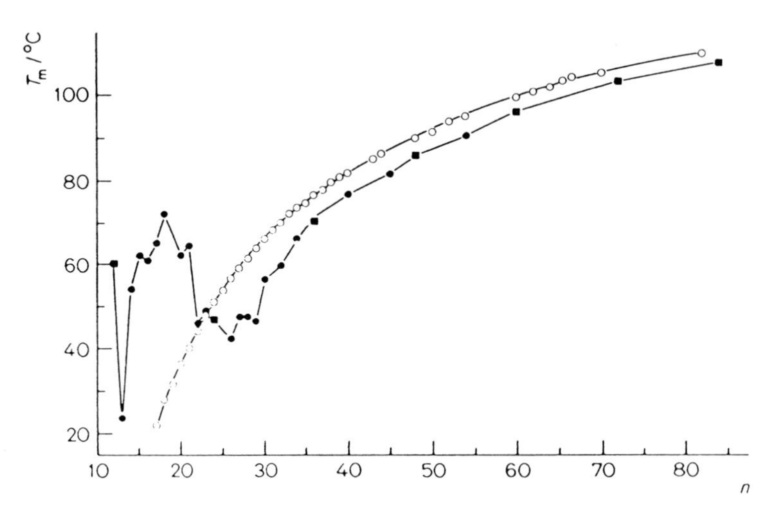

Fig. 5.5: Melting points of cycloalkanes (●, data of compounds from different sources; ■, data of compounds from metathesis reaction of cyclododecene and hydrogenation) as compared with those of n-alkanes (○) as a function of the number of C atoms, n.

content of 20 %, the decamer of 45 %. The boiling points increase from 415 K for cis-cyclooctene to 760 K for the pentamer (isomeric mixture, 35 % trans) of cyclooctene. The index of refraction, n = 1.46 for cyclooctene, quickly approaches the limiting value of 1.49. The proton resonances exhibit an upfield shift going from the monomer to the dimer and then remain constant.

The mass spectra of the oligomers of cyclooctene and cyclododecene have been studied in detail [8]. They show high intensity molecular ion peaks. The most prominent fragments have the general formula $C_nH_{2(n-p)-1}$ with $0 \leq p \leq x$, where x is the degree of polymerization of the respective oligomer (or the number of double bonds in it) and p represents the number of double bonds in the fragment.

After GPC separation the individual oligomers are readily hydrogenated to form the respective cycloalkanes. In Fig.5.5, the melting points of the cycloalkanes are shown as a function of ring size and compared with those of n-alkanes [9]. Formally, the melting points of the cycloalkanes are represented by an empirical equation proposed by Broadhurst [10] for n-alkanes

$$T_m = T\,(n + a)/(n + b) \qquad \begin{aligned} T &= 414.3\ \text{K} \\ a &= -1.4\ (-1.5) \\ b &= 5.75\ (5.0) \end{aligned}$$

where the numbers in parentheses refer to n-alkanes.

5.5 Cyclic Oligomers of Cycloacetals

It is long known that during the cationic ring-opening polymerization of 1,3-dioxolane *(4)* cyclic oligomers are formed [11, 12]. A Jacobson-Stockmayer plot shows that there is a close agreement between experiment and theory for cyclics with 25 or more skeletal bonds (that means above cyclic pentamers).

The question was whether higher homologues of dioxolane are also polymerizable and whether simultaneous formation of macrocycles takes place. Therefore the following compounds 1,3,6-trioxocane *(5)*, 1,3,6,9-tetraoxacycloundecane *(6)*, 1,3,6,9,12-pentaoxacyclotetradecane *(7)* and 1,3,6,9,12,15-hexaoxacycloheptadecane *(8)* have been prepared [13].

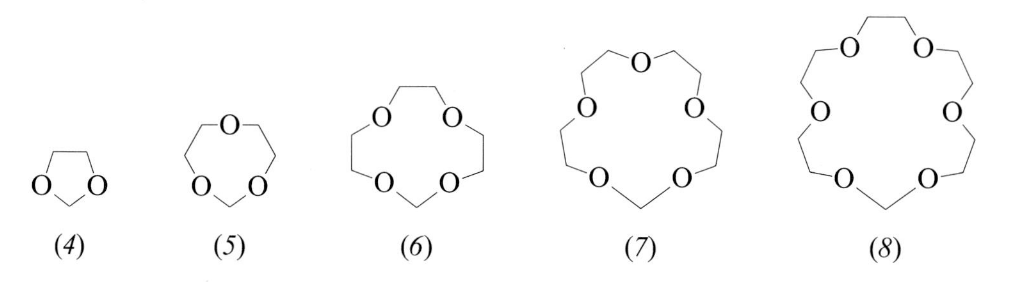

Indeed, all of them can be polymerized by cationic catalysts under formation of regular polymers with high molecular weight.

$$\overbrace{(CH_2CH_2O)_b\ OCH_2} \longrightarrow [-(OCH_2)-(OCH_2CH_2)_b-]_x$$

$$b = 1, 2, 3, 4, 5$$

By GPC of the reaction mixtures in all cases besides the polymers cyclic oligomers could be detected. In particular in the case of *(6)*, in the early stages of the reaction (in CH_2Cl_2 with CF_3SO_3H as initiator) exclusively cyclic oligomers are observed. This behaviour is characteristic of an end-biting reaction [14].

The product distribution of the cationic polymerization of *(6)* is shown in Fig. 5.6a. Fig. 5.6b shows the Jacobson-Stockmayer plot of the respective equilibrium oligomer distribution.

Furthermore, it was possible to isolate the first members of the cyclic oligomers up to the octamer by preparative GPC.

$$\text{cyclo-}(-OCH_2-(OCH_2CH_2)_3)_n \qquad (6\,a-g)$$

$$n = 1-8$$

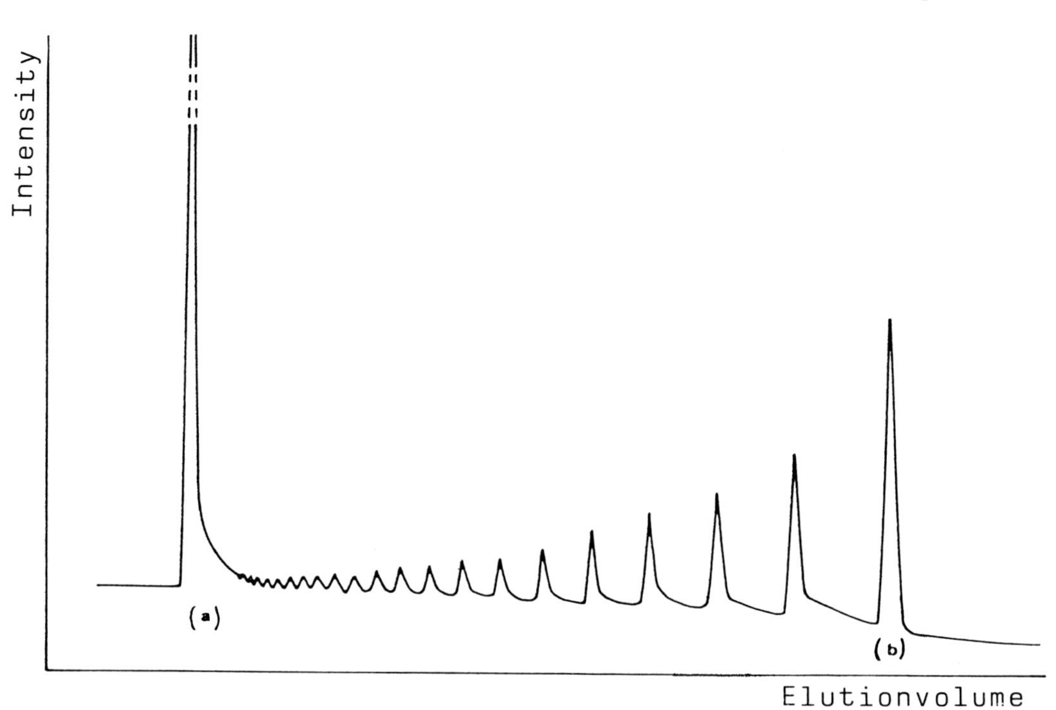

Fig. 5.6 a: GPC of a polymerization product of *(6)*. [monomer] = $2.5 \cdot 10^{-4} mol \cdot l^{-1}$; solvent: CH_2Cl_2; temp.: 0 °C, (a) polymer fraction; (b) dimer.

The melting points show a strong odd-even-effect with decreasing tendency. The high polymer fraction have a crystalline melting point around +6 °C [15].

n	1(=(6))	2	3	4	5	6	7	8
mp. °C	27	88	28	55	18	35	20	28

In spite of the structural similarity of the compounds *(5)–(8)* and their corresponding cyclic oligomers with the well-known crownethers non of these compounds form cage complexes with cations.

In a similar manner the cationic ring-opening polymerization of an unsaturated cycloacetal (4H, 7H-1,3 dioxepin *(9)*) with BF_3- OEt_2 between -20° and + 25 °C was achieved [16].

Beside the linear polymer *(9a)*

$$\longrightarrow -OCH_2-O-CH_2-CH=CH-CH_2- \quad (9a)$$

(9)

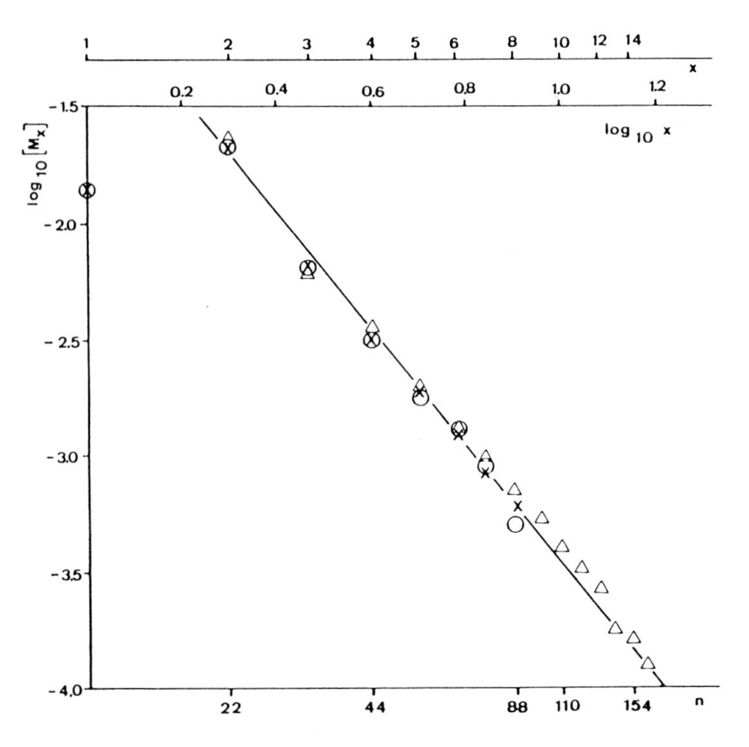

Fig. 5.6b: Jacobson-Stockmayer plot of equilibrium cyclic oligomer concentration of (*6*) on degree of polymerization from three (△ x ○) different experiments. The straight line has a slope of −2.5.

cyclic oligomers at least up to the hexamer have been identified. The cyclic dimer *(9b)* was isolated in high yield (m. p. 117 °C).

(*9b*)

The same compound had been prepared previously by a quite different method [17].

The bulk polymerization of the bicyclic unsaturated acetal *(10)* (3,5-dioxa-bicyclo[5.4.0] undec-9-ene) (m. p. 1 °C) leads to a colourless waxlike polymer (T_g= −11 °C) *(10a)*. During the polymerization oligomers are also formed whose cyclic structure up to the pentamer can be proved by HP-GPC [18]. The cyclic dimer *(10b)* (m. p. 234 °C) is produced in higher quantities. It is also formed as a byproduct during the synthesis of the monomer *(10)*. It should be mentioned that the corresponding trans-isomer is not polymerizable [19]. In this case only the cyclic dimer is formed.

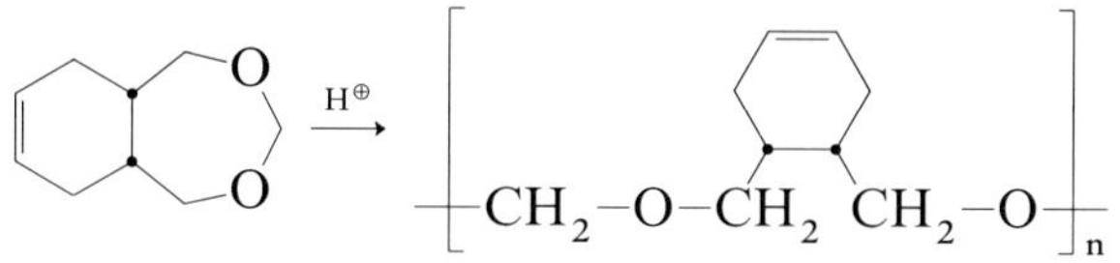

$\xrightarrow{H^{\oplus}}$ $\left[-CH_2-O-CH_2 \quad CH_2-O- \right]_n$

(*10*) (*10a*)

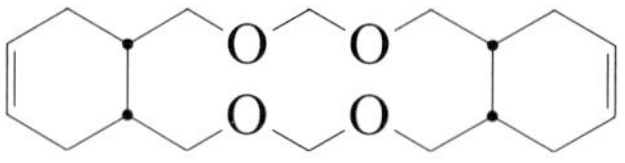

(*10b*)

5.6 Synthesis of Cyclophanes via Polycombination Reaction Induced by Electron Transfer to Divinylidene Compounds

Since the work of Szwarc et al. [20] it has been known that 1,1-diphenylethylene (DPE) upon electron transfer forms the monomeric radical anion which immediately and exclusively forms the dimeric dianion. The homopolymerization of DPE, on the other hand, does not occur, very probably because of steric reasons. Molecules which are bifunctional with respect to the characteristic group of DPE react in a corresponding way upon electron transfer:

$$2\ H_2C=C(C_6H_5)-C_6H_4-C(C_6H_5)=CH_2 \xrightarrow{2e^{\ominus}}$$

$$2\ H_2C=C(C_6H_5)-C_6H_4-\overset{\ominus}{\overline{C}}(C_6H_5)-\dot{C}H_2 \longrightarrow$$

$$\begin{array}{l} H_2C=C(C_6H_5)-C_6H_4-\overset{\ominus}{\overline{C}}(C_6H_5)-CH_2 \\ \qquad\qquad\qquad\qquad\qquad\qquad\qquad\quad | \\ H_2C=C(C_6H_5)-C_6H_4-\underset{\ominus}{\underline{C}}(C_6H_5)-CH_2 \end{array}$$

They, however, possess another vinylidene group which is still eligible for an electron transfer reaction. Whether this electron transfer reaction occurs with a similar rate as the first one or at a much lower rate depends on the substitution mode of -C_6H_4-. If it is a p-phenylene (or 4.4'-biphenylene) group, the negative charge formally placed at the neighbouring carbon atom is delocalized over the whole molecule, the colour of the solution (in THF) is deep violet, and the electron affinity of the residual double bond is so low that, with lithium as an electron transfer reagent, no further reaction occurs at all; the dimeric α,ω-diene may be obtained in nearly 100 % yield. With sodium as an electron transfer reagent, the dimer is formed as well; but in the further course of the reaction, further electron transfer occurs by which new radical species are formed which combine. Thus, the even members of the homologous series of the oligomers are obtained.

In contrast, if the middle group is a m-phenylene group (or

$$-p-C_6H_4-(CH_2)_n-p-C_6H_4-$$

or

$$-p-C_6H_4-CH(C_6H_5)-(CH_2)_n-CH(C_6H_5)-p-C_6H_4-)$$

the two double bonds react individually, i. e. the electron affinity of a double bond is independent of whether the twin double bond has been converted to a carbanionic species already or not. Consequently, intramolecular reactions (combination reaction/radical addition reaction) become possible by which cyclic oligomers, cyclophanes, are formed. The colour of the THF solution is red.

All molecules may react intermolecularly to build up cyclic oligomer with a certain degree of polymerization. Once they have reacted intramolecularly they do not react further; the cyclics are formed irreversibly [21]. It has to be mentioned that the divinylidene compound containing the m-phenylene unit is an exceptional case since the dimer highly favours the intramolecular reaction. Thus the cyclic dimer (*11* in its protonated form) is isolated in more than 90 % yield [22].

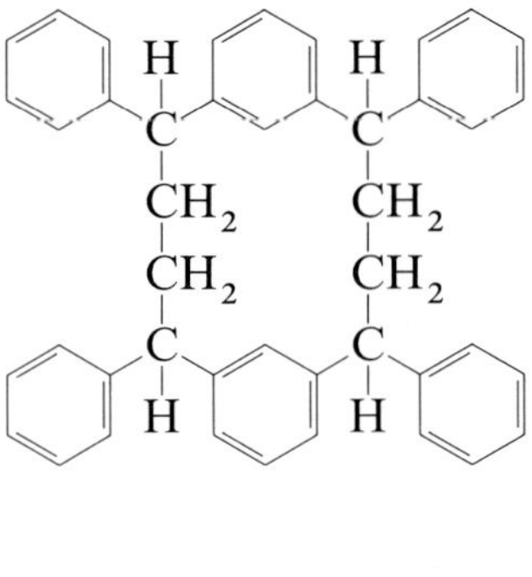

(*11*)

Molecules with a similar geometrical structure behave in a similar way, i. e. divinylidene compounds which contain the 1,5-naphthylene group or the 1,4-pyridylene group [23]. The divinylidene compounds *(12)* and *(13)* form a

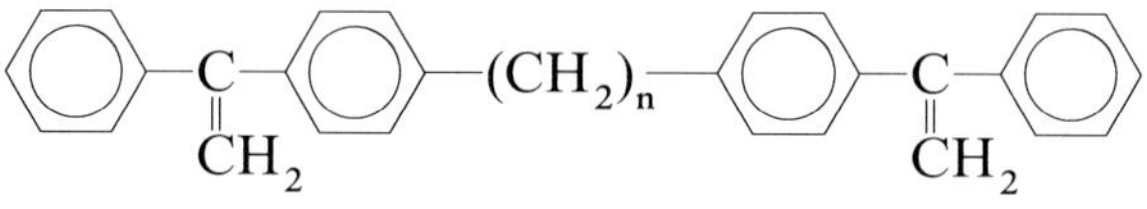

$C(=CH_2)$ … $(CH_2)_n$ … $C(=CH_2)$ (*12*)

$H_2C{=}C$ … $CH{-}CH_2{-}CH_2{-}CH$ … $C{=}CH_2$ (*13*)

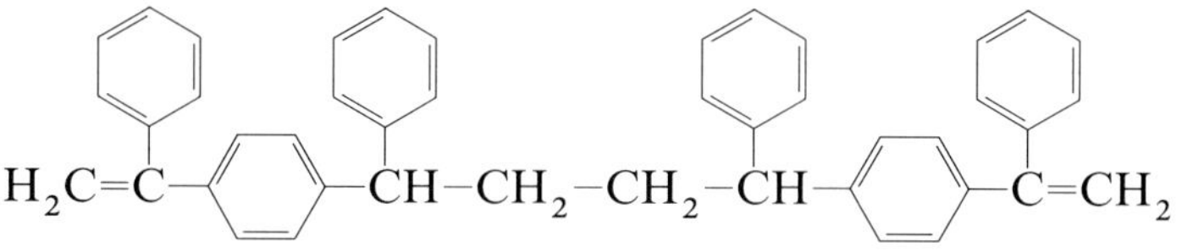

homologous series of oligomers without favouring a certain oligomer. The formation of a cyclic monomer may be observed for $n \geq 3$ and reaches a maximum for $n = 6$. The individual oligomers (degree of polymerization x) contain 2x centres of asymmetry, which for the cyclic monomers and dimers is clearly reflected in the NMR spectra. The larger the rings, the more the chiral centres are decoupled and the less is the effect on the nuclear magnetic resonances, i. e. the less is the number of lines [23].

5.7 Macrocyclic Polystyrene

Cyclic oligomers are formed by ring-opening polymerization of cyclic monomers as well as during polycondensation and polyaddition reactions. In these systems the cyclics' concentration corresponds to a ring-chain equilibrium. The concentration of high molecular weight cyclics is assumed to be low (although indications are present that high molecular weight cyclics are formed as well). The preparation of high molecular weight cyclics was first achieved by Brown and Slusarczuk [24] and later by Bannister and Semlyen [25]. They obtained high molecular weight cyclic poly(dimethylsiloxane)s.

With advantage high molecular weight cyclic polymers may be obtained using bifunctional "living" polymers obtained by anionic polymerization in conjunction with bifunctional terminating agents (electrophiles X-R-X).

As the initiator for the anionic polymerization of styrene sodium naphthalene may be used in tetrahydropyran as the solvent and α,α'-dichloro-p-xylene as the electrophile. The deep red solution of the bifunctional living polymer and the colourless solution of the bifunctional electrophile are slowly added to a certain volume (e. g. 2 l) of the solvent following the principle of mutual titration to guarantee an equimolar ratio of the reactants. At the end of the cyclization reaction, an excess of the electrophile should be provided to end-cap the acyclic molecules with X-groups. Then the polymer may be isolated, dissolved and reacted with high molecular weight living polystyrene. By this reaction, the acyclic material is fixed to the high molecular weight polystyrene and the cyclic material may be isolated by fractionation [26].

$$2\,[\text{naphthalene}]^{\dot{\ominus}}\ Mt^{\oplus} + 2\,CH_2{=}CH(C_6H_5) \longrightarrow$$

$$2\,\text{naphthalene} + \overset{\ominus}{C}H(C_6H_5){-}CH_2{-}CH_2{-}\overset{\ominus}{C}H(C_6H_5)\quad 2\,Mt^{\oplus} \xrightarrow{\text{Styrene}}$$

$$\overset{\ominus}{C}H(C_6H_5){-}CH_2 \cdots CH_2{-}\overset{\ominus}{C}H(C_6H_5)\quad 2\,Mt^{\oplus} \xrightarrow{X{-}R{-}X}$$

$$\overbrace{CH(C_6H_5){-}CH_2 \cdots CH_2{-}CH(C_6H_5)}^{R} + 2\,MtX$$

Mt: Na, K X: Cl, Br

By this method, we have obtained cyclic polymers up to molecular weights of 25,000. For a number of reasons cyclic polymers with a higher molecular weight are desirable. They were prepared in the mean time by Roovers and Toporowski [27].

The GPC elution volume V_e of cyclics is larger than that of acyclic molecules. The reason for this behaviour is to be found in the smaller molecular dimensions of cyclics as compared with acyclic molecules. In the case of polystyrene the so-called calibration curve, that is the log M vs. V_E plot of cyclic molecules, is, within the limits of experimental error, parallel to that of acyclic molecules but shifted to higher elution volumes (Fig. 5.7). The ratio of molecular weights of cyclic and acyclic polymers eluted at a certain elution volume is close to 1.4 in THF as eluting solvent.

Since GPC is sensitive to the hydrodynamic volume of the respective molecules, according to Benoit [28], a "universal" calibration curve is obtained if log M [η] is plotted instead of log M, [η] being the intrinsic viscosity.

The intrinsic viscosity reflects most obviously the hydrodynamic volume of polymer molecules in dilute solution (Fig. 5.8). Measured in cyclohexane at 307 K the ratio of the intrinsic viscosity of cyclic and acyclic polystyrene was found to be 0.66. This value is in close agreement with the theoretically expected value [29].

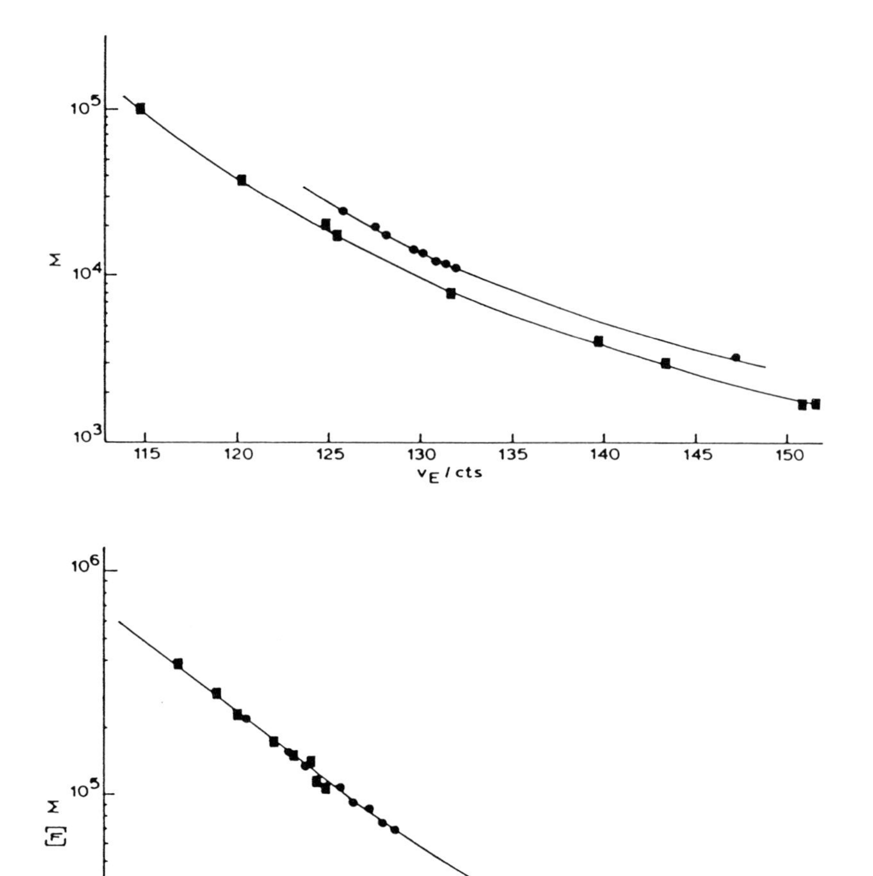

Fig. 5.7: GPC calibration (a) and universal calibration (b) curve for cyclic (●) and acyclic (■) polystyrenes.

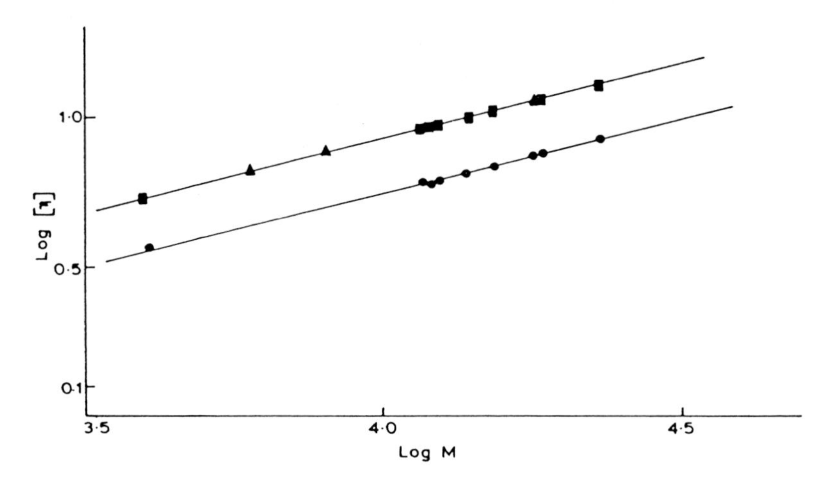

Fig. 5.8: Log [η] vs. log M curves for cyclic (●) and acyclic (▲, ■) polystyrene.

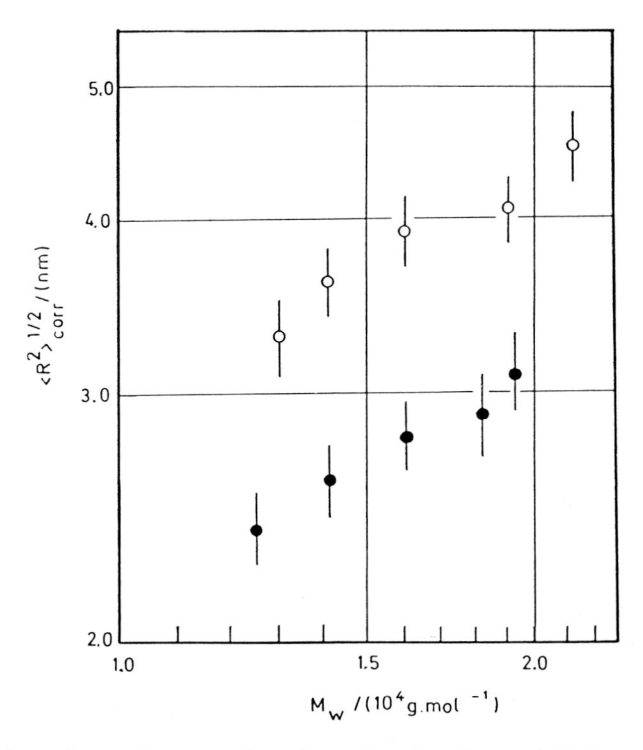

Fig. 5.9: Radius of gyration as a function of molecular weight for cyclic (●) and acyclic (○) polystyrene as obtained from low angle neutron scattering in toluene-d_8.

The most direct evidence for the hydrodynamic volume of both cyclic and acyclic molecules is the radius of gyration which was obtained from small angle neutron scattering in toluene-d_8 [30]. Although toluene is a thermodynamically good solvent, the theoretically expected ratio of the mean square radii of gyration of cyclic and acyclic molecules, i. e., a value of 0.5 (as expected for Θ-conditions), was found. The dependence of the radius of gyration on the molecular weight is shown in Fig. 5.9.

5.8 Concluding Remarks

Cyclics play an important role in polymer chemistry. Very often they are unavoidable associates of linear macromolecules and may disturb but may also improve the properties of the respective polymers. In particular, if the cyclics are of high molecular weight, they very well may improve the mechanical and thermal properties of polymers since they affect the diffusion behaviour and hence the rheological properties of the bulk material.

5.9 References

[1] L. Rossa, F. Vögtle: Topics in Current Chemistry 113 (1983) 1.
[2] J. A. Semlyen: Adv. Polym. Sci. 21 (1976) 41.
[3] J. A. Semlyen: Pure Appl. Chem. 53 (1981) 1797.
[4] H. Jacobson, W. H. Stockmayer: J. Chem. Phys. 18 (1950) 1600.
[5] I. L. Herisson, Y. Chauvin: Makromol. Chem. 141 (1970) 161.
[6] L. Reif, H. Höcker: Macromolecules 17 (1984) 952.
[7] H. Höcker, R. Musch: Makromol. Chem. 175 (1974) 1395.
[8] H. Höcker, K. Riebel: Makromol. Chem. 179 (1978) 1765.
[9] H. Höcker, K. Riebel: Makromol. Chem. 178 (1977) 3101.
[10] M. G. Broadhurst: J. Chem. Phys. 36 (1962) 2578.
[11] S. Penczek, P. Kubisa, K. Matyjaszewski: Adv. Polym. Sci. 68/69 (1985) 35.
[12] J. A. Semlyen: Adv. Polym. Sci. 21 (1976) 59.
[13] K. Albrecht, D. Fleischer, C. Rentsch, H. Yamaguchi, R. C. Schulz: Makromol. Chem. 178 (1977) 3191.
[14] C. Rentsch, R. C. Schulz: Makromol. Chem. 178 (1977) 2535; 179 (1978) 1403.
[15] K. Albrecht, D. Fleischer, A. Kane, C. Rentsch, Q.V. Tran Thi, H. Yamaguchi, R. C. Schulz Makromol. Chem. 178 (1977) 881.
[16] W. Hellermann, R. C. Schulz: Makromol. Chem., Rapid Commun. 2 (1981) 585.
[17] J.U. R. Nielson, S. E. Jorgensen, N. Frederiksen, R. B. Jensen, O. Dahl, O. Burchardt, G. Schroll: Acta Chem. Scand. Ser. B 29 (1975) 400; B 33 (1979) 197.
[18] Ch. Mletzko, R. C. Schulz: Makromol. Chem., Rapid Commun. 4 (1983) 445.

[19] E. Schwarzenbach, R. C. Schulz: unpublished results.
[20] G. H. Monteiro, M. Levy, M. Szwarc: Trans. Faraday Soc. 58 (1962) 1809; M. Matsuda, J. Jagur-Grodzinski, M. Szwarc: Proc. Roy. Soc. (London) Ser. A 288 (1965) 212; J. Jagur-Grodzinski, M. Szwarc: Proc. Roy. Soc. (London) Ser. A 288 (1965) 224; E. Ureta, J. Smid, M. Szwarc: J. Polymer Sci. Part A-1, 4 (1966) 2219.
[21] H. Höcker, G. Lattermann: J. Polym. Sci. 54 (1976) 316.
[22] H. Höcker, G.G. H. Schulz: Makromol. Chem. 178 (1977) 2589.
[23] T. Bastelberger, C. Tran Thu, H. Höcker: Makromol. Chem. 185 (1984) 1565.
[24] J. F. Brown, G.M. J. Sluzarczuk: J. Amer. Chem. Soc. 87 (1965) 931.
[25] D. J. Bannister, J. A. Semlyen: Polymer 22 (1981) 381.
[26] G. Hild, A. Kohler, P. Rempp: Europ. Polym. J. 16 (1980) 525.
[27] J. Roovers, P. M. Toporowski: Macromolecules 16 (1983) 843.
[28] H. Benoit, Z. Grubisic, P. Rempp, D. Decker, B. J. Zilliox: J. Chem. Phys. Phys.-Chem. Biol. 63 (1966) 1507; Z. Grubisic, P. Rempp, H. Benoit: J. Polym. Sci., Part B, 5 (1967) 735.
[29] V. Bloomfield, B. H. Zimm: J. Chem. Phys. 44 (1965) 315; M. Fukatsu, M. Kurata: J. Chem. Phys. 44 (1966) 4539.
[30] M. Ragnetti, D. Geiser, H. Höcker, R. C. Oberthür: Makromol. Chem. 186 (1985) 1701.

6 Kinetics of Copolymerization

Dietrich Braun and Wojciech K. Czerwinski*

Abstract

On the basis of various models of rate equations for binary copolymerizations the classical ternary rate equation of Melville et al. was developed. New equations were derived for systems characterized by significant viscosity differences between the monomers and the viscosity control of termination is considered. Rate equations presented here were examined by measurements with several monomer systems in bulk at 60 °C with AIBN as initiator.

6.1 Introduction

The development and verification of reaction models at the end of the 40s was one of the most important stages on the field of free radical copolymerization [1–3]. Because of interesting properties of binary, ternary and more component copolymers respective copolymerization models were developed [4–7]. Their basic applicability was confirmed by means of experimental investigations. For many systems deviations of the copolymer composition and/or of the polymerization rate from the classical behaviour were detected. These observations gave raise to development of supplementary reaction models.

Some authors have postulated such effects as the influence of the penultimate monomer unit on the reactivity of the final growing radical (the penultimate effect) [8, 9], the significance of the initiation dependency on monomer feed composition [10], effect of medium viscosity [11], of degradative chain transfer [12] or CT-complexation between the reacting components [13–15] on polymerization. Recently several workers reexamined the classical reaction model [16–18].

* Deutsches Kunststoff-Institut, Schloßgartenstraße 6, D-6100 Darmstadt

It follows from a study of different rate equations [19] that none of them is particularly advantageous in comparison with the classical rate equation [3]. In another work the copolymerization of styrene with a wide number of comonomers was reviewed in terms of the classical copolymerization model [20]. For 72% of monomer systems under consideration a good or very good description was achieved.

The aim of the present work was to obtain systematic knowledges of copolymerization and rate behaviour of several binary systems. Further it was attended to describe quantitatively measurements by means of the existing models [4–6, 21]. Basing on the above study [22–27] detailed investigations of two ternary systems were performed: styrene (St,M_1)/N-vinylpyrrolidone (NVP,M_2)/methyl methacrylate (MMA,M_3) – system I – and diethyl maleate (MSE,M_1)/MMA (M_2)/St (M_3) – system II. The kinetical analysis comprised the composition of terpolymers, dilatometric contraction constant, overall rate of polymerization, initiation rate, rate constant of initiator dissociation and efficiency of the primary radicals as functions of the monomer feed composition.

6.2 Experimental Methods

Copolymers were prepared in glass ampoules under N_2 atmosphere in bulk with AIBN as initiator. The substrates were degassed by threefold freeze-thawing cycles and then heated in a water bath at 60 °C for a given reaction time. The copolymers were reprecipitated two- or threefold and dried under vacuum. Copolymer composition was determined by elemental analysis or by means of IR- and NMR-spectroscopy.

Rate measurements were performed dilatometrically under the same conditions. For initiation rate investigations stabile Banfield radical and triphenylverdazyle were applied as inhibitors [28, 29].

6.3 Binary Copolymerization

6.3.1 Classical Rate Model

For binary copolymerizations several kinetic models were described in the literature. The classical rate model developed by Melville et al. [3] is based on the assumption that the reactivity of the growing radical is controlled by the reactivity of the terminal chain component only. Under these conditions the following scheme of propagation steps is valid:

$$-M_1\cdot + M_1 \xrightarrow{k_{p11}} -M_1\cdot \tag{1}$$

$$-M_1\cdot + M_2 \xrightarrow{k_{p12}} -M_2\cdot \tag{2}$$

$$-M_2\cdot + M_1 \xrightarrow{k_{p21}} -M_1\cdot \tag{3}$$

$$-M_2\cdot + M_2 \xrightarrow{k_{p22}} -M_2\cdot \tag{4}$$

Under steady state conditions and supposing that Equ. (5) is valid:

$$k_{p12}[-M_1\cdot][M_2] = k_{p21}[-M_2\cdot][M_1] \tag{5}$$

Melville, Noble, and Watson [3] derived the following rate equation:

$$v_{Br} = \frac{(r_1[M_1]^2 + 2[M_1][M_2] + r_2[M_2]^2) \cdot v_i^{1/2}}{(r_1^2\delta_1^2[M_1]^2 + 2\Phi\delta_1\delta_2 r_1 r_2[M_1][M_2] + r_2^2\delta_2^2[M_2]^2)^{1/2}} \tag{6}$$

$$r_i = \frac{k_{pii}}{k_{pij}} \qquad \delta_i = \frac{k_{tii}^{1/2}}{k_{pii}}$$

$$\Phi = \frac{k_{t12}}{(k_{t11} \cdot k_{t22})^{1/2}} \quad \text{cross termination constant}$$

$$v_i = 2k_d f \cdot [I] \text{ rate of initiation}$$

Some rational simplifications of this model have been proposed by Abkin [10]. The termination rate was assumed by this author to be constant for various monomer feed compositions and therefore only the initiation rates for the homopolymerizations are required. Under this prerequisite time-consuming initiation rate measurements in the monomer mixtures can be omitted.

6.3.2 Model of Diffusion Controlled Termination

Many experimental investigations showed that these rather simple assumptions can only be applied in some special cases. Therefore other authors have discussed more complex models. As the so-called cross termination constant in Equ. (6) in many experimental investigations was found not to be constant, several authors took into consideration mainly the termination steps. North and Postlethwaite [11, 30] have postulated a reaction scheme with diffusion control:

$$-M_1\cdot + \cdot M_2- \underset{k_{-1}}{\overset{k_1}{\rightleftharpoons}} -M_1\cdot\cdot M_2- \tag{7}$$

$$-M_1\cdot\cdot M_2- \underset{k_{-2}}{\overset{k_2}{\rightleftharpoons}} -M_1 : M_2- \tag{8}$$

$$-M_1 : M_2- \xrightarrow{k_c} P \tag{9}$$

reaction (7): Growing radicals diffuse towards one another
reaction (8): Sequential motion permits a direct approach of reactive centres
reaction (9): Termination step

The overall rate constant of termination (k_t) depends on the viscosity of the reaction medium:

$$k_t = \frac{k_t^\circ}{\eta_L} \tag{10}$$

k_t° = proportionality factor

Applying this fact the rate equation can bc modificd as follows:

A: for homopolymerizations

$$v_{br} = \frac{k_p}{(k_t^\circ)^{0.5}} \eta_L^{0.5} \, (2k_d f[I])^{0.5}[M] \tag{11}$$

B: for copolymerizations (Atherton and North [11])

$$v_{br} = \frac{(r_{ij}[M_i]^2 + 2[M_i][M_j] + r_{ji}[M_j]^2) \cdot v_i^{0.5}}{\left(\frac{k_t^\circ}{\eta_L}\right)^{0.5} \cdot \left(\frac{[M_i]}{k_{pij}} + \frac{[M_j]}{k_{pji}}\right)} \tag{12}$$

By substituting

$$d_i = (k_t)^{0.5}/k_{pij}$$

and $$d_j = (k_t)^{0.5}/k_{pji}$$

Equ. (13) follows :

$$v_{br} = \frac{(r_{ij}[M_i]^2 + 2[M_i][M_j] + r_{ji}[M_j]^2) \cdot v_i^{0.5} \cdot \eta_L^{0.5}}{d_i[M_i] + d_j[M_j]} \tag{13}$$

The parameters d_i and d_j are calculated as fitting factors for given values of v_{br}, η_L, r_{ij} and r_{ji}.

6.3.3 Penultimate Model

The diffusion rate of the growing chain end depends not only on the medium viscosity but also on the conformational characteristics of this radical chain end. Regarding this effect a reaction model has been developed assuming, in first approximation, that the length of the rearranging segment is given by the last four carbon atoms (penultimate effect). With this simplification Merz, Alfrey, and Goldfinger [8] derived a modified copolymer composition equation. A rate equation corresponding to this model was developed by Russo and Munari [9] and was successfully applied to describe the copolymerization rate of several binary systems. If this assumption is made, ten different reactions must be regarded [9]:

$$\begin{array}{l} \sim M_i - M_j^{\cdot} + {}^{\cdot}M_i - M_j\sim \xrightarrow{k_{tijij}} \\ \sim M_i - M_j^{\cdot} + {}^{\cdot}M_j - M_j\sim \xrightarrow{k_{tijjj}} \end{array} P \tag{14}$$

Using the simplification:

$$\begin{aligned} k_{tijij} &= 2(k_{tijji} \times k_{tjiij})^{0.5} \\ k_{tijjj} &= 2(k_{tijji} \times k_{tjjjj})^{0.5} \end{aligned} \tag{15}$$

expression (16) for the rate of copolymerizations was obtained:

$$v_{br} = \frac{[M_1 + M_2]v_i^{0.5}}{x+1} \cdot \frac{r_{12}x^3 + 3r_{12}x^2 + 2x + r_{12}r_{21}x + r_{21}}{r_{12}^2x^2\delta_1 + r_{12}x\delta_{21} + \frac{r_{21}x(r_{12}r_{21}x+1)}{r_{21}+x}\delta_{12} + \frac{r_{21}^2(r_{12}x+1)}{r_{21}+x}\delta_2} \tag{16}$$

for $x = [M_1]/[M_2]$, r_{12}, r_{21}, δ_1 and δ_2 see Equ. (6) and

$$\delta_{12} = \frac{k_{t1221}^{\;0.5}}{k_{p22}} \qquad \delta_{21} = \frac{k_{t2112}^{\;0.5}}{k_{p11}}$$

Recent reexamination of the classical binary model by Inagaki et al. [17, 31–34] shows however that the velocity of free-radical copolymerizations is yet not comprehensively understood in terms of the elementary process of propagation and termination. Measurements of absolute rate constants k_p and k_t reveal that particularly the constant k_p is entirely different from what the model predicts. The authors therefore indicate the necessity of reinterpretation of the propagation step in copolymerization reactions.

6.3.4 Supplementary Concepts

The individual growing radicals may react in different energy states before they give up their reaction heat. Assuming only the lowest (cold) and highest (hot) extreme states the "theory of hot radicals" was developed [16]. The theory leads to the conclusion that the rate constants of chain propagation reactions, hence the reactivity ratios, may depend on the composition of the monomer mixture or the dilution.

In some cases the classical binary model may not be valid, e.g., if the monomers are capable of association, owing to secondary valence forces. In this regard the donor acceptor interaction between molecules has a very remarkable effect. Models considering the interactions between radical and monomer or solvent [35] or monomer-monomer [36] (complex-models) were developed. For the second case a rather complicated rate equation [15] results. A simplier method by Georgiev and Zubov [37] was succesfully applicated by many authors [38–41] to describe the reaction rate of some binary systems exhibiting a significant monomer-monomer complex formation.

6.3.5 Experimental Results

For the verification of copolymer composition and rate models for binary copolymerizations some experimental results are presented in the following.
The investigated systems may be ordered in systems behaving classically:

p-chlorostyrene/n-butyl methacrylate	[42–44]
styrene/N-phenyl maleimide	[45]
styrene/1-vinyl naphthalene	[46]
methyl methacrylate/1-vinyl naphthalene	[46]
acrylonitrile/1-vinyl naphthalene	[46]
styrene/N-vinyl pyrrolidone	[20, 24, 27, 49]
styrene/methyl methacrylate	[22–26]
styrene/diethyl maleate	[47]
methyl methacrylate/diethyl maleate	[48]

and systems showing non classical properties:

styrene/N-p-tolylcitraconimide	[45] (polymer composition)
styrene/methyl α-cyanocinnamate	[45] (polymer composition)
methyl methacrylate/N-vinyl pyrrolidone	[49] (reaction rate)

For all systems monomer feed and copolymer composition data were transformed by means of the Kelen-Tüdös method [16] yielding values for classical relative reactivity ratios r_{ij}. r-values found are summarized in Table 6.1. For systems styrene/N-p-tolylcitraconimide and styrene/methyl α-cyanocinnamate, for which the classical model is not adequate, the relative reactivity ratios (r_1 and r_1') were evaluated in terms of the simplified penultimate model [8]. For both monomer systems the expectation of the penultimate model permitted to linearize the data points and to calculate the relative reactivity ratios. The magnitude of the penultimate effect may be expressed by means of the ratio r_1'/r_1. From table 6.1 it results to 1.8 for the system St/N-p-tolylcitraconimide and 6 for the system St/methyl α-cyanocinnamate. As discussed in [50], the penultimate effect may be caused by the very polar structure of the comonomers. Owing to the strong electron withdrawing character of both substituents, comonomers considered are able at the penultimate position to influence the reactivity of the final radical unit. This is a common property of a wide number of disubstituted vinyl monomers [51]. In the case of monosubstituted monomers investigated in the present work no deviations from the classical copolymerization model were observed.

As Fig. 6.1 shows the polymerization reactivities of styrene and N-vinyl pyrrolidone (NVP) are very different. The copolymerization can be described in a very satisfying way and with good accuracy by means of the classical rate

Table 6.1: Relative reactivity ratios calculated in terms of the terminal copolymerization model (r_1, r_2) or of the penultimate model (r_1, r_1') for several monomer systems (bulk polymerizations at 60 °C with AIBN).

Monomer system		Relative reactivity ratios terminal model (1, 2)		References
M_1	M_2	r_1	r_2	
p-chlorostyrene	n-butyl methacrylate	1.036 ± 0.100	0.517 ± 0.100	41–44
styrene	N-phenyl maleimide	0.024 ± 0.008	0.098 ± 0.106	45
styrene	1-vinyl naphthalene	0.688 ± 0.433	0.978 ± 0.365	46
methyl methacrylate	1-vinyl napthalene	0.347 ± 0.150	0.640 ± 0.102	46
acrylo-nitrile	1-vinyl napthalene	0.095 ± 0.089	0.511 ± 0.300	46
styrene	N-vinyl pyrrolidone	20.51 ± 5.13	0.030 ± 0.151	20, 24, 27, 49
methyl methacrylate	N-vinyl pyrrolidone	4.163 ± 0.715	0.103 ± 0.102	49
styrene	diethyl maleate	6.592	0.001	47
methyl methacrylate	diethyl maleate	370	0	48
styrene	N-p-tolyl citraconimide	0.083	0	45[a)]
styrene	methyl α-cyanocinnamate	0.402	0	45[a)]
		penultimate model (8)		
		r_1	r_1'	
styrene	N-p-tolyl citraconimide	0.061 ± 0.022	0.108 ± 0.017	50[a)]
styrene	methyl α-cyanocinnamate	0.229 ± 0.047	1.230 ± 0.225	50[a)]

[a)] in toluene as solvent

Equ. (6). The reason is that due to the big differences in the reactivity of the two monomers involved under practical conditions only few active chains with NVP units as radical end groups are present; therefore, one kind of termination reaction (with styrene at the growing chain ends) predominates which is in agreement with the prerequisit of this model.

For the system styrene/methyl methacrylate the polymerization reactivities of the two monomers are also quite different. But as shown in Fig. 6.2, the copolymerization rate cannot be exactly described by means of the classical model (fine-dashed line). This fact is well known from the literature and was explained by Russo and Munari [9] as an effect of the penultimate monomer unit. Another possibility for describing this system is to use variable cross termination parameters Φ in the classical model (solid line in Fig. 6.2). The change of Φ with the monomer feed is also indicated in Fig. 6.2 by a long-dashed line.

As can be seen in Fig. 6.3 the reaction rate of the system N-vinyl pyrrolidone/methyl methacrylate cannot be described by means of the classical copolymerization model (neither with $\Phi = 1$ or with $\Phi = 0.0002$ as the best value (dashed lines)). Here, interactions between NVP and MMA can be observed using ^{1}H- and ^{13}C-NMR spectroscopy [49]. The observed signal shifts can be explained in analogy to [52, 53] by an interaction between the carbonyl atom of MMA and the nitrogen atom by NVP.

Starting from earlier investigations [26] on the MMA homopolymerization in N-methyl pyrrolidone (NMP) as a nonpolymerizable model of NVP a series

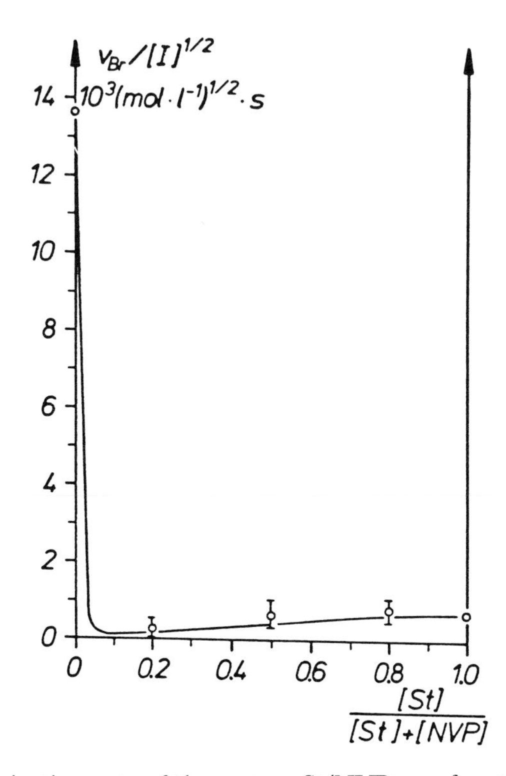

Fig. 6.1: Copolymerization rate of the system St/NVP as a function of the monomer feed (60 °C, AIBN; ○ measured points, —— calculated (Equ. 6), $\Phi = 0.033$ ($r_{12} = 20.51$, $r_{21} = 0.030$).

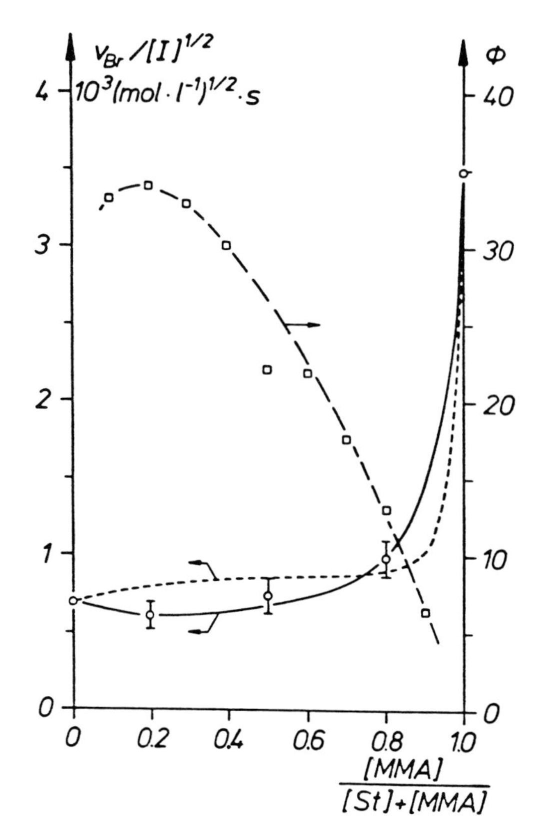

Fig. 6.2: Copolymerization rate of the system St/MMA and the value of Φ (Equ. 6) as a function of the monomer feed (60 °C, AIBN); ○ measured values, ------ calculated (Equ. 6), Φ = 15.73, —— calculated (Equ. 6), variable Φ-values (r_{13} = 0.522, r_{31} = 0.482), -□-□- Φ-values calculated from the reaction rate.

of polymerizations was carried out with increasing dilution, where the monomer concentrations were varied in a wide range. From the values of the overall rate constant k_{br} the ratio $\delta = k_{tii}^{0.5}/k_{pii}$ was calculated supposing the initiation rate constant $2k_df$ to have the same value as in bulk homopolymerization.

This condition is not exactly fulfilled [54]. The experimentally found values, therefore, are indicated as δ'. The change of $\delta'_{3(MMA)}$ with increasing dilution is represented in Fig. 6.3 (dotted-dashed line) for [NMP] = [NVP]. Since this dependence is very pronounced, the variation was introduced as a function of the ratio [NVP] : [MMA] into the classical rate Equ. (6). The obtained rate curve fits the experimental points quite well confirming the importance of the monomer-monomer interaction in the copolymerization kinetics of this monomer system.

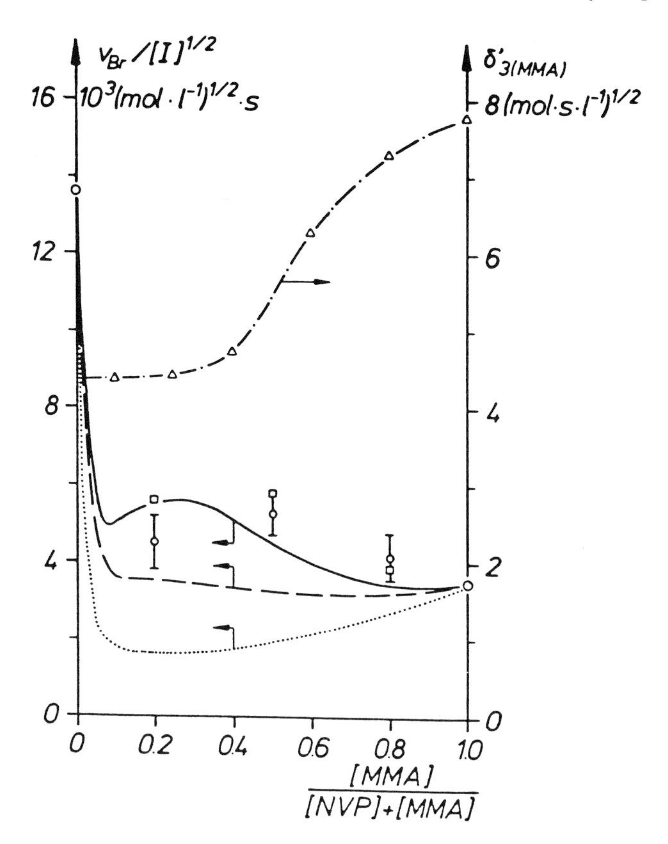

Fig. 6.3: Copolymerization rate of the system NVP/MMA and δ'_3 as a function of the monomer feed (60 °C, AIBN); ○ measured values, □ literature data (26) (same reaction conditions), ········ calculated (Equ. 6), $\Phi = 1$, ------ calculated (Equ. 6), $\Phi = 0.0002$ as best value (nonlinear estimation), —— calculated (Equ. 6), $\Phi = 3$ and variable δ'_3-values ($r_{23} = 0.103$, $r_{32} = 4.163$), --·△---△-- δ'_3, calculated with $2k_d f = 0.91 \cdot 10^{-5} s^{-1}$.

The polymerization rate of the system diethyl maleate (MSE)/MMA [55] is shown in Fig. 6.4 (full points). The monomer reactivity ratios were determined to $r_{MSE} = 0$ and $r_{MMA} = 370$ [48]. From these values follows that MSE participates only to a very little extend in the propagation step; practically it is not homopolymerizable. Starting from this fact, the copolymerization of the system MSE/MMA may be regarded with a very good approximation as homopolymerization of MMA in MSE as the solvent. Under these conditions the copolymerization rate can be interpreted by Equ. (11). Taking into account the viscosity of the reaction medium which changes between $\eta_L = 0.37$ cp for MMA and $\eta_L = 1.33$ cp for MSE with this simple model results a rather good agreement between the measured and the calculated values [48].

The copolymerization rates of the systems diethyl maleate/styrene and diethyl maleate/methylmethacrylate show a quite similar feature [47, 55]. From the copolymerization reactivity ratios of the system MSE/St (r_{MSE} = 0.001 and r_{St} = 6.592) results, however, that MSE is more frequently incorporated into the polymer chains during the copolymerization with styrene as in the system MSE/MMA. Therefore, the copolymerization rate can be better interpreted by means of a binary copolymerization model using the change of the viscosity of the reaction medium from η_L = 0.47 cp for styrene to η_L = 1.34 cp for MSE, which is nearly the same interval as for MSE/MMA. For this reason the reaction model of Atherton and North [11] (Equ. 12) could be used to describe the measured rate values (see Fig. 6.5). The fitting factors d_1 and d_3 were estimated with Equ. (13) to 135.7 and 186.6 $(mol \cdot s \cdot cp \cdot l^{-1})^{0.5}$, respectively [47].

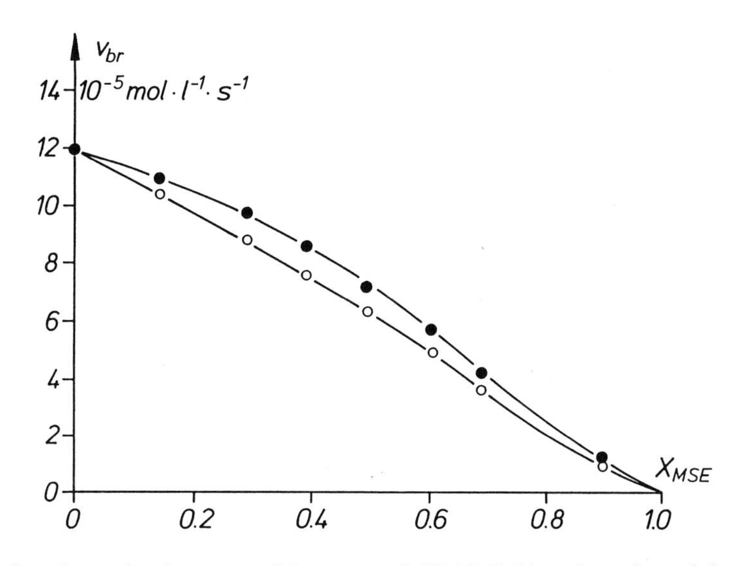

Fig. 6.4: Copolymerization rate of the system MSE/MMA as function of the monomer feed (60 °C, [AIBN] = $2 \cdot 10^{-3}$ mol · l^{-1}); ● measured values, ○ calculated with Equ. (11).

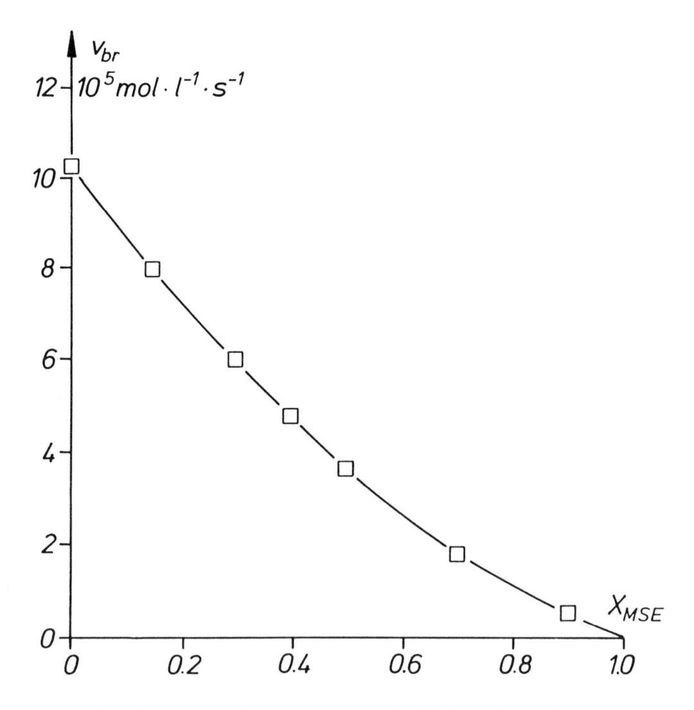

Fig. 6.5: Copolymerization rate of the system MSE/St as function of the monomer feed (60 °C, [AIBN] = $2 \cdot 10^{-2}$ mol · l^{-1}); □ measured values, —— calculated with Equ. (13) [11] and d_1 = 135.7 (mol · s · cp · $l^{-1})^{0.5}$, d_3 = 186.6 (mol · s · cp · $l^{-1})^{0.5}$, r_{13} = 0.001 and r_{31} = 6.592.

6.4 Ternary Copolymerization

As mentioned above, up to now there are no systematic investigations on the rate of ternary copolymerizations in the literature. From the experiences with binary copolymerization systems for an exact evaluation of the rate equation for ternary systems the knowledge of the initiation rates seems to be necessary; in this regard a linear interpolation of the initiation rate for binary and ternary monomer systems is not sufficient. Since the required values are missing, the applicability of the classical ternary rate equation of Blackley, Melville, and Valentine [6] is essentially unknown.

6.4.1 Classical Rate Model for Three Homopolymerizable Monomers

For three homopolymerizable monomers the propagation steps of a ternary copolymerization are much more complicated than in the case of binary copolymerizations as the propagation is composed of nine reactions:

$$-M_1\cdot + M_1 \xrightarrow{k_{p11}} -M_1\cdot \tag{17}$$

$$-M_1\cdot + M_2 \xrightarrow{k_{p12}} -M_2\cdot \tag{18}$$

$$-M_1\cdot + M_3 \xrightarrow{k_{p13}} -M_3\cdot \tag{19}$$

$$-M_2\cdot + M_1 \xrightarrow{k_{p21}} -M_1\cdot \tag{20}$$

$$-M_2\cdot + M_2 \xrightarrow{k_{p22}} -M_2\cdot \tag{21}$$

$$-M_2\cdot + M_3 \xrightarrow{k_{p23}} -M_3\cdot \tag{22}$$

$$-M_3\cdot + M_1 \xrightarrow{k_{p31}} -M_1\cdot \tag{23}$$

$$-M_3\cdot + M_2 \xrightarrow{k_{p32}} -M_2\cdot \tag{24}$$

$$-M_3\cdot + M_3 \xrightarrow{k_{p33}} -M_3\cdot \tag{25}$$

On this basis a ternary rate equation was derived by Melville et al. [6] supposing the validity of the steady-state condition:

$$v_{Br} = \frac{\left\{\sum\limits_{1,2,3}\left([M_1]r_{12} + [M_2] + [M_3]\frac{r_{12}}{r_{13}}\right)\Delta_1\right\} \cdot v_i^{1/2}}{\left\{\sum\limits_{1,2,3} D_{11} + 2\cdot \sum\limits_{1,2,3} D_{12}\right\}^{1/2}} \tag{26}$$

$$\sum_{1,2,3} D_{11} = \delta_1^2 r_{12}^2 \Delta_1^2 + \delta_2^2 r_{23}^2 \Delta_2^2 + \delta_3^2 r_{31}^2 \Delta_3^2$$

$$\sum_{1,2,3} D_{12} = \Phi_{12}\delta_1\delta_2 r_{12} r_{23}\Delta_1\Delta_2 + \Phi_{23}\delta_2\delta_3 r_{23} r_{31}\Delta_2\Delta_3$$

$$+ \Phi_{31}\delta_3\delta_1 r_{31} r_{12}\Delta_3\Delta_1$$

$$\Delta_1 = \begin{vmatrix} \frac{r_{23}}{r_{21}}[M_1] & [M_1] \\ -\left\{\frac{r_{23}}{r_{21}}[M_1] + [M_3]\right\} & \frac{r_{31}}{r_{32}}[M_2] \end{vmatrix}$$

$$\Delta_2 = -\begin{vmatrix} -\left\{[M_2] + \frac{r_{12}}{r_{13}}[M_3]\right\} & [M_1] \\ [M_2] & \frac{r_{31}}{r_{32}}[M_2] \end{vmatrix}$$

$$\Delta_3 = \begin{vmatrix} -\left\{[M_2] + \frac{r_{12}}{r_{13}}[M_3]\right\} & \frac{r_{23}}{r_{21}}[M_1] \\ [M_2] & -\left\{[M_3] + \frac{r_{23}}{r_{21}}[M_1]\right\} \end{vmatrix}$$

Melville et al. [6] have verified this equation for the monomer system styrene/p-methoxystyrene/MMA. With the aid of interpolated initiation rate values a good agreement between the calculated and the measured rate was obtained for three different monomer feeds.

The assumptions of Abkin [10] concerning the initiation and termination steps are particularly interesting in the case of ternary copolymerizations. They would allow to reduce considerably the number of experiments permitting the application of the classical model. In this connection Braun et al. [56] have applied Abkin's simplifications to the classical ternary reaction model as a first approximation.

Setting λ_{nm} as ratio of the initiation rate constants (k_i) for monomers n and m

$$\lambda_{nm} = \left(\frac{k_{i_n}}{k_{i_m}}\right)^{0.5} \text{ and}$$

$$v_{br,n} = \frac{v_{br,homo,n}}{[I]^{0.5}}$$

($v_{br,homo,n}$ – absolute homopolymerization rate of the monomer M_n at the initiator concentration [I])

the following rate equation results:

$$v_{Br} = -\frac{d[M]}{dt} = \frac{v_{Br,3} \cdot T^{1/2}}{[M]_o^{3/2}} \cdot \left\{ \frac{\frac{a}{r_{32}}[M_1][M_2] + \frac{[M_1]}{r_{31}} \cdot Z}{ab[M_1][M_2] - x \cdot z} \left(W + \frac{e[M_2] \cdot Y}{Z} \right) + \right.$$

$$\left. + \frac{[M_2] \cdot Y}{r_{32} \cdot Z} + U \right\} \qquad (27)$$

$$T = \lambda_{13}^2[M_1] + \lambda_{23}^2[M_2] + [M_3]$$

$$U = \frac{[M_1]}{r_{31}} + \frac{[M_2]}{r_{32}} + [M_3]$$

$$W = c[M_1] - b[M_2] + d[M_3]$$

$$X = \frac{\lambda_{13}[M_1]}{r_{31}v_{Br,1}} + \frac{[M_2]}{r_{12}v_{Br,3}} + \frac{[M_3]}{r_{13}v_{Br,3}}$$

$$Y = a[M_1] + f[M_2] + g[M_3]$$

$$Z = \frac{[M_1]}{r_{21}v_{Br,3}} + \frac{\lambda_{23}[M_2]}{r_{32}v_{Br,2}} + \frac{[M_3]}{r_{23}v_{Br,3}}$$

$$a = \frac{1}{r_{21}v_{Br,3}} + \frac{\lambda_{23}}{r_{31}v_{Br,2}} \qquad e = \frac{\lambda_{13}}{r_{32}v_{Br,1}} - \frac{1}{r_{12}v_{Br,3}}$$

$$b = \frac{1}{r_{12}v_{Br,3}} - \frac{\lambda_{13}}{r_{32}v_{Br,1}} \qquad f = \frac{1}{v_{Br,3}} - \frac{\lambda_{23}}{r_{32}v_{Br,2}}$$

$$c = \frac{\lambda_{13}}{r_{31}v_{Br,1}} - \frac{1}{v_{Br,3}} \qquad g = \frac{1}{r_{23}v_{Br,3}} - \frac{\lambda_{23}}{v_{Br,2}}$$

$$d = \frac{\lambda_{13}}{v_{Br,1}} - \frac{1}{r_{13}v_{Br,3}}$$

For the calculations using Equ. (27) the following constants are required: monomer reactivity ratios r_{ij}, the three individual homopolymerization rates $v_{br,n}$ and only three initiation rate constants k_i. The results correspond directly to the values divided by $[I]^{0.5}$.

6.4.2 Model of Diffusion Controlled Termination

The ternary copolymerization kinetics can be simplified if one of the comonomers is not able to homopolymerize. Supposing the diffusion control of the chain termination [11, 30] (Equ. 10) the following ternary rate equation can be developed in case that M_1 cannot homopolymerize:

$$v_{br} = \{2k_d f(I)\eta_L\}^{0.5} \cdot \frac{(S_1' + S_2' + S_3')}{(V_1 + V_2 + V_3)} \tag{28}$$

$$S_1' = \{[M_2] + R[M_3]\}\Delta_1'$$

$$S_2' = \left\{ \frac{[M_1]}{r_{21}} + [M_2] + \frac{[M_3]}{r_{23}} \right\} \Delta_2' r_{23}$$

$$S_3' = \left\{ \frac{[M_1]}{r_{31}} + \frac{[M_2]}{r_{32}} + [M_3] \right\} \Delta_3' r_{31}$$

$$R = \frac{k_{p13}}{k_{p12}}$$

$$V_1 = d_1 \Delta_1' \cdot R \qquad d_1 = \frac{(k_t^\circ)^{0.5}}{k_{p13}}$$

$$V_2 = \delta_2^\circ \cdot \Delta_2' \cdot r_{23} \qquad \delta_2^\circ = \frac{(k_t^\circ)^{0.5}}{k_{p22}}$$

$$V_3 = \delta_3^\circ \cdot \Delta_3' \cdot r_{31} \qquad \delta_3^\circ = \frac{(k_t^\circ)^{0.5}}{k_{p33}}$$

$$\Delta_1' = \begin{vmatrix} \frac{r_{23}}{r_{21}}[M_1] & [M_1] \\ -[M_3] & [M_1] + \frac{r_{31}}{r_{32}}[M_2] \end{vmatrix} \qquad \Delta_2' = \begin{vmatrix} \frac{r_{31}}{r_{32}}[M_2] & [M_1] \\ -[M_2] & [M_2] + R[M_3] \end{vmatrix}$$

$$\Delta_3' = \begin{vmatrix} R[M_3] & [M_2] \\ -[M_3] & [M_3] + \frac{r_{23}}{r_{21}}[M_1] \end{vmatrix}$$

Considering the penultimate effect for monomers M_2 and M_3 according to Russo et al. [9] Equs. (14) and (15) and regarding the steady-state condition Equ. (29) results:

$$v_i = v_t = \{(k_{t2222})^{0.5}(\sim M_2{-}M_2^{\cdot}) + (k_{t2332})^{0.5}(\sim M_2{-}M_3^{\cdot}) + (k_{t3333})^{0.5}(\sim M_3{-}M_3^{\cdot}) + (k_{t3223})^{0.5}(\sim M_3{-}M_2^{\cdot})\}^2 \quad (29)$$

In a system with a nonhomopolymerizable monomer M_1 the following three assumptions may be made:

a) $[-M_1{-}M_1^{\cdot}] = 0$

b) The radicals $-M_2{-}M_1^{\cdot}$, $-M_3{-}M_1^{\cdot}$, $-M_1{-}M_2^{\cdot}$ and $-M_1{-}M_3^{\cdot}$ do not show any penultimate effect and

c) the viscosity of the medium has no influence on the termination of all growing radicals.

Under these conditions Equ. (29) changes to Equ. (30):

$$v_i = v_t = \frac{1}{\eta_L}[(k^{\circ}_{t2222})^{0.5}(\sim M_2{-}M_2^{\cdot}) + (k^{\circ}_{t3223})^{0.5}(\sim M_3{-}M_2^{\cdot}) + (k^{\circ}_{t3333})^{0.5}(\sim M_3{-}M_3^{\cdot}) + (k^{\circ}_{t2332})^{0.5}(\sim M_2{-}M_3^{\cdot}) + (k^{\circ}_{t})^{0.5}\{(M_1^{\cdot}) + (\sim M_1{-}M_2^{\cdot}) + (\sim M_1{-}M_3^{\cdot})\}]^2 \quad (30)$$

With

$$k_{12}[-M_1\cdot] : k_{23}[-M_2\cdot] : k_{31}[-M_3\cdot] = \Delta_1 : \Delta_2 : \Delta_3 \quad (31)$$

$$\frac{(\sim M_i - M_j^{\cdot})}{(\sim M_j - M_j^{\cdot})} = \frac{k_{ij}(M_i^{\cdot})}{k_{jj}(M_j^{\cdot})} \quad (32)$$

and the Equs. (28) and (30) the final rate Equ. (33) results:

$$v_{br} = \{2k_d f(I)\eta_L\}^{0.5} \cdot \frac{(S_1' + S_2' + S_3')}{\left[J + \frac{L}{M} + \frac{N}{P}\right]} \tag{33}$$

s_i' – see Equ. (28) and

$$J = d_1 R \Delta_1' \quad M = \Delta_1' + r_{23}\Delta_2' + \frac{r_{31}}{r_{32}} \Delta_3' \quad P = r\Delta_1' + \Delta_2' + r_{31}\Delta_3'$$

$$L = \left(\delta_2^\circ(r_{23}\Delta_2' + \Delta_1') + \delta_{32}^\circ \frac{r_{31}}{r_{32}} \Delta_3'\right) r_{23}\Delta_2'$$

$$N = (\delta_3^\circ(r_{31}\Delta_3' + R\Delta_1') + \delta_{23}^\circ \Delta_2') r_{31}\Delta_3'$$

d_1, δ_2°, δ_3° – see Equ. (28)

$$\delta_{23}^\circ = \frac{k_{t2332}^\circ}{k_{p33}} \qquad \delta_{32}^\circ = \frac{k_{t3223}^\circ}{k_{p22}}$$

6.4.3 Experimental Results

The results of rate measurements of the polymerization of the system styrene/N-vinyl pyrrolidone/methyl methacrylate are shown in Fig. 6.6. A detailed analysis of the polymerization rate of this system is given in [56]. As can be seen, the homopolymerization rates of these three monomers are very different. The binary systems styrene/N-vinyl pyrrolidone and styrene/methyl methacrylate as well as nearly all investigated ternary systems polymerize with a very similar rate to that of the styrene homopolymerization, that means, that in this case the slowest polymerizing component determines the overall rate of the ternary system.

The experimentally measured values can be interpreted by means of the classical rate Equ. (26). All necessary kinetic constants have been determined by homopolymerization and binary copolymerization experiments. Φ_{13} and δ_3 were taken as functions of the ratios [St] : [MMA] and [NVP] : [MMA], respectively. The measured and calculated rate values are summarized in Table 6.2; the comparison shows a very good agreement between the calculated and the experimentally determined rates. Therefore in the present case the classical rate equation can be applied.

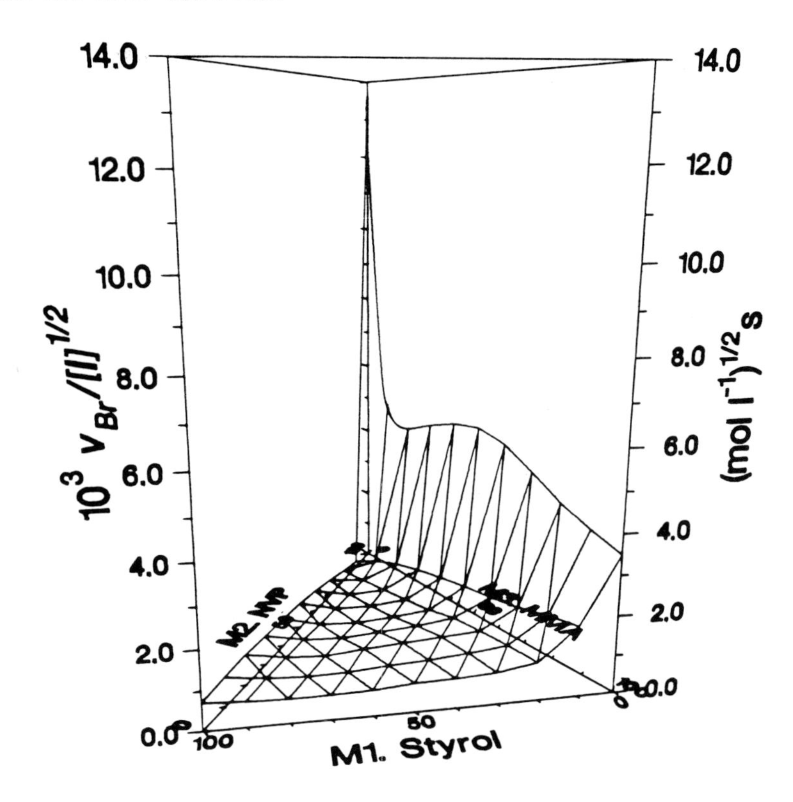

Fig. 6.6: Experimentally determined ternary copolymerization rate of the system St/NVP/MMA as a function of the monomer feed (60 °C, AIBN, in bulk).

An interpretation with the aid of the simplified ternary model (Equ. 27) yields only approximative rate values (Table 6.2). It is, however, possible in this case to estimate quickly the copolymerization rate for different monomer feeds without any knowledge of the initiation rates of the monomer mixture.

In a similar way the experimentally obtained rate values for the system MSE/MMA/St are shown in Fig. 6.7 [29]. Also in this case, the component with the slowest homopolymerization rate (MSE) determines the copolymerization rate of the ternary system over a wide range of monomer feed.

For the interpretation of the experimental results both Equs. (28) and (33) were used; measured and calculated values are compared in Table 6.3.

As can be seen Equ. (28) is able to describe the experimentally determined rates in a very satisfying manner. Nevertheless, considering a penultimate effect in the case of the binary system styrene/MMA the applied rate model (Equ. 33) permits to describe the polymerization rate of the ternary system even more precisely.

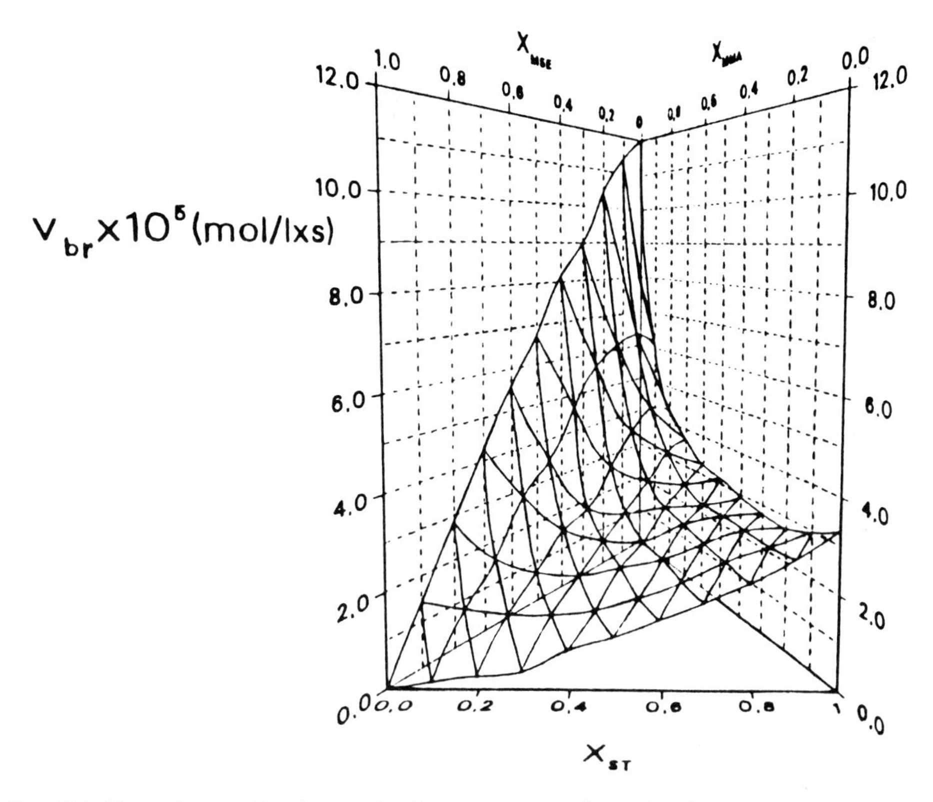

Fig. 6.7: Experimentally determined ternary copolymerization rate of the system MSE/MMA/St as a function of the monomer feed (60 °C, AIBN, in bulk).

6.5 Conclusions

In the present paper some systematic investigations on the rate of several binary and ternary copolymerization systems are presented and compared with theoretically developed rate equations based on various kinetic models. The here discussed systems can be described using the classical copolymerization model by Melville et al. respectively by some newly developed models being modifications of the classical model. The rather good agreement between calulations and experimental results permits the conclusion that the rate of binary and ternary copolymerizations may be interpreted on the basis of Melville's model, but that it is necessary to take into consideration some deviations which result from some simplifications of the model. The here for the first time presented systems show also, that Melville's model must be adopted to the individual nature of the considered monomers, especially in

Table 6.2: Measured and calculated polymerization rate of the ternary system ST/NVP/MMA.

System I St : NVP : MMA	$v_{br}/[AIBN]^{0.5} \cdot 10^4$ $(mol \cdot l^{-1})^{0.5}s$ 60 °C in bulk, AIBN		
(mol-%)	measured	calculated (Equ. 26)	calculated (Equ. 27)
10 : 10 : 80	16.672	13.778	25.084
40 : 10 : 50	9.727	6.941	15.676
80 : 10 : 10	6.982	6.773	8.057
20 : 20 : 60	9.311	7.870	18.259
40 : 20 : 40	7.745	6.626	13.108
60 : 20 : 20	6.310	5.940	9.184
10 : 40 : 50	11.803	9.316	16.868
50 : 40 : 10	5.105	4.849	6.052
10 : 60 : 30	9.183	6.666	10.794
30 : 60 : 10	4.315	3.579	4.903
10 : 80 : 10	5.236	3.180	4.348

Table 6.3: Measured and calculated polymerization rate of the ternary system MSE/MMA/St.

System II MSE : MMA : St	$v_{br} \cdot 10^5$ $mol \cdot l^{-1} \cdot s^{-1}$ 60 °C in bulk, AIBN		
(mol-%)	measured[a)]	calculated (Equ. 28)	calculated (Equ. 33)
10.3 : 9.8 : 79.9	2.69	3.66	2.70
45.6 : 9.7 : 44.7	1.45	2.00	1.49
29.4 : 10.1 : 10.5	0.65	0.49	0.43
23.0 : 23.0 : 54.0	2.42	3.88	2.24
33.6 : 33.1 : 33.3	2.37	4.00	2.14
55.8 : 22.5 : 21.7	1.58	1.83	1.18
9.6 : 45.4 : 45.0	3.36	7.38	3.31
22.9 : 53.8 : 23.3	3.84	5.93	2.81
44.8 : 45.0 : 10.2	2.98	2.94	1.74
9.3 : 81.3 : 9.4	5.83	7.78	4.36

a) $[AIBN] = 2 \cdot 10^{-3}$ $[mol \cdot l^{-1}]$

case of big differences in the polymerization reactivities of the involved monomers or in case of specific interactions between them.

Acknowledgement

The authors thank the Deutsche Forschungsgemeinschaft/Sonderforschungsbereich 41 for support of this work.

6.6 References

[1] T. Alfrey, G. Goldfinger: The mechanism of copolymerization, J. Chem. Phys. 12 (1944) 205–209.
[2] F. R. Mayo, F. M. Lewis: Copolymerization I. A basis for comparing the behaviour of monomers in copolymerization; the copolymerization of styrene and methyl methacrylate, J. Am. Chem. Soc. 66 (1944) 1594–1601.
[3] H.W. Melville, B. Noble, W. F. Watson: Copolymerization I. Kinetics and some experimental considerations of a general theory, J. Polym. Sci. 2 (1947) 229–245.
[4] T. Alfrey, G. Goldfinger: Copolymerization of systems of three and more components, J. Chem. Phys. 12 (1944) 322.
[5] T. Alfrey, G. Goldfinger: Copolymerization of systems containing three components, J. Chem. Phys. 14 (1946) 115–116.
[6] D.C. Blackley, H.W. Melville, L. Valentine: Rates of polymerization in the system styrene-methoxystyrene and methyl methacrylate, Proc. Roy. Soc. A 227 (1954) 10–21.
[7] C. Walling, E. R. Briggs: Copolymerization III. Systems containing more than two monomers, J. Am. Chem. Soc. 67 (1945) 1774–1778.
[8] E.T. Merz, T. Alfrey, G. Goldfinger: Intramolecular reactions in vinyl polymers as a means of investigation of the propagation step, J. Polym. Sci. 1 (1946) 75–82.
[9] S. Russo, S. Munari: A model for the termination stage of some radical copolymerizations, J. Macromol. Sci., Chem. A2 (1968) 1321–1332.
[10] A. D. Abkin: Kinetics and mechanism of copolymerization, Dokl. Akad. Nauk SSSR 75 (1950) 403–406.
[11] J. Atherton, A. North: Diffusion-controlled termination in free radical copolymerization, Trans. Faraday Soc. 58 (1962) 2049–2057.
[12] J. Ulbricht: Habil. Thesis, TH Merseburg 1964.
[13] P. D. Bartlett, K. Nozaki: The polymerization of allyl compounds. III. The peroxide-induced copolymerization of allyl acetate with maleic anhydride, J. Am. Chem. Soc. 68 (1946) 1495–1504.
[14] R. G. Farmer, D. J.T. Hill, J. H. O'Donnell: Study of the rate of charge-transfer complexes in some bulk-phase free-radical polymerizations, J. Macromol. Sci., Chem. A14 (1980) 51–68.

[15] D. Braun, W. Czerwinski: Anwendung des Komplexmodells zur Beschreibung der Bruttopolymerisationsgeschwindigkeit bei Copolymerisationen mit ausgeprägter Beteiligung von EDA-Komplexen, Makromol. Chem. 184 (1983) 1071–1082.
[16] F. Tüdös, T. Kelen, T. Földes-Berezsnich: Copolymerization and the hot radical theory, J. Polym. Sci., Symp. 50 (1975) 109–132.
[17] T. Fukuda, Y. D. Ma, H. Inagaki: The failure of terminal-model kinetics of free-radical copolymerization, Polym. Bull. 10 (1983) 288–290.
[18] J.-F. Kuo, C.-Y. Chen, C.-W. Chen, T.-C. Pan: The apparent rate constant model for kinetics of radical chain copolymerization: chemical-controlled process, Polym. Eng. Sci. 24 (1984) 22–29.
[19] O. Prochazka, P. Kratochvil: Termination in radical copolymerization: correlation of experimental data, J. Polym. Sci., Polym. Chem. Ed. 21 (1983) 3269–3279.
[20] D. Braun, W. Czerwinski, G. Disselhoff, F. Tüdös, T. Kelen, B. Turcsányi: Analysis of the linear methods of determining copolymerization reactivity ratios, VII. A critical reexamination of radical copolymerizations of styrene, Angew. Makromol. Chem. 125 (1984) 161–205.
[21] G. Disselhoff: Dilatometric investigations of the terpolymerization of benzyl methacrylate, styrene and methyl methacrylate, Polymer 19 (1978) 111–114.
[22] G. Mott: Kinetische Untersuchungen zur Homo-, Co- und Terpolymerisation von Benzylmethacrylat, Styrol und Methylmethacrylat, Ph. D. Thesis, TH Darmstadt 1973.
[23] D. Braun, G. Disselhoff: Kinetics of copolymerization, 1. Dilatometric investigation of the copolymerizations of benzyl methacrylate, styrene and methyl methacrylate, Polymer 18 (1977) 963–966.
[24] D. Braun, G. Disselhoff, F. Quella: Kinetik von Copolymerisationen, 2. Dilatometrische Untersuchungen der Copolymerisationen von N-Vinyl-2-pyrrolidon, Styrol und Methylmethacrylat, Makromol. Chem. 179 (1978) 1239–1248.
[25] D. Braun, F. Quella: Bestimmung der Geschwindigkeitskonstanten und der Effektivitäten beim Zerfall von 2,2-Azoisobutyronitril in Styrol, N-Vinyl-2-pyrrolidon und Methylmethacrylat, Makromol. Chem. 179 (1978) 387–394.
[26] D. Braun, G. Disselhoff, F. Quella: Kinetik von Copolymerisationen, 3. Geschwindigkeit der Copolymerisation von N-Vinyl-2-pyrrolidon, Styrol und Methylmethacrylat, Makromol. Chem. 182 (1981) 2951–2959.
[27] F. Quella: Untersuchungen zur Geschwindigkeit binärer Copolymerisationen, Ph. D. Thesis, TH Darmstadt 1976.
[28] D. Braun, W. K. Czerwinski: Kinetik der Startreaktion der radikalischen Homo- und Copolymerisation von Styrol, N-Vinyl-2-pyrrolidon und Methylmethacrylat, Makromol. Chem. 188 (1987) 2371–2387.
[29] D. Braun, G. Cei: Kinetics of the free-radical terpolymerization of diethyl maleate with methyl methacrylate and styrene, Makromol. Chem. 188 (1987) 171–187.
[30] A. North, D. Postlethwaite, in T. Tsuruta, K. O'Driscoll (eds.): Structure and Mechanism in Vinyl Polymerization. Marcel Dekker Inc., New York 1959, 99.
[31] T. Fukuda, Y.-D. Ma, H. Inagaki: Re-examination of free-radical copolymerization kinetics, Makromol. Chem. Suppl. 12 (1985) 125–132.
[32] T. Fukuda, Y.-D. Ma, H. Inagaki: Free-radical copolymerization, 1. Reactivity ratios in bulk copolymerization of p-chlorostyrene and methyl acrylate, Polym. J. 14 (1982) 705–711.

[33] T. Fukuda, Y.-D. Ma, H. Inagaki: Free-radical copolymerization, 3. Determination of rate constants of propagation and termination for the styrene/methyl methacrylate system. A critical test of terminal model kinetics, Macromolecules 18 (1985) 17–26.
[34] Y.-D. Ma, T. Fukuda, H. Inagaki: Free-radical copolymerization, 4. Rate constants of propagation and termination for the p-chlorostyrene/methyl acrylate system, Macromolecules 18 (1985) 26–31.
[35] G. Henrici-Olivé, S. Olivé: Über den Lösungsmitteleinfluß bei der Radikalpolymerisation: I. Makromol. Chem. 58 (1962) 188–194; II. Makromol. Chem. 68 (1963) 219–222; III. Z. Phys. Chem. 47 (1966) 286–298; IV. Z. Phys. Chem. 48 (1966) 35–50; V. Z. Phys. Chem. 48 (1966) 51–60.
[36] D. J.T. Hill, J. H. O'Donnell, P.W. O'Sullivan: The role of donor-acceptor complexes in polymerization, Prog. Polym. Sci. 8 (1982) 215–275.
[37] G. S. Georgiev, V. P. Zubov: Mechanism of alternating copolymerization, I. Kinetic method for the determination of the ratio of constants for the addition of donor-acceptor complexes and free monomers to the propagating centre, Eur. Polym. J. 14 (1978) 93–100.
[38] B. Tizianel, G. Caze, C. Loucheux: Studies of radical alternating copolymerization, III. Kinetics of the copolymerization of citraconic anhydride and vinyl acetate: a new method of evaluating the kinetics constant, J. Macromol. Sci., Chem. A 22 (1985) 1477–1494.
[39] K. Fujimori: Rate of copolymerization of maleic anhydride with butyl vinyl ether in $CHCl_3$ and relative reactivity of free monomers and the complex, Polym. Bull. 13 (1985) 459–462.
[40] K. Fujimori, P. P. Organ, M. J. Costigan, I. E. Craven: Relative reactivity of free monomers and donor-acceptor complex in alternating copolymerization of isobutyl vinyl ether with maleic anhydride from the rate of polymerization, J. Macromol. Sci., Chem. A 23 (1986) 647–655.
[41] K. Fujimori, A. S. Brown: Overall rate of alternating copolymerization of vinyl acetate with maleic anhydride in methyl ethyl ketone. Relative reactivity of the complex and free monomers, Polym. Bull. 15 (1986) 223–226.
[42] E. Manger: Untersuchungen zur Copolymerisation von p-Chlorstyrol und n-Butylmethacrylat, Diplomarbeit, TH Darmstadt 1985.
[43] D. Braun, E. Manger: Radikalische Copolymerisation von p-Chlorstyrol und n-Butylmethacrylat, Colloid Polym. Sci. 264 (1986) 494–497.
[44] D. Braun, E. Manger: Alternierende Copolymerisation mit Lewissäuren. Copolymerisation von p-Chlorstyrol und n-Butylmethacrylat mit Ethylaluminiumsesquichlorid, Angew. Makromol. Chem. 145/146 (1986) 101–124.
[45] R. Bednarski: Über die Copolymerisation von Styrol mit nicht homopolymerisierbaren Monomeren. Diplomarbeit, TH Darmstadt 1985.
[46] A. Essmail Pour: Untersuchungen zur binären radikalischen Copolymerisation von 1-Vinylnaphthalin. Diplomarbeit, TH Darmstadt 1987.
[47] D. Braun, G. Cei: Kinetics of the free-radical copolymerization of diethyl maleate with styrene, Makromol. Chem. 187 (1986) 1713–1726.
[48] D. Braun, G. Cei: Kinetics of the free-radical copolymerization of diethyl maleate with methyl methacrylate, Makromol. Chem. 187 (1986) 1699–1711.
[49] D. Braun, W. K. Czerwinski: Kinetische Analyse der Copolymerisationsgeschwindigkeit von N-Vinyl-2-pyrrolidon mit Styrol und Methylmethacrylat, Makromol. Chem. 188 (1987) 2389–2401.

[50] D. Braun, F. Tüdös, T. Kelen, W. K. Czerwinski, R. Bednarski: Analysis of copolymerization of styrene with methyl α-cyancinnamate and fumaroyldipyrrolidone; analysis of copolymerizations of styrene and methylmethacrylate with N-p-tolylcitraconimide and fumaronitrile, in preparation.
[51] D. Braun, F. Tüdös, T. Kelen, B. Turcsanyi, W. K. Czerwinski: Analysis of the linear methods of determining copolymerization reactivity ratios, XIII; Non classical binary copolymerization, a critical application of penultimate model to styrene radical copolymerization data, in preparation.
[52] T. Ishida, S. Kondo, K. Tsuda: Free-radical polymerization of methyl methacrylate initiated by N,N-dimethylaniline, Makromol. Chem. 178 (1977) 3221–3228.
[53] K. Tsuda, S. Kondo, K. Yamashita, K. Ito: Initiation mechanism of free-radical polymerization of methyl methacrylate by p-substituted N,N-dimethylanilines, Makromol. Chem. 185 (1984) 81–89.
[54] M. Kamachi, D. J. Liaw, S. Nozakura: Solvent effect on radical polymerization of methyl methacrylate, Polym. J. 13 (1982) 41–50.
[55] D. Braun, G. Cei: Measurement of conversion in free-radical copolymerization from dilatometric data, Makromol. Chem. 188 (1987) 189–199.
[56] F. Quella, W. K. Czerwinski, D. Braun: Zur theoretischen Beschreibung der Kinetik der radikalischen Terpolymerisation von Styrol mit N-Vinyl-2-pyrrolidon und Methylmethacrylat anhand des klassischen Copolymerisationsmodells, Makromol. Chem. 188 (1987) 2403–2415.

7 Free Radical Polymerization: From Spontaneous Initiation up to the Glass Transition State

Günther Meyerhoff*, Günter V. Schulz*, and Jürgen Lingnau**

7.1 Introduction

High molecular weight compounds play an outstanding role among chemical industrial products. Among these compounds polymers obtained by free radical polymerization processes are of special importance. This led to remarkable cumulation of engineering data to perform polymerizations on large scale with the necessary accuracy to obtain reproducible materials and to run the reactions in a safe way. While this descriptive approach has its merits for practice there remains a lack of understanding of quite a number of elementary reaction steps of the polymerization reaction during its various stages.

In contrast to low molecular weight organic chemistry there are two aspects which are typical for polymerization reactions. First, due to the enhancement by the chain reaction of polymerization, elementary reactions believed to be of otherwise minor importance can account for great effects seen through the magnifying glass of polymerization. And second, increase of the viscosity during the polymerization process strongly influences the kinetics of diffusion controlled reactions as well as local reaction conditions. As for radical reactions in general these aspects are combined with the diversity of radical-radical reactions, like primary radical termination [1–4] and polymer-polymer termination [5, 6], which again are complicated by the influence of the degree of polymerization of the respective radicals. Especially for a group specialized in the fields of polymerization kinetics and the determination of polymer molecular weight distributions these aspects are a challenge and it seemed worthwhile to contribute to a better understanding of these elementary reactions.

* Institut für Physikalische Chemie der Universität Mainz, Jakob-Welder-Weg, D-6500 Mainz
** Hoechst AG, Werk Kalle-Albert, Geschäftsbereich Technische Spezialprodukte, D-6200 Wiesbaden-Biebrich

7.2 Experimental Approach

Elucidation of reaction mechanisms is based on knowledge of the structure of products under consideration and their build-up with time. This includes characterization of side-products as well as information on the polymer molecular weight distributions as a function of polymerization conditions.

7.2.1 Kinetic Experiments

For the determination of polymerization kinetics of azo-compound initiated polymerization dilatometry proved to be the method of choice. For the high conversion experiments the reactions were performed in PET-bags within the dilatometer, and the shrinkage was followed via water/methanol as embedding medium [7–9]. Since side reactions leading to oligomers may not be disregarded in the case of the spontaneous polymerization, dilatometry had to be ruled out for this purpose [10, 11]. Gravimetry yielded the necessary data on polymerization kinetics, supplying also polymer samples to follow trends in molecular weight distribution with conversion. In practice, not the methods themselves proved to be the crucial point, but rather the necessary purification steps for monomer as well as solvents to avoid any side reaction paths influencing especially the slow initiation in the spontaneous polymerization. This had to include monomer purification by repeated distillation under nitrogen, removal of peroxide traces by column chromatography over alumina, and prepolymerization already under high vacuum in sealed glassware. For the gravimetric runs the monomer and solvents then were high vacuum distilled into brown reaction ampoules, which had to be surface-modified to avoid contributions to initiation by the glass surface, which otherwise led to irreproducible results. The purification thus strongly reminds the requirements of anionic polymerization.

For the determination of the rate constants of propagation and termination, k_p and k_t, the "classical" combination of steady-state and non-steady state experiments was applied. For the latter intermittent illumination by rotating sector is applicable in the lower conversion range, while for higher conversions determination of the post-effect of polymer build-up after discontinuation of the illumination is the method of choice [7–9].

7.2.2 Identification and Kinetics of Formation of Side Products [10, 12–14]

For the determination of oligomers gas (GC) and liquid chromatography (HPLC) proved to be the best tools to follow the kinetics, while the combination of GC with mass spectroscopy (GC-MS), and the combination of preparative GC and HPLC with ^{1}H-NMR- and IR-spectroscopy were applied for the characterization. GC also proved to be useful for the determination of thermolytic or photolytic decay of azo compounds [15, 16].

7.2.3 Determination of Molecular Weight Distributions

For the determination of molecular weight distributions, a variety of methods was applied. Gel permeation chromatography (GPC, SEC) (low and high pressure) was used for most of the samples. For samples with especially high molecular weights as those obtained e.g. by spontaneous polymerization at lower temperatures or the initiated polymerization at high conversions, the molecular weight average values had to be checked by alternative methods [10, 17]. This also applied for non-standard polymers like Poly-2-chloroethyl-methacrylate and Poly-methyl-α-chloro-acrylate, for which neither narrow MWD standards nor universal calibration data were available [16, 18].

One approach is based on the combination of GPC and viscosity within one experimental setup by inclusion of a viscosity detector in a GPC arrangement [19]. Measuring the viscosity values of fractions separated by GPC directly leads to universal calibration, as long as extrapolation to $[\eta]$ is assured. For high molecular weights, i. e. high viscosity numbers $[\eta] > 500$, the velocity gradient within the viscometer may not further be neglected, since it leads to deformation of the polymer coil. The gradient may be eliminated by the relatively simple concept of a gradient viscometer [20,21] consisting of a U-type viscometer with automatic meniscus pursuit. In contrast to standard viscosity measurements, where only the time for a given volume to pass the capillary is measured, gradient viscometry determines the volume flow as a function of the applied pressure difference, which may be related to both, the calculation of viscosity and maximum gradient by:

$$[\eta] = \frac{\varrho g h r^4}{4 l R^2} \frac{t}{\ln(h_o/h)} \tag{1}$$

$$G_{max} = 4R^2 \Delta h/(r^3 \Delta t) \tag{2}$$

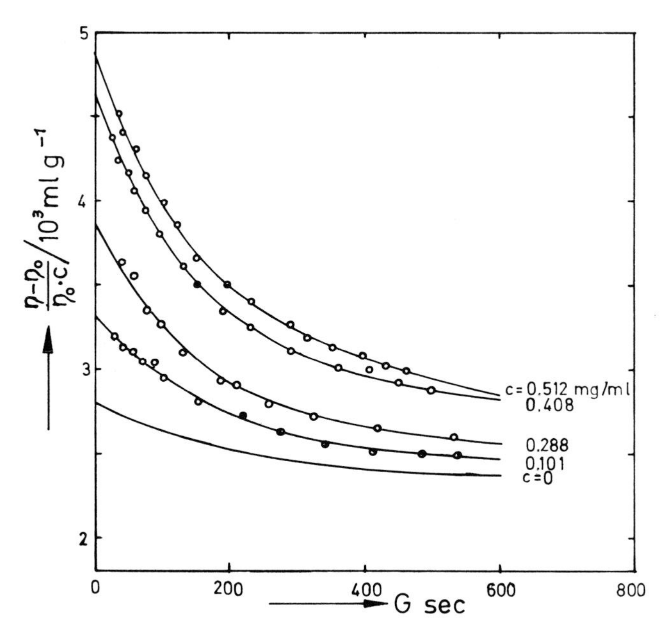

Fig. 7.1: Shear dependence of the reduced viscosity for a polystyrene fraction ($\overline{M}_w$ = 35 x 10^6) in toluene at 20 °C. The lowest value (c = 0) represents the intrinsic viscosity number [η] [21].

where r and l are radius and length of the capillary, R is the tube radius and h the height above equilibrium. Extrapolation to G = 0 and c = 0 yields the Staudinger Index and allows calculation of $\overline{M}_\eta$ in the familiar way (Fig. 7.1). The experiments clearly indicate that – at least for polystyrene – no deviation from the Kuhn Mark Houwink parameters established for medium molecular weights could be observed, at least up to $\overline{M}_\eta = 4 \times 10^7$ [21].

Ultracentrifuge measurements [10, 22, 23] based on both, sedimentation velocity and sedimentation equilibrium [22, 23] provide an additional tool for the determination of molecular weight distributions, again especially useful for high molecular weights. The application of both types especially to PMMA was investigated in detail, the results ($\overline{M}_w$, U) are in good agreement to data obtained by GPC and light scattering. Application of density gradient experiments to copolymers offer an easy access to the determination of their chemical and physical inhomogenities [22, 24].

The characterization of molecular weight distributions by means of light scattering [25, 26] has been developed for the classical and the dynamic light scattering. The first depends on the molecular weights, the radius of gyration and the concentration of the polymers in solution. After extrapolation to zero concentration the angular dependent scattering function contains valuable information on the different chain lengths contributing to this experimentally observed angular dependence. Comparison with the theoretically calculated scattering function allows the determination of the width of the molecular weight distribution as expressed by the polydispersity $\overline{M}_w/\overline{M}_n$ [27] (Fig. 7.2). Of course this method is restricted to molecular weights high enough to show angular dependence of the scattered light. The dynamic light scattering basically measures the diffusion coefficient of polymers and thus may be used also for lower molecular weight samples, but needs still more experimental precision than its classical counterpart and also more computational work for the evaluation. All these problems could be overcome [28–30], and presently the characterization of almost all possible molecular weight distributions can be performed. Table 7.1 compares molecular weights and polydispersities determined by light scattering and other techniques for radically polymerized polystyrene samples [27, 28]. It should be mentioned here that for the small polydispersities resulting from well-performed anionic polymerization (cf. Müller et al., this Volume, Chapter 8) gel chromatography is at its limits especially for higher $\overline{M}_w$s. This is the field where classical light scattering has its merits. Furthermore the experiences in light scattering and other MWD determination techniques have been applied to the characterization of polycondensates, which were obtained in the course of kinetic investigations of step polymerization ([31–34], Höcker and Schulz, this Volume, Chapter 5).

Back in 1940 Ruska and Husemann already demonstrated that macromolecules of glycogenes could be made visible by electron microscopy (EM) [35].

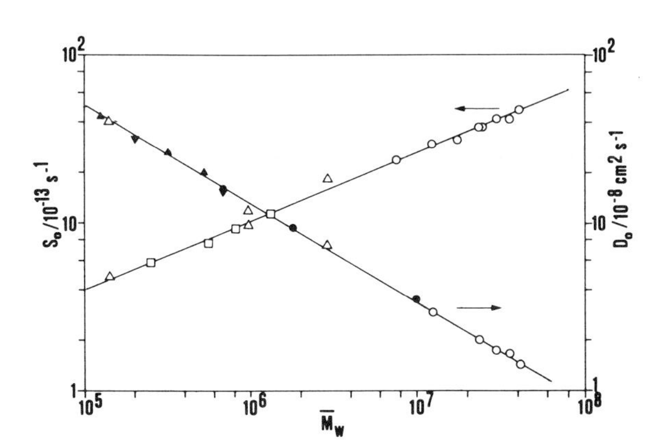

Fig. 7.2: Sedimentation coefficient S_o and diffusion coefficient D_o as a function of $\overline{M}_w$ up to 4.4×10^7 for polystyrene in toluene at 20 °C. (o) data of [17], (▼) data of [28]. Further details in [17].

Table 7.1: Molecular weight $\overline{M}_w$ and polydispersities $\overline{M}_w/\overline{M}_n$ determined by Dynamic Light Scattering as compared with results from Classical Light Scattering, GPC, and data given by NBS[a]. (F = fractions) [28].

Sample	Dyn. LS $\overline{M}_w/10^6$	Dyn. LS $\overline{M}_w/\overline{M}_n$	Cl. LS $\overline{M}_w/10^6$	GPC $\overline{M}_w/10^6$	GPC $\overline{M}_w/\overline{M}_n$	NBS $\overline{M}_w/10^6$	NBS $\overline{M}_w/\overline{M}_n$	
NBS 706	0.284	2.23	0.278	0.259	2.28	0.258		Cl. LS[a]
						0.288	2.10	Sed. Equil.[a]
ThA60	1.62	2.01	1.59	1.69	1.98	1.64	1.91	Cl. LS [26, 27]
ThA20	4.52	1.98	4.43	4.79	2.33			
ThA60 F	0.881	1.33	0.919	0.833	1.29			
ThA60 F	1.37	1.25	1.39	1.44	1.22			
ThA60 F	2.00	1.27	1.94	2.14	1.27			

By applying the method of inclined sputtering it became possible to determine the volume of single molecules [36, 37]. The respective molecular weight $M_i = N_L \cdot \varrho V_i$ of particle i results from ist volume V_i and its density ϱ_i. Assuming the particle shape to be a cut sphere, V_i is given by:

$$V_i = (\pi/6)(H^3 + 3\,R^2H)\ (H > R) \tag{3}$$

$$V_i = (\pi/6)(6RH^2 - 2\,H^3)\ (H < R) \tag{4}$$

Height H and radius R of the particle are calculated from shadow length S and magnification factor q.

Fig. 7.3 shows single molecules of PMMA, which were obtained by low conversion polymerisation at 80 and 20 °C, respectively. The pictures indicate the validity of the above assumptions on particle shape. To rule out association effects, concentration variation for the casting process is considered essential. A more detailed investigation of the data shows a standard deviation for the single particle of about 30 %, which may be reduced by averaging. Sorting the molecules with respect to quantity and molecular weight contribution directly yields the known average values $\overline{M}_n$ and $\overline{M}_w$. Cross check of MWDs obtained by electron microscopy (Fig. 7.4) with data obtained by GPC, light scattering and ultracentrifuge show good agreement within a few percent [37, 38]. By principle the EM method is especially useful for high molecular weight polymers.

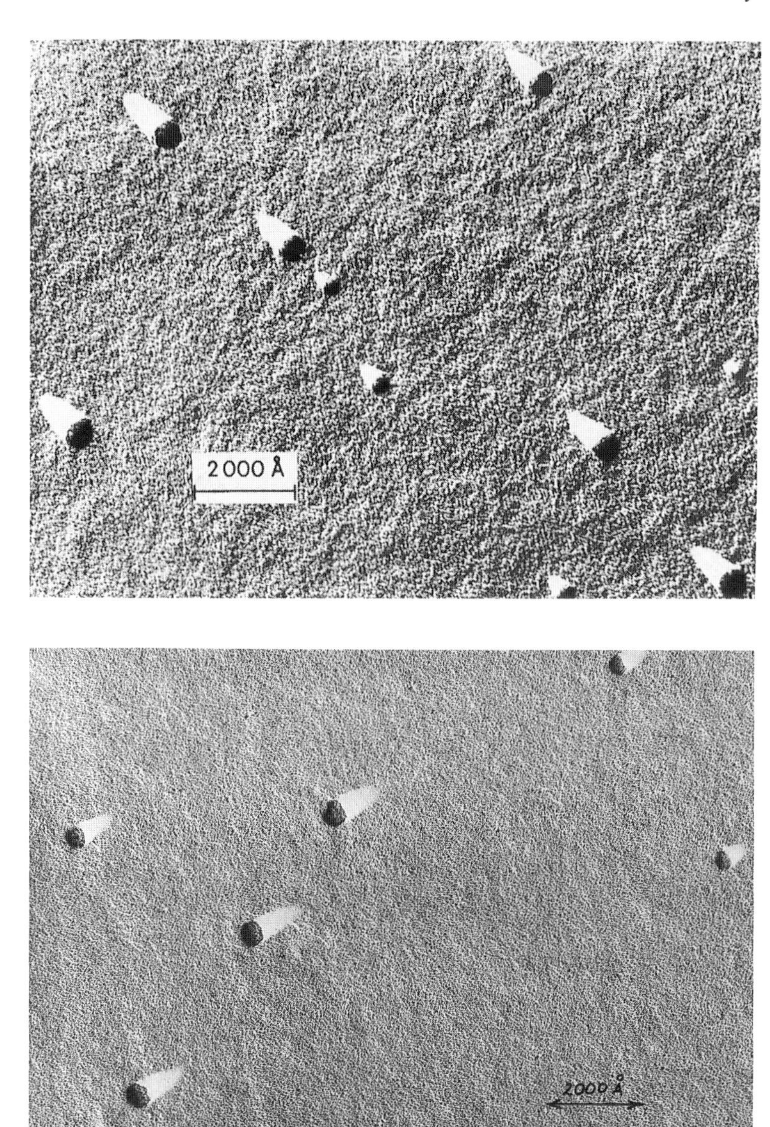

Fig. 7.3: Single molecules and/or associates of poly(methyl methacrylate) polymerized (a) at 80 °C, (b) at 20 °C. Precipitated from aceton solution of 2.2×10^{-6} g, cm^{-3} [37].

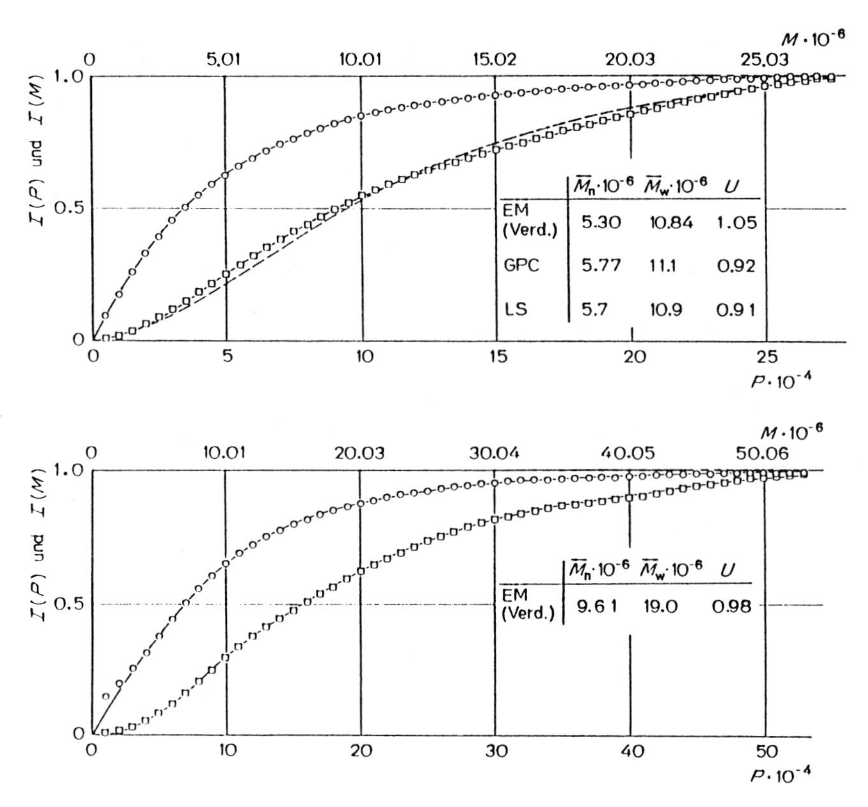

Fig.7.4: Molecular weight distributions of the sample of Fig. 7.3 (-o- number distribution and – weight distribution) by electron microscopy after evaporation of the solvent; –□– M_{GPC}; LS = classical light scattering; (a) sample polymerized at 80 °C, (b) at 20 °C [37].

7.3 Spontaneous Polymerization

The normal way to perform the radical polymerization of monomers is the initiation by some type of added free-radical source, like azo or peroxy compounds, at appropriate temperature or radiation [39–43]. Some monomers, however, undergo free-radical polymerization without addition of initiators, even under strict exclusion of any external radical source like oxygen or radiation. This type of polymerization of pure monomers is called „spontaneous" or „thermal" polymerization [44, 45].

7.3.1 The Spontaneous Polymerization of Styrene

Since long styrene is known to polymerize spontaneously, a process which even finds practical application in the industrial polymerization of this monomer. After extensive investigation [20, 46–49] the initiation path is now understood to take place via the extended „Diels-Alder-Mechanism“ outlined in Scheme *(1)*, where two styrene molecules – one acting as diene, the other as dienophile – first cyclicize under formation of bicyclo-compounds AHA and AHE, which then may disproportionate to yield the initiating radicals, which – upon cage recombination or disproportionation – yield the dimers and trimers also determined in the reaction mixtures. The Diels-Alder-Mechanism has been supported by the evaluation of reaction orders, by spectroscopic proof of the Diels-Alder-adducts and an extensive investigation of oligomer structures and build-up. Still the formation of cyclobutane derivatives may be correlated to a minor second reaction path via a biradical, which originally was discussed as alternative initiation path.

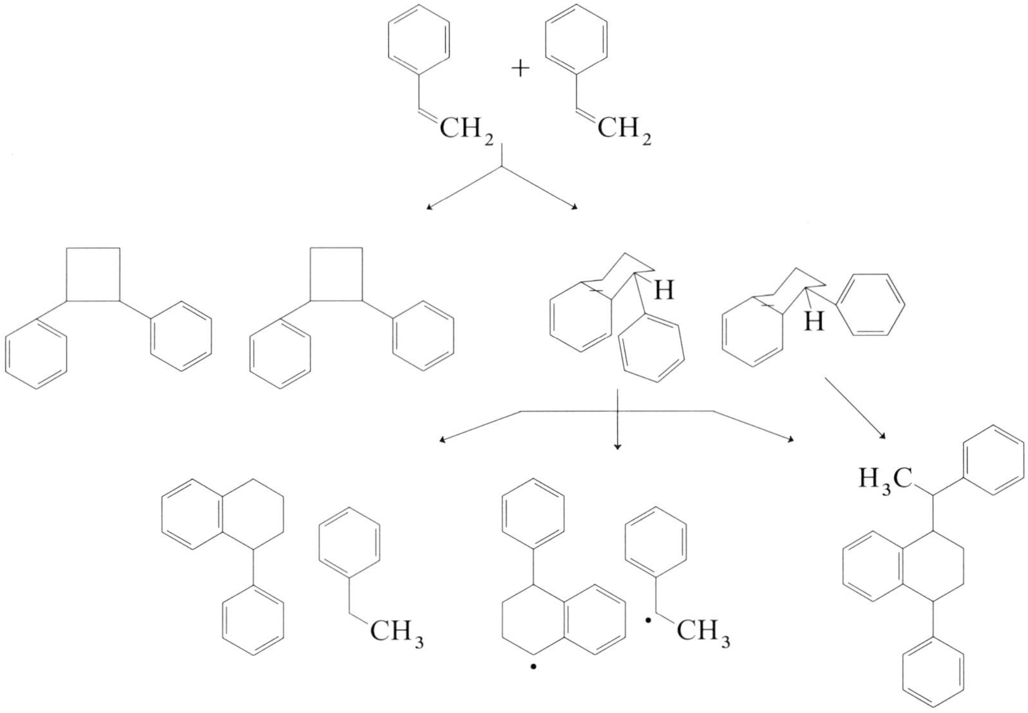

CH_2 + CH_2

H H

CH_3 CH_3 H_3C

Scheme 1: Initiation and oligomerization mechanism of the spontaneous polymerization of styrene.

7.3.2 The Spontaneous Polymerization of Methacrylates

7.3.2.1 General Aspects

For methyl methacrylate experimental data on its spontaneous polymerization were conflicting (see [10, 11] and Table 7.2), indicating at least that the overall reaction rate is much slower than for styrene. Severe purification steps as outlined in the experimental approach were the condition to obtain reproducible results. These proved to be much below the values published before, indicating that the initiation step of the spontaneous polymerization may be easily overlaid by initiation already caused by traces of impurities. The high molecular weights even at low conversions, the inhibition by benzoquinone, the insensitivity towards addition of hydroquinone (in the absence of oxygen) as well as molecular weight distributions and the way these could be influenced by radical transfer agents clearly indicated the radical character of the polymerization reaction, ruling out addition type polymer formation and ionic mechanisms.

Polymerization in benzene solution revealed that the initiation step is second order in monomer concentration, its activation parameters are ΔH^{+} =

Table 7.2: Overall rates of polymerization of the spontaneous polymerization of methyl methacrylate.

Ref.	Temp. (°C)	$R_p \times 10^6$ (mol/l sec)	$R_i \times 10^{14}$ (mol/l sec)	$\overline{M}_n \times 10^6$
[39]	70	0.21		6.67
	100	4.13		4.07
[50]	100	0.62		2.52
	131	3.25		1.56
	150	8.32		1.05
[51]	80	2.69		7.20
	100	5.3		4.75
	120	14.7		3.1
[52]	40	0.7		
[11]	0	0.0015	0.0037	95
	60	0.0075	0.0067	14.6
	100	0.105	0.383	6.69
	130	0.5	4.05	3.22
Styrene [20]	60	2.02		1.81

143 kJ/mol and $\Delta S^{\ddagger} = -178$ J/mol/K, $\Delta H^{\ddagger}$ being in good agreement to a value of 142 kJ/mol as estimated from thermochemical data for the formation of a dimeric biradical. For energetic reasons ($\Delta H^{\ddagger} = 226$ kJ/mol) a mechanism via a monomeric biradical would be unfavourable [10, 13, 14].

7.3.2.2 The Oligomers

As in the case of styrene additional indication for the reaction mechanism comes from the analysis of oligomeric side compounds. Two groups of oligomers could be identified: A linear dimer H-1 is formed in amounts about equal to the amount of polymer [10, 12], which – by its double bond – can participate in the polymerization. Experiments with added H-1 revealed copolymerization parameters of $r_{MMA} = 1.79$ and $r_{H\text{-}1} = 0.33$ at 60 °C, while homopolymerization of H-1 is extremely slow [51, 52]. Besides H-1 a further linear dimer, P-1, and three trimers could be analyzed. While the reaction order for the formation of both H-1 and P-1 is two in monomer concentration, the formation of the trimers depends not only on monomer concentration, but also on the

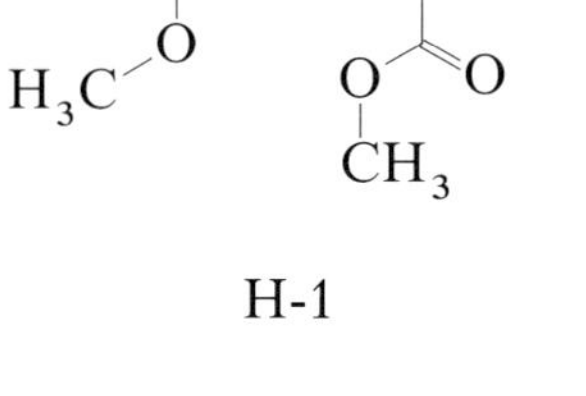

H-1 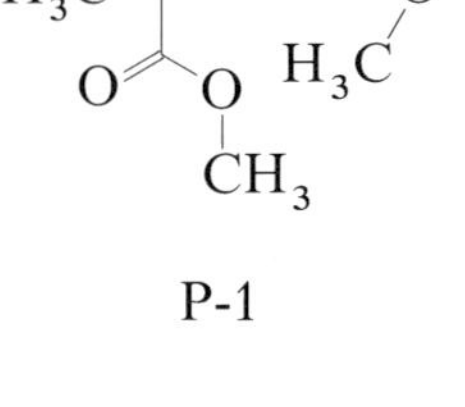 P-1 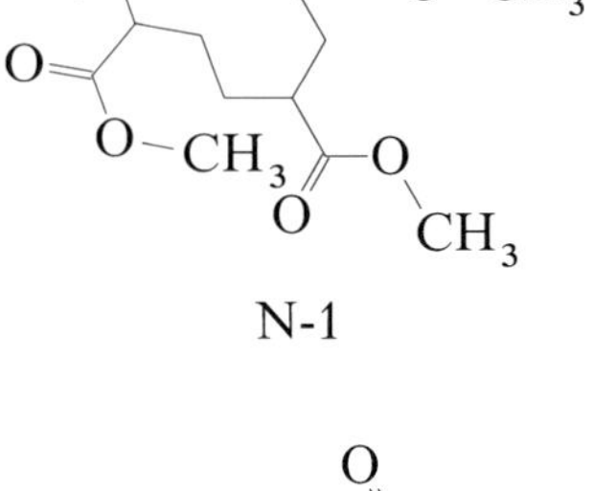 N-1

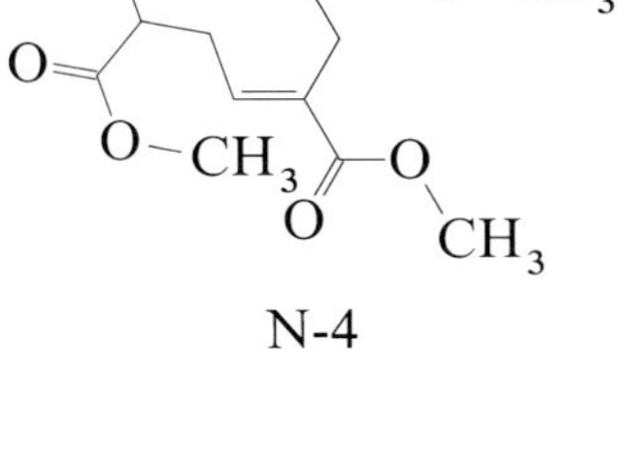

N-4 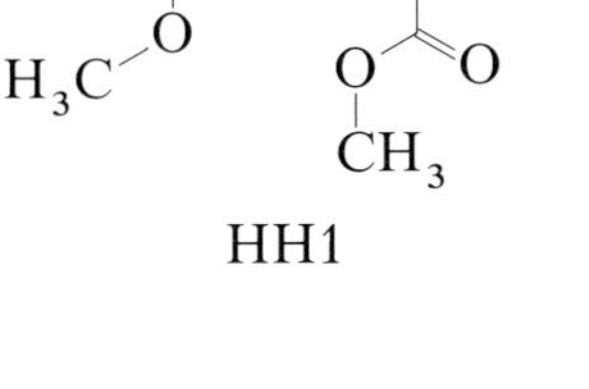 HH1 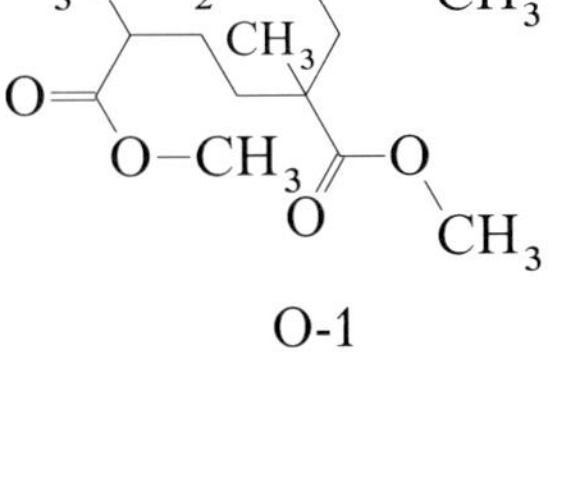 O-1

concentration of H-1, indicating, that formation of trimers proceeds via a step mechanism rather than via a radical chain mechanism [12–14]. Determination of the activation parameters (Table 7.3) also suggests a different reaction path as compared with initiation. The choice of trimers isolated may be explained by a concerted mechanism like:

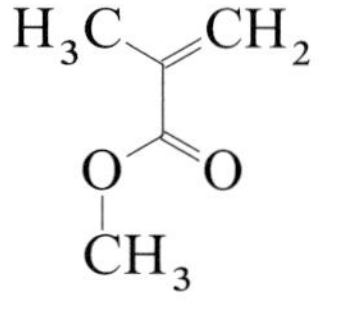

MMA 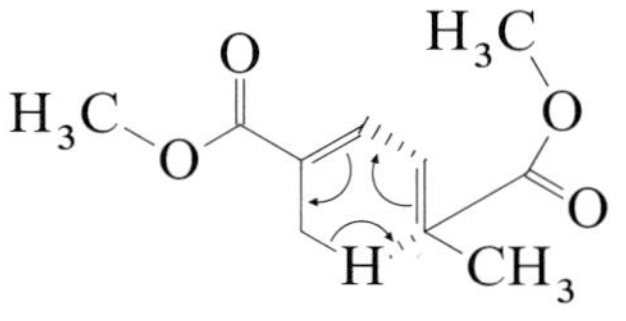 $\quad$ H_{zwi} 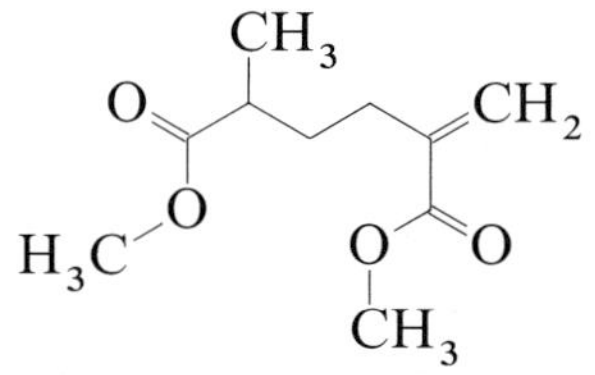 $\quad$ H-1

which for different relative positions of the reactants results in exactly this manifold of dimers and trimers.

As for styrene, cyclobutane derivatives could be isolated and identified, having structures cCB and tCB [13, 14]. The activation parameters of their formation (Table 7.3) fit fairly well with those of initiation giving further proof for an initiation reaction via biradicals, since formation of cyclobutanes by a thermal concerted mechanism is symmetry forbidden, at least in first approximation.

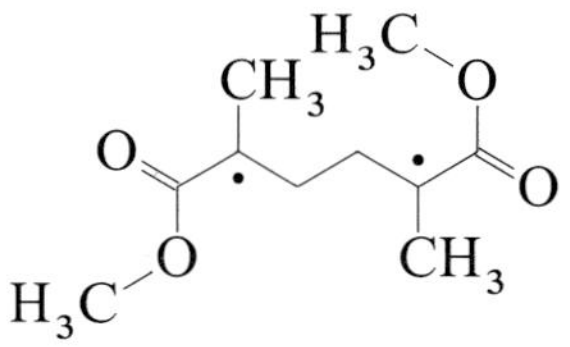

$\cdot M_2 \cdot$ 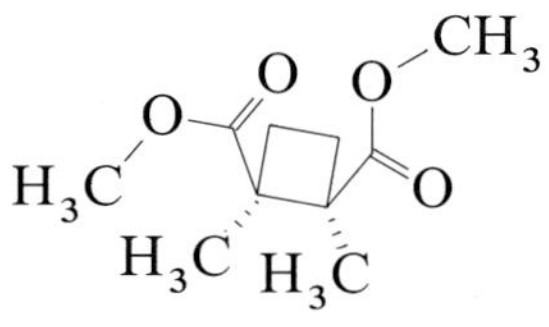 $\quad$ cCB 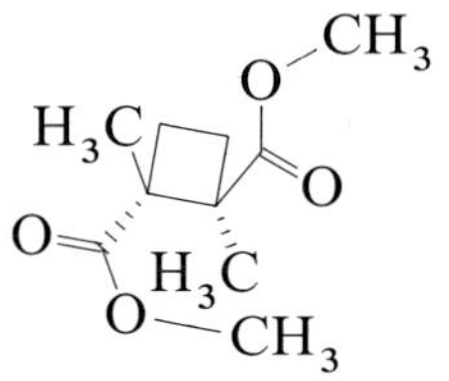 $\quad$ tCB

Table 7.3: Enthalpies and entropies of activation for the initiation step and the formation of dimers and trimers.

	$\Delta H^{\ddagger}$(kJ/mol)	$\Delta S^{\ddagger}$(J/mol.K)
Initiation	143	–178
Formation of cCB	141	–102
Formation of tCB	126	–140
Formation of H-1	102	–150
Formation of O-1	80	–228
Formation of N-1 + N-4	81	–205

7.3.2.3 Solvent Effects

While benzene as the "most neutral" solvent to think of does not influence the polymerization behaviour [10], change to solvents with higher hydrogen transfer power leads to an unusual kinetic behaviour [16, 55]: Addition of small amounts of solvent S leads to an increase in the polymerization rate, while due to decreasing monomer concentration, the rate is decreased upon further addition of solvent (analogous to Fig. 7.5). Linearization was possible for $R_p^2/[M]^4$ vs [S], yielding a slope which was linearly correlated with the transfer constant of the respective solvents (Table 7.4). Since the dependence of R_p and thus of R_i on $[M]^2$ remains, participation of a transfer step at the biradical level is a reasonable assumption. Consequently heating the monomer in presence of large amounts of strong transfer agents like thiophenol should yield "double-transferred" product HH1, which indeed could be identified by GC-MS in comparison with a reference sample obtained by hydrogenation of unsaturated dimer H-1.

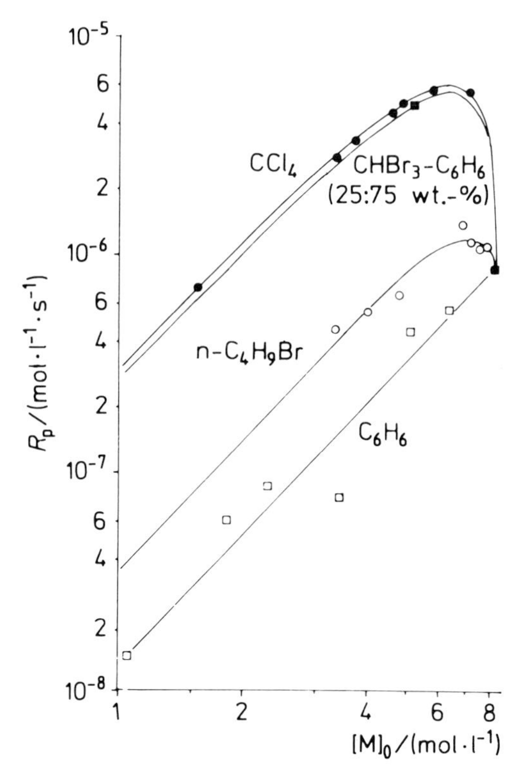

Fig. 7.5: Effect of halogen containing solvents on the overall rate of polymerization of the spontaneous polymerization of methyl methacrylate [16].

Table 7.4: Comparison of transfer constant of solvents with their influence on the initiation rate of the spontaneous polymerization of methyl methacrylate.

Solvent	$C_S \times 10^4$ (130 °C)	Slope of plot $R_p^2 k_t/[M]^4 k_p^2$ vs. [S] x 10^{15}
Benzene	0.12	ca. 0
MMA	0.67	ca. 0
Isobutyric acid methyl ester	1.49	0.39
i-Propanol	2.04	1.89
N,N-Dimethyl-aniline	18.5	5.7
Tri-n-butyl-amine	33.4	5.7
Nitromethane	181	17.5
Thiophenol	428000	22400

The use of halogen containing solvents [16, 56] yields rather fast overall polymerization rates, which cannot be explained on the basis of a pure transfer effect. Again an increase is observed on addition of small amounts of solvent (Fig. 7.5), and again, linearization is possible for $R_p^2/[M]^4$ vs [S] (Table 7.5). The dependence on the atomic number of the halogen contained in the solvent reminds parallels to the photochemical behaviour of 1-chloro-naphthalene in the presence of halogencontaining (solvent-) glasses [57]. It suggests a necessary distinction between radicals of different multiplicity. Since from the thermal formation mechanism the dimeric biradical discussed so far must have singlet multiplicity, the heavy atom assisted intersystem crossing to yield the triplet biradical seems to enhance the initiating probability, which otherwise is in the range of $1:10^4$ based on the amount of cyclobutanes formed. While the transfer effect is temperature dependent, there is no further activation energy for the reaction step, in which the heavy atom containing solvents participate.

Table 7.5: Comparison of the heavy atom effects on the photochemical behaviour of 1-chloronaphthalene and on the initiation of the spontaneous polymerization of methyl methacrylate.

Solvent	Slope of plot $R_p^2 k_t/[M]^4 k_p^2$ vs. [S] x 10^{13}	Integr.absorp. of 1-chloro-naphthalene [55]
Carbontetrachloride	1.35	1.60
Bromoform	3.24	9.24
Methyliodide	12.40	27.70

7.3.2.4 Other Acrylates [16, 18]

As in the case of photochemistry the heavy atom effect may be further enhanced by including the heavy atom into the active species, i.e. the monomer. While change from MMA to the ethyl ester (EMA) has no influence, a pronounced increase in the initiation constant is observed upon Cl-substitution in the rather remote 2-position of EMA (CEMA). Bringing the heavy atom closer to the reactive center as in Methyl α-chloroacrylate (MCA) yields another increase in the initiation constant k_i. Again the temperature dependence reveals that this increase is almost completely temperature independent.

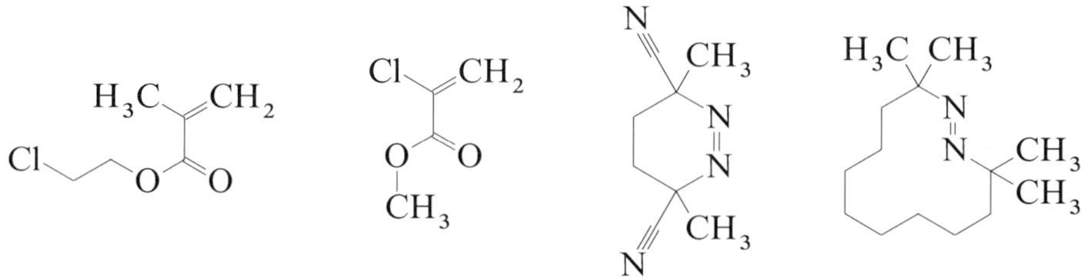

CEMA MCA DN TM-12

Finally the influence of the otherwise highly inert Xenon [16, 52] as inert gas in comparison with nitrogen was investigated, which also yielded a linear context between R_p^2 vs p_{Xe} as expected from the above, while variation of p_{N2} does not affect the polymerization rate.

Since formation of the cyclobutanes is the dominant reaction path for $.M_2.$ neither transfer nor heavy atom effect can be observed for the formation of the cyclobutane dimers [16].

7.3.2.5 "Chemical Simulation" [16]

To further proof the participation of a dimeric biradical of type $.M_2.$ "chemical simulations" were performed. As model substances cyclic azo compounds were used. The six-membered ring DN can be produced from the saturated ring by in-situ oxidation with various oxidizing agents. The analysis of the decay products showed that the biradical besides forming the cyclobutanes decays to monomer, while an H-1 analogous product could not be detected:

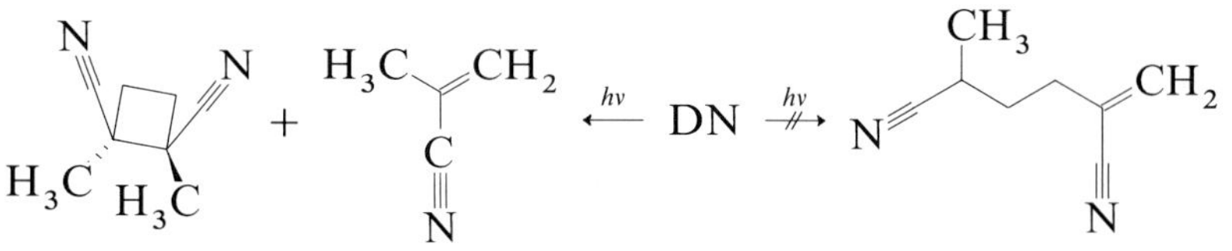

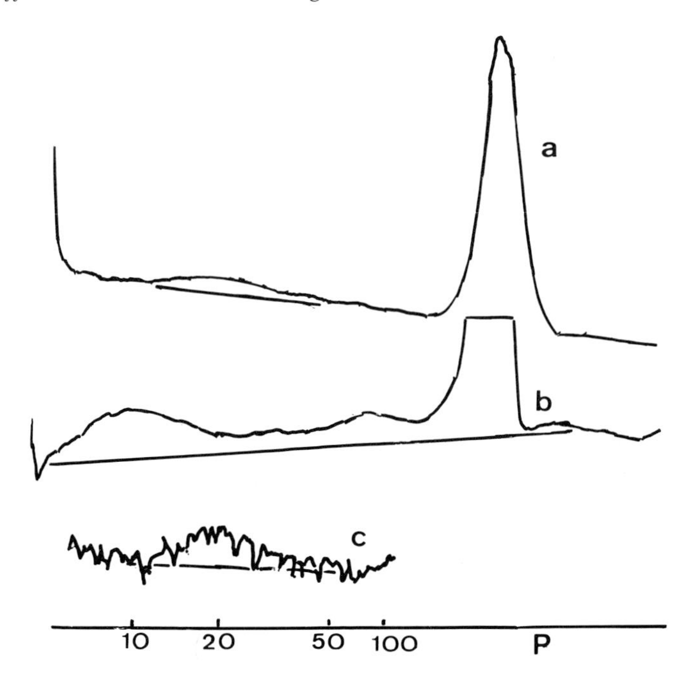

Fig. 7.6: Oligomers of DP 10 to 50, sample (a) MMA/TM-12/130 °C (x1), (b) MMA/CBr_4/130 °C/spontaneous polymn (x0.5), (c) MMA/130 °C/spontaneous polymn (x8).

Experiments to elucidate the initiating properties of DN were performed in MMA at room temperature. Polymer formation was checked by injecting samples of that solution directly into a GPC working at the exclusion limit. For the un-assisted reaction this led to an efficiency of about 2.5×10^{-10} for initiation. Addition of tetrabromomethane, acting both as strong transfer agent and an additive with high heavy atom content, led to an increase in the efficiency f to about 5×10^{-7}/mole CBr_4. Besides high polymer there are reasonable amounts of oligomers formed having degrees of oligomerization between 10 and 50, which may also be found in samples obtained by spontaneous polymerization (Fig. 7.6).

Switching from the labile cis-6-ring azo compound to the much more stable trans-12-ring azo compound TM-12 (sample courtesy CIBA-Geigy) yields a much higher efficiency, which for bulk polymerization at 130 °C is 3×10^{-3}. The system follows the square root dependence on initiator concentration as does the parallely tested azo-bis-isobutane. Molecular weights are as expected for a chain length dominated by chain transfer, i.e. $\overline{P}_n = 17.000$ at 130 °C. As in the case of the spontaneous initiation an increase in the overall polymerization rate is observed upon addition of tetrachloromethane or methyliodide.

7.3.2.6 The Mechanism and its Formal Description

The experimental data result in a rather complex initiating scenario, which covers a series of reactions which otherwise would be regarded as very improbable, but which do not seem to be unreasonable for a reaction with steady state radical concentrations in the range of 10^{-12} mol/l at 130 °C and a polymerization rate yielding 5 % conversion after 1 week at 130 °C in bulk.

Biradicals play a central role in this mechanism as outlined in Scheme *(2)*.

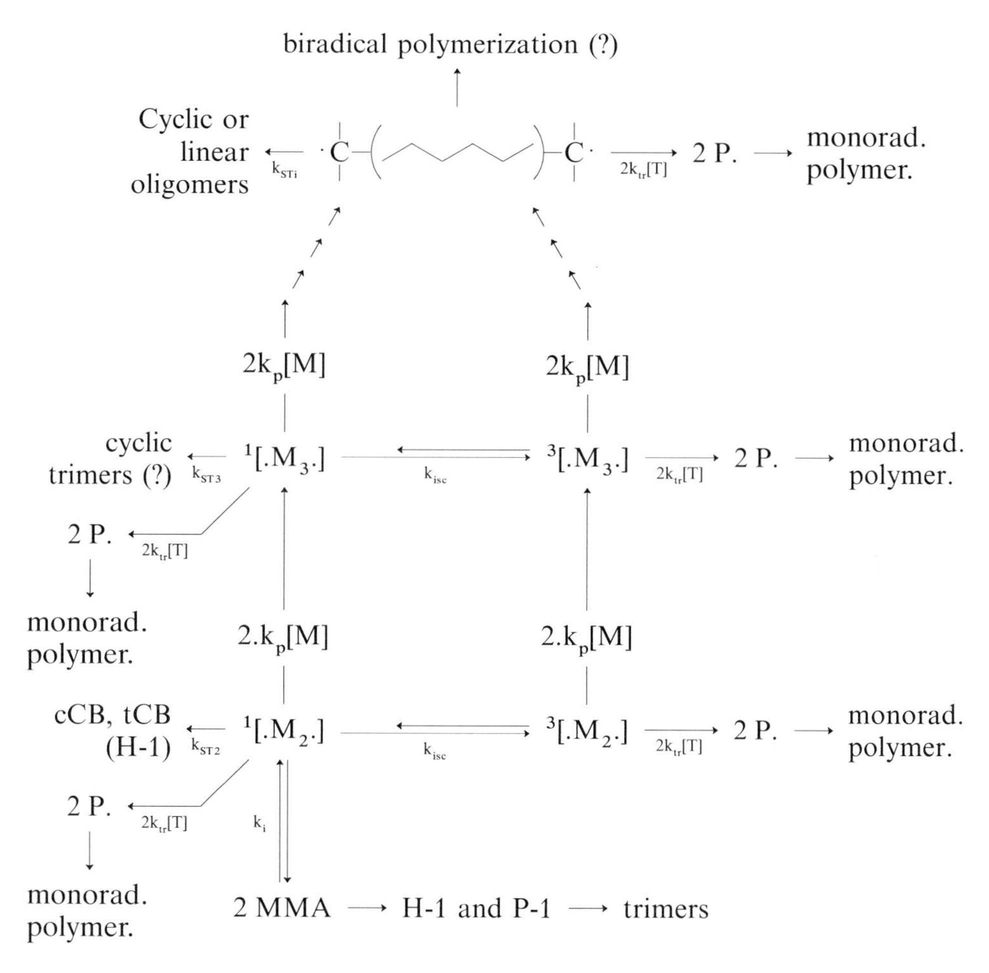

Scheme 2: Initiation and oligomerization mechanism of the spontaneous polymerization of methyl methacrylate.

Formation of the singlet dimeric biradical, $.M_2.$, from two monomer molecules is regarded as the primary step, which mainly deactivates to cyclic dimers of the cyclobutane type. A minor part, however, may undergo intersystem crossing to the triplet analog, which should have much better chances for one of the other possible reaction alternatives: propagation and transfer. While the first yields the respective biradical, $.M_3.$, containing one more monomer unit, transfer yields two monoradicals, which undergo "normal" propagation and termination reactions without the high probability of selftermination [16].

Two attempts have been undertaken to get an estimate of the various kinetic constants of such a system. In a first simulation [58] emphasis was laid on the consequences of the build-up of oligomers on the reaction kinetics and product distribution for high conversion. Especially the linear dimer and its implications on the properties of the resulting polymers are of practical interest (Fig. 7.7).

The second approach [16] tried to describe more details within the initiation scheme, covering especially the cyclic oligomers of various degrees of polymerization and the effects of transfer agents and isc-catalysts on the probability of mono- and biradical propagation and survival. It yields a number of interesting implications, e.g. a narrowing of the molecular weight distribution for biradically initiated polymerizations, indications of which have already been observed in the preliminary experiments with cyclic initiators [16]. However, much more detailed investigations will be necessary to finally proof the proposed mechanism of a biradical initiated polymerization.

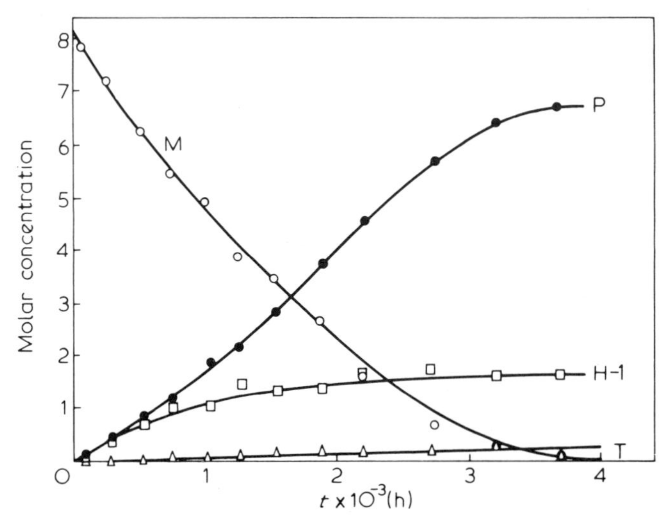

Fig. 7.7: Experimental data and computed curves for the molar concentrations of reactant and products for the thermal polymerization and oligomerization of MMA at 130 °C. M = monomer, H-1 = dimer, T = mixture of the trimers N-1 and N-4. P = polymer (in base mole units).

7.4 Kinetics for Medium and High Conversions

In industrial practice polymerization processes are almost ever performed up to the highest possible conversion to obtain maximum output and to cut down undesired residual monomer influencing both polymer performance and toxicological aspects. The understanding of the kinetics of the polymerization elementary reactions at high conversions is limited. Full descriptions are at best semiempirical. There have not been determinations of the absolute rate constants of propagation, k_p, and termination, k_t, for the free radical polymerization since the early paper of Hayden and Melville [59], who measured the heat of polymerization of methyl methacrylate up to high conversion in order to evaluate k_p and k_t. An update and completion using the different approach of applying steady state and non-steady state experimental techniques on free radical polymerization up to highest conversion seemed essential in view of theoretical considerations published since then [60].

7.4.1 Kinetics of the Initiated Polymerization below the Onset of the Gel Effect

The first stage of a polymerization process, the so-called ideal polymerization, is covered by a vast literature, mainly restricted to the determination of the ratio k_p^2/k_t and other parameters available from steady state experiments. A great number of experiments, however, has also been directed towards the measurement of the efficiency f and the ratio k_p/k_t and thus of the rate constants k_p and k_t at low conversion. In this region the three parameters have almost always been accepted to be constant. Mahabadi and O'Driscoll [61] reported a few years ago that the termination rate constant has a chain length dependence with a more rapid decrease for shorter chains and a slow decrease for longer chains. No further verifications have been reported since, but some theoretical backing of the decrease [62].

Proceeding to higher conversion the first almost ideal stage is followed by a stage where polymer coils touch and then entangle each other, which means the begin of the gel or Trommsdorff effect [63].

7.4.2 Kinetics of the Initiated Polymerization beyond the Onset of the Gel Effect up to the Glass Transition Point

Due to the increasing polymer concentration the termination reaction turns from a chemical to a diffusional control because of the growing hindrance of the movement of the radical chain ends. This results in an increase of the rate of polymerization R_p and the ratios k_p^2/k_t and k_p/k_t, and consequently an increase in the overall rate of polymerization as well as the chain length.

Upon further increase of the conversion the glass transition point of the system is reached at which both R_p and k_p become diminishing low. The chain length is reported to increase or to remain constant up to this point. k_p was accepted to be constant from conversion $x = 0$ up to shortly before the glass point x_{gl}, beyond which it was assumed to fall steeply [64]. Newer considerations, however, make it more reasonable to assume a decrease of k_p already far below x_{g1}. Authors calculating k_p/k_{po} via the free volume theory find a decrease of k_p/k_{po} due to the decreasing self diffusion coefficient of the monomer from the very beginning of the polymerization as shown e.g. by Stickler [65].

All experiments leading to this conclusion have been performed under steady state conditions not giving a direct access to k_p. Non-steady state experiments were necessary for medium and high conversions to yield f, k_p, and k_t as a function of x and to allow deeper insight into the growth and termination mechanisms over the full conversion range.

Selection of the Monomer and the Polymerization Conditions

The experiments had to be done with a well characterized polymer showing negligible transfer reaction with an initiator decaying almost completely by photodegradation. Methyl methacrylate and 1,1'-azo bis(1-cyclohexane-nitrile) (ACN) at 0 °C fulfill these conditions, but result in rather high molecular weight for low initiator concentrations. $f.k_p^2/k_t$ was determined at steady state from R_p, and k_p/k_t at non-steady state as described in Chapter 7.1.

For the determination of the efficiency f the instantaneous degrees of number and weight averages had to be known. Measurements by size exclusion chromatography and/or light scattering yield only cumulative degrees $\overline{\overline{P}}_n$ and $\overline{\overline{P}}_w$ as averages covering the full conversion range from 0 to x. The transformation is done according to G.V. Schulz [66] by:

$$\overline{P}_n = \overline{\overline{P}}_n \left(1 + \frac{x}{\overline{\overline{P}}_n} \frac{d\overline{\overline{P}}_n}{dx}\right)^{-1} ; \quad \overline{P}_w = \overline{\overline{P}}_w + \frac{d\overline{\overline{P}}_w}{dx} \tag{5}$$

From these values the fraction w_2 of chains resulting from termination by combination as compared with chains resulting from any type of termination can be calculated as:

$$w_2 = (4-2\,\overline{P}_w/\overline{P}_n)^{1/2} \tag{6}$$

Together with R_p, [I], and k_d as rate constant for the decay of the initiator at conversion x the efficiency f follows as:

$$f = \frac{2}{2-w_2}\;\frac{R_{p,x}}{2\,\overline{P}_n\,k_d\,[I]_x} \tag{7}$$

7.4.3 Results of the Polymerization Experiments

Fig. 7.8 shows the cumulative and the instantaneous degrees of the weight averages for three initiator concentrations. Of course the instantaneous values are appreciably higher than the cumulative ones. But already the cumulative values for the highest conversions demonstrate a distinct decrease of $\overline{P}_w$ which has so far not yet been observed.

The determination of the efficiency for ACN and methyl methacrylate at 0 °C resulted in:

$$f = 0.386\,(1-1.31\,x) \tag{8}$$

As shown in Fig. 7.9 there is a linear relation from the value of f at $x = 0$ to $f = 0$ at about the glass transition point of the system. This linearity is unexpected and, since the determination of $\overline{P}_n$ is difficult, it is desirable to test whether the experimental finding is reasonable in the light of the definition of f. The efficiency f is the fraction of radical pairs escaping the cage to become free radicals as compared with the radical pairs which react within the cage to form inert products. With k_{sep} and k_{rec} for the rate constants of separation and recombination, respectively, f is given by:

$$f = k_{sep}/(k_{sep} + k_{rec}) \tag{9}$$

In this relation k_{sep} is replaced by the self-diffusion coefficient D_s of the single radical and k_{rec} by a corresponding coefficient for the recombination k'_{rec}. k'_{rec} should be independent of the amount of polymer present and thus of conversion. D_s on the other side strongly depends on conversion. There are no diffusion data available for radicals in a mixture of monomers and polymers. But for solvents diffusing in a polymer matrix D_s determinations from spin-echo experiments have been published for poly methyl methacrylate and the sol-

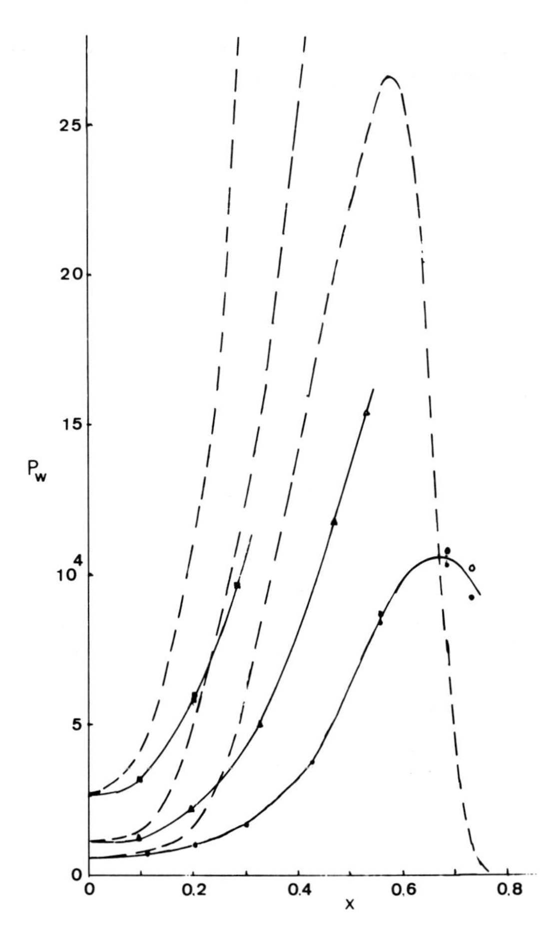

Fig. 7.8: Cumulative (———) and instantaneous (– – –) weight average degrees of polymerization, $\overline{P}_w$ and $\overline{\overline{P}}_w$, as a function of conversion x for methyl methacrylate at 0 °C.

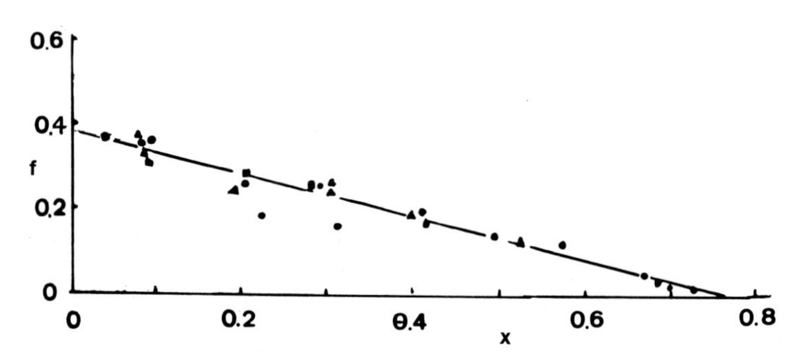

Fig. 7.9: Experimental values of efficiency f as a function of conversion x for ACN and methyl methacrylate at 0 °C.

vents acetone, benzene and methyl ethyl ketone [67, 68]. If k'_{rec} is calculated from $f_o = 0.386$ and the D_s value for $c = 0$, the experimental D_s values directly yield f as a function of the polymer concentration. Independent of the type of solvent data used an even better linearity for f is observed with slopes close to the slope 1.31 of Equ. (8).

From $f.k_p^2/k_t$, k_p/k_t and f the single rate constants can be calculated. Fig. 7.10 shows the conversion dependence of the termination rate constant for $x = 0.04$ up to the glass transition point conversion of about 0.73. In the lowest conversion range k_t is almost constant or even increases slightly. No effect of the different initiator concentrations or the different chain lengths could be observed as our chain lengths are too long as compared with those of Mahabadi and O'Driscoll [61] and thus should show only a very slight influence of $\overline{P}_n$ on k_t. With the onset of the gel effect k_t starts to decrease. The onset is shifted to lower conversions as the degrees of polymerization increase due to lower initiator concentrations. These observations are valid up to $x = 0.45$ where scattering of the k_t values becomes quite large and interpretation becomes difficult.

The behaviour of the rate of propagation is demonstrated in Fig. 7.11. Instead of k_p the product $f.k_p$ is shown for better comparison with the data of

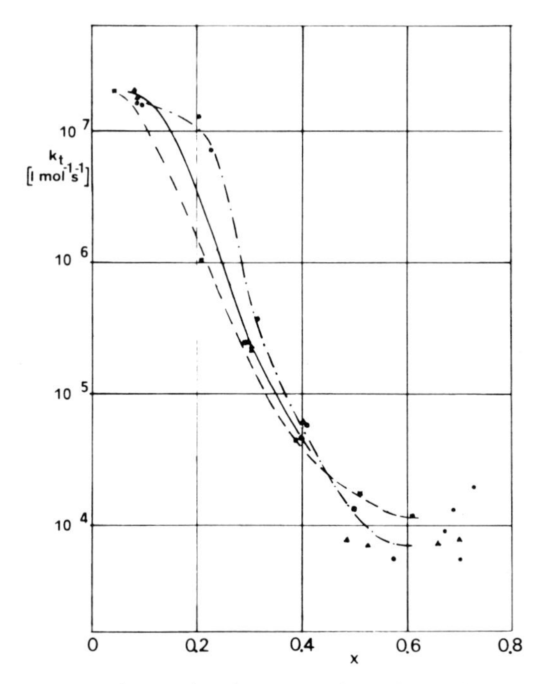

Fig. 7.10: Rate constants of termination as a function of conversion x for methyl methacrylate at 0 °C and ACN concentrations of (■) 0.4, (▲) 1.6, and (●) 6.4×10^{-3} mol l^{-1}.

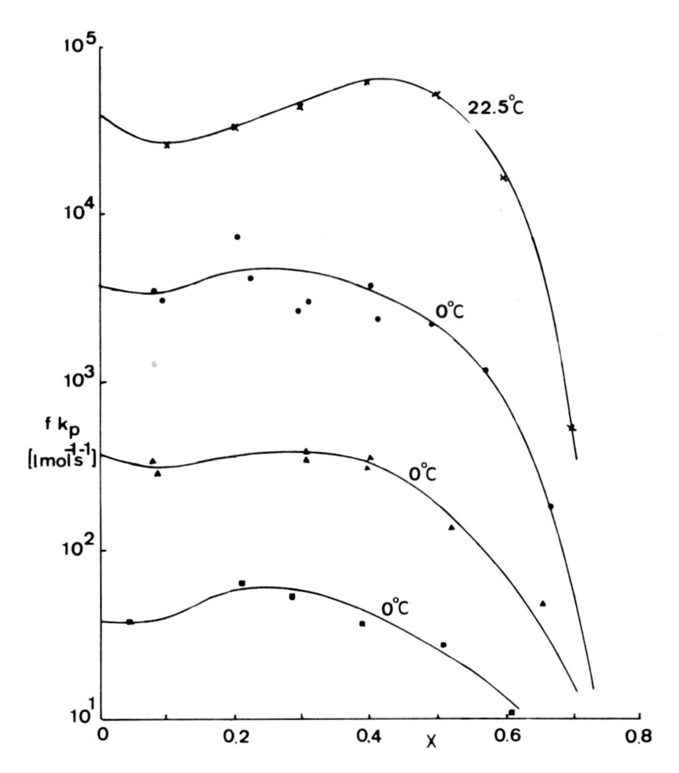

Fig. 7.11: Products f.k_p from Hayden and Melville [58] for poly (methyl methacrylate) at 22.5 °C and from [7–9] with initiation rates: 22.5 °C: R_i = 2.8 x 10^{-8} mol $l^{-1}s^{-1}$; 0 °C: (●) R_i = 1.0 x 10^{-8}, (▲) 2.1 x 10^{-9}, (■) 3.5 x 10^{-10} mol $l^{-1}s^{-1}$.

Hayden and Melville [59] who did not determine the efficiency. For f.k_p – and also for k_p – at low conversions a slight decrease or almost constant value is observed, followed by an increase first observed at about the onset of the entanglement. f.k_p then runs through a weak maximum for our 0 °C experiments and a more pronounced maximum for the 22.5 °C experiments [59]. Since f is a linearly declining function of x the maxima of k_p are even more pronounced.

7.4.4 Rate Constants and Diffusion Mechanisms

For the rate constant of termination it is well known that after the gel point there is a strong decrease of k_t. Generally the termination reaction can take place only if two radical chain ends have an encounter close enough for reaction. In dilute solution an unhindered translational selfdiffusion of the polymer and a segmental or rotational diffusion of the chain ends occur simultaneously. It is reasonable to accept the segmental diffusion as the slower and rate determining process. The behaviour should change with increasing entanglements which result in a strong decrease of the translational mobility. This means, for higher conversions the translational self-diffusion coefficient of the polymer $D_{s,p}$ will determine k_t.

Measurements of segmental diffusion coefficients of polymers D_{seg} have only been reported for poly(vinyl pyrrolidone) in ethyl alcohol by Buchachenko and Wassermann [69]. These authors report a qualitative agreement with the k_t-data of [57], which may also be concluded to be valid for the presented k_t-values up to high conversions.

The translational self-diffusion coefficient of polymers can be measured by Forced Rayleigh Scattering. Especially for polystyrene a number of data has been reported [70] which show a rapid decrease of $D_{s,p}$ with increasing chain length and concentration starting at $x = 0$. For poly(methyl methacrylate) the same behaviour is to be expected. This is in contrast to the experimental $k_t = f(x)$ function with an almost constant k_t for low conversions x. For higher conversions, however, the observed chain length dependence of k_t (Fig. 7.10) fits in the experimental $D_{s,p}$-data for polymers. Since there are no experimental data on D_{seg} of poly(methyl methacrylate) we cannot yet decide if D_{seg} together with $D_{s,p}$ can properly describe the k_t dependence on conversion for all polymerization stages.

The rate constant of propagation k_p during the initial stages of conversion is determined chemically as long as there are monomer molecules readily available close to the radical chain ends. Formerly it had been assumed that k_p is constant up to close to the glass point of the reaction mixture [64]. But theoretical considerations and the data of Fig. 7.11 clearly show a k_p influenced by the restricted diffusion of monomers due to the presence of polymers already far below the glass transition point. Thus the translational self-diffusion of the monomer $D_{s,m}$ will govern the k_p behaviour at least from the maximum of the k_p curve on.

After the first stages of conversion at about the onset of entanglement not only R_p increases but also k_p shows a slight increase. In contrast to the accepted theory [63] that the decrease of k_t alone effects the increase of R_p now a slight contribution from the increasing k_p seems reasonable, especially when taking into account the conversion dependence of the efficiency. Unfortunately, so far there is no physical interpretation of this behaviour.

7.4.5 Chain Length and Conversion

The molecular weights of the polymers start to increase with the onset of the gel effect and finally decrease. The decrease so far not yet reported concerns both the number and the weight average of the molecular weight and is observed in the experiments between high conversion ($x > 0.57$) and the glass transition point (x = 0.73). The effect may be described by descreasing lengths of the terminating radicals in this regime. Up to the maximum of the degrees of polymerization the average length of two terminating chains grows and is about the same for the two radicals. With decreasing mobility and content of the monomers the average length of the reacting chains becomes shorter until at least one of the reaction partners is an oligomer or even primary radical.

7.5 Conclusion

During the period of support within the Sonderforschungsbereich 41 a reasonable effort has been undertaken to elucidate what is going on at various stages of polymerization. Of course the presentation given above cannot include all work, which has been done, in the extent necessary to cover all aspects believed to be of importance. These have to be considered in detail in the publications quoted within this article. The compilation should, however, give a summary of what can be contributed from kinetic experiments and the determination of molecular weights to come to a better understanding of the diverse elementary reactions involved even in such "simple" reaction schemes like that of radical polymerizations.

The approach has so far been limited to the most common monomeric species. And it will be a future task to extend all these aspects to the variety of polymeric species of homogenous or mixed composition. Work has also been done on copolymerization kinetics (cf. Braun and Czerwinski, this Volume, Chapter 6). But there remains a lot to be done to proceed to a complete understanding of what has so successfully been applied in practice – work, which in consequence will lead to even better, "taylor-made" polymers.

7.6 References

[1] P. C. Deb, G. Meyerhoff: Primary radical termination and chain transfer in vinyl polymerization, J. Polym. Sci., Polym. Phys. Ed. 12 (1974) 2162–2167.
[2] P. C. Deb, G. Meyerhoff: Primary radical termination: evaluation of the characteristic constant, Europ. Polym. J. 10 (1974) 709–715.
[3] K. C. Berger, P. C. Deb, G. Meyerhoff: Radical vinyl polymerization. Reactions of benzoyl peroxide during bulk polymerization with labelled initiator, Macromolecules 10 (1977) 1075–1080.
[4] H. K. Mahabadi, G. Meyerhoff: A new method for evaluation of the characteristic constant for primary radical termination in free radical polymerization, Europ. Polym. J. 15 (1979) 607–613.

[5] K. C. Berger, G. Meyerhoff: Disproportionierung und Kombination als Abbruchsmechanismen bei der radikalischen Polymerisation von Styrol, 1: Versuche mit ^{14}C-markiertem AIBN, Makromol. Chem. 176 (1975) 1983–2003.
[6] K. C. Berger, G. Meyerhoff: Disproportionierung und Kombination als Abbruchsmechanismen bei der radikalischen Polymerisation von Styrol, 2: Analyse der Temperaturabhängigkeiten, Makromol. Chem. 176 (1975) 3575–3592.
[7] R. Sack: Bestimmung der absoluten Geschwindigkeitskonstanten k_w und k_{ab} und der Radikalausbeute f der radikalischen Polymerisation von Methylmethacrylat bei hohen Umsätzen, Diss. Mainz 1984.
[8] R. Sack, G.V. Schulz, G. Meyerhoff: Free radical polymerization of methyl methacrylate up to the glassy state. Rates of propagation and termination, Macromolecules 21 (1988) 3345–3352.
[9] R. Sack-Kouloumbris, G. Meyerhoff: Radikalpolymerisation von Methylmethacrylat im gesamten Umsatzbereich. Stationäre und instationäre Experimente zur Bestimmung von Wachstums- und Abbruchgeschwindigkeit, Makromol. Chem. 190 (1989) 1133–1152.
[10] M. Stickler: Die Kinetik der spontanen thermischen Polymerisation von Methylmethacrylat, Diss. Mainz 1977.
[11] M. Stickler, G. Meyerhoff: Die thermische Polymerisation von Methylmethacrylat, 1: Polymerisation in Substanz, Makromol. Chem. 179 (1978) 2729–2745.
[12] M. Stickler, G. Meyerhoff: Die thermische Polymerisation von Methylmethacrylat, 2: Bildung des ungesättigten Dimeren, Makromol. Chem. 181 (1980) 131–147.
[13] J. Lingnau: Zur thermischen Polymerisation des Methylmethacrylats: Oligomerenbildung und Polymerisation in Lösung, Diplomarbeit, Mainz 1978.
[14] J. Lingnau, M. Stickler, G. Meyerhoff: The spontaneous polymerization of methyl methacrylate, IV: Formation of cyclic dimers and linear trimers, Europ. Polym. J. 16 (1980) 785–791.
[15] J. A. Benson, C. D. Duncan, G. C. O'Connell, M. S. Platz: Kinetic confirmation and synthetic circumvention of the cascade mechanism for population of the triplet ground state of a trimethylenemethane, J. Amer. Chem. Soc. 98 (1976) 2358–2361.
[16] J. Lingnau: Der Startvorgang bei der spontan-thermischen Polymerisation von Methylmethacrylat, Diss. Mainz 1982.
[17] B. Appelt, G. Meyerhoff: Characterization of polystyrenes of extremely high molecular weights, Macromolecules 13 (1980) 657–662.
[18] J. Lingnau, G. Meyerhoff: Spontaneous Polymerization of methyl methacrylate, 8: Polymerization kinetics of acrylates containing chlorine atoms, Macromolecules 17 (1984) 941–945.
[19] G. Meyerhoff: Extension of GPC techniques, Separation Sci. 6 (1971) 239–248.
[20] B. Appelt: Zum Mechanismus der thermischen Polymerisation von Styrol. Kinetik und Molekulargewichtsverteilung, Diss. Mainz 1977.
[21] G. Meyerhoff, B. Appelt: A low shear viscometer with automated recording and application to high molecular weight polystyrene solutions, Macromolecules 12 (1979) 968–971.
[22] R. Härzschel: Molekulargewichtsverteilungen von Polymeren durch Ultrazentrifugenmessungen, Diss. Mainz 1985.
[23] R. Härzschel, G. Meyerhoff: Sedimentation equilibrium in a density gradient and characterization of the molecular weight distribution of poly(methyl methacrylate), Makromol. Chem. 191 (1990) 3139.

[24] P. Munk: Measurements of density gradient of a solvent mixture in an ultracentrifuge, Macromolecules 15 (1982) 500–505.
[25] J. Raczek: Charakterisierung von Molekulargewichtsverteilungen gelöster Polymerer aus Autokorrelationsfunktionen und Frequenzspektren der Lichtstreuung, Diss. Mainz 1980.
[26] H. Hack: Streulichtmessungen zur Bestimmung der Verteilungsbreite des Molekulargewichts, Diss. Mainz 1977.
[27] H. Hack, G. Meyerhoff: Streulichtmessungen an Polymerlösungen zur Bestimmung der Verteilungsbreite des Molekulargewichts, Makromol. Chem. 179 (1978) 2475–2485.
[28] J. Raczek, G. Meyerhoff: Determination of molecular weight distributions from frequency-analysing (dynamic) light scattering, Macromolecules 13 (1980) 1251–1254.
[29] J. Raczek: Theory of polydispersity correction procedures for the determination of molecular weight dependences, Europ. Polym. J. 18 (1982) 351–357.
[30] J. Raczek: Examples for the determination of molecular weight distributions by light scattering, Europ. Polym. J. 18 (1982) 863–873.
[31] J. Herold: Molekulargewichte und deren Verteilung bei Polyamiden am Beispiel Poly-2,4,4-Trimethylhexamethylen-terephthalamid, Diss. Mainz 1977.
[32] J. Herold, G. Meyerhoff: Solution properties of poly-trimethylhexamethylene-terephthalamide (Trogamid), Europ. Polym. J. 15 (1979) 525–532.
[33] K. Weißkopf: Charakterisierung N-Trifluoracetylierter Polyamide und Kinetik der Polykondensation in Lösung am Beispiel von Trogamid, Diss. Mainz 1983.
[34] K. Weißkopf: Determination of molecular weight averages and molecular weight distributions by g.p.c. of N-trifluoroacetylated polyamides, Polymer 26 (1985) 1187–1190.
[35] E. Husemann, H. Ruska: Die Sichtbarmachung von Molekülen des Parajodbenzoylglycogens, Naturwissenschaften 28 (1940) 534–535.
[36] G. Koszterszitz, W.K. R. Barnikol, G.V. Schulz: Die Bestimmung der Molekulargewichtsverteilung von nichtkristallisierenden Polymeren mit dem Elektronenmikroskop, 4: Weiterentwicklung der Präparationsmethode durch Gefriertrocknung, Makromol. Chem. 178 (1977) 1133–1148.
[37] G. Koszterszitz, G.V. Schulz: Die Bestimmung der Molekulargewichtsverteilung von nichtkristallisierenden Polymeren mit dem Elektronenmikroskop, 7: Präparation aus Lösung durch Verdampfung des Lösungsmittels, Makromol. Chem. 178 (1977) 2437–2450.
[38] G. Koszterszitz, G. S. Greschner, G.V. Schulz: Die Bestimmung der Molekulargewichtsverteilung von nichtkristallisierenden Polymeren mit dem Elektronenmikroskop, 6: Fehlerkorrektur der experimentellen Zahlenverteilungsfunktion, Makromol. Chem. 178 (1977) 1169–1185.
[39] G.V. Schulz, K. Blaschke: Orientierende Versuche zur Polymerisation des Methacrylsäuremethylesters – Über die Kinetik der Kettenpolymerisation, 11, Z. Physik. Chem. B 50 (1941) 305–322.
[40] M. Kouloumbris: Die Polymerisation von Methylmethacrylat mit Azoisobuttersäuremethylester – Ein Modellsystem der radikalischen Polymerisation, Diss. Mainz 1984.
[41] K. Neubecker: Kinetik der gestarteten Substanzpolymerisation von n-Butylmethacrylat im Umsatzbereich von 0 bis 100 %, Diplomarbeit, Mainz 1980.

[42] G. Mayer, G.V. Schulz: Über die Elementarreaktionen der radikalischen Polymerisation von Benzylmethacrylat, Makromol. Chem. 173 (1973) 101–112.
[43] U. Naust: Der Einfluß von alpha-Polymeren auf die Kinetik der radikalischen Polymerisation von Styrol, Diplomarbeit, Mainz 1977.
[44] W. A. Pryor, L. D. Lasswell: Advances in Free Radical Chemistry, Vol. V, Academic Press, New York 1975.
[45] V. A. Kurbatov: Thermal polymerization and oligomerization of monomers, Russian Chem. Revs. 56 (1987) 505.
[46] K. Kirchner, F. Patat: Die Startreaktion der unkatalysierten Styrolpolymerisation, Makromol. Chem. 37 (1960) 251–253.
[47] K. Kirchner, H. Riederle: Thermal polymerization of styrene – The formation of oligomers and intermediates, 1, Angew. Makromol. Chem. 111 (1983) 1–16.
[48] H. F. Kauffmann, O. F. Olaj, J.W. Breitenbach: Spectroscopic measurements on spontaneously polymerizing styrene, 1. Evidence for the formation of the two Diels-Alder isomers of different stability, Makromol. Chem. 177 (1976) 939–945.
[49] O. F. Olaj, H. F. Kauffmann, J.W. Breitenbach: Spectroscopic measurements on spontaneously polymerizing styrene, 2. The estimation of the reactivity of the two Diels-Alder isomers towards polymer radicals, Makromol. Chem. 178 (1977) 2707–2717.
[50] C. Walling, E. R. Briggs: The thermal polymerization of methyl methacrylate, J. Amer. Chem. Soc. 68 (1946) 1141–1149.
[51] G. Henrici-Olivé, S. Olivé: Kettenübertragung bei der radikalischen Polymerisation, Fortschr. Hochpolym. Forsch. 2 (1961) 494–577.
[52] R. S. Lehrle, A. Shortland: A study of the purification of methyl methacrylate suggests that the "thermal" polymerization of this monomer is initiated by adventitious peroxides, Eur. Polym. J. 24 (1988) 425–429.
[53] E. Brand: Der Einfluß des dimeren Methylmethacrylats auf die Polymerisation von Methylmethacrylat, Diplomarbeit, Mainz 1977.
[54] E. Brand, M. Stickler, G. Meyerhoff: Die thermische Polymerisation von Methylmethacrylat, 3: Verhalten des ungesättigten Dimeren bei der Polymerisation, Makromol. Chem. 181 (1980) 913–921.
[55] J. Lingnau, G. Meyerhoff: The spontaneous polymerization of methyl methacrylate, 6: Polymerization in solution; participation of transfer agents in the initiation reaction, Polymer 24 (1984) 1473–1478.
[56] J. Lingnau, G. Meyerhoff: The spontaneous polymerization of methyl methacrylate, 7: External heavy atom effect on the initiation, Makromol. Chem. 185 (1984) 587–600.
[57] A. A. Lamola, G. S. Hammond: Mechanisms of photochemical reactions in solution. 33. Intersystem crossing efficiencies, J. Chem. Phys. 43 (1965) 2129.
[58] M. Stickler, G. Meyerhoff: The spontaneous thermal polymerization of methyl methacrylate, 5: Experimental study and computer simulation of the high conversion reaction at 130 °C, Polymer 22 (1981) 928–933.
[59] P. Hayden, Sir H. Melville: The kinetics of methyl methacrylate, I: The bulk reaction, J. Polymer Sci. 43 (1960) 201–214.
[60] J. N. Cardenas, K. F. O'Driscoll: High conversion polymerization, I: Theory and application to methyl methacrylate, J. Polym. Sci., Polym. Chem. Ed. 14 (1976) 883–897.

[61] K. H. Mahabadi, K. F. O'Driscoll: Absolute rate constants of free-radical polymerization, III: Determination of propagation and termination rate constants for styrene and methyl methacrylate, J. Macromol. Sci., Chem. A-11 (1977) 967.
[62] O. F. Olaj, G. Zifferer: General treatment of a kinetic scheme with chain length dependent termination, Makromol. Chem., Symposia 10,11 (1987) 165.
[63] E. Trommsdorff, H. Köhle, P. Lagally: Zur Polymerisation des Methacrylsäuremethylesters, Makromol. Chem. 1 (1947) 168.
[64] A. F. Moroni, G.V. Schulz: Ein Modell zur Beschreibung bimolekularer Reaktionen zwischen reaktiven Gruppen, die in zwei Polymerknäuel eingebaut sind, Makromol. Chem. 118 (1968) 313–323.
[65] M. Stickler: Free-radical polymerization kinetics of methyl methacrylate at very high conversions, Makromol. Chem. 184 (1984) 2563.
[66] G.V. Schulz, G. Harborth: Über den Mechanismus des explosiven Polymerisationsverlaufs des Methacrylsäuremethylesters, Makromol. Chem. 1 (1947) 106.
[67] R. Kosfeld, J. Schlegel: Zur Selbstdiffusion von Benzol und Aceton in Polymethylmethacrylat, Angew. Makromol. Chem. 29/30 (1973) 105.
[68] D. Hwang, C. Cohen: Diffusion and relaxation in polymer-solvent-systems, II: Poly(methyl methacrylat)/methyl ethyl ketone, Macromolecules 17 (1984) 2890.
[69] A. L. Buchachenko, A. M. Wassermann: The structure and dynamics of macromolecules in solution as studied by ESR and NMR techniques, Pure and Appl. Chem. 54 (1982) 507.
[70] J. A. Wesson, I. Noh, T. Kitano, H. Yu: Self-diffusion of polystyrene by forced Rayleigh scattering, Macromolecules 17 (1984) 782.

8 Kinetics and Mechanisms of Anionic and Group Transfer Polymerization

Axel H. E. Müller*, Hartwig Höcker**, and Günter V. Schulz*

8.1 Introduction

"Living" polymerization renders the possibility of preparing a multitude of model polymers of narrow molecular weight distribution which are well-defined in terms of topology (e.g. block copolymers, star polymers, cyclic polymers, networks) and functualization (e.g. telechelics, macromonomers).

"Living" polymerization is characterized by the absence of termination and transfer reactions. Since its discovery by M. Szwarc [1] in 1956 it has been associated with anionic polymerization and only recently various other polymerization mechanisms (e.g. cationic polymerization [2], group transfer polymerization (GTP) [3], metathesis [4]) have proven to be "living" in character in certain cases. As most of the work done in Mainz dealt with anionic polymerization this article will mainly focus on the elucidation of the kinetics and mechanism of this type of polymerization. Only in a later part more recent investigations on group transfer polymerization will be reviewed.

Due to the absence of side reactions "living" polymerizations consist of only two principal reactions:

initiation:

$$I^* + M \xrightarrow{k_i} P_1^* \qquad (1)$$

* Institut für Physikalische Chemie der Universität Mainz, Jakob-Welder-Weg 11–15, D-6500 Mainz

** Lehrstuhl für Textilchemie und Makromolekulare Chemie der RWTH Aachen, Veltmanplatz 8, D-5100 Aachen

and propagation:

$$P_i^* + M \xrightarrow{k_p} P_{i+1}^* \tag{2}$$

Here, I^* denotes initiator, M monomer, P_i^* an active polymer chain of degree of polymerization i.

In most cases it is also possible to find initiators which are reactive enough to give instantaneous initiation, i.e.

$$k_i \geq k_p \tag{3}$$

This implies that the concentration of active centres $c^* = \Sigma[P_i^*]$ remains constant during the polymerization process. For such an "ideal" type of polymerization, only propagation has to be taken into account. Thus, for the rate of polymerization simple pseudo first-order kinetics can be applied:

$$R_p = -\frac{d[M]}{dt} = k_p \cdot [M] \cdot c^* = k_{app} \cdot [M], \tag{4}$$

where k_{app} is the "apparent" pseudo first-order rate constant. Integration leads to

$$\ln \frac{[M]_0}{[M]} = k_p \cdot c^* \cdot t = k_{app} \cdot t. \tag{5}$$

In such a case the number-average degree of polymerization, P_n, is a linear function of the monomer conversion, x_p:

$$P_n = \frac{\text{concentration of polymerized monomers}}{\text{concentration of polymer chains}} = \frac{[M]_0 \cdot x_p}{c_{tot}}. \tag{6}$$

Here, $c_{tot} = c^*/k$ is the total concentration of polymer chains, *including terminated ones,* and k is the number of active centres per polymer chain.

The molecular weight distribution (MWD) for a "living" polymerization with fast initiation was calculated by Flory [5] as early as 1940. It is identical with a Poisson distribution. The polydispersity ratio M_w/M_n is given by

$$M_w/M_n = 1 + \frac{P_n - 1}{P_n^2} \approx 1 + 1/P_n \approx 1. \tag{7}$$

This shows that "living" polymerization has the capability to form polymers of extremely narrow MWD.

As the half-lives of the polymerizations investigated range from fractions of seconds to hours, it was necessary to use different types of reactors for the kinetic measurements:

- a flow tube for fast reactions ($0.05 \leq t_{1/2} \leq 2$ s) [6]
- a stirred tank reactor equipped with a solenoid valve allowing for the withdrawal of samples at intervals of 1 s or larger for reactions having half-lives ≥ 2 s. [7]

The advantage of both methods is that for all conversions a polymer sample is obtained which can be characterized for

- molecular weight averages,
- molecular weight distribution,
- activity of chains (after labelling of active chains, e.g. by $^3H^+$),
- microstructure, etc.

8.2 Polymerization of Styrene

Most of the early work on the mechanism of anionic polymerization was done on styrene and α-methyl styrene in ethereal solvents like THF, 1,2-dimethoxyethane (DME), tetrahydropyrane (THP) or dioxane. These monomers are perfect for kinetic investigations because the active centres are stable for longer times than necessary for complete polymerization. Already in 1958 it was shown by Worsfold and Bywater [8] that the polymerization of α-methylstyrene in THF leads to linear first-order plots (cf. Equ. 5), the apparent (i.e. pseudo first-order) rate constant being a linear function of c^*. During the 60s and early 70s thorough investigations, mainly of the groups of M. Szwarc (Syracuse) and G.V. Schulz (Mainz), gave insight not only into the elementary reactions of anionic polymerization, but also into the properties of ions and ion pairs in non-aqueous solvents [9, 10].

8.2.1 Ions and Ion Pairs

Independently, the groups of Szwarc [11, 12] and Schulz [13, 14] reported that in the anionic polymerization of styrene in THF the propagation rate constant k_p decreases with increasing concentration of active centres c^*. This was attributed to the coexistence of two different kinds of active species, namely free anions and ion pairs, propagating at different rates (cf. Scheme 1)

$$\begin{array}{ccc} P_i^-,\ Mt^+ & \xrightleftharpoons{K_D} & P_i^- + Mt^+ \\ +M \downarrow k_\pm & & +M \downarrow k_- \\ P_{i+1}^-,\ Mt^+ & \rightleftharpoons & P_{i+1}^- + Mt^+ \\ \text{ion pair} & & \text{free anion} \end{array}$$

Scheme 1

Thus, the experimental rate constant is given by

$$k_p = \alpha \cdot k_- + (1-\alpha) \cdot k_\pm = k_\pm + (k_- - k_\pm) \cdot \alpha, \tag{8}$$

α being the degree of dissociation. Due to the low dielectric constants of the ethereal solvents, the dissociation constants are very low ($K_D < 10^{-6}$ mol/l) and consequently $\alpha \ll 1$. Thus, the mass action law can be simplified, rendering:

$$\alpha = (K_D/c^*)^{1/2}, \tag{9}$$

leading to

$$k_p = k_\pm + (k_- - k_\pm) \cdot K_D^{1/2} \cdot (c^*)^{-1/2}. \tag{10}$$

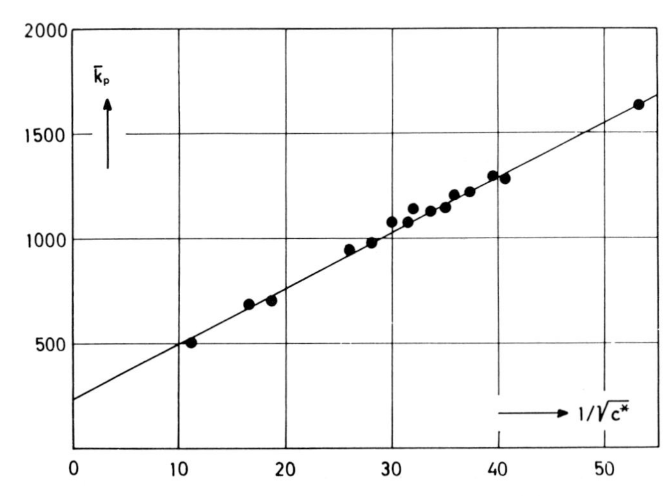

Fig. 8.1: Dependence of propagation rate constants on the concentration of active centres in the anionic polymerization of styrene in THF using Na^+ as the counterion at 25 °C [9].

Fig. 8.1 shows that a plot of k_p versus $(c^*)^{-1/2}$ is indeed linear and has a positive intercept, allowing for the determination of $k_\pm$.

Two different methods were used to determine K_D and k_-. The Syracuse group used conductivity measurements in order to determine K_D. A different approach was used by the Mainz group. Addition of a common ion salt having a dissociation constant much larger than that of the polymeric ion pairs (e.g. tetraphenyl borates) decreases the degree of dissociation of the latter and thus leads to lower average rate constants [15]:

$$k_p = k_\pm + (k_- - k_\pm) \cdot K_D/[Mt^+]. \tag{11}$$

Fig. 8.2 shows that a plot according to Equ. (11) has a good linearity. The values of $k_\pm$ determined from Figs. 8.1 and 8.2 agree within limits of experimental error. Combination of the slopes of both plots renders K_D and k_-.

It was found that the propagation rate constant for the free anions is nearly invariable of the solvent ($k_- \approx 1.3 \cdot 10^5$ l mol^{-1} s^{-1}). In contrast there is a strong increase of $k_\pm$ with increasing polarity of the solvent, especially for the smaller cations. This is a first indication that the nature of the solvent influences the structure of the ion pair. The dependence of $k_\pm$ on the size of the counterion for a given solvent is also peculiar: For the less polar solvents like dioxane $k_\pm$ increases with increasing ionic radius. This is expected, because a large interionic distance in the ion pair favours the charge separation in the transition state and thus leads to an increase in reaction rate. However, for solvents of higher polarity this dependence is reversed. The reason for this observation is discussed below.

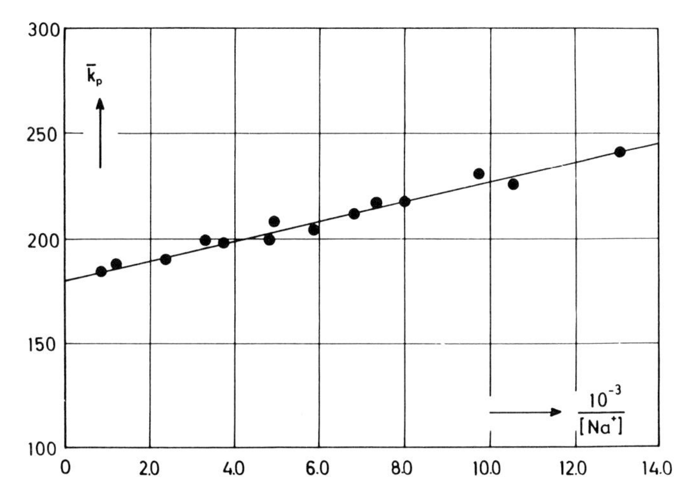

Fig. 8.2: Dependence of propagation rate constants on the concentration of added sodium ions in the anionic polymerization of styrene in THF at 25 °C [9].

8.2.2 Contact and Solvent-Separated Ion Pairs

The most conclusive information on the nature of the ion pairs is obtained from the temperature dependence of the propagation rate constants. As can be seen in Fig. 8.3 the rate constants for the free anion give one straight line in the Arrhenius plot, irrespective of counterion and solvent [15] ($\log A_{-} = 8.0$; $E_{a,-} = 16.7$ kJ mol^{-1}).

An Arrhenius plot for the ion pair rate constants (counterion Na^+) is shown in Fig. 8.4. Here, the Arrhenius lines significantly deviate from linearity. For dimethoxyethane (DME) and THF in certain temperature intervals this even leads to "negative activation energies". The reason for this observation is the coexistence of two different types of ion pairs, i.e. contact (or tight) and

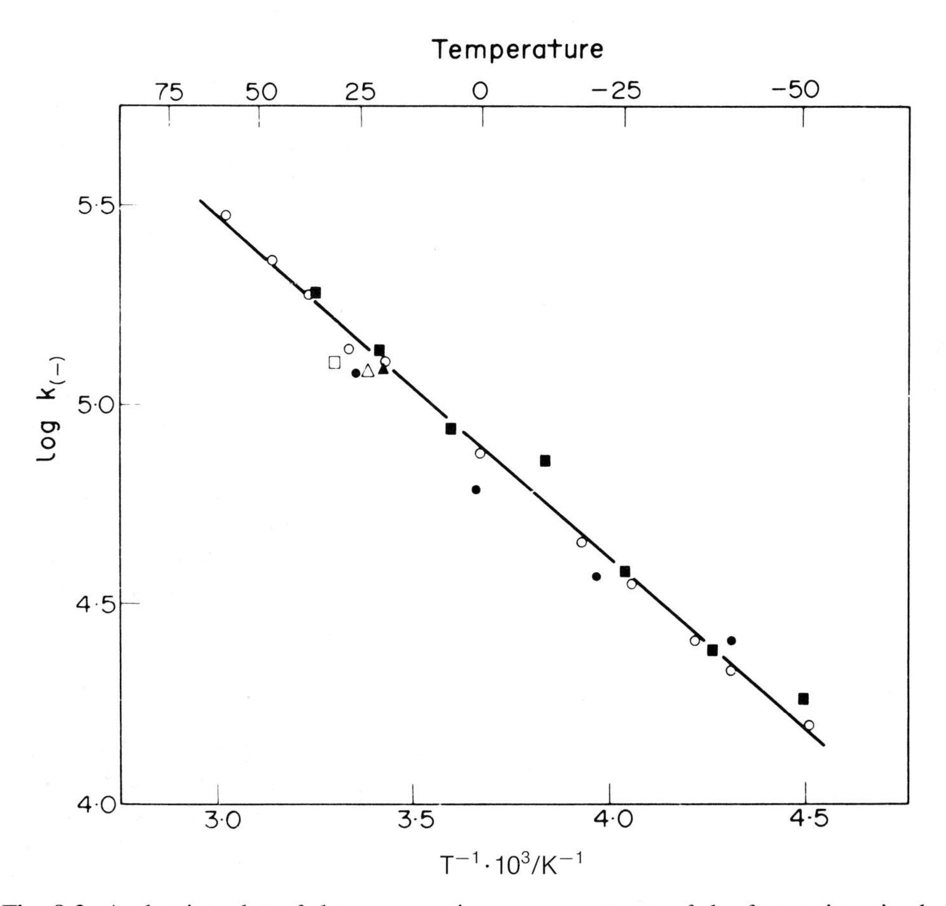

Fig. 8.3: Arrhenius plot of the propagation rate constants of the free anions in the anionic polymerization of styrene in THF using various solvents [15].

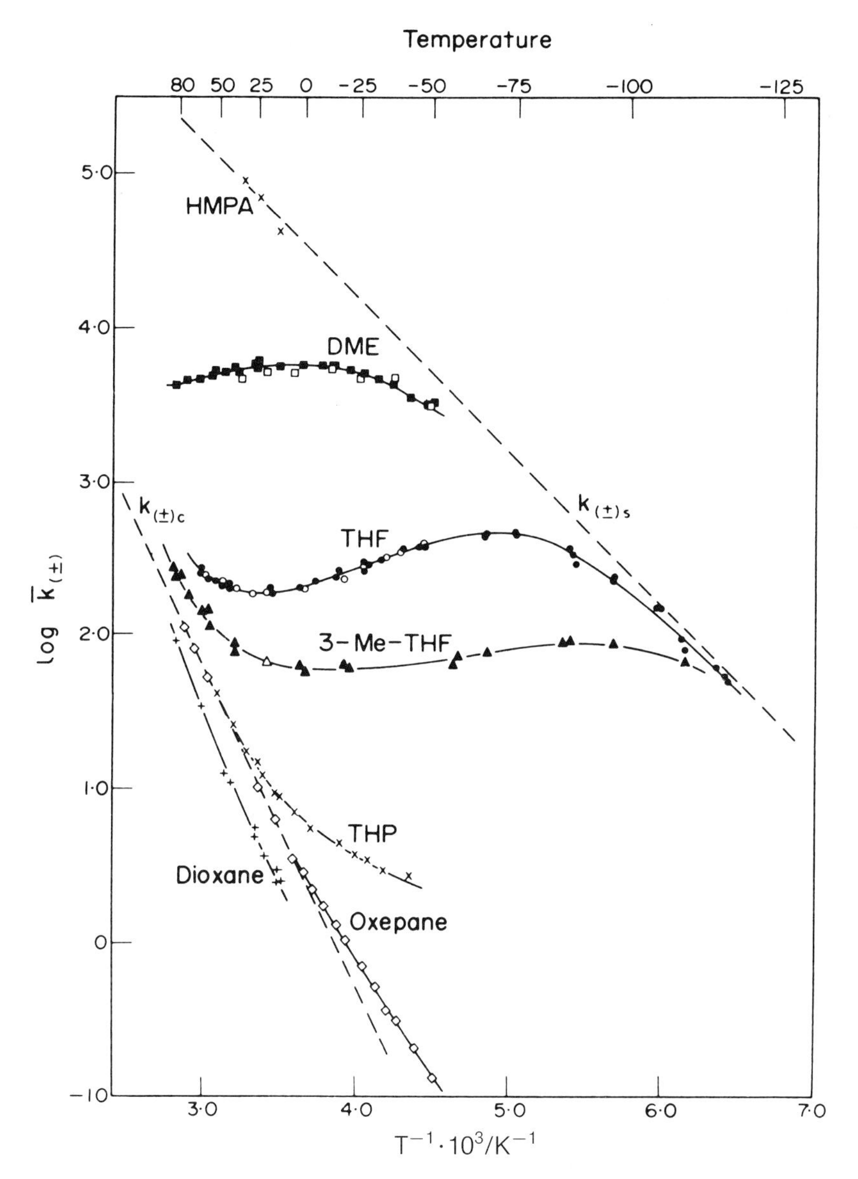

Fig. 8.4: Arrhenius plot of the propagation rate constants of the ion pairs in the anionic polymerization of styrene. Counterion; Na^+ [15].

solvent-separated (or loose) ion pairs [16]. This results in a three-state mechanism (cf. Scheme 2).

$$
\begin{array}{ccccc}
P_i^-, Mt^+ + n\,S & \underset{}{\overset{K_{cs}}{\rightleftharpoons}} & P_i^-, S_n, Mt^+ & \overset{K_D^*}{\rightleftharpoons} & P_i^- + Mt^+_, S_n \\
+M \downarrow k_{\pm,c} & & +M \downarrow k_{\pm,s} & & +M \downarrow k_- \\
P_{i+1}^-, Mt^+ + n\,S & \rightleftharpoons & P_{i+1}^-, S_n, Mt^+ & \rightleftharpoons & P_{i+1}^- + Mt^+_, S_n
\end{array}
$$

contact ion pair — solvent-separated ion pair — free anion

Scheme 2

In the solvent-separated ion pair the cation is completely surrounded by solvent molecules, S, and thus is separated from the carbanion. Consequently, its interionic distance and reactivity are largely increased.

From the equilibrium constant of contact and solvent-separated ion pairs

$$K_{cs} = [P_{\pm,s}]/[P_{\pm,c}] \tag{12}$$

the fraction of solvent-separated ion pairs can be calculated:

$$\alpha_s = [P_{\pm,s}]/[P_\pm] = \frac{K_{cs}}{1+K_{cs}} \tag{13}$$

The overall ion pair propagation rate constant $k_\pm$ is given by

$$k_\pm = \alpha_s \cdot k_{\pm,s} + (1-\alpha_s) \cdot k_{\pm,c} \tag{14}$$

$$= \frac{k_{\pm,s} \cdot K_{cs} + k_{\pm,c}}{1+K_{cs}} \tag{15}$$

Here, $k_{\pm,c}$ and $k_{\pm,s}$ denote the propagation rate constants of contact and solvent-separated ion pairs, respectively. Both constants exhibit straight lines in the Arrhenius plot.

As $k_\pm$ is composed of three different constants it depends on temperature in a very complex way:

$$k_\pm = \frac{A_s \cdot \exp(\Delta S_{cs}/R) \cdot \exp[-(E_s + \Delta H_{cs})/RT] + A_c \cdot \exp(-E_c/RT)}{1+\exp[(T\Delta S_{cs} - \Delta H_{cs})/RT)]} \tag{16}$$

Here, A_s and A_c denote the frequency factors of the Arrhenius equations for $k_{\pm,s}$ and $k_{\pm,c}$, respectively, E_s and E_c denote the corresponding activation energies, and ΔS_{cs} and ΔH_{cs} denote the parameters of the van't Hoff equation for K_{cs}. By using a curve-fitting technique, it was possible to determine all thermodynamic and activation parameters [17]. The existence of solvent-separated ion pairs was later confirmed by spectroscopic [18] and conductometric techniques [19].

8.3 Polymerization of Acrylic Monomers

Acrylic monomers are one of the most important classes of vinyl monomers. Especially methacrylates, acrylates and acrylonitrile are of high industrial interest. Moreover, acrylic polymers have some advantages compared to non-polar ones:

- tacticity can be controlled by the polymerization conditions,
- polymers are easily converted to other interesting polymers, e.g. hydrolysis of poly(alkyl acrylate)s rendering poly(acrylic acid),
- polymers are UV-degradable, e.g. poly(vinyl ketones).

8.3.1 Potential Problems

In contrast to non-polar monomers the polymerization mechanism is complicated by the presence of the polar side group (e.g. ester, ketone, nitrile), which gives rise to chemical and physical interactions with the living chain end.

The most important potential complications are:

a) Side reactions of the initiator or living end with the carbonyl group of the monomer or polymer. This results in low initiator efficiency or chain termination, e.g:

$$R^{\ominus} + CH_2{=}C(CH_3){-}C(=O){-}OR' \longrightarrow CH_2{=}C(CH_3){-}C(=O){-}R + R'O^{\ominus}$$

b) Activation of protons in α position to the carbonyl group leading to chain transfer:

$$\sim\!\!\sim CH^{\ominus}(C{=}O)(OR) + RO{-}\underset{\|}{\overset{}{C}}(O){-}C{-}H \longrightarrow \sim\!\!\sim CH_2(C{=}O)(OR) + RO{-}\underset{\|}{C}(O){-}C^{\ominus}$$

c) Due to the bidentate character of the active centres, they may attack the monomer not only by the carbanion (1,2 addition) but also by the enolate oxygen (1,4 addition):

$$\left[\ \sim\!\!\sim C^{\ominus}(CH_3){-}C(R){=}O \longleftrightarrow \sim\!\!\sim C(CH_3){=}C(R){-}O^{\ominus}\ \right] \equiv \sim\!\!\sim \overset{\delta^-}{C}(CH_3){\cdots}C(R){\cdots}\overset{\delta^-}{O}$$

carbanion enolate

$$\sim\!\!\sim \overset{\delta^-}{C}(CH_3){\cdots}C(R){\cdots}\overset{\delta^-}{O} + CH_2{=}C(CH_3){-}C(R){=}O$$

$$\xrightarrow{1,2} \sim\!\!\sim C(CH_3)(R{-}C{=}O){-}CH_2{-}\overset{\delta^-}{C}(CH_3){\cdots}C(R){\cdots}\overset{\delta^-}{O}$$

$$\xrightarrow{1,4} \sim\!\!\sim C(CH_3){=}C(R){-}O{-}CH_2{-}\overset{\delta^-}{C}(CH_3){\cdots}C(R){\cdots}\overset{\delta^-}{O}$$

d) The carbonyl group may co-ordinate with the counterion of the living chain end ("intramolecular solvation") or may lead to association of ion pairs.

e) The chain end may interact with electron donors.

As a result of these complications (especially in non-polar solvents) non-ideal polymerizations of acrylic monomers have been reported frequently (for reviews, see [10, 20, 21, 22]).

Thus, our primary aim was to find conditions for an "ideal" anionic polymerization of these monomers characterized by simple kinetics and narrow molecular weight distributions. Additionality, it was of interest to determine the factors which influence the kinetics, the amount and nature of side products, the molecular weight distribution and the microstructure of the polymers. Our second aim was to understand these findings on the basis of mechanisms and the structures of the active centres involved.

Although acrylates and vinyl ketones also have been investigated in recent years, the results reported here will focus only on methacrylates, which have been the most thoroughly investigated class of acrylic monomers.

8.3.2 Methyl Methacrylate

Most of the mechanistic work reported on acrylic monomers has been done on methyl methacrylate (MMA). In 1965 Figueruelo et al. [23, 24] showed that anionic polymerization of MMA in THF at −78 °C leads to narrow molecular weight distributions. First kinetic investigations stating "living" kinetics for MMA polymerization were published independently by Mita et al. [25] and by Löhr and Schulz [26, 27] in 1973. From the concentration dependence of the propagation rate constants, the coexistence of ions and ion pairs was shown. The number-average degree of polymerization is a linear function of conversion. During the past 15 years the kinetics of methacrylates have been investigated in detail using different counterions, solvents, additives, and monomers [28, 29].

In THF as the solvent, the kinetics of MMA polymerization were investigated using Cs^+ [30], K^+ [31], Na^+ [30, 32], and Li^+ [31] as counterions. After complexation of the sodium ion by the bicyclic cryptand 222 ("Na^+, 222") it was possible to independently determine the rate constants of the cryptated ion pairs and the free anions [33]. Fig. 8.5 shows an Arrhenius plot of the rate constants for all counterions investigated [31]. This plot reveals several informations:

- the linearity indicates only one kind of active species,
- except for Na^+, K^+, and Cs^+, a distinct dependence of the rate constants on the counterion is found. This indicates that the active species are contact ion pairs, because a large interionic distance in the contact ion pairs favours the charge separation in the transition state and thus leads to an increase in reaction rate (cf. 8.2.1).

This is seen clearly in Fig. 8.6, which plots log k_p *vs.* the reciprocal interionic distance, *a* (calculated as the sum of the crystal radius of the cation and the electrostatic radius of the anion which is assumed to be 1.5 A). The deviation of Na^+ and K^+ from the straight line indicates that the interionic distances are larger than the values expected from the crystal radii.

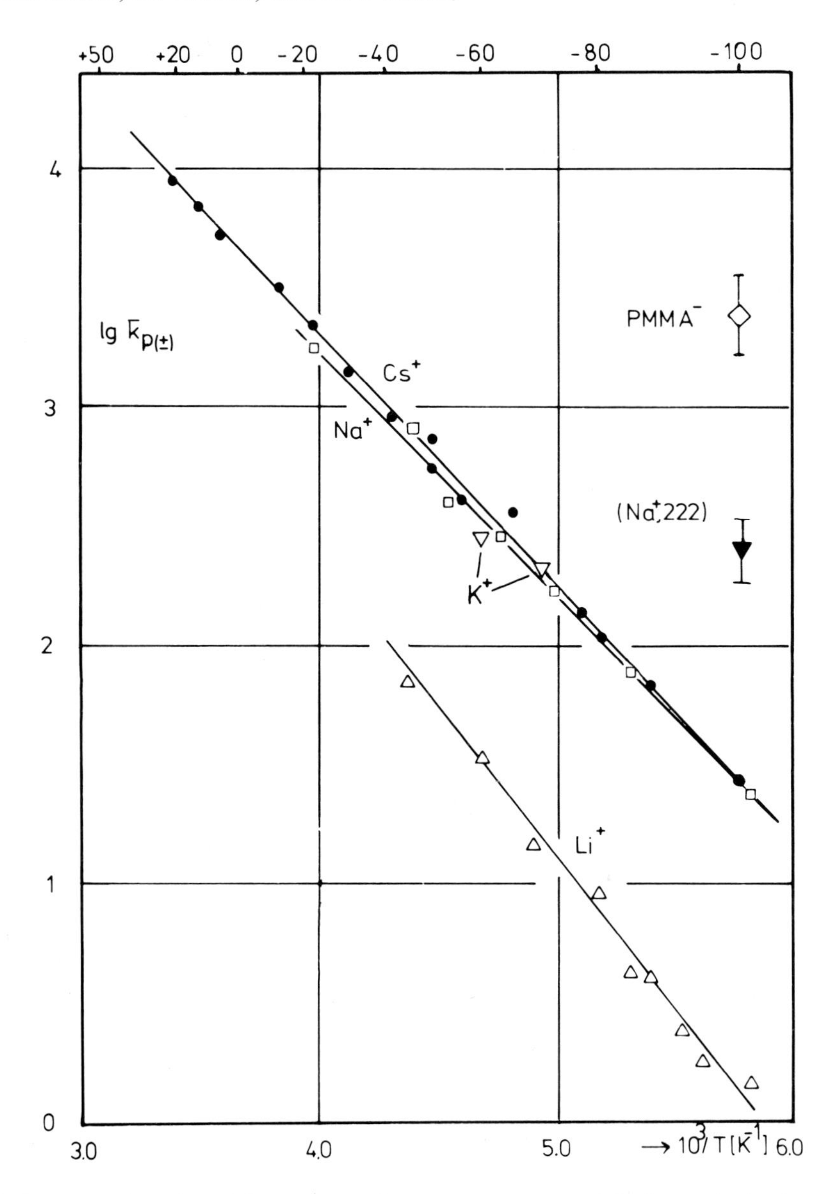

Fig. 8.5: Arrhenius plot of the propagation rate constants of the ion pairs in the anionic polymerization of MMA in THF [31].

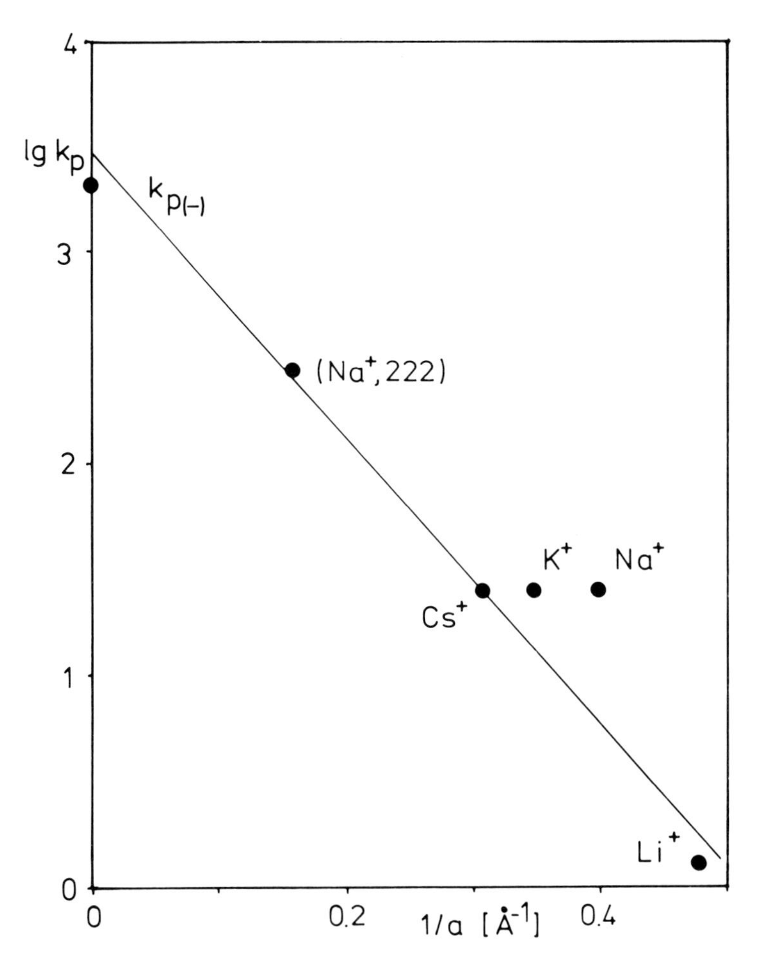

Fig. 8.6: Dependence of the propagation rate constants on the interionic distance in the anionic polymerization of MMA in THF at $-100\,°C$ [31].

These results conform to a model of the active centre described as a contact ion pair peripherally solvated by THF molecules (S denotes solvent):

$$
\begin{array}{l}
\qquad\qquad\qquad CH_3 \\
\qquad\qquad\qquad\ |\ \delta^- \\
\sim\sim CH_2-C \quad\quad\quad CH_3 \\
\qquad\qquad\ \ \vdots \quad C-O \\
\qquad S\cdots Mt^{\oplus}\ \ \| \\
\qquad\qquad\ \ \vdots \quad O_{\delta^-} \\
\qquad\qquad\ \ S
\end{array}
$$

The increased interionic distance of Na^+ and K^+ ion pairs seems to be caused by peripheral solvation spreading the positive charge over a larger area and thus weakening the electrostatic interaction of the cation with the α-carbon and the partially enolized carbonyl oxygen. On the other hand, Li^+ is on the straight line again because of a strong p-π overlap leading to a very tight bond between anion and cation.

It is interesting to note that due to the strong interaction of the counterions with the enolate oxygen, no solvent-separated ion pairs are found.

At higher concentrations the ion pairs tend to form dimeric or higher associates of low reactivity [34]. This leads to decreasing rate constants. The kinetic scheme for such a process is given in Scheme 3, assuming dimeric associates only:

$$\begin{array}{ccc} 2\,P^*_{\pm} & \overset{K_A}{\rightleftharpoons} & (P^*)_2 \\ +M \downarrow k_{\pm} & & +M \downarrow k_a \end{array}$$

Scheme 3

The dependence of the rate constant k_p on the concentration of active centres is given by the rate constant of non-associated ion pairs, $k_{\pm}$, the rate constant of the associates, k_a, and the fraction of non-associated ion pairs, α, which decreases with increasing concentration of active centres:

$$k_p = \alpha \cdot k_{\pm} + [(1-\alpha)/2] \cdot k_a = \tfrac{1}{2}k_a + (k_{\pm} - \tfrac{1}{2}k_a) \cdot \alpha. \qquad (17)$$

$$\alpha = [-1 + (1 + 8 \cdot K_A \cdot c^*)^{1/2}]/(4 \cdot K_A \cdot c^*). \qquad (18)$$

For $K_A \cdot c^* \gg 1$

$$\alpha = (2 \cdot K_A \cdot c^*)^{-1/2} \qquad (19)$$

and for $K_A \cdot c^* \ll 1$

$$\alpha = 1 - 2 \cdot K_A \cdot c^*. \qquad (20)$$

Direct evidence for the existence of associates was obtained from viscosity measurements on "living" and terminated polymers [35] and from vapour pressure osmometry measurements on oligomers [36].

When working with bifunctional initiators (e.g. oligo-α-methylstyryl sodium), the chain ends are very close to each other. Even at very low initiator concentrations associates were found in these systems. The degree of association was observed to decrease with increasing degree of polymerization [37, 38].

Studies on the oligomerization of MMA indicate that coordination of the counterion with the penultimate or ante-penultimate ester groups leads to an active species different to the peripherally solvated contact ion pair [34, 39]:

CH_3 CH_2 δ− CH_3
~~~ CH_2–C C
C $Mt^{\oplus}$ C–OCH_3
CH_3O O S $O_{\delta-}$

⇌

H_3C $COOCH_3$
C–CH_2 δ− CH_3
CH_2 C
C–C C–OCH_3
~~~ CH_2 O··$Mt^{\oplus}$
H_3C OCH_3 S $O_{\delta-}$

Similar structures were proposed by Fowells et al. [40] on the basis of the stereostructure of partially deuterated poly(acrylates) and methacrylates. There is IR-spectroscopic evidence on the existence of these structures [41, 42]. However, NMR [43] and conductivity [44] studies on living oligomers conducted at *lower* temperatures did not give unequivocal results. It appears that "intramolecular solvation" is an endothermic process, due to the lower solvating power of the ester group compared to THF. Thus, the driving force seems to be the entropy gain due to desolvation of THF favouring this process at higher temperatures only.

Mechanism of monomer addition

When working in more polar solvents, (e.g. 1,2-dimethoxyethane; DME), external solvation of the ion pairs presumably increases. Accordingly, the interionic distance in ion pairs should increase, leading to higher rate constants. The opposite is expected for less polar solvents, (e.g. tetrahydropyran; THP). This is true for Na^+ and Cs^+ as counterions [45, 46] but not for Li^+ [47]. Here, the rate constants are even smaller in DME, as compared to THF.

In order to understand this effect, it is necessary to have a closer look at the mechanism of monomer addition. Two possible mechanisms are discussed below. In mechanism I, (cf. Scheme 4) the monomer carbonyl oxygen first coordinates with the counterion displacing a solvent molecule (only one is shown in the scheme). Ion pair-monomer complexes of this type have been proposed for various monomer/solvent systems in order to explain the formation of highly stereoregular polymers [40, 48]. In the transition state the counterion is moved from the active centre to the newly formed one; the counterion is solvated again:

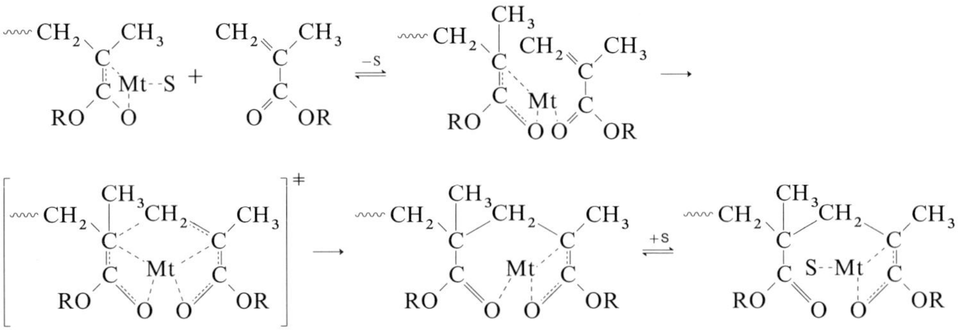

Scheme 4

Li^+ strongly coordinates with the anion as well with the peripheral solvent molecules. Thus, it is difficult for the incoming monomer to displace the solvent, especially for DME which is a bidentate ligand. In this case an alternative mechanism II (cf. Scheme 5) is more likely. Here, the monomer vinyl group directly attacks the carbanion without coordination with the shielded counterion. After the transition state the solvated counterion is transferred to the newly formed anion:

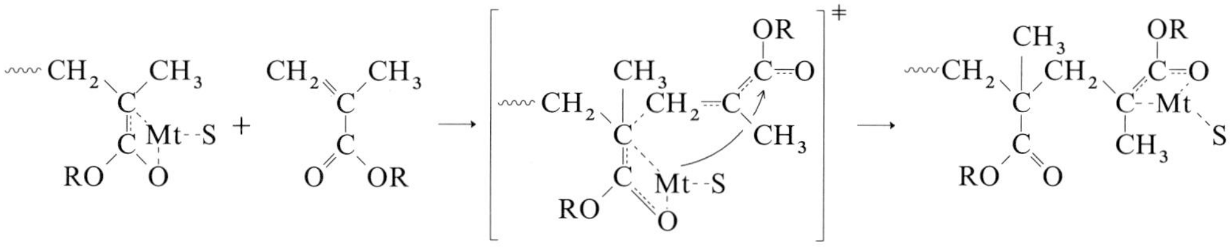

Scheme 5

This step is energetically unfavourable. However, mechanism II is the only choice for polymerizations *via* free anions or ion pairs involving large, nonsolvated counterions, like cryptated sodium (Na^+, 222). Evidence from polymer microstructure *(vide infra)* indicates that this is also the preferred mechanism for Li^+.

Stereochemistry of monomer addition

Mechanisms I and II should differ significantly in their stereoregulating effect. Fowells et al. [40] proposed that mechanism I should lead to a meso placement. For mechanism II (which is valid for Li^+, cryptated Na^+ and the free anion) the stereochemistry is not pre-determined. Because of steric reasons, predominantly racemic placements are found (cf. Fig. 8.7). This is also observed in radical polymerization [49].

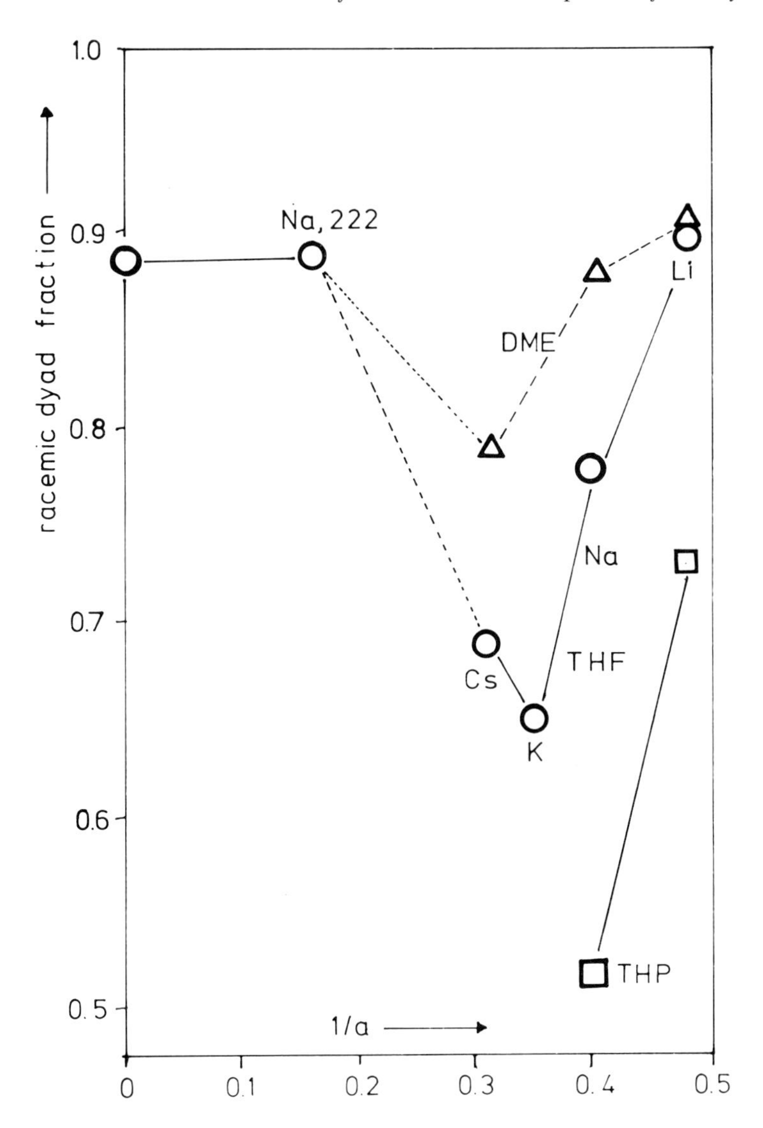

Fig. 8.7: Fractions of racemic dyads in PMMA prepared in various solvents and using different counterions at ca. $-50\,°C$ (free anion and cryptated Na^+ at $-100\,°C$) [29].

For the larger alkali cations the fraction of racemic dyads decreases with increasing interionic distance as well as with decreasing solvating power of the solvent [29]. It is thus probable that in these systems both mechanisms I and II are effective to varying extents, the fraction of monomer additions *via* mechanism I increasing with interionic distance and decreasing with polarity of solvent. These assumptions are corroborated by the fact that the statistics of the tactic placements deviate from a Bernoullian one in the same order.

By terminating the oligomerization of MMA with $^{13}CH_3I$ or $ClSi(CH_3)_3$ it was possible to trap the stereochemistry of the end group. $^{13}CH_3$ methylation allows for the determination of the end group dyads and triads of placements by ^{13}C NMR [50, 51]. Comparison with the main chain tacticity gives information on the applicability of Bernoullian or Markovian statistics.

Silylation of the "living" oligomers occurs selectively at the enolate oxygen and leads to silyl ketene acetals. Thus, E and Z structures are stabilized and their ratio can be determined by ^{13}C NMR [52]. The fine structure of the peaks even allows for the separate determination of the dyads and triads of placements for the E and Z end groups. The fraction of Z end groups drastically changes from 99.7 % (Li^+) to 3 % (Cs^+) on a sharp sigmoidal curve which has its inflection for K^+. For the same cation the meso fraction of the main chain and the deviation from Bernoullian statistics show a maximum, too. This indicates that tacticity is determined by the E/Z stereoisomerism of the chain end.

Side reactions

Side reactions are frequently observed when working at temperatures above −75 °C. They manifest in a broadening of the molecular weight distributions. A downward curvature of the first-order time-conversion curves (cf. Equ. 4) is observed which indicates a decrease in the concentration of active centres c* due to chain termination. Generally, P_n remains proportional to monomer conversion (cf. Equ. 6), indicating the absence of transfer reactions.

Based on these effects it was seen qualitatively that the relative amount of termination increases with decreasing polarity of the solvent and decreasing size of the counterion [28]. In the polymerization of *tert*-butyl methacrylate (TBMA) no termination was detected even at ambient temperature [53].

Three termination reactions have been discussed widely in the literature.

They were first proposed by Schreiber [54]:

a) a reaction of the active centre with the carbonyl group of the monomer resulting in a vinyl ketone ("monomer termination"):

$$\sim\!\!\sim C^{\ominus}(CH_3)(COOR) + CH_2{=}C(CH_3){-}C({=}O){-}OR \longrightarrow CH_2{=}C(CH_3){-}C({=}O){-}C(CH_3)(COOR)\!\sim\!\!\sim + RO^{\ominus}$$

b) reaction of the living end with the ester group of another polymer resulting in chain coupling ("intermolecular polymer termination"):

$$\sim\!\!\sim C^{\ominus}(CH_3)(COOR) + RO{-}\underset{\|}{\overset{}{C}}(O){-}C{-}CH_3 \longrightarrow \sim\!\!\sim C(CH_3)(COOR){-}C(=O){-}C{-}CH_3 + RO^{\ominus}$$

c) reaction of the active centre with the ante-penultimate ester group of its own polymer chain resulting a cyclic β-ketoester structure ("intramolecular polymer termination"; "backbiting"):

$$\begin{array}{c} H_3C\text{–}C(COOR)\text{–}CH_2\text{–}C^{\ominus}(CH_3)(COOR) \\ CH_2\text{–}C(H_3C)(\sim\!\!\sim CH_2)\text{–}C(=O)OR \end{array} \longrightarrow \begin{array}{c} H_3C\text{–}C(COOR)\text{–}CH_2\text{–}C(CH_3)(COOR) \\ CH_2\text{–}C(H_3C)(\sim\!\!\sim CH_2)\text{–}C(=O) \end{array} + RO^{\ominus}$$

All these reactions produce alkoxide as a by-product. Wiles and Bywater [55, 56, 57] and Mita et al. [25] showed that most of the methoxide is produced in the *initial* stage of the polymerization of MMA. This can be explained in two ways. Either the methoxide is formed by the initiator attacking the carbonyl group of the monomer *or* by an increased tendency of living *oligomers* to undergo any type of termination. Both explanations are correct, as will be seen below.

The vinyl ketone resulting from the attack of initiator onto the carbonyl group of the monomer (a), as well as the carbinol formed by subsequent addition of a second initiator molecule to the keto group was detected by Schreiber [54], Kawabata and Tsuruta [58] and Hatada et al. [59, 60, 61]. The vinyl ketone may as well be added to a living polymer *via* its vinyl group. This decreases the reactivity of the active centre leading to a "dormant" species which only eventually will be reactivated by adding another methacrylate monomer [60, 62].

Consequently, the choice of the initiator is of utmost importance. Highly reactive initiators, e.g. n-butyllithium lead to side products due to the attack of the monomer carbonyl group. Thus, initiators of higher delocalization and steric hindrance, like 1,1-diphenylhexyllithium or metalloesters (being models of the active centre [63]) are more favourable.

Whereas there is no evidence for *inter*molecular termination (b), the product of *intra*molecular termination (or "backbiting"), i.e. the cyclic β-ketoester end group could be detected by its characteristic IR absorption [64, 65]. The termination product of the living trimer was isolated [63, 66]. This makes "backbiting" the most probable termination mechanism.

By kinetically analyzing the product distribution of linear and cyclic oligomers formed in the oligomerization of MMA using Li^+ as the counterion in THF at 25 °C, the rate constants of cyclization were determined as a function of the degree of polymerization [39]. It was found that the rate of cyclization drops from the trimer to the tetramer by a factor of ca. 50 and does not change significantly for the higher oligomers. This was attributed to steric effects of the pendant monomer units in the formation of the cyclic β-ketoester structure. This effect partially explains the above observation that most of the methoxide is formed in the initial stage of the polymerization.

Earlier investigations showed that addition of alkoxides, especially *tert*-butoxide, decreases the effect of termination [56, 57]. Recently, it was shown that this effect is due to a decrease of the cyclization rate constants by a factor of ca. 100 and of the polymerization rate constants by a factor of only 10 [68]. It was assumed that alkoxide addition changes the structure of the active centre, i.e. the Li^+ cation may change from intramolecular coordination to coordination with lithium alkoxide molecules.

Warzelhan et al. [32] investigated the kinetics of termination by labelling the active polymers with tritiated acetic acid (CH_3COO^3H). A first-order decrease of c^* was found during polymerization using Na^+ as the counterion in THF at temperatures below −40 °C. This is in accordance with intramolecular termination.

A very thorough re-investigation and extension of Warzelhan's kinetic work by Gerner et al. [69] showed that the termination mechanism at temperatures below −40 °C is more complicated. Determination of c^* for long times after complete monomer consumption unexpectedly revealed that only a fraction of the polymer chains becomes terminated. This fraction, as well as the rate of termination depends on the *initial* monomer concentration. None of the termination mechanisms discussed above (including combinations thereof) is able to explain these results. A new mechanism for termination was proposed which is based on the assumption that a deactivating species is formed in the initial step of the polymerization. This species reacts with the living polymers during polymerization, leading to termination in a second-order reaction. The corresponding Arrhenius plots are linear and do not significantly differ for both THF and THP as solvents, indicating that the higher extent of termination in the latter solvent is due to the lower propagation rate constants.

Unfortunately, the exact nature of the deactivating species is unknown. Its initial concentration is directly proportional to the initial concentrations of both monomer and active centres. GPC and GC/MS analysis of oligomers prepared at −35 °C and end-labelled with ^3H revealed the existence of several by-products for each regular oligomer, one of which possibly being the product of termination by the deactivating species [46, 70]. From the GC/MS data it was not possible to assign a molecular structure to this series. However it seems to be linear and to contain one initiator fragment per chain.

There is very little evidence for *transfer reactions* occuring in methacrylate polymerization. Müller et al. [71] reported that the caesium methoxide formed in the termination of PMMA-Cs in THF at ambient temperature is able to initiate the polymerization of MMA. The initiation rate, however, is very low, so that this effect only gains importance when all the polymer chains started initially have been terminated before complete monomer conversion. The newly initiated polymer chains again can suffer from termination, producing methoxide, *etc.* This is equivalent to a spontaneous transfer reaction.

8.3.3 Other Methacrylates

8.3.3.1 Homopolymerization

Ester groups different from methyl may exert steric and/or electronic effects on the rates of polymerization. Whereas steric effects should always lead to lower rate constants, the electronic effects are difficult to predict. Electron-donating groups, e.g. *tert*-butyl, decrease the polarization of the monomer and hence its reactivity. On the other hand, they increase the charge density on the carbanion increasing its reactivity. The latter effect may be partially cancelled by ion pairing effects; the increase in charge density in turn may lead to smaller interionic distances which decreases rate constants again (cf. 8.2.1).

Table 8.1 shows the rate constants and Arrhenius parameters for the polymerization of different methacrylates in THF using Na^+ as the counterion. Taft's σ^* parameter [72] is given as a measure for the electron-withdrawing (–I) effect of the substituent.

Table 8.1: Propagation rate constants and Arrhenius parameters in the anionic polymerization of methacrylates in THF using Na^+ as the counterion [29, 74].

Substituent	σ^*	E_a / kJ/mol	log A	k_p^{198} / l/mol · s	Symbol
decyl	−0.12	15.8	6.5	211	✧
methyl [30]	0.00	19.5	7.3	145	□
n-butyl	−0.13	23.6	8.3	128	⬣
ethyl	−0.10	16.5	6.4	100	○
benzyl	+0.23	19.1	7.0	81	●
i-propyl	−0.20	19.5	7.0	72	△
2-ethoxyethyl	−0.17	26.0	8.4	38	▼
t-butyl [53]	−0.32	31.2	8.7	3.0	■
phenyl	+0.60	15.1	4.1	1.4	⬡

The poor correlation between the relative rate constants and Taft's σ^* parameter indicates that the substituent effect is too complex to be described by one parameter only. On the other hand *tert*-butyl methacrylate (TBMA) has the highest frequency exponent (i.e. least steric requirements in the transition state). This indicates that steric effects seem to be of minor importance. Moreover, the relatively bulky decyl methacrylate has even higher rate constants than MMA.

In order to gain further information it is necessary to determine separately the relative reactivities of monomers towards a given polymer anion and the relative reactivities of different polymer anions towards a given monomer. This can be done either by direct determination of the rate constants k_{12} and k_{21} [73] or by copolymerization experiments.

8.3.3.2 Copolymerization

Assuming the "terminal" model for the kinetics of copolymerization, the copolymerization parameters

$$r_1 = k_{11}/k_{12} \text{ and } r_2 = k_{22}/k_{21}$$

give information only for the relative reactivities of *monomers* adding to a given "living" polymer. The relative reactivities of the *active centres* are given by the parameters

$$q_1 = k_{11}/k_{21} \text{ and } q_2 = k_{22}/k_{12} \,,$$

which cannot be calculated from the copolymerization parameters alone. The homopolymerization rate constants are necessary in order to calculate these parameters.

A novel method was used in order to determine k_{12} and k_{21} [74]. The differential equations for the consumption of the two monomers

$$-\frac{d[M_1]}{dt} = (k_{11} \cdot [P_1^*] + k_{21} \cdot [P_2^*]) \cdot [M_1] \tag{21}$$

$$-\frac{d[M_2]}{dt} = (k_{22} \cdot [P_2^*] + k_{12} \cdot [P_1^*]) \cdot [M_2] \tag{22}$$

and for the change in the concentration of active centres P_1^*

$$\frac{d[P_1^*]}{dt} = k_{21} \cdot [M_1] \cdot c^* - (k_{21} \cdot [M_1] + k_{12} \cdot [M_2]) \cdot [P_1^*] \tag{23}$$

were integrated numerically. By varying k_{12} and k_{21}, the calculated time-conversion curves were fitted to the experimental data. Thus, it is possible to investigate the copolymerization kinetics up to high conversions.

All experiments were performed in THF at $-75\,°C$, using Na^+ as the counterion. For most monomers, the "terminal" model is sufficient, i.e. the time conversion curves can be described by four rate constants. Only for *iso*-propyl methacrylate and TBMA the terminal model gives poorer fits probably due to penultimate effects. However, our data are not sufficient to calculate the eight rate constants describing such a model.

Fig. 8.8 shows the Hammett plots for the relative reactivities of the monomers towards $PMMA^-Na^+$ ($1/r_1 = k_{12}/k_{11}$) and of the living polymers towards MMA ($1/q_1 = k_{21}/k_{11}$) as a function of Taft's σ^* parameter:

$$\log(1/r_1) = \log(k_{12}/k_{11}) = \varrho^*_{PMMA-Na} \cdot \sigma^*$$
$$\log(1/q_1) = \log(k_{21}/k_{11}) = \varrho^*_{MMA} \cdot \sigma^*$$

Obviously, there is a good correlation for the relative reactivity of the monomers adding to $PMMA^-Na^+$ ion pairs ($\varrho^*_{PMMA-Na} = 2.2$) indicating that the influence of the ester group on the reactivity of the monomer is predominantly electronic.

For the relative reactivity of the active centres towards MMA the correlation is worse. $PTBMA^-Na^+$ deviates by one order of magnitude from the straight line ($\varrho^*_{MMA} \approx -2.0$) indicating that the influence of the ester group on the reactivity of the active centre, though mainly electronic, is more complex.

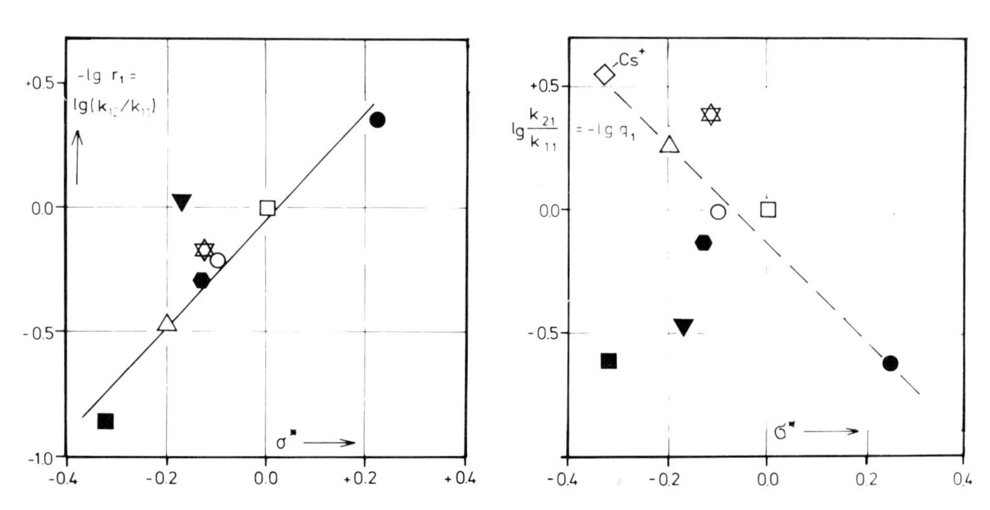

Fig. 8.8: Dependence of the relative reactivities of (a) different monomers and (b) different polymer anions on Taft's σ^* parameter in the anionic copolymerization of different methacrylates with MMA in THF at $-75\,°C$ using Na^+ as the counterion [29, 74]. Symbols, cf. Table 8.1.

Except for TBMA, the effects of the ester group on monomer and active centre roughly cancel each other. This explains the rather small influence of the ester group on the homopolymerization kinetics for most methacrylates.

Further experiments showed that the reactivity ratios do not vary much with temperature but strongly depend on the counterion. Selectivity increases with increasing ionic radius. This is unexpected since the reactivity of the ion pairs also shows the same trend. This effect, as well as the unexpectedly low reactivity of the $PTBMA^{-}Na^{+}$ ion pairs, may be due to different mechanisms and stereochemistry of monomer addition. Further work is necessary to clarify this point.

8.3.4 Polymerization in Non-Polar Solvents

The investigation of the kinetics of polymerization in non-polar solvents is of special interest, because only in these media isotactic polymers can be prepared. However, it is known that in non-polar systems, like toluene, the polymerization of MMA is very complex, resulting from severe side reactions and a multiplicity of active species (for reviews, see [10, 20, 21, 22]). Consequently, very broad, multimodal molecular weight distributions were reported [57, 75]. Only the use of some Grignard initiators leads to narrow distributions with MMA [61, 76]. However, these initiators are not favourable for non-polar monomers making the preparation of block copolymers problematic.

On the other hand it was found that *tert*-butyl methacrylate (TBMA) completely lacks any side reactions during its polymerization in THF [53]. Thus it is no problem to prepare polymers or block copolymers of high molecular weight and narrow molecular weight distribution (cf. Fig. 8.9 a, b). Furthermore, PTBMA is easily hydrolized to poly(methacrylic acid) thus rendering the possibility to prepare narrowly distributed polyelectrolytes or copolymers having electrolyte and non-electrolyte blocks. These polymers have very interesting properties, e.g. as emulsifiers or membrane precursors.

Consequently TBMA appears to be a more suitable monomer for non-polar solvents than MMA. The polymerization of TBMA in toluene in fact is very different from that of MMA. When using diphenylhexyl-Li or ethyl α-lithio isobutyrate as initiators first-order kinetics with respect to monomer and initiator and narrow molecular weight distributions ($M_w/M_n \geq 1.15$; cf. Fig. 8.9c) were found [29, 77]. The rate constants are *higher* by one order of magnitude than those in THF, indicating that all species exist as non-associated, unsolvated contact ion pairs lacking the steric hindrance of peripheral solvation. Consequently, addition of lithium *tert*-butoxide which is presumably breaks associates by complexation with the ion pairs [67, 78], has no effect on kinetics, MWD, and tacticity.

In the contrast to MMA which results only ca. 90 % isotactic dyads in the polymers under these conditions, the PTBMAs prepared are 100 % isotactic. The latter observation conforms with mechanism I brought forward in section 3.2 (Scheme 4), i.e. complex formation between monomer and cation prior to addition.

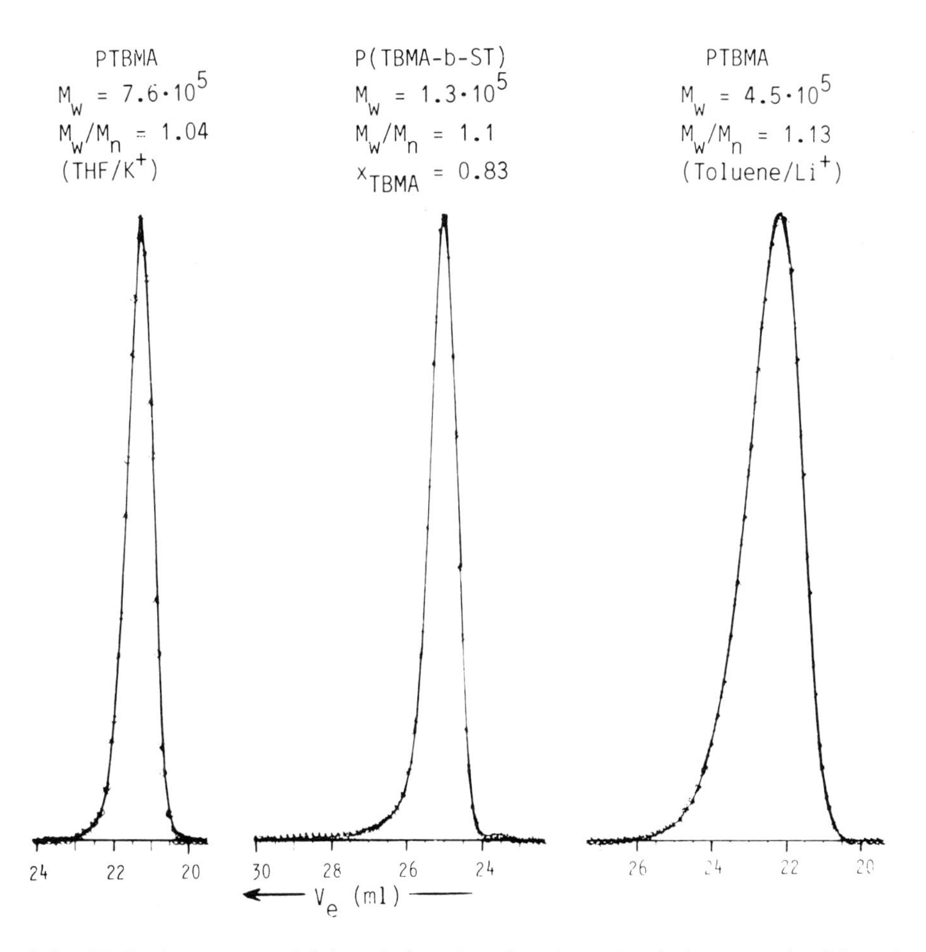

Fig. 8.9: GPC eluograms of (a) poly(*tert*-butylmethacrylate) (PTBMA), (b) poly-(styrene-*b*-TBMA) prepared in THF, (c) PTBMA prepared in toluene. Counterions; (a, c) Li^+, (b) K^+ [29, 77].

A. H. E. Müller, H. Höcker, and G.V. Schulz

8.4 Group Transfer Polymerization of Methacrylates

8.4.1 Introduction

Group transfer polymerization (GTP) is a new polymerization process discovered and patented by E. I. du Pont de Nemours & Company in 1983 [3, 79,

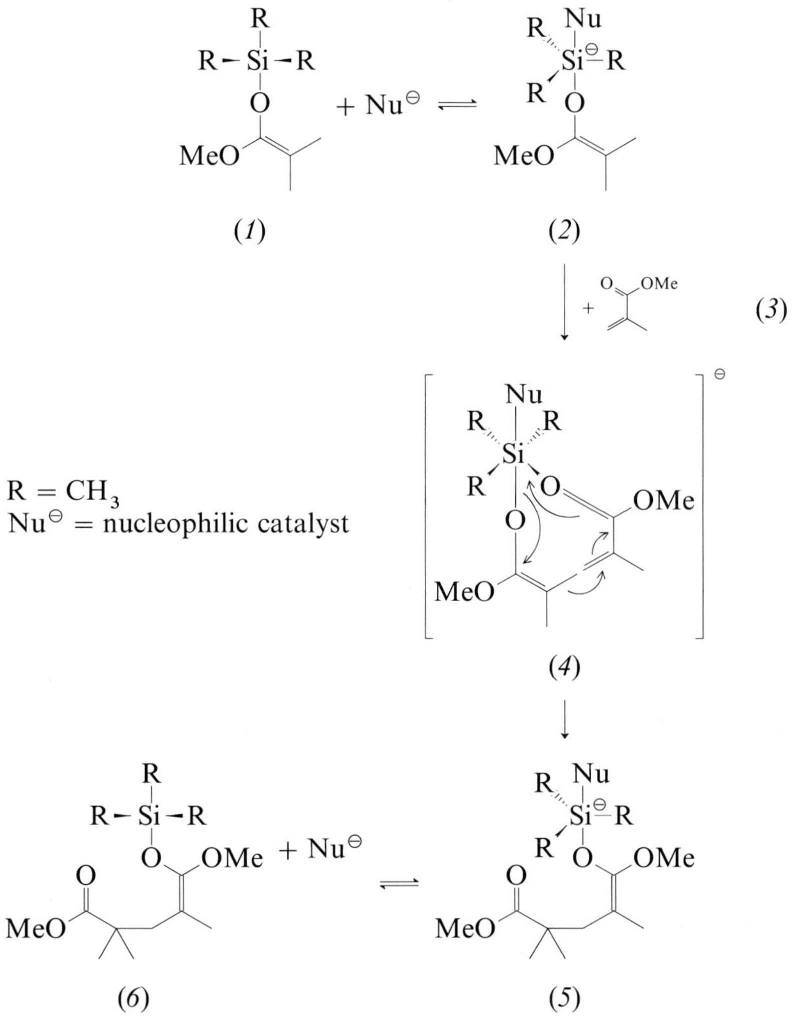

R
R – Si – R
O + $Nu^{\ominus}$ ⇌
MeO

R Nu
$Si^{\ominus}$ R
R O
MeO

(*1*) (*2*)

+ O OMe (*3*)

Nu
R R
Si
R O
O OMe
MeO
$\ominus$

$R = CH_3$
$Nu^{\ominus}$ = nucleophilic catalyst

(*4*)

R
R – Si – R
O OMe + $Nu^{\ominus}$
O
MeO

⇌

R Nu
$Si^{\ominus}$ R
R O OMe
O
MeO

(*6*) (*5*)

Scheme 6

80]. It allows the polymerization of polar aprotic monomers, such as methacrylates and acrylates by initiation with silyl compounds, such as ketene silyl acetals, in the presence of a nucleophilic or electrophilic catalyst.

Especially for methacrylates, GTP combines the important advantages of "living" polymerization with the added advantage that the reaction can be carried out at ambient or above temperatures.

Only few studies have been published, so far, on the mechanism of nucleophilic catalysis. An associative mechanism was proposed by the du Pont group [3, 81, 82]. It is outlined in Scheme 6.

The nucleophilic catalyst (e.g. HF_2^-) coordinates to the silicon atom of the initiator, 1-methoxy-1-(trimethylsiloxy)-2-methyl-1-propene 1 (MTS) to provide a pentacoordinate species 2. The activated initiator and monomer 3 are proposed to form a hypervalent silicon intermediate 4. A new C-C bond is created between initiator and monomer and the trimethylsilyl group is transferred to the carbonyl oxygen of the monomer under formation of a new Si-O bond and cleavage of the old Si-O bond. This transfer regenerates a structure similar to that of the initiator. The propagation reaction 2 ⟶ 5 should proceed as long as monomer is present. However, there is no direct evidence for a "concerted" mechanism symbolized by structure 4.

8.4.2 General Considerations

A kinetic scheme for chain propagation which is consistent with the mechanism is shown in Scheme 7.

$$
\begin{array}{ccc}
C + P_i & \overset{K^*}{\rightleftharpoons} & P_i^* \\
 & & \quad \downarrow k_p \; (+M) \\
C + P_{i+1} & \overset{K^*}{\rightleftharpoons} & P_{i+1}^*
\end{array}
$$

P_i = "dormant" polymer of DP = i, P_i^* = activated polymer,
C = catalyst, M = monomer.

Scheme 7

As was shown in section 8.1 (Equs. 4 and 5), the rate of polymerization is first order with respect to monomer concentration and the apparent rate constant k_{app} (i.e the slope of a first-order plot of conversion vs. time) is given by

$$k_{app} = k_p [P^*]. \tag{24}$$

However, in GTP the concentration of active centres [P*] is not equal to the initiator concentration $[I]_0$, because the concentration of catalyst is much lower than that of the initiator and thus only a small fraction of the chains is active.

Application of the mass action law renders [83]

$$[P^*] = \frac{K^* \cdot [I]_0}{1 + K^* \cdot [I]_0} \cdot [C]_0 \tag{25}$$

and

$$k_{app} = k_p \cdot \frac{K^* \cdot [I]_0}{1 + K^* \cdot [I]_0} \cdot [C]_0. \tag{26}$$

Thus the reaction is expected to follow first-order kinetics with respect to catalyst concentration. The kinetic order with respect to initiator is fractional and depends on the product $K^* \cdot [I]_0$. Two limiting cases can be discussed:

a) $K^* \cdot [I]_0 \gg 1$, i.e. the activation equilibrium is shifted to the right. This leads to

$$[P]^* = [C]_0$$

and

$$k_{app} = k_p \cdot [C]_0. \tag{27a}$$

Here, the reaction is of zeroth order with respect to initiator concentration.

b) $K^* \cdot [I]_0 \ll 1$, i.e. the activation equilibrium is shifted to the left. This leads to

$$[P]^* = K^* \cdot [I]_0 \cdot [C]_0$$

and

$$k_{app} = k_p \cdot K^* \cdot [I]_0 \cdot [C]_0 \cdot = k_p' \cdot [C]_0 \tag{27b}$$

Here, the reaction is first order with respect to initiator and the "pseudo" rate constant of propagation, k_p', is not equal to the "true" rate constant, k_p:

$$k_p' = k_{app}/[C]_0 = k_p \cdot K^* \cdot [I]_0 \ll k_p. \tag{28}$$

8.4.3 Kinetic Investigations

The kinetics as well as the tacticities of the resulting polymers in the GTP of methyl methacrylate (MMA) and *tert*-butyl methacrylate (TBMA) were investigated in THF using 1-methoxy-1-(trimethylsiloxy)-2-methyl-1-propene (MTS, 1) as the initiator. Tris(dimethylamino)sulfonium bifluoride (TAS HF_2), tetrabutyl-ammonium bibenzoate (TBA BB), and TAS benzoate (TAS B) were used as catalysts.

Experiments using MMA as the monomer and TAS HF_2 as the catalyst [83, 84, 85] showed that the curves for monomer conversion vs. time, as plotted for a first-order dependence, are linear at ambient temperature (cf. Fig. 8.10). Depending on reaction conditions (most pronounced at low temperatures) "induction periods" are observed. At higher conversions, the rate of polymerization often decreases.

The propagation reaction is first order with respect to both initial monomer and catalyst concentration and nearly zeroth order with respect to initiator concentration. This is in accordance with Scheme 6, the equilibrium being shifted to the right-hand side ($K^* \cdot [I]_0 \gg 1$).

An Arrhenius plot for the GTP of MMA is given in Fig. 8.11. The data are compared to those obtained in anionic polymerization. The activation parameters for the propagation rate constants [84] as well as the tacticities [49, 86] are very similar to those in anionic polymerization using bulky counter-ions [29] (cf. 8.3.2). It was concluded that the mechanism of monomer addition is similar to that in anionic polymerization, i.e. the monomer vinyl group is first added to the carbon α to the enolized carbonyl group, then the trimethylsilyl group is transferred to the newly formed chain end (cf. Scheme 8).

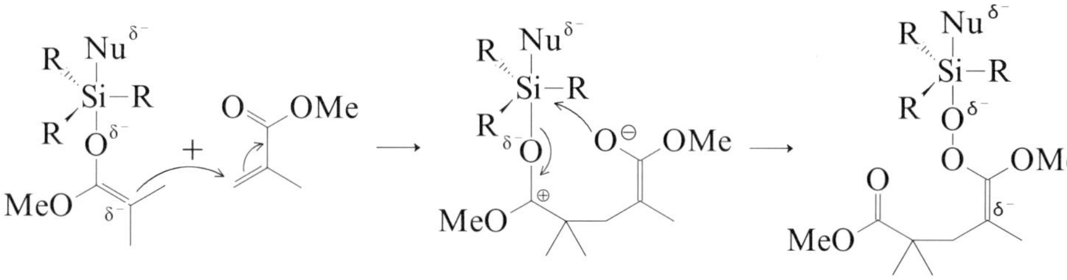

Scheme 8

This mechanism of monomer addition is now generally accepted [86, 87].

The GPT of TBMA is very similar to that of MMA. However, beside being slower by a factor of ca. 2, it differs in two details. "Introduction periods" are observed at very low temperatures only, but the first-order time-conversion plots strongly deviate from linearity. Depending on catalyst concentration, this may lead to incomplete monomer conversion [88, 89].

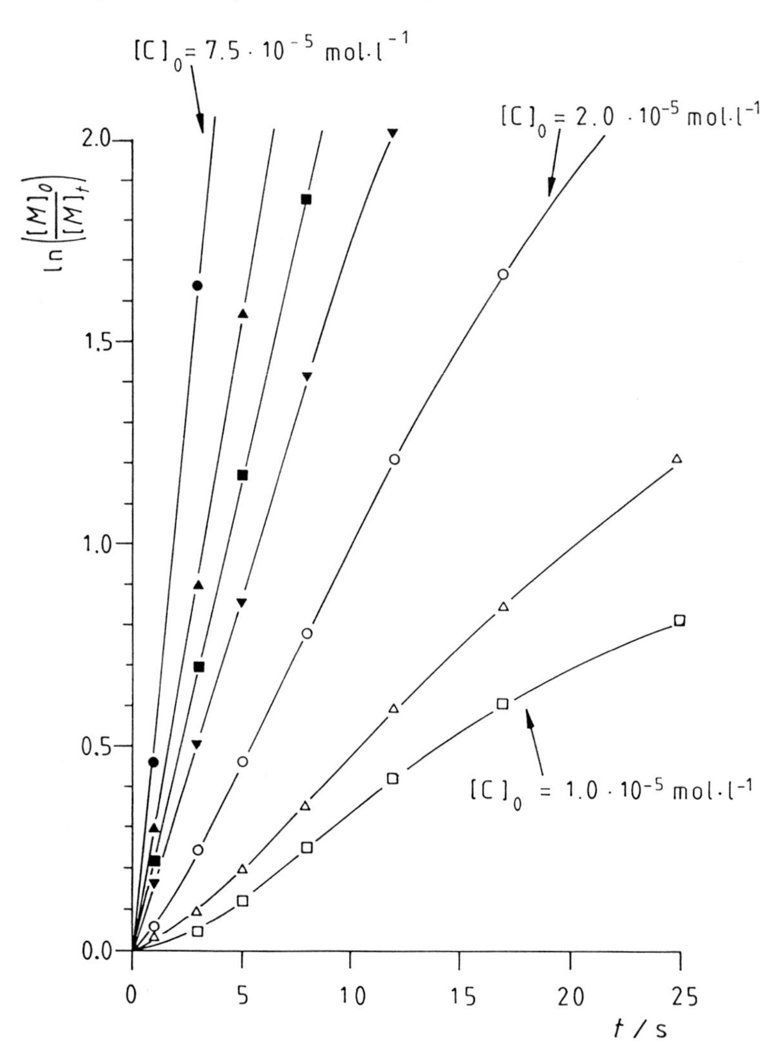

Fig. 8.10: First-order time-conversion plots for the group transfer polymerization (GTP) of MMA as a function of TAS HF_2 catalyst concentration at 20 °C [83].

This "termination" effect is due to a deactivation of catalyst by either the silyl ketene acetal ended initiator, polymer, or both. Thus, addition of a second catalyst dose reactivates an apparently "terminated" polymerization (cf. Fig. 8.12).

The oxyanion-catalyzed GTP of MMA [89] seems to be little disturbed by termination. However, "induction periods" are very pronounced here. The polymerization is first order with respect to monomer concentration, but the reaction order with respect to catalyst concentration is complex. It ranges from

−0.3 to +1.0, depending on reaction conditions and nature of catalyst. The reaction order with respect to initiator concentration is 1.0, indicating that the activation equilibrium is shifted to the left-hand side ($K^* \cdot [I]_0 \ll 1$). Due to the smaller number of activated polymer chains the reaction is considerably slower as compared to bifluoride catalysis. Presently, the rate constants of propagation cannot be determined because the equilibrium constants K^* are not known and due to the variations in the reaction order with respect to catalyst.

The "induction periods" found in bifluoride and benzoate catalysis possibly are due to slow initiation. When using the dimer 6 as the initiator linear time-conversion plots are observed [90, 91]. However, it is not clear whether this effect is due to the rate constant, k_i, or activation equilibrium constant, K_I^*, being lower than the corresponding constants for propagation. This question is currently under investigation.

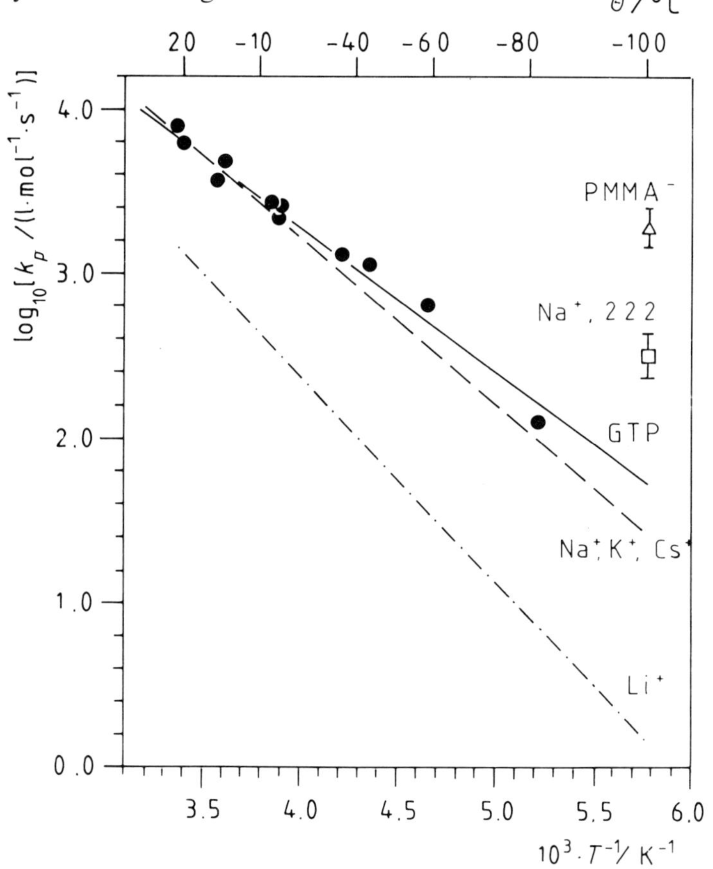

Fig. 8.11: Arrhenius plot for the polymerization rate constants in the group transfer polymerization of MMA with MTS and $TASHF_2$ in THF (●) [84] and in the anionic polymerization of MMA in THF using different counterions [29].

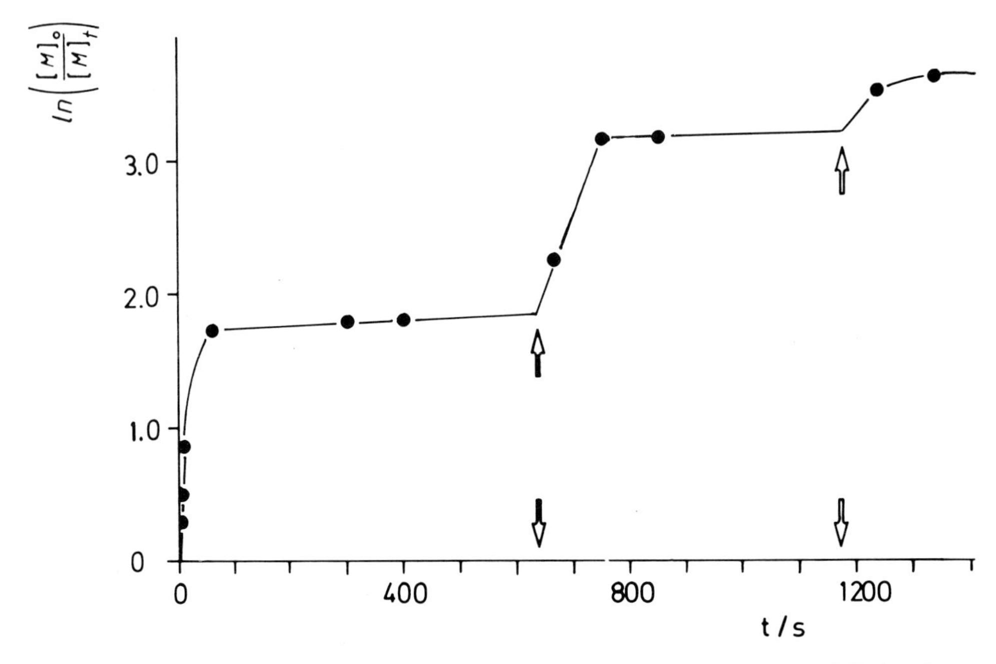

Fig. 8.12: Time-conversion curve for the GTP of TBMA catalyzed by TAS HF_2, where additional catalyst solution is added at 645 and 1190 seconds [88, 89].

8.4.4 Control of Molecular Weight Distribution

During our kinetic investigation, it was observed that the resulting molecular weight distribution (MWD) is influenced to a large extent by the mode of monomer addition into the reaction system [85, 89]. Du Pont's procedure, which involves a slow monomer feed, has claimed polydispersities as low as $M_w/M_n = 1.05$ [79, 3]. Such introduction of monomer guarantees a low monomer concentration throughout the course of the reaction.

In order to carry out kinetic studies, it is necessary that all reagents are thoroughly mixed over the shortest time span possible (the stirred tank reactor used in our laboratory ensures mixing times shorter than 0.4 s). Under such conditions the monomer concentration is initially large but decreases with consumption. The resulting MWD's for the polymers obtained from the kinetic investigation were consistently broader ($M_w/M_n \geq 1.25$) than those reported by du Pont. Moreover, the polydispersity drastically decreases with monomer conversion, x_p (cf. Fig. 8.13). There is also a significant deviation from theory in the plot of the number-average degree of polymerization, P_n, versus x_p.

This indicates that the mechanism of polymerization is more complicated than that of anionic polymerization. It was proposed that these effects are due to rate of catalyst exchange between "dormant" and "living" polymers [85]:

$$C + P_i \underset{k_{-1}}{\overset{k_1}{\rightleftharpoons}} P_i^* \qquad K^* = k_1/k_{-1}$$

$$P_i^* + M \xrightarrow{k_p}$$

$$C + P_{i+1} \underset{k_{-1}}{\overset{k_1}{\rightleftharpoons}} P_{i+1}^*$$

Scheme 9

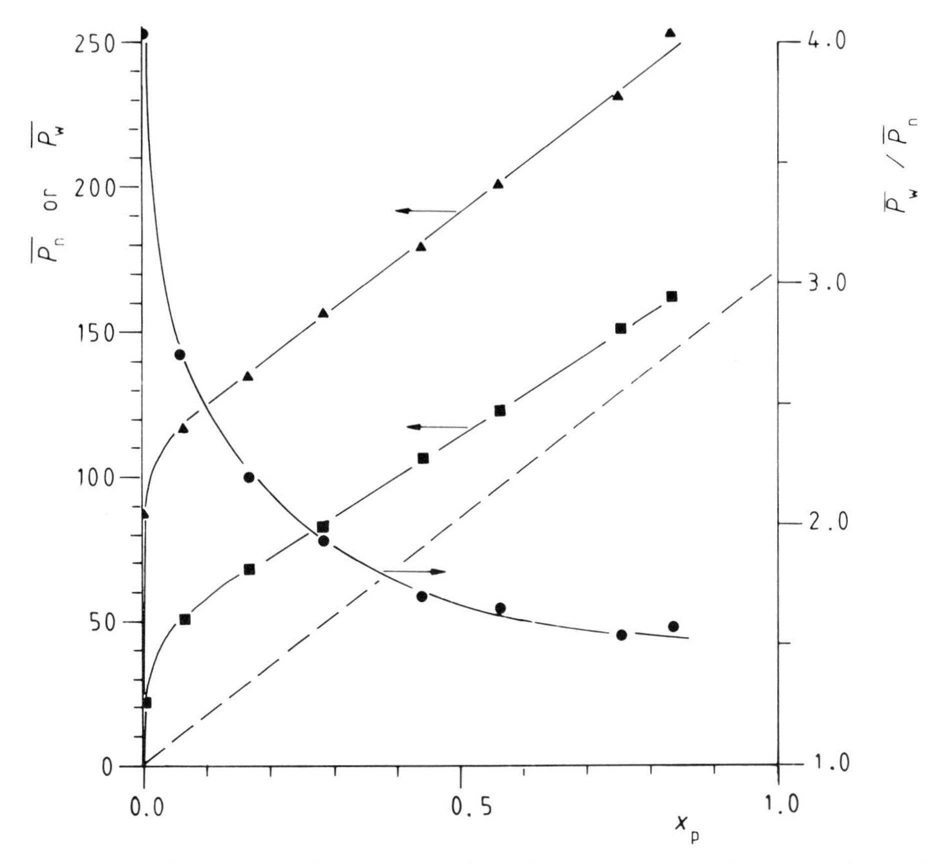

Fig. 8.13: Number and weight averages of the degree of polymerization and polydispersity as a function of monomer conversion in the GTP of MMA using TAS HF_2 as the catalyst (---; expected for P_n) [85, 89].

It was demonstrated [85] that the MWD depends on the ratio $k_p \cdot [M]/k_{-1}$. This means it is dependent on the *actual* monomer concentration. Narrow MWD's are only obtained for $k_p \cdot [M] \ll k_{-1}$, i.e. especially at low monomer concentrations.

When monomer is added using the du Pont procedure, the given equilibrium involving the catalyst exchange between dormant and activated silyl ketene acetal end groups is established rapid enough to ensure that all chains propagate in an even manner. This accounts for the reported narrow MWD's. In our kinetic experiments, we believe the high initial monomer concentration literally swamps the activated chain ends resulting in the addition of several monomer units before the catalyst can equilibrate to another dormant end group. Over the course of the reaction, this effect would diminish due to the contiual consumption of monomer. However, it is the initially uneven incorporation of monomer that results in the observed broadening of the MWD.

8.5 References

[1] M. Szwarc: Living polymers, Nature (London) 178 (1956) 1168.

[2] M. Miyamoto, M. Sawamoto, T. Higashimura: Living polymerization of isobutyl vinyl ether with the hydrogen iodide/iodine initiating system, Macromolecules 17 (1984) 265.

[3] O.W. Webster, W. R. Hertler, D.Y. Sogah, W. B. Farnham, T.V. RajanBabu: Group transfer polymerization, 1: A new concept for addition polymerization with organosilicon initiators, J. Am. Chem. Soc. 105 (1983) 5706.

[4] L. R. Gillion, R. H. Grubbs: Titanocyclobutanes derived from strained cyclic olefins: The living polymerization of norbonene, J. Am. Chem. Soc. 108 (1986) 733.

[5] P. J. Flory: Molecular size distribution in ethylene oxide polymers, J. Am. Chem. Soc. 62 (1940) 1561.

[6] G. Löhr, B. J. Schmitt, G.V. Schulz: Das Strömungsrohr als kontinuierlicher Reaktor zur kinetischen Untersuchung schneller Polymerisationsprozesse, Z. Phys. Chem. (Frankfurt am Main) 78 (1972) 177.

[7] V. Warzelhan, G. Löhr, H. Höcker, G.V. Schulz: An automatically controlled reaction vessel suitable for the kinetic investigation of polymerization processes with half life periods of 2 seconds and greater, Makromol. Chem. 179 (1978) 2211.

[8] D. J. Worsfold, S. Bywater: Anionic polymerization of α-methylstyrene, II: Kinetics, Can. J. Chem. 36 (1958) 1141.

[9] L. L. Böhm, M. Chmelir, G. Löhr, B. J. Schmitt, G.V. Schulz: Zustände und Reaktionen des Carbanions bei der anionischen Polymerisation des Styrols, Adv. Polym. Sci. 9 (1972) 1.

[10] M. Szwarc: Living polymers and mechanisms of anionic polymerization, Adv. Polym. Sci. 49 (1983) 1.

[11] D. N. Bhattacharyya, C. L. Lee, J. Smid, M. Szwarc: The absolute rate constants of anionic polymerization by free ions and ion pairs of living polystyrene, Polymer 5 (1964) 54.
[12] D. N. Bhattacharyya, C. L. Lee, J. Smid, M. Szwarc: Reactivities and conductivities of ions and ion-pairs in polymerization processes, J. Phys. Chem. 69 (1965) 612.
[13] H. Hostalka, R.V. Figini, G.V. Schulz: Zur anionischen Polymerisation des Styrols in THF, Makromol. Chem. 71 (1964) 198.
[14] H. Hostalka, G.V. Schulz: Zur Kinetik anionischer Polymerisationsvorgänge, II: Bestimmung der Geschwindigkeitskonstanten für die Monomer-Addition an das Polystyryl-Anion und das Ionenpaar Polystyryl-Natrium in THF, Z. Phys. Chem. (Frankfurt am Main) 45 (1965) 286.
[15] B. J. Schmitt, G.V. Schulz: The influence of polar solvents on ions and ion pairs in the anionic polymerization of styrene, Eur. Polym. J. 11 (1975) 119.
[16] W. K. R. Barnikol, G.V. Schulz: Zur Kinetik anionischer Polymerisationsvorgänge, III: Die Polymerisation von Styrol in THP, Z. Phys. Chem. (Frankfurt am Main) 47 (1965) 89.
[17] B. J. Schmitt, G.V. Schulz: Vollständige Kinetik der anionischen Polymerisation von Polystyrylnatrium in THF, Makromol. Chem. 142 (1971) 325.
[18] T. E. Hogen-Esch, J. Smid: Studies of contact and solvent-separated ion pairs of carbanions, I: Effect of temperature, counterion and solvent, J. Am. Chem. Soc. 88 (1966) 307.
[19] M. Chmelir, G.V. Schulz: Leitfähigkeitsmessungen an Polystyrylnatrium in DME und THF, Ber. Bunsenges. Phys. Chem. 75 (1971) 830.
[20] D. M. Wiles: Polymerization of α,β-unsaturated carbonyl compounds, in T. Tsuruta, K. F. O'Driscoll: Structure and Mechanism in Vinyl Polymerization, Marcel Dekker Inc., New York 1969, 223.
[21] M. Szwarc: Ion and ion pairs in ionic polymerization, in M. Szwarc: Ions and Ion Pairs in Organic Reactions, Vol. 2. Wiley Interscience, New York 1974, p. 375.
[22] S. Bywater: Anionic polymerization of olefins in C. H. Bamford, C. F. H. Tipper: Comprehensive Chemical Kinetics, Vol 15, Elsevier, Amsterdam 1976, 1.
[23] A. Roig, J. E. Figueruelo, E. Llano: Monodisperse stereoregular poly(methyl methacrylate) by anionic polymerization, J. Polym. Sci., Part B 3 (1965) 171.
[24] A. Roig, J. E. Figueruelo, E. Llano: Monodisperse stereoregular poly(methyl methacrylate) by anionic polymerization II., J. Polym. Sci., Part C 16 (1965) 4141.
[25] I. Mita, Y. Watanabe, T. Akatsu, H. Kambe: Anionic polymerization of methyl methacrylate in THF, Polym. J. 4 (1973) 271.
[26] G. Löhr, G.V. Schulz: Zur Kinetik der anionischen Polymerisation von Methylmethacrylat in THF bei −75 °C, Makromol. Chem. 172 (1973) 137.
[27] G. Löhr, G.V. Schulz: Kinetics of anionic polymerization of methyl methacrylate with caesium and sodium as conter-ions in THF, Eur. Polym. J. 10 (1974) 121.
[28] A. H. E. Müller: Present view of the anionic polymerization of methyl methacrylate and related esters in polar solvents, in J. E. McGrath: Anionic Polymerization. Kinetis, Mechanisms and Synthesis. ACS Symposium Series No. 166, Washington 1981, 441.
[29] A. H. E. Müller: Kinetics and mechanisms in the anionic polymerization of methacrylic esters, in T. Hogen-Esch, J. Smid: Recent Advances in Anionic Polymerization. Elsevier Science Publishing Co. Inc., New York – Amsterdam – London 1987, p. 205.

[30] R. Kraft, A. H. E. Müller, V. Warzelhan, H. Höcker, G.V. Schulz: On the structure of the active species in the anionic polymerization of methyl methacrylate in THF, Macromolecules 11 (1978) 1095.
[31] H. Jeuck, A. H. E. Müller: Kinetics of the anionic polymerization of methyl methacrylate in tetrahydrofuran using lithium and potassium as counterions, Makromol. Chem., Rapid Commun. 3 (1982) 121.
[32] V. Warzelhan, H. Höcker, G.V. Schulz: Kinetic studies of the anionic polymerization of methyl methacrylate, Makromol. Chem. 179 (1978) 2221.
[33] C. Johann, A. H. E. Müller: Kinetics of the anionic polymerization of methyl methacrylate using cryptated sodium as the counterion in THF, Makromol. Chem., Rapid Commun. 2 (1981) 687.
[34] Ch. B. Tsvetanov, A. H. E. Müller, G.V. Schulz: Dependence of the propagation rate consatants on the degree of polymerization in the initial stage of the anionic polymerization of methyl methacrylate in tetrahydrofuran, Macromolecules 18 (1985) 863.
[35] C. Johann: Untersuchung der Wachstumskinetik der anionischen Polymerisation von tert-Butylvinylketon in Tetrahydrofuran, Diss. Univ. Mainz 1985.
[36] V. Halaska, L. Lochmann: Aggregation of α-lithio esters of carboxylic acids, Collect. Czech. Chem. Commun. 38 (1973) 1780.
[37] V. Warzelhan, G.V. Schulz: On a new active species in the anionic polymerization of methyl methacrylate in THF using a bifunctional initiator with sodium as counterion, Makromol. Chem. 177 (1976) 2185.
[38] V. Warzelhan, H. Höcker, G.V. Schulz: The anionic polymerization of methyl methacrylate with a bifunctional initiator, Makromol. Chem. 181 (1980) 149.
[39] A. H. E. Müller, L. Lochmann, J. Trekoval: Equilibria in the anionic polymerization of methyl methacrylate, 1: Chain-length dependence of the rate and equilibrium constants, Makromol. Chem. 187 (1986) 1473.
[40] W. Fowells, C. Schuerch, F. A. Bovey, F. P. Hood: Solvation control in the anionic polymerization of stereospecifically deuterated acrylate and methacrylate esters, J. Am. Chem. Soc. 89 (1967) 1396.
[41] L. Lochmann, D. Lím: Preparation and properties of pure lithio esters of some carboxylic acids, J. Organomet. Chem. 50 (1973) 9.
[42] L. Lochmann, J. Trekoval: Metalloesters, 8: Esters of Diacids and Oligo(carboxylic acid)s (oligomers of methyl methacrylate) substituted in α-position with an alkali metal. Their stability and IR spectra, Makromol. Chem. 183 (1982) 1361.
[43] L. Vancea, S. Bywater: ^{13}C-NMR studies on anion pairs related to acrylate polymerization, 2: Dimer models, Macromolecules 14 (1981) 1776.
[44] C. B. Tsvetanov, D.T. Dotcheva, D. K. Dimov, E. B. Petrova, I. M. Panayotov: Donor-acceptor interactions of the active centers of chain propagation in the anionic polymerization of acrylonitrile and methacrylonitrile and some other polar vinyl monomers, in T. Hogen-Esch, J. Smid: Recent Advances in Anionic Polymerization. Elsevier Science Publishing Co. Inc., New York – Amsterdam – London 1987, 155.
[45] R. Kraft, A. H. E. Müller, H. Höcker, G.V. Schulz: Kinetics of anionic polymerization of methyl methacrylate in 1.2-dimethoxyethane (DME), Makromol. Chem., Rapid Commun. 1 (1980) 363.
[46] F. J. Gerner: Untersuchungen zu Mechanismus und Kinetik der Abbruchreaktion bei der anionischen Polymerisation von Methacrylat in polaren Lösungsmitteln. Diss. Univ. Mainz 1982.

[47] P. Kilz: Untersuchungen zur Kinetik der anionischen Polymerisation von Methylmethacrylat mit Lithium als Gegenion in Dimethoxyethan. Diplomarbeit, Univ. Mainz 1982.
[48] A. Soum, M. Fontanille: Living anionic stereospecific polymerization of 2-vinylpyridine, in J. E. McGrath: Anionic Polymerization. ACS Symposium Series No. 166, Washington D.C. 1981, 239–269.
[49] M. A. Müller, M. Stickler: On tacticity of poly(methyl methacrylate) prepared by group transfer polymerization, Makromol. Chem., Rapid Commun. 7 (1986) 575.
[50] F. Gores: Untersuchungen zur Stereochemie der anionischen Polymerisation von Methylmethacrylat in Tetrahydrofuran. Diplomarbeit, Univ. Mainz 1986.
[51] R. A. Volpe, T. E. Hogen-Esch, A. H. E. Müller, F. Gores: Stereochemistry of oligometrization of MMA and correlation with endgroup stereochemistry of PMMA, Polym. Prepr., ACS Div. Polym. Chem. 28(2) (1987).
[52] J. L. Baumgarten, A. H. E. Müller, T. E. Hogen-Esch: Determination of E/Z and meso/racemic end group stereochemistry in the anionic polymerization of methyl methacrylate in THF, Macromolecules, 24 (1990) 353–359.
[53] A. H. E. Müller: Kinetics of the anionic polymerization of tert.-butyl methacrylate in THF, Makromol. Chem. 182 (1981) 2863.
[54] H. Schreiber: Über die Abbruchsreaktionen bei der anionischen Polymerisation von Methylmethacrylat, Makromol. Chem. 36 (1960) 86.
[55] D. M. Wiles, S. Bywater: Methoxide ions in the anionic polymerization of methyl methacrylate, Chem. Ind. (London) 1963 (1963) 1209.
[56] D. M. Wiles, S. Bywater: The butyl-lithium initiated polymerization of methyl methacrylates III. Effect of lithium alkoxides, J. Phys. Chem. 68 (1964) 1983.
[57] D. M. Wiles, S. Bywater: Polymerization of methyl methacrylate initiated by 1.1-diphenyl-hexyl-lithium, Trans. Faraday Soc. 61 (1965) 150.
[58] N. Kawabata, T. Tsuruta: Elementary reactions of metal alkyl in anionic polymerization, I: Reaction mode of n-butyllithium in the initiation step of methyl acrylate and methyl methacrylate polymerization, Makromol. Chem. 80 (1965) 231.
[59] K. Hatada, T. Kitayana, K. Fujikawa, K. Ohta, H. Yuki: Studies on the anionic polymerization of methyl methacrylate with butyl-lithium using perdeuterated monomer, Polym. Bull. (Berlin) 1 (1978) 103.
[60] K. Hatada, T. Kitayama, M. Nagakura, H. Yuki: Structure of methyl methacrylate oligomers formed in the polymerization with butyl-lithium in toluene, Polym. Bull. (Berlin) 2 (1980) 125.
[61] K. Hatada, T. Kitayama, K. Fumikawa, K. Ohta, H. Yuki: Studies on the anionic polymerization of methyl methacrylate initiated with butyllithium in toluene by using perdeuterated monomer, in J. E. McGrath: Anionic Polymerization. ACS Symposium Series No. 166, Washington D.C. 1981, 327–341.
[62] A.T. Bullock, G. G. Cameron, J. M. Elsom: ESR studies of spin-labelled synthetic polymers, 12: The polymerization of methyl methacrylate initiated by butyl-lithium, Eur. Polym. J. 13 (1977) 751.
[63] L. Lochmann, M. Rodová, J. Petránek, D. Lím: Reactions of models of the growth center during anionic polymerization of methacrylate esters, J. Polym. Sci., Polym. Chem. Ed. 12 (1974) 2295.
[64] W. E. Goode, F. H. Owens, W. L. Myers: Crystalline acrylic polymers, II: Mechanism studies, J. Polym. Sci. 47 (1960) 75.

[65] D. L. Glusker, I. Lysloff, E. Stiles: The mechanism of the anionic polymerization of methyl methacrylate, II: The use of molecular weight distribution to establish a mechanism, J. Polym. Sci. 49 (1961) 315.
[66] F. H. Owens, W. L. Myers, F. E. Zimmermann: The reaction of acrylates and methacrylates with organo-magnesium compounds, J. Org. Chem. 26 (1961) 2288.
[67] L. Lochmann, J. Kolarik, D. Doskocilova, S. Vozka, J. Trekoval: Metalloesters VII. Stabilizing effect of sodium tert-butoxide on the growth enter in the anionic polymerization of methacrylic esters, J. Polym. Sci., Polym. Chem. Ed. 17 (1979) 1727.
[68] L. Lochmann, A. H. E. Müller: Equilibria in the anionic polymerization of methyl methacrylate, 2. Effect of lithium tert-butoxide on the rate and equilibrium constants, Makromol. Chem. 191 (1990) 1657.
[69] F. J. Gerner, H. Höcker, A. H. E. Müller, G.V. Schulz: On the termination reaction in the anionic polymerization of methyl methacrylate in polar solvents, 1: Kinetic studies, Eur. Polym. J. 20 (1984) 349.
[70] F. J. Gerner, A. H. E. Müller, H. Höcker, G.V. Schulz: Termination Reactions in the Anionic Polymerization of Methyl Methacrylate in Polar Solvents, Prepr., IUPAC Intl. Symp. on Macromol., Strasbourg (1981) 213.
[71] A. H. E. Müller, V. Warzelhan, H. Höcker, G. Löhr, G.V. Schulz: Termination and Transfer Reactions in the Anionic Polymerization of Methyl Methacrylate in THF, Prepr., IUPAC Intl. Symp. on Macromol., Dublin, (1977) 31.
[72] R.W. Taft jr.: Separation of polar, steric, and resonance effects in reactivity, in: Steric Effects in Organic Chemistry; R.W. Taft jr., M. S. Newman, Ed., Wiley, New York 1956, p. 556.
[73] J. Smid, M. Szwarc: Kinetics of anionic copolymerization. Determination of absolute rate constants, J. Polym. Sci. 61 (1962) 31.
[74] H. Jeuck: Kinetik der anionischen Homo- und Copolymerisation von Methacrylaten in Tetrahydrofuran, Diss. Univ. Mainz 1985.
[75] K. E. Piejko: Untersuchung der anionischen Polymerisation von Methylmethacrylat in unpolaren Lösungsmitteln unter besonderer Berücksichtigung von Nebenreaktionen und Molmassenverteilungen, Diss. Univ. Bayreuth 1982.
[76] K. Hatada, K. Ute, K. Tanaka, T. Kitayama, Y. Okamoto: Preparation of highly isotactic poly(methyl methacrylate) of low polydispersity, Polym. J. 17 (1985) 977.
[77] P. Kilz: Diss. Univ. Mainz 1991.
[78] L. Lochmann, M. Rodová, J. Trekoval: Anionic polymerization of methacrylate esters initiated with esters of α-metallocarboxylic acids, J. Polym. Sci., Polym. Chem. Ed. 12 (1974) 2091.
[79] O.W. Webster, W. B. Farnham, D.Y. Sogah: Living Polymers and Process for their Preparation, E. P. O. Patent 68887 (E. I. du Pont de Nemours & Co.) (1983).
[80] D.Y. Sogah, O.W. Webster: Telechelic polymers by group transfer polymerization, J. Polym. Sci., Polym. Lett. Ed. 21 (1983) 927.
[81] D.Y. Sogah, W.B. Farnham: Group transfer polymerization. Mechanistic studies, in H. Sakurai: Organosilicon and Bioorganosilicon Chemistry, Wiley, New York 1986, p. 219.
[82] W. B. Farnham, D.Y. Sogah: Group transfer polymerization. Mechanistic studies, Polym. Prepr., ACS Div. Polym. Chem. 27(1) (1986) 167.

[83] P. M. Mai, A. H. E. Müller, Kinetics of group transfer polymerization of methyl methacrylate in tetrahydrofuran, 1: Effect of concentrations of catalyst and initiator on reaction rates, Makromol. Chem., Rapid Commun. 8 (1987) 99.
[84] P. M. Mai, A. H. E. Müller: Kinetics of group transfer polymerization of methyl methacrylate in tetrahydrofuran, 2: Effect of monomer concentration and temperature on reaction rates, Makromol. Chem., Rapid Commun. 8 (1987) 247.
[85] A. H. E. Müller: Kinetics of group transfer polymerization, in M. Fontanille, A. Guyot: Recent Advances in Mechanistic and Synthetic Aspects of Polymerization. D. Reidel, Dordrecht/NL 1987, 23.
[86] D.Y. Sogah, W. R. Hertler, O.W. Webster, G. M. Cohen: Group transfer polymerization. Polymerization of acrylic monomers, Macromolecules 20 (1987) 1473.
[87] O.W. Webster, D.Y. Sogah: Recent advances in the controlled synthesis of acrylic polymers by group transfer polymerization, in M. Fontanille, A. Guyot: Recent Advances in Mechanistic and Synthetic Aspects of Polymerization. D. Reidel Publishing Company, Dordrecht – Boston – Lancaster – Tokyo 1987, 3–21.
[88] M. A. Doherty, A. H. E. Müller: Kinetics of the group transfer polymerization of tert-butyl methacrylate in tetrahydrofuran, Makromol. Chem. (1988).
[89] M. A. Doherty, F. Gores, P. M. Mai, A. H. E. Müller: Mechanism and molecular weight distribution in the group transfer polymerization of methacrylates, Polym. Prepr. (ACS Div. Polym. Chem.) 29(2) (1988) 73.
[90] A. H. E. Müller: Group transfer and anionic polymerization: A critical comparison, Makromol. Chem., Macromol. Symp. 32 (1990) 87.
[91] F. Gores: Diss. Univ. Mainz 1991.

9 Molecular Engineering of Liquid Crystalline Polymers*

Helmut Ringsdorf**, Ingrid Voigt-Martin***,
Joachim Wendorff****, Renate Wüstefeld**,
and Rudolf Zentel**

9.1 Introduction: Self Organizing Systems

Polymer science within the scope of the Sonderforschungsbereich 41 was, at its inception, devoted to classical problems in chemistry and physics of macromolecules, e. g. polymerization kinetics, solution properties and partially crystalline polymers. From this starting point, research in Mainz has evolved and expanded into the areas of material and life sciences. Polymeric liquid crystals and polymeric membrane models have played significant roles in this evolution. Investigations of these materials have led to a view of polymer science as an interdisciplinary field, with the areas of polymeric liquid crystals and polymeric membrane models (cellular liquid crystals) as two important research areas within this field. This relationship is schematically represented in Fig. 9.1.

Within the perspective of the Sonderforschungsbereich 41, the investigation of polymeric liquid crystals and membrane models was incorporated into the study of *self-organizing supramolecular structures*. These systems are unique in that they combine *order and mobility* as well as *function via organization* (see Fig. 9.2). Polymeric thermotropic liquid crystals and liposomes are the two extremes within this framework. These areas of investigation are linked by

* The investigation of liquid crystalline polymers was done in close cooperation between chemists and physicists. So this article is divided in one section that deals with the molecular architecture of liquid crystalline polymers (section 9.2 by H. Ringsdorf, R. Wüstefeld and R. Zentel) and one section that deals with the physical properties (section 9.3 by I. Voigt-Martin and J. Wendorff).

** Institut für Organische Chemie der Universität Mainz, J.-J.-Becher-Weg 18–20, D-6500 Mainz

*** Institut für Physikalische Chemie der Universität Mainz, Jakob-Welder-Weg, D-6500 Mainz

**** Deutsches Kunststoff-Institut, Schloßgartenstraße 6, D-6100 Darmstadt

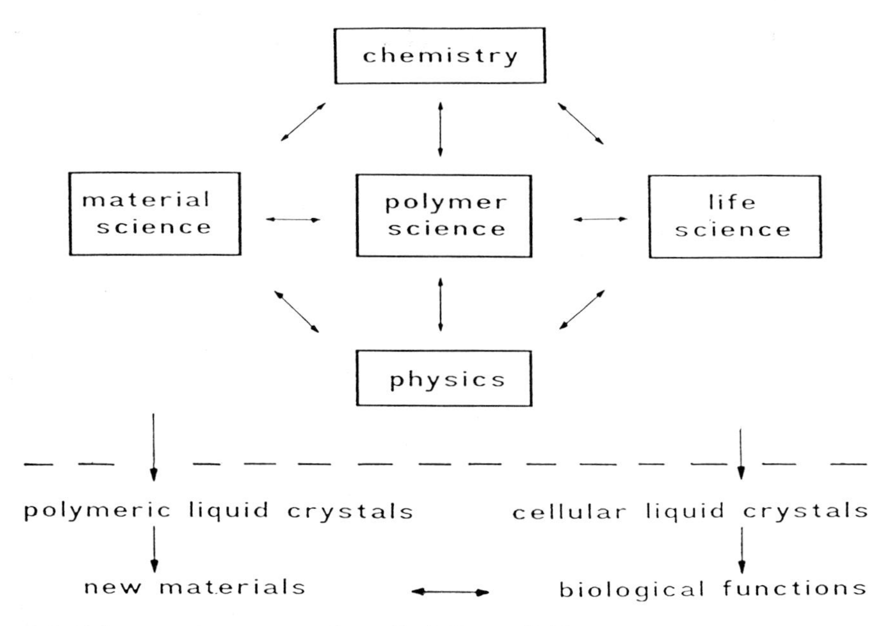

Fig. 9.1: Polymer science as an interdisciplinary field.

the investigation of polymeric surfactants. Some of these surfactants are used to build mono- and multilayers, while others aggregate in solution to form micelles of different shapes and lyotropic liquid crystalline phases. Since investigations on polymeric amphiphiles have been reviewed quite recently [1], this chapter will concentrate on the synthesis, structures and properties of thermotropic liquid crystalline polymers.

9.2 Molecular Architecture of Liquid Crystalline Polymers

Liquid crystals [2–4] combine the properties of liquids and crystals – most characteristically, those of mobility and order. Liquid crystalline phases or mesophases ("meso" = inbetween, here the liquid and crystalline state) occur in a particular temperature interval between the well ordered crystalline phase and the mobile isotropic liquid. Whether a liquid crystalline phase is formed or not, as well as its type, is closely related to molecular structure. Typical of liquid crystalline molecules, also called mesogens, is their strong formanisotropy. They are in general composed of an inner rigid core which is rod-like or

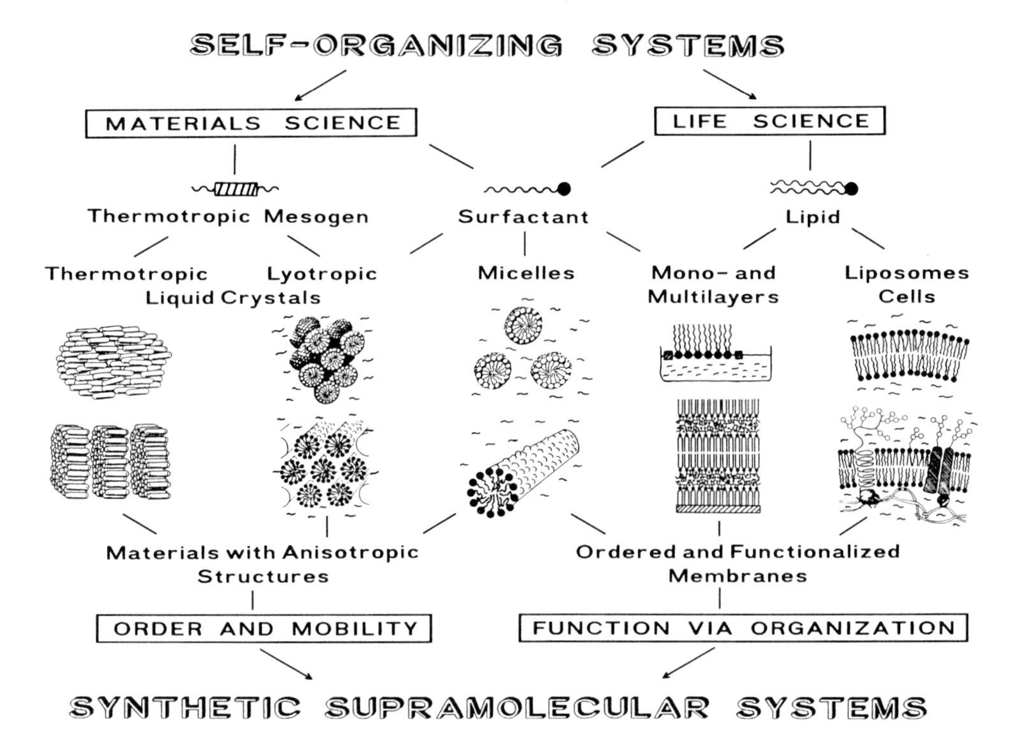

Fig. 9.2: Self-organization and supramolecular systems [1] in material science and life science. The supramolecular structures range from thermotropic liquid crystals to liposomes.

disc-like [5, 6] in shape and often substituted with alkyl chains (the so-called "tails" of the mesogen). Fig. 9.3 shows the formation of different thermotropic liquid crystalline phases starting from rod-like mesogens (for the formation of liquid crystalline phases from disc-like mesogens see section 9.2.2).

The *nematic phase* (n) is usually the high temperature phase. The mesogens are arranged in such a way that their centres of mass display a short range order and their long axes lie preferentially in one direction *(long range orientational order)*. This long range orientational order describes the characteristic difference with respect to the isotropic melt, where the centres of mass are also isotropically distributed but a preferred orientation of the molecules is not found over long distances. At lower temperatures different *smectic* phases may occur. The mesogens now arrange in layers. As in nematic phases, their long axes still lie preferentially in one direction (long range orientational order). Their centres of mass, however, are no longer isotropically distributed but ordered (in layers), giving the smectic phase an additional element of order with the *long range positional order.*

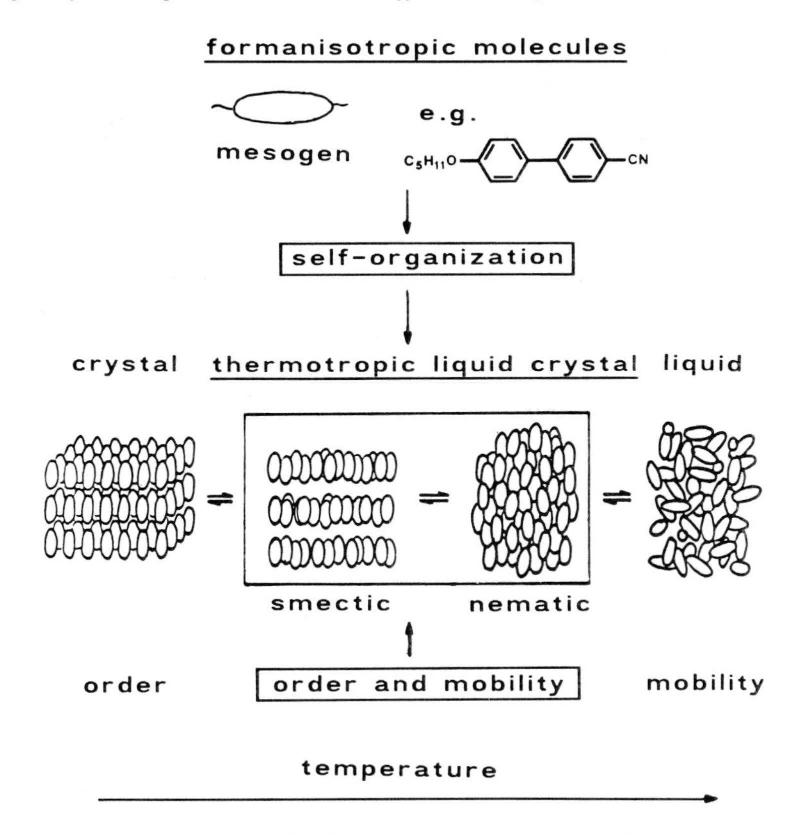

Fig. 9.3: Structure and phase behaviour of thermotropic liquid crystals from rod-like molecules.

Eleven types of smectic mesophases have been identified, ranging from smectic A (S_A) to smectic K (S_K). These differ in the arrangement of the mesogens within each individual layer (e.g. no order within the layer in S_A, hexagonal order in S_B) and/or their orientation within the layer plane (e.g. orthogonal in S_A, tilted in S_C) [7].

Rapid developments in the field of liquid crystalline (LC) polymers [8] occured after the realization of *rigid rod macromolecules* in which the main chain, as a whole, functions as the mesogen. Such polymers form liquid crystalline phases either in solution (lyotropic) [9] or in the bulk at high temperatures close to the decomposition temperature (thermotropic) [10]. At present they are used as high tensile strength fibers (e.g. Kevlar®) or as thermoplastically processable, self-reinforcing plastics (e.g. Xydar®, Vectra®, Ultrax®).

Parallel to these industrial developments, numerous studies have been carried out in areas of basic research during the past fifteen years. The original goal was to prepare thermotropic LC polymers which form liquid crystalline phases at moderate temperatures, primarily for kinetic investigations in oriented systems [11]. For this purpose, rod-like mesogenic groups, well known from low molar mass liquid crystals, were incorporated into flexible polymers. The crucial point in the systematic synthesis of these polymers was the combination of two opposing tendencies: Formanisotropic mesogens with their tendency towards an anisotropic arrangement had to be linked to polymer chains with the tendency to adopt a statistical coil conformation. This problem was solved by the use of a *flexible spacer* [12, 13] to ensure a *partial decoupling* between polymer chains and mesogenic groups (see Fig. 9.4).

Starting from the first – recognized as "classical" – liquid crystalline (LC) polymers prepared according to this concept (the side chain polymers *A* and the main chain polymers *B*, shown in Fig. 9.5), the *molecular architecture* of liquid crystalline polymers has been varied extensively [14]. This is illustrated schematically in Fig. 9.5.

On the one hand, it is possible to realize discotic phases, known from low molar mass liquid crystals, by using disc-like mesogens in side group [15] (type *C*) and main chain [16] (type *D*) polymers. On the other hand, it is, of course, also possible to incorporate rod-shaped mesogens in ways different from the two classical types *A* and *B*: examples are side group polymers [17–18] (type *E*) as well as main chain polymers [18] (type *F*) with laterally fixed mesogens. Contrary to expectations, these strange-looking architectures do not prevent the formation of liquid crystalline phases. As far as polymers of type *E* are concerned, the limited rotation of the mesogenic groups around their longitudinal axes leads to the formation of biaxial nematic phases [17]. A third type of structural variation makes use of a principle that is well known in polymer chemistry but has rarely been used in the field of liquid crystalline materials: the combination of different structural elements and different building principles within one molecule. This includes different combinations of rod- and disc-shaped mesogens [19–22] (types *G*, *H* and *I*) and the combination of structural principles of the two classical types of liquid crystalline polymers (type *A* and *B*) in combined main chain/side group polymers [23–24] (types *J* and *K*).

Furthermore, polymers containing cross-shaped mesogens [18] (type *M*) should be noted, as well as the idea of combining laterally and terminally connected mesogenic groups (type *L*).

Parallel to this variation in molecular architecture, LC polymers have been *functionalized* in different ways. For this purpose, dye containing groups [25, 26], groups undergoing chemical [27] or photochemical reactions [28, 29] and groups carrying chirality and strong dipole moments [30, 31] (see section 9.2.3 and Fig. 9.6) have either been incorporated into the mesogens or were added into the polymer chain as comonomers.

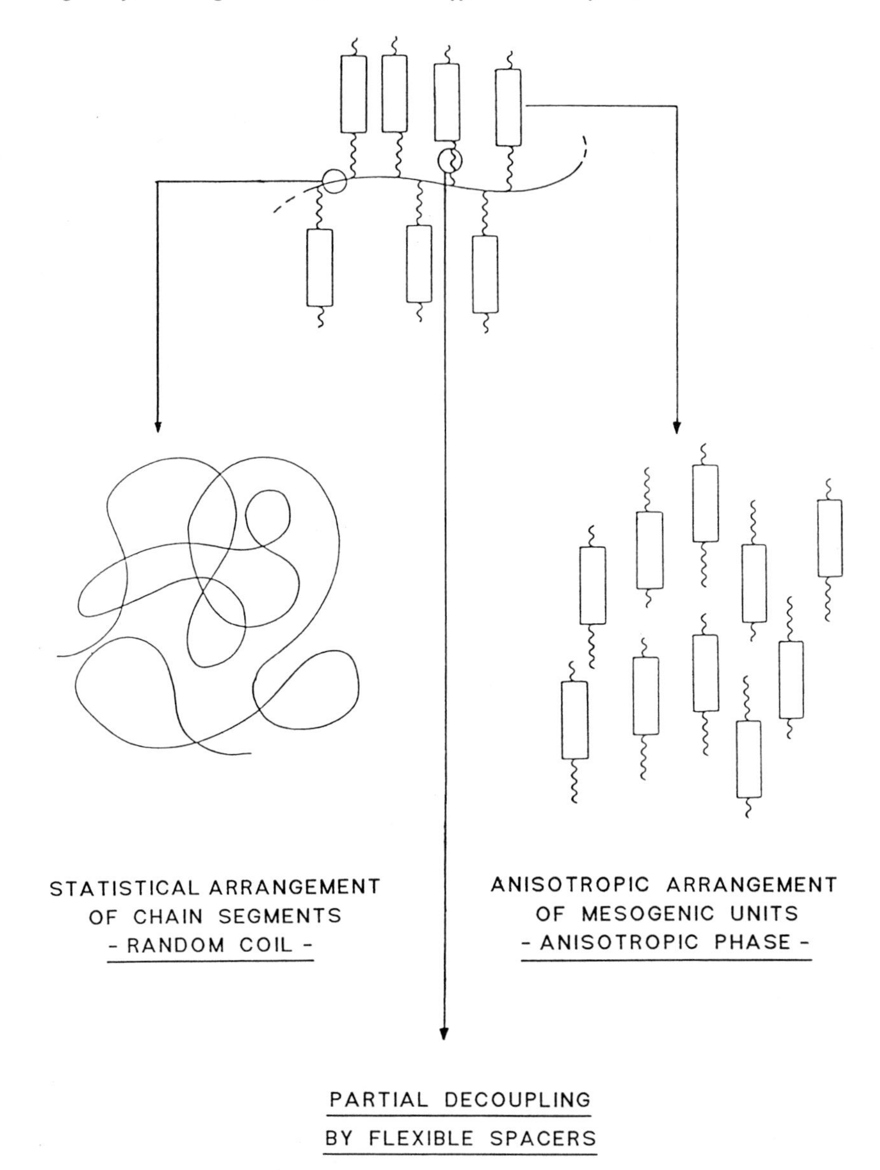

Fig. 9.4: The concept for the synthesis of LC polymers: Mesogens with their tendency towards anisotropic arrangement and polymer chains with their tendency towards random coil formation are partially decoupled via flexible spacers.

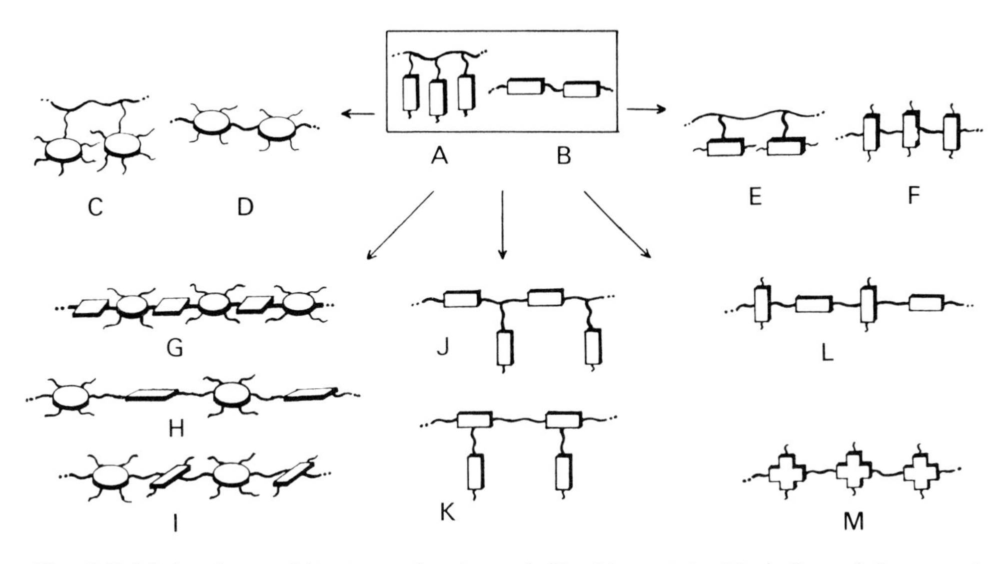

Fig. 9.5: Molecular architecture of polymeric liquid crystals: Variation of shape and arrangement of the mesogens.

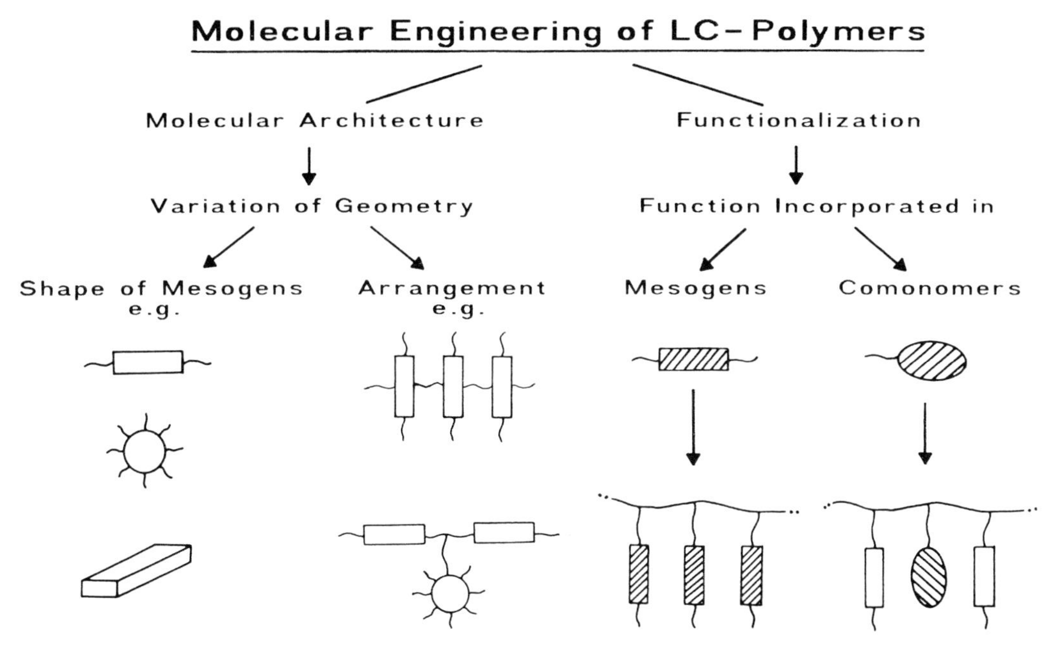

Fig. 9.6: Molecular engineering of LC polymers can be achieved either by varying the molecular geometry or by incorporating functionalized groups in mesogens or comonomer units.

9.2.1 Liquid Crystalline Polymers with Rod-like Mesogens

9.2.1.1 Synthesis and Structure-Property Relations of LC Side Group Polymers

The concept of linking mesogenic groups via a flexible spacer to the polymer chain to produce LC side group polymers (type *A* of Fig. 9.5) was first realized with polymethacrylates using benzoic acid phenylesters as mesogens [12] (see Table 9.1). Later, biphenyls [32, 33], phenylcyclohexanes [25, 34] and phenylpyrimidines [35] were also used – in Mainz – as mesogens, with polyacrylates [36], polysiloxanes [33, 37], polychloroacrylates [38], and polyitaconates [34] as polymer chains. For a comprehensive summary of the systems synthesized so far see references [39] and [40]. The structural variation of polymer subunits – mesogenic groups, spacer groups, and polymer chains – enables the influence of each of these constituents on the properties of the liquid crystalline polymers to be determined. The influence of variations of the mesogenic groups and the spacer length on the phase behaviour is summarized in Table 9.1.

The *influence of varying the mesogenic structure* is demonstrated by comparing polymer *1* and *2*. A polyacrylate with 6 methylene units as a spacer and benzoic acid phenylester as mesogen (polymer *1*) shows a glass transition temperature at 35° C, a smectic A phase up to 97° C and a nematic phase that becomes isotropic at 123 °C. A change of the mesogenic group, in this case the exchange of the benzoic acid phenylester with a phenylcyclohexane (polymer *2*), keeping the rest of the polymer structure constant, leads to drastic changes in the phase behaviour: The glass transition temperature drops to −10° C and only a nematic phase is found for this polymer. This is in accordance to what is known from low molar mass liquid crystals: the phenylcyclohexanes [41] are known for their low viscosities, low melting temperatures and a preference for nematic phases. The influence of a decrease in spacer length while keeping the mesogen constant is revealed by a comparison of polymer *1* and polymer *3*. In this case, the glass transition temperature rises and only a nematic phase is observed for the polymer with the short spacer. This again is not astonishing because a decrease in the spacer length corresponds to a decrease in the length of the tails of the mesogens and it is well-known from low molar mass liquid crystals [3–5,7] that this favours the formation of nematic phases. In conclusion, the main effect of varying the mesogenic groups and spacer length is both a change in the mesophase type and the clearing temperature as the upper limit of the liquid crystalline phase.

In contrast, the main *influence of varying the polymer chain* is summarized in Table 9.2 [37].

All three polymers in Table 9.2 show, independent of the polymer chain, a nematic phase. However, the glass transition temperatures are very different.

Table 9.1: The influence of variations of the mesogens and the spacer length on the phase behaviour demonstrated by comparing polymers *1–3* [34, 36].

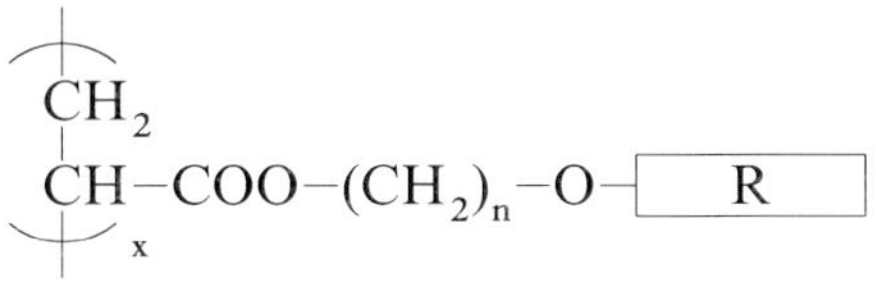

$$\left[CH_2-CH(-COO-(CH_2)_n-O-R) \right]_x$$

No.	n	R	Phase transitions/°C
1	6	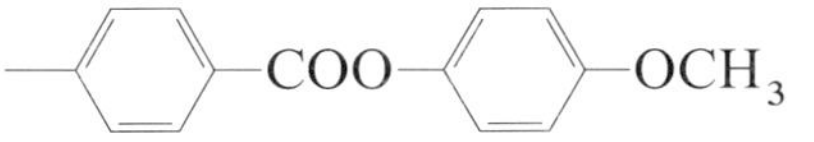 $-C_6H_4-COO-C_6H_4-OCH_3$	g 35 s_A 97 n 123 i
2	6	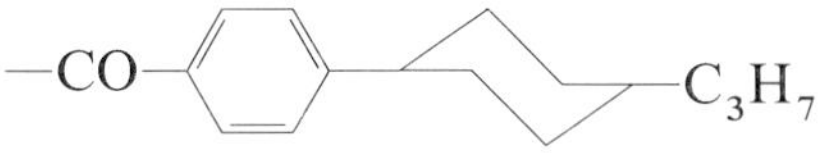 $-CO-C_6H_4-C_6H_{10}-C_3H_7$	g −10 n 20 i
3	2	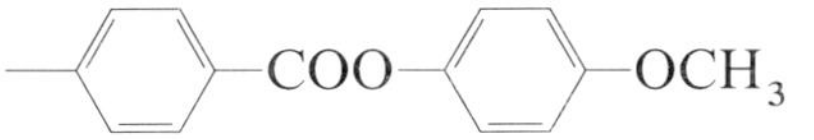 $-C_6H_4-COO-C_6H_4-OCH_3$	g 62 n 116 i

They are as high as 97° C for the polymethacrylate *4*, and only 15° C for the polysiloxane *5*. This certainly reflects the different mobilities of polymethacrylate, polyacrylate and polysiloxane chains (T_g-values are 105° C for PMMA, 10° C for PMA and −120° C for PDMS homopolymers, respectively [42]). The fact that the glass transition of the polysiloxane *5* (15° C) is still much higher than the glass transition temperature of the pure polymer backbone (−120° C), demonstrates that the mobilities are, of course, modified by the mesogenic groups. Nevertheless, the main effect of a variation in the polymer chains is a variation of the mobility and the glass transition temperature as the lower limit of the liquid crystalline phase.

So far, the discussion has focused on structure-property relations of LC-polymers. *Characterization of the liquid crystalline phases* is addressed in the next section. The polymers prepared so far show different liquid crystalline phases, which can be classified in complete analogy to low molar mass liquid crystals [39]. In this way, nematic, cholesteric [43], and different smectic phases can be clearly identified. They show the optical textures typical for low molar mass liquid crystals and the corresponding X-ray patterns [44–47]. The order parameter of the mesogenic groups was determined, for example, by ESR- (see Fig. 9.7) and ^{2}H-NMR-spectroscopy [48a–c, 49] (see Spiess and Sillescu,

Table 9.2: The influence of variations of the polymer chain on the phase behaviour demonstrated by comparing polymers *3–5* [33, 36, 37].

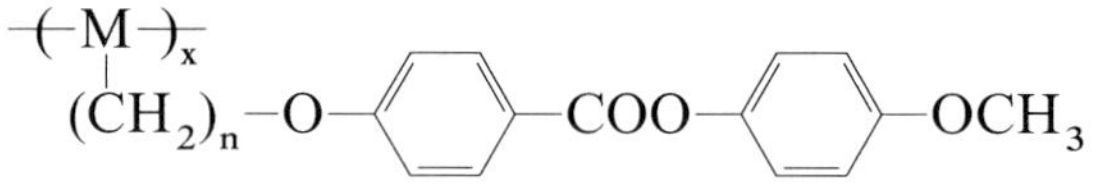

$$-(M)_x- \quad | \quad (CH_2)_n-O-C_6H_4-COO-C_6H_4-OCH_3$$

No.	M	n	Phase transitions/°C
4	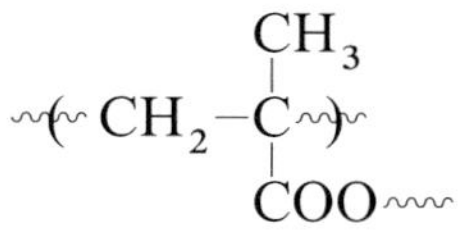 $-(CH_2-C(CH_3)(COO\sim))-$	2	g 97 n 120 i
3	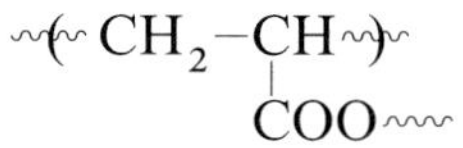 $-(CH_2-CH(COO\sim))-$	2	g 62 n 116 i
5	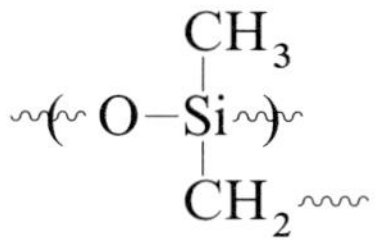 $-(O-Si(CH_3)(CH_2\sim))-$	2	g 15 n 61 i

Chapter 12, this Volume) and has been found to be in good agreement with the values measured for low molar mass liquid crystals. The results of ^{2}H-NMR-measurements [48a] proved, for example, that all mesogens participate in the liquid crystalline ordering. That means these polymers do not show a partial "degree of liquid-crystallinity" in analogy to partially crystalline polymers [48d]. Fig. 9.7 summarizes the results of ESR-measurements with a spin-probe on the polymers *1* (nematic and smectic) and *3* (nematic) of Table 9.1 [49]. For both polymers, the order parameter jumps at the clearing temperature from 0 (isotropic phase) to a finite value in the nematic phase. With decreasing temperature, it increases continuously, until it becomes constant around the glass transition temperature. For polymer *1* the transition from the nematic to the smectic A phase is accompanied by an additional jump of the order parameter. For this polymer an order parameter as high as 0.9 is frozen in the glassy state.

Well oriented samples of the LC polymers can be obtained by using electric and magnetic fields [39, 46–52] as well as mechanical stretching [16, 23, 27, 39, 44, 45]. Because of their orientability in electric fields, these polymers

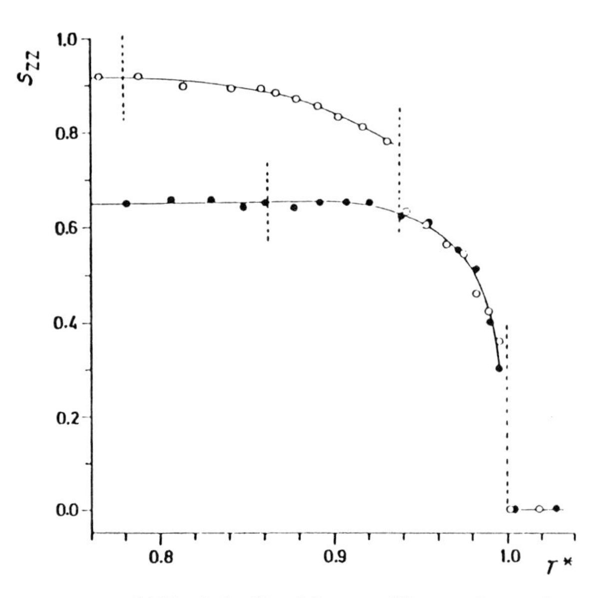

Fig. 9.7: Order parameters [49] of the liquid crystalline polyacrylates *1* (○) and *3* (●) (see Table 9.1) as function of the reduced temperature $T^* = T/T_{ni}$. (T_{ni}: transition temperature nematic-isotropic). The dashed lines indicate the phase transitions.

have been tested for use in liquid crystal displays (see Fig. 9.8). Fig. 9.8 shows schematically the reorientation of the mesogenic groups upon application of an electric field. Without an electric field, the mesogens are oriented parallel to the glass plates, giving rise to a birefringent texture which transmits light between crossed polarizers. Upon application of an electric field, the mesogens orient perpendicular to the glass plates and give rise to a pseudoisotropic, dark texture. Fig. 9.8 shows the switching behaviour for one polymer [50]. With rise and decay times around 200 msec, this polymer switches fairly fast, but only at high temperatures. As the measuring temperature approaches the glass transition, the reorientation times increase to infinity.

Properties of LC side group polymers and low molar mass liquid crystals differ mainly in the dynamic aspects mentioned above. This is demonstrated also in dielectric relaxation behaviour [53–57]. Dielectric relaxation measurements show both the dynamic glass process of the polymer chains (α-relaxation) and 180° jumps of the long axes of the mesogens (δ-relaxation; see Fig. 9.9) [54]. The δ-relaxation is orders of magnitude slower than in low molar mass liquid crystals and freezes at the glass transition temperature. In addition, up to 3 local relaxation processes are found below the glass transition temperature (see Fig. 9.9) which can be attributed to relaxations of the mesogenic groups (β), the spacer (γ_1) and the end group (γ_2) [54]. The results obtained for polymer *1* of Table 9.1 are presented in Fig. 9.9b. It is interesting

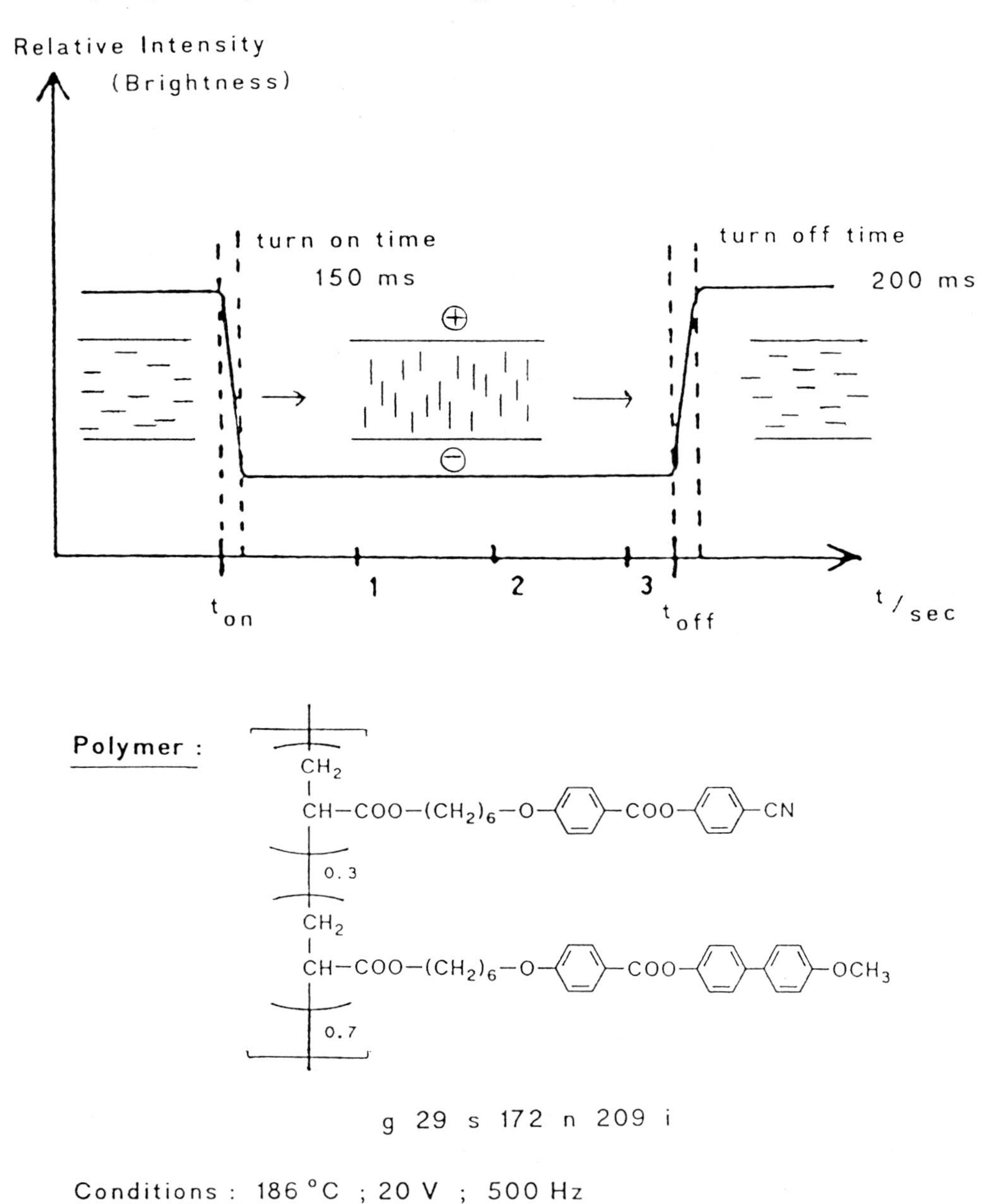

Fig. 9.8: The switching of a liquid crystalline polymer in an electric field [50] changes the transmittance of the sample between two crossed polarizers.

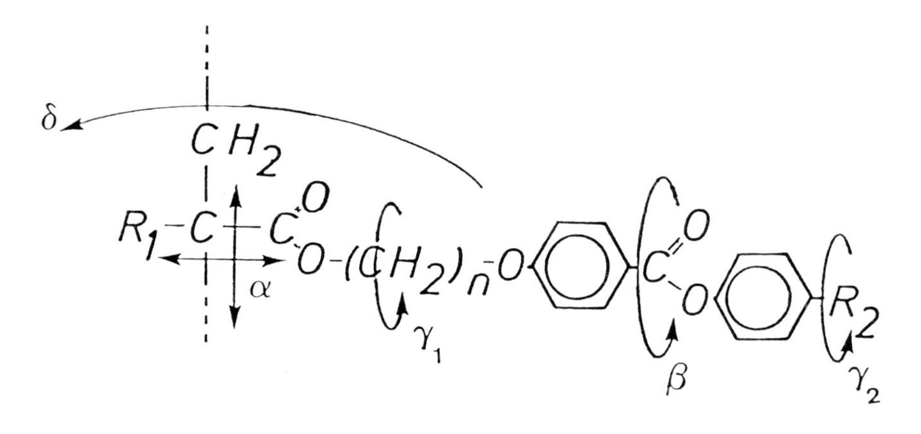

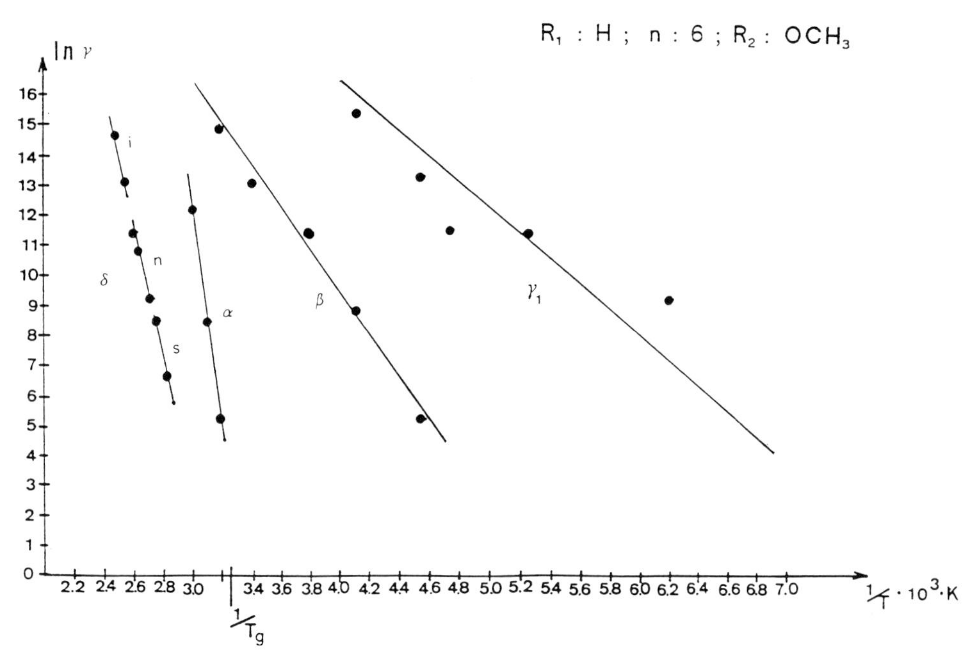

Fig. 9.9: Dielectric relaxation measurements [54] have revealed 5 different relaxation processes for LC polyacrylates (a). The plot of the natural logarithm of the relaxation frequency versus 1/T (b) shows the temperature dependence for 4 of the relaxation processes for polymer *1* (see Table 9.1).

to note that the ß-relaxation of the polymers *1* and *3* agrees completely with the dynamics of 180° phenylflips determined for the same polymers by ^{2}H-NMR-spectroscopy [48c] (see Spiess and Sillescu, Chapter 12, this Volume).

9.2.1.2 Synthesis and Structure-Property Relationships of Main Chain Polymers

Classical rigid-rod polymers [10] show thermotropic liquid crystalline phases only at very high temperatures close to the temperature of thermal decomposition. The incorporation of flexible spacer segments, however, allows the synthesis of semiflexible main chain polymers (type *B* of Fig. 9.5) that show liquid crystalline phases at moderate temperatures [13]. Within the scope of the Sonderforschungsbereich 41, the research interests in semiflexible main chain polymers was aimed at a further decrease of the lower limit of the liquid crystalline temperature range by the use of very flexible spacer segments. For this purpose siloxane spacers were used. Fig. 9.10 shows the results obtained for one series of these siloxane spacer-containing main chain polymers [58]. With increasing length of the siloxane spacer, the glass transition temperature drops drastically and is as low as −100° C at a spacer length of 13 repeating units, while the phase width of the liquid crystalline phase is kept constant.

9.2.1.3 Molecular Engineering of LC Polymers: Side Group and Main Chain LC Systems

In connection with the structural variation of side group polymers, it was especially interesting to *vary the number of mesogenic units per repeat unit* in the polymer chain. In order to increase the concentration of the mesogens, polyitaconate diesters and polysiloxanes substituted with allylmalonic acid diesters (*"dimesogenic" polysiloxanes*, see Figs. 9.11, 9.12) were utilized [34]. Both systems allow the fixation of two mesogens per repeat unit. In order to decrease the concentration of the mesogens, copolysiloxanes were used in which only some of the siloxane units carried mesogenic groups.

These copolymers have found interest for preparation of LC polymers with low glass transition temperatures. If, in these copolymers, each mesogen is fixed separately to the polymer chain, then the liquid crystalline phase is lost at a ratio of 1 mesogen per more than 5 dimethylsiloxane units (see Fig. 9.12). If, however, „dimesogenic" polysiloxanes are used [34, 59], then the ratio can be as low as two mesogens (fixed together) per more than 30 dimethylsiloxane units (see Fig. 9.12). These copolymers show low glass transitions and broad liquid crystalline phases, the phase width of which is comparable to the homopolymers.

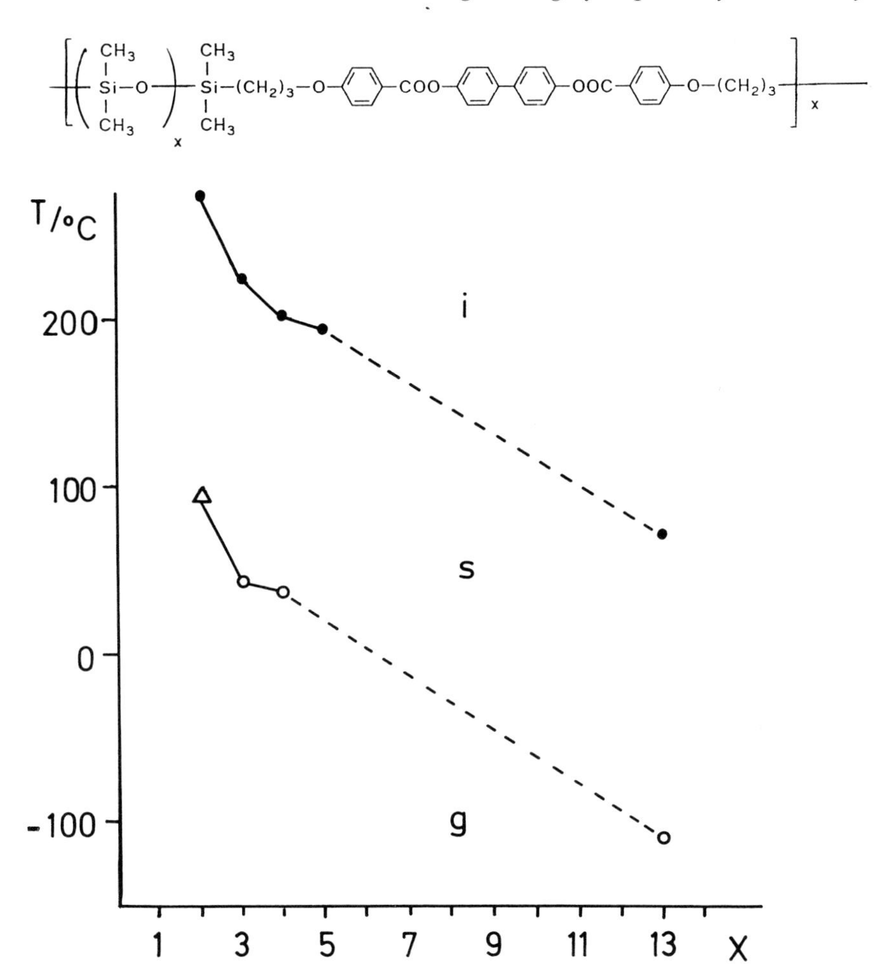

Fig. 9.10: Dependence of the transition temperatures of semiflexible LC main chain polymers on the length of the siloxane spacer [58].

The fact that „dimesogenic" polysiloxanes form liquid crystalline phases, even if only each 30th repeating unit is linked to mesogens, can be understood with a model derived from X-ray measurements of homologous series of these polymers [59] (see Fig. 9.13). The model assumes that mesogenic groups pack densely and reject the polysiloxane chains, which are incompatible with them. Thus, two sublayers are formed (microphase separation), one consisting of the disordered polysiloxane chains and the other of the mesogenic groups, which are in a smectic A arrangement.

However, if each mesogen is linked to the polymer chain separately, or if the polymer chains are more compatible (e.g. polyacrylates) with the mesogens,

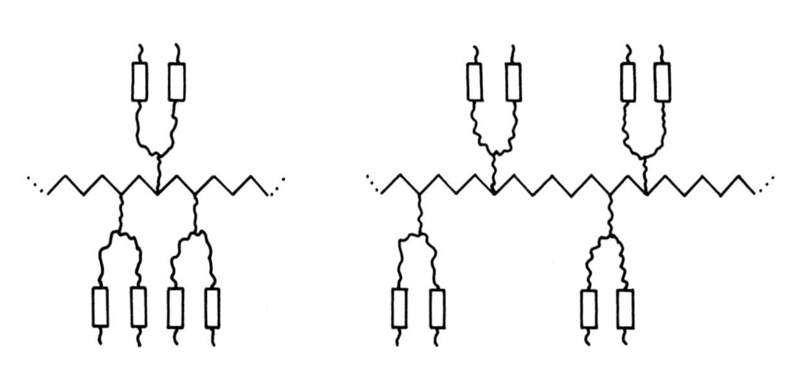

Fig. 9.11: Schematic structure of "dimesogenic" polysiloxanes. In the case of the homopolymers, two mesogens are fixed per each repeating unit. In the case of the copolymers, the mesogens are statistically distributed along the polymer chain.

then the tendency towards microphase separation is reduced. In these cases, the incorporation of nonmesogenic units into the polymer chain reduces the interaction among the mesogens and destroys the liquid crystalline phase.

Another structural variation of LC polymers led to the synthesis of *combined main chain/side group polymers* [23, 24] (types *J* and *K* of Fig. 9.5) that combine the structural principles of the „classical" main chain and side group polymers (types *A* and *B* of Fig. 9.5). These polymers show very broad liquid crystalline phases that are, in most cases, broader than the liquid crystalline phases of the corresponding pure main chain or side group polymers. This suggests that the mesogenic groups in the main chain and the side groups are not oriented perpendicular to each other as suggested by the chemical formula in Fig. 9.14. They orient parallel to each other (see Fig. 9.14) to produce an uniaxial mesophase, as confirmed by X-ray measurements on oriented fibres [60]. This structure, in which both types of mesogens are oriented parallel to each other, can also explain the differences in the phase behaviour of polymers of type *J* and *K* [24]. For polymers of type *J*, a smectic layer structure is preformed locally and consequently mostly smectic phases are observed. For polymers of type *K*, however, the fixation of mesogenic side groups to the middle of the mesogenic groups in the main chain favours nematic phases (see Fig. 9.14).

By the use of chiral end groups, polymers of type *J* with cholesteric and chiral smectic C* phases can be prepared [30, 31] (see also Fig. 9.29 in Section 9.2.3.3). As far as the dynamics of these „combined" polymers is concerned, it is especially interesting that the reorientation of the long axes of the mesogenic side groups (δ-relaxation, see Fig. 9.9) is still possible [61].

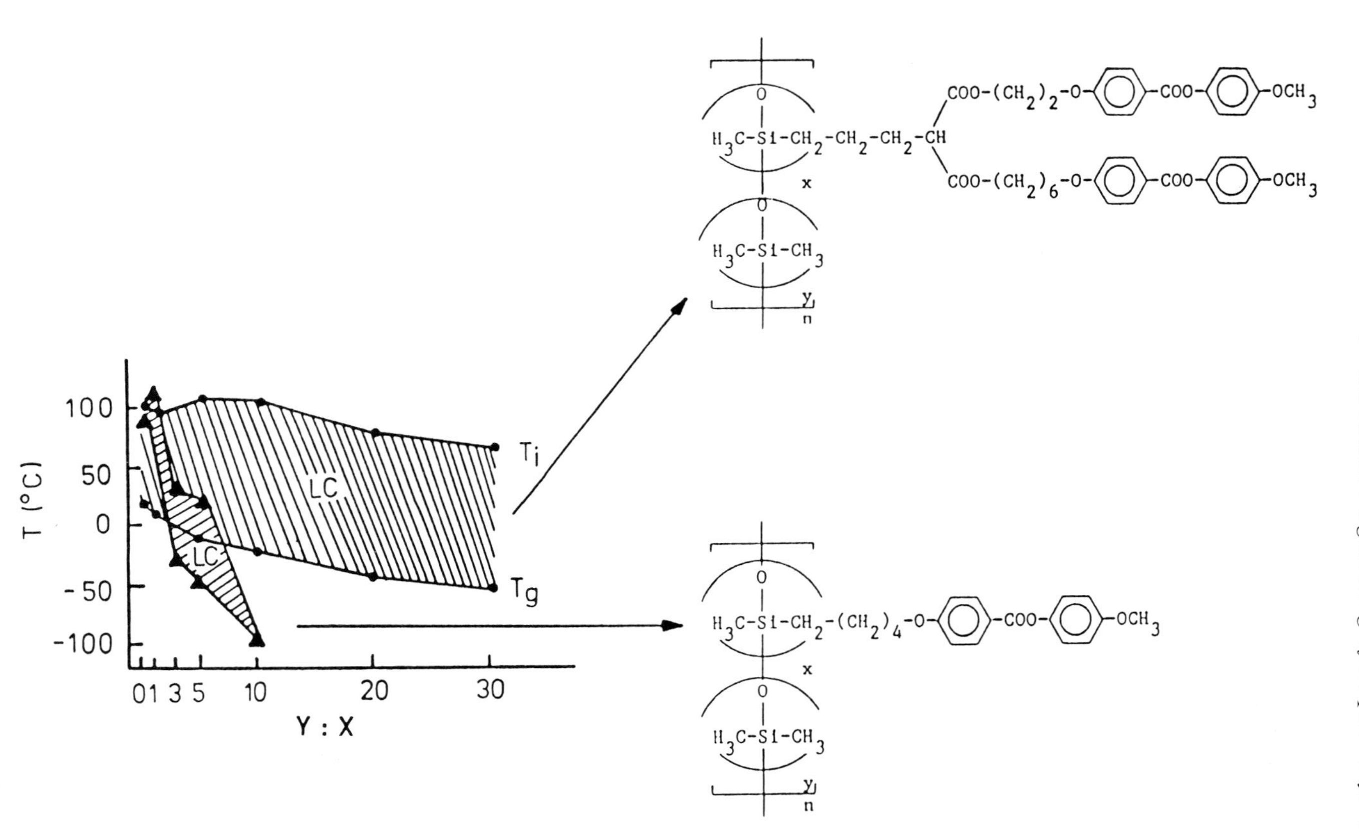

Fig. 9.12: Phase behaviour of "monomesogenic" and "dimesogenic" copolymers versus the ratio of repeating units carrying no mesogens (Y) to units carrying mesogens (X). For the "dimesogenic" copolymers the lc-phase width remains broad and constant up to 30 dimethylsiloxane units per "dimesogenic" unit (Y : X = 30). In contrast, for the "monomesogenic" copolymers the lc-phase width shrinks and is lost above 10 dimethylsiloxane units per "monomesogenic" unit (Y : X ≥ 10).

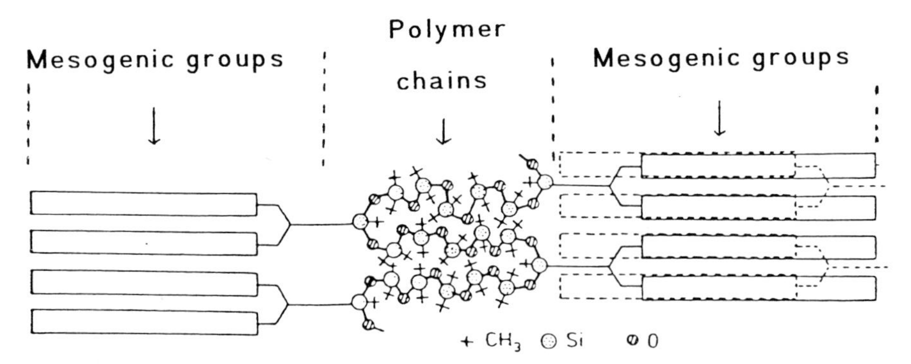

Fig. 9.13: Schematic model of the layer structure for "dimesogenic" polysiloxanes [59]. The mesogenic groups drawn in dashed lines are positioned above the drawing plane.

9.2.2 Liquid Crystalline Polymers with Disc-like Mesogens

Calamitic mesophases formed from rod-like mesogens are well-known and have been realized in many polymeric systems. This is not the case with discotic mesophases formed from disc-like mesogens. Predicted theoretically in 1974 [2] discotic mesophases remained undiscovered until 1977 [5] (see Fig. 9.15).

Analogously to calamitic phases which are divided into nematics (n) or smectics (s), discotic mesophases [62] can be either nematic-discotic (N_D) involving long range orientational order of the mesogens only, or columnar (e.g. D_{ho}) involving both long range orientational and long range positional order. In the latter columnar phases disc-like molecules are stacked, regularly (ordered) or irregularly (disordered), in columns. These arrange in various lattices of which the hexagonal lattice is most commonly observed (D_{ho} = **d**iscotic **h**exagonal **o**rdered mesophase). As with calamitic systems, *discotic polymers* of both the side group [15] and main chain type [16] have been synthesized analogously by linking disc-like mesogens through flexible spacers (see Fig. 9.16).

A third type of fundamentally different molecular structure is a board-like shape (see Fig. 9.15). Boards do not show rotational symmetry about any axes at all, neither about the long axis as for rods nor about the short axis as for discs. Molecular (polymeric) engineering has recently succeeded in making board-like structures by linking disc-like mesogens rigidly in a main chain polymer (see Fig. 9.16). For the resulting *sanidic polymers*, both phase types, sanidic nematic [66] and the higher ordered sanidic phases [19, 69], are found.

J

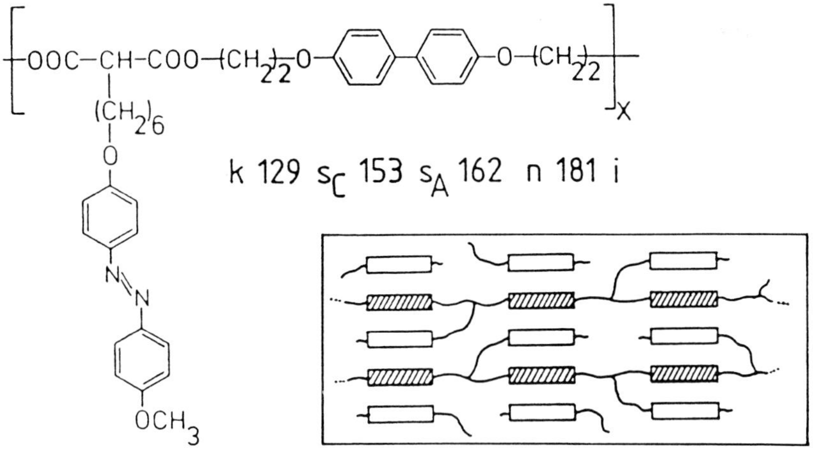

k 129 s_C 153 s_A 162 n 181 i

K

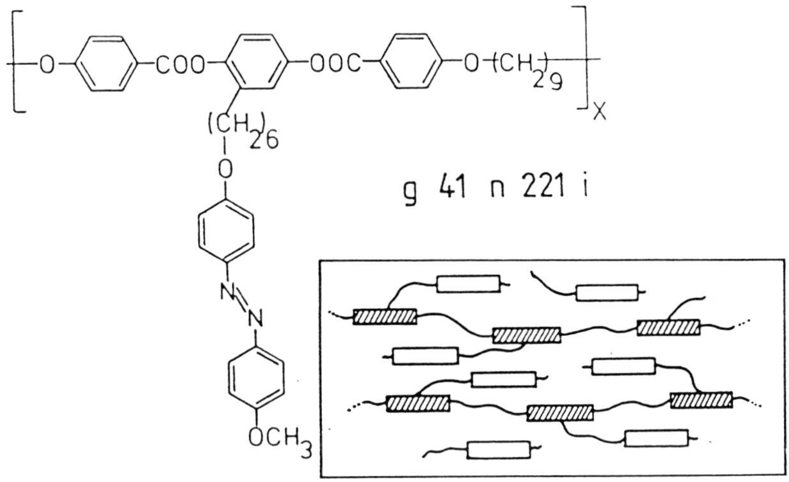

g 41 n 221 i

Fig. 9.14: Chemical structure, phase behaviour and structural models of the mesophases of combined main chain/side group polymers [24].

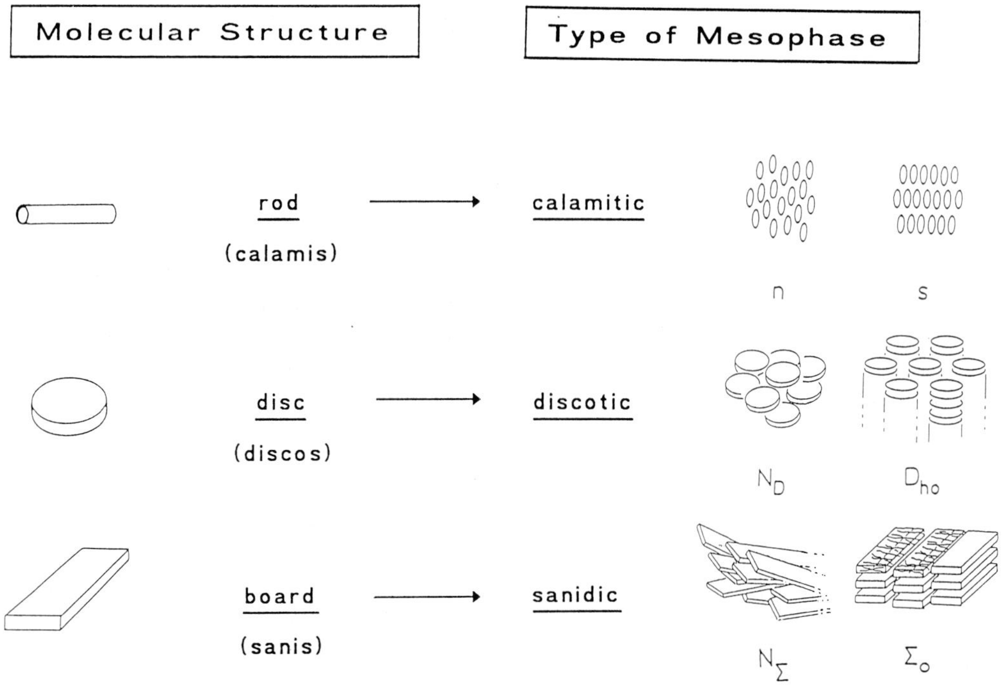

Molecular Structure | Type of Mesophase

rod (calamis) → calamitic — n, s

disc (discos) → discotic — N_D, D_{ho}

board (sanis) → sanidic — N_Σ, Σ_o

Fig. 9.15: Relationship between molecular structure and mesomorphic structure for rod-like, disc-like and board-like mesogens. The greek expressions are in brackets.

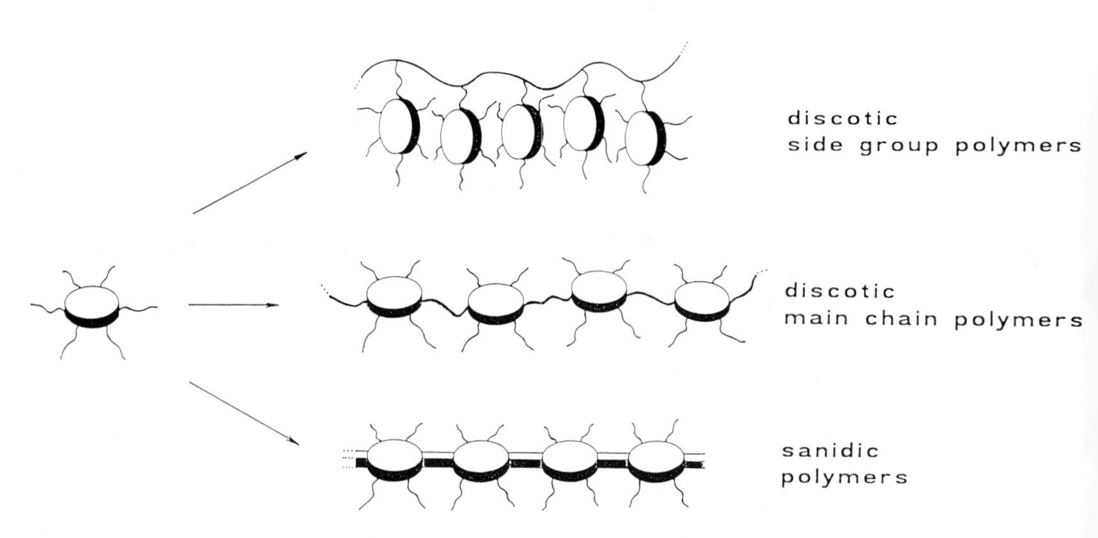

Fig. 9.16: Different ways to prepare discotic and sanidic polymers.

9.2.2.1 Polymers with Discotic Phases

Linking disc-like molecules through *flexible* spacers leads to discotic polymers, i.e. polymers with discotic mesophases. The first discotic polymers were side group systems incorporating hexapentyloxytriphenylene as the disc-like mesogen [15]. More recently some more polymers with disc-like mesogens [16, 64a–d] have been synthesized. Structural variations, however, are still restricted and not to be compared to those of calamitic polymers. The disc-like mesogens so far incorporated are triphenylene ethers [15, 16, 64a–d] and benzene esters [16a], all highly substituted with six long alkyl chains; and alkylated phthalocyanines [65] (only rigid main chain polymers). The backbones are polysiloxanes [15], polymethacrylates [64a], polyacrylates [64c], and polymalonates [64b] for side chain polymers and polyphenylesters [15, 16, 64a–d] for main chain polymers.

Molecular engineering of discoid polymers has so far indicated some *differences with respect to calamitic polymers*. Polymerization generally increases or stabilizes the order for rod-like mesogens, bringing about, for example, smectic phases from nematic monomers or thermodynamically stable nematic mesophases from monotropic nematic monomers (monotropic mesophase = metastable mesophase only observed on cooling). This is not necessarily the case for disc-like mesogens: The discoid side group polymethacrylate in Fig. 9.17, for instance, is not liquid crystalline and amorphous ($T_g = 30°$ C), the preceding monomer, however, shows a highly ordered monotropic D_{ho}-mesophase. Likewise, the influence of the spacer in calamitic and discoid polymers is different. Long spacers do not prevent the formation of smectic mesophases for calamitic polymers, but lead to non liquid crystalline amorphous systems for discoid polymers [16a, 64a]. Thus, the discoid main chain polyester in Fig. 9.17 exhibits a D_{ho} – mesophase for spacer lengths $n = 10$ and $n = 14$, but is non liquid crystalline and amorphous for the higher spacer length $n = 20$.

Recent experiments have shown that if the intracolumnar interactions are enhanced, such as through charge-transfer interactions, discoid polymers can stand more variations of molecular structure (longer spacers, bulkier polymer backbones) without loosing the liquid crystalline phase. In this way, many amorphous discoid polymers with electron rich triphenylene mesogens can be transformed into liquid crystals by adding electron acceptors: e.g. the amorphous side group polymethacrylate and the amorphous main chain polyester ($n = 20$) in Fig. 9.17 by adding 2,4,7-trinitrofluorenone [64c].

To discuss a general *relationship between polymeric structure and mesomorphic properties*, the number of discotic polymers is far too small. For hexapentyloxytriphenylene containing polymers the following relationships are found. Polymerization or a change in polymeric structure do not bring about any change in the phase type but influence the mesomorphic ranges, clearing temperatures and lower transition temperatures (melting point or glass-transi-

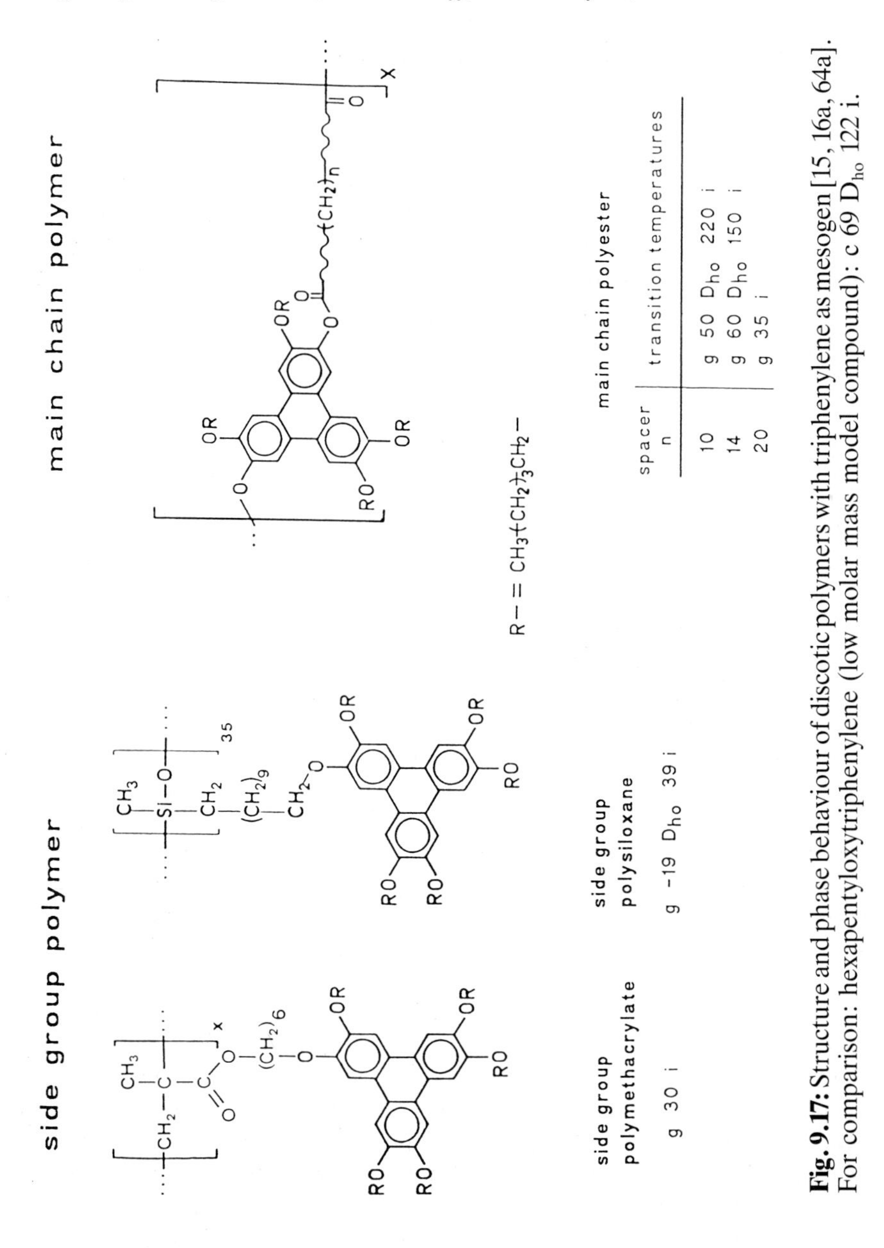

Fig. 9.17: Structure and phase behaviour of discotic polymers with triphenylene as mesogen [15, 16a, 64a]. For comparison: hexapentyloxytriphenylene (low molar mass model compound): c 69 D_{ho} 122 i.

tion) [15, 16a–c] (see Fig. 9.17). The hexapentyloxytriphenylene, its side group polysiloxane and its main chain polyester in Fig. 9.17 all exhibit one identical mesophase type. The temperature range of the D_{ho}-mesophase however varies from 50° C for the hexapentyloxytriphenylene to 70° C for the side group polysiloxane and 170° C for the main chain polyester (n = 10). The clearing temperature is the lowest for the side group polysiloxane ($T_c = 39°$ C) and the highest for the main chain polyester ($T_c = 220°$ C).

The side group polysiloxane and main chain polyester also differ in the intermolecular distances between the mesogens in their liquid crystalline state and most distinctly in their *alignment upon stretching* [66, 16a] (see Fig. 9.18). The side group polysiloxane orients like low molar mass discotics with the column axes *parallel* to the direction of strain [66], whereas the main chain polyester (n = 14) orients in the opposite way with the column axes directed *perpendicular* to the direction of strain [16a]. Which factors are responsible for the orientation of the main chain polyester is not fully investigated yet. Probably, the alignment results from the arrangement of the polymer chain which links discs of neighbouring columns rather than those of the same column. This is supported by ^{2}H-NMR measurements of core- and tail-deuterated derivatives of the main chain polyester, the corresponding dimer and the monomer [67]. There are further factors which may also play an important role in the alignment of main chain discotic polymers, such as molecular weight (as found for smectic polymers [60]) as well as the relative flexibility of backbones and mesogenic tails.

In order to answer the above and many additional open questions in the field of discoid polymers, the scope of structural variations has to be expanded. This will be one task for the synthetic chemist in the near future. The major tasks, though, will be to synthesize functionalized and less viscous polymeric discotic systems, and also to look at some potential applications that have so far only been discussed for low molar mass discotics [68] (e.g. as one-dimensional conducting materials, as stationary phases for liquid chromatography). To lower the viscosity, polymer backbones and side chains can be modified, or polymers with highly fluid nematic-discotic phases should be made. To introduce functionality, many approaches are possible; one of the easiest is certainly by inserting functionalized low molar mass dopants [64c].

9.2.2.2 Polymers with Sanidic Phases

Linking disc-like mesogens *rigidly in the main chain* leads to sanidic polymers, i.e. polymers with sanidic mesophases. The first sanidic polymer was a fully aromatic polyamide obtained through condensation of tetrasubstituted p-phenylenediamine and disubstituted terephthalic acid dichloride [19a]. The mesomorphic structure, as determined by X-ray investigations of oriented fibres, is neither calamitic nor discotic, but shows characteristics of both systems: the well defined layered structure of smectic (calamitic) phases and the one-dimensional packing in stacks of columnar (discotic) phases (see Fig. 9.19). The layer spacing (2.85 nm) corresponds to the maximum cross-section of the macromolecules perpendicular to their backbones.

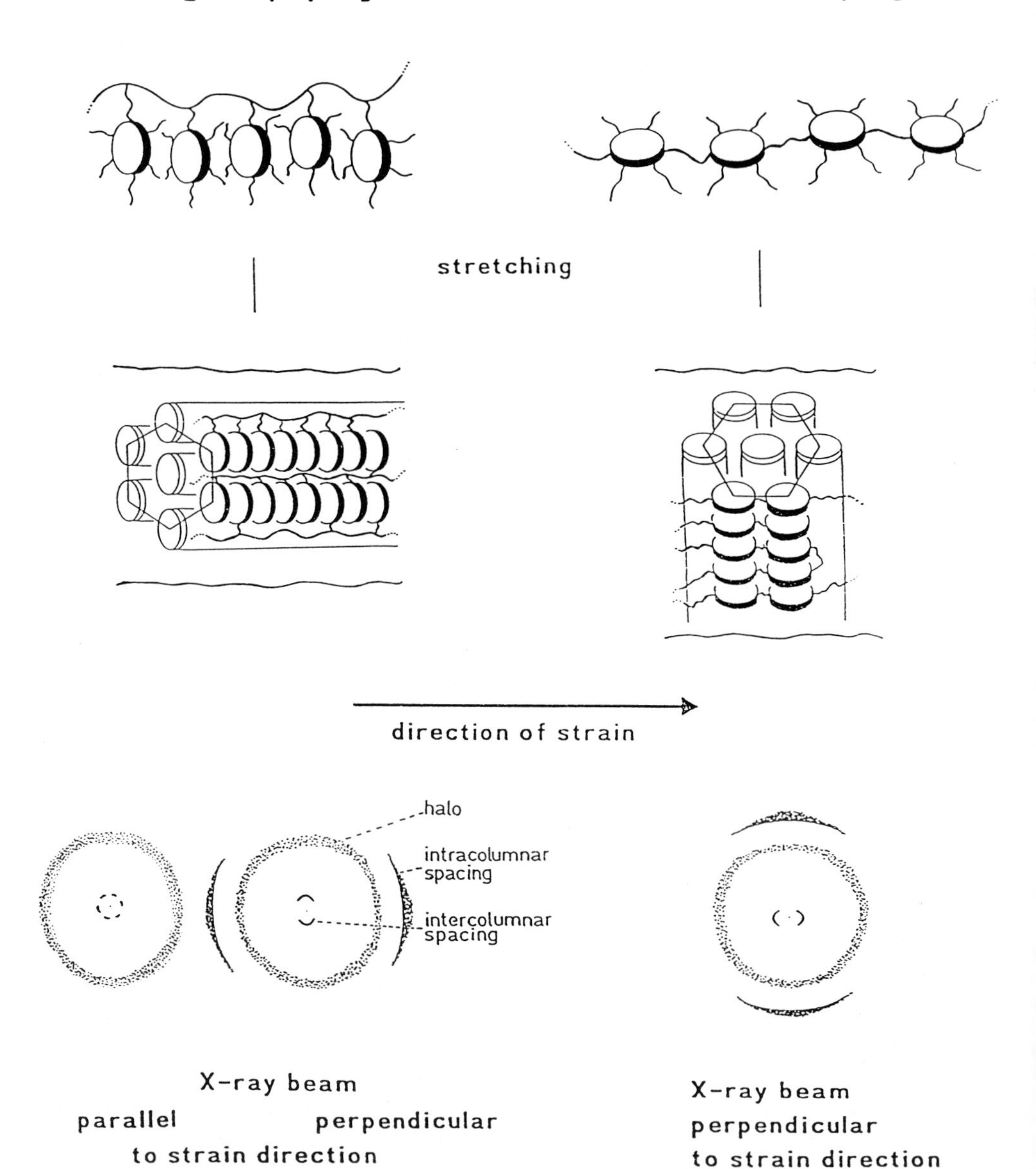

Fig. 9.18: Alignment of a discotic side group polysiloxane [66] and a discotic main chain polyester [16b] on stretching.

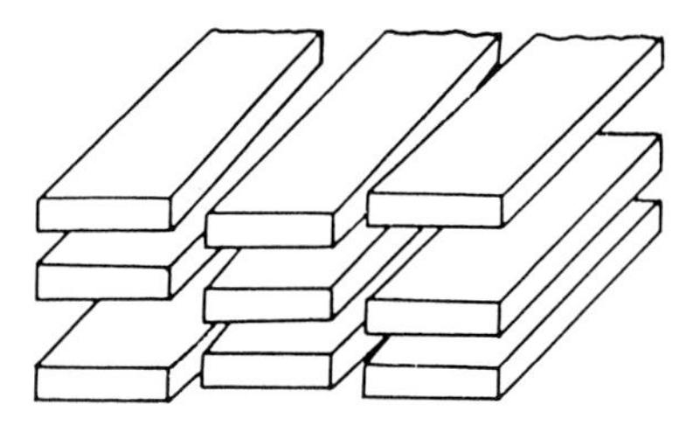

$R_1 = -\overset{O}{\overset{||}{C}}OC_{12}H_{25}$

$R_2 = -OC_{12}H_{25}$

c 17 Σ_{d_1} 70 Σ_{d_2} 130 i

fibre axis

halo 0.63 nm 2.85 nm

X-ray pattern
2.85 nm = layer spacing
0.63 nm = regularity of the stiff backbone

Fig. 9.19: Chemical structure, X-ray pattern and structural model for a sanidic disordered phase Σ_d of a fully aromatic polyamide [19a].

This new mesomorphic structure called *"sanidic disordered phase"* has also been found for many similarly structured polyamides and -esters [19b, 20, 22]. It is explained by the new geometric shape of the mesogens acquired only through polymerization. This is a board-like structure on the average since the long densely packed alkyl chains hinder rotation around the main chain axis and thus prevent the macromolecule from adopting a rod-like shape.

Meanwhile, molecular engineering has produced two further types of sanidic phases: the higher ordered *"sanidic ordered phase"* [69] and the less – merely orientationally ordered – *"sanidic nematic phase".* The former differs from the *"sanidic disordered phase"* (in Fig. 9.19) in that the spacing between the boards of each layer is constant. The latter is obtained for a polyamide dissymmetrically substituted with ethylene oxide chains [63] (see Fig. 9.20).

The sanidic nematic phase is of special theoretical interest due to its simultaneous nematic and biaxial character. Contrary to the ordinary uniaxial nematics, orientational order now exists in two directions, giving the phase

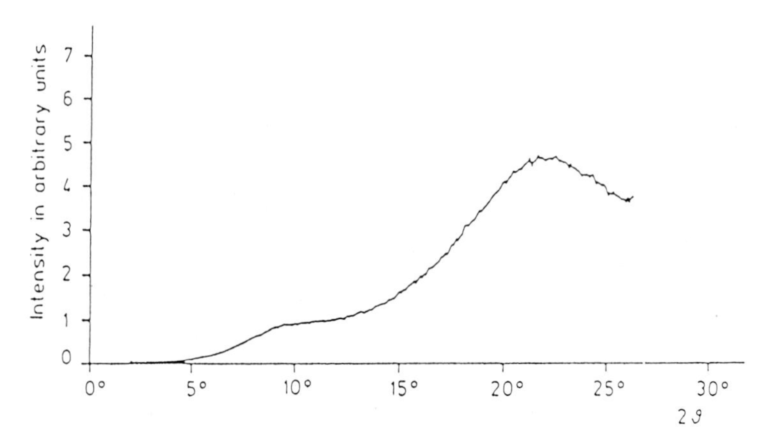

wide-angle X-ray diagram of the biaxial nematic phase

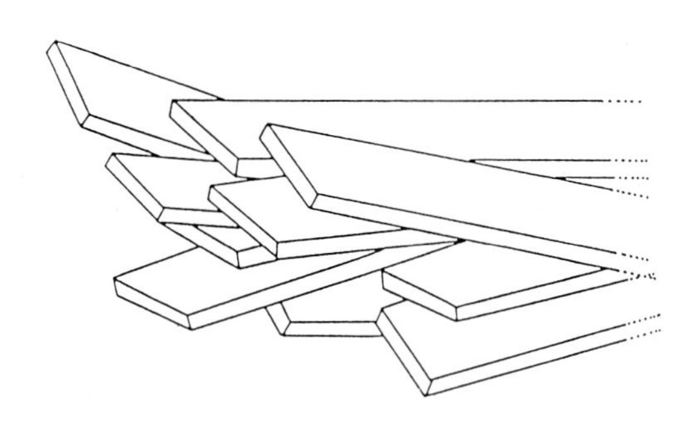

$R = -O(CH_2CH_2O)_2CH_3$

$R' = -OCH_3$

c 131 N_{Σ} 216 i

two directors. Biaxiality has already been reported for some other nematic systems [17b,c, 70], however, only on the grounds of conoscopic investigations. For the sanidic polyamide [63] further evidence for present biaxiality is provided by X-ray investigations (see Fig. 9.20): two halos are found, one corresponding to the packing of the boards along their normal, the other corresponding to the packing in a direction perpendicular to both their normal and their main chain axis.

Future work on sanidic polymers will certainly, as is the case for discotic polymers, concentrate on introducing functionality to make the sanidics more attractive to applied chemists.

9.2.3 Functionalized Liquid Crystalline Polymers

Besides the variations in molecular architecture of liquid crystalline polymers which has led to the use of mesogens of different shape (rods, discs and boards) and to different arrangements of mesogens (see Fig. 9.6), *functionalized LC polymers* have been synthesized and investigated. These polymers were functionalized by dye containing groups [25, 26] (see section 9.2.3.1), or by mesogens undergoing photoreactions [28, 29]. This can lead subsequently to a formation or destruction of the liquid crystalline phase (see section 9.2.3.2). Alternatively liquid crystalline polymers were functionalized by groups capable of undergoing chemical reaction [27] and by chiral groups [30, 31]. This led to the formation of chiral LC elastomers (see section 9.2.3.3).

9.2.3.1 Dye Containing Polymers

The incorporation of dichroic dyes into liquid crystalline polymers enables the formation of coloured LC copolymers [25]. Due to the covalent linkage of dyes and mesogens via the polymer chain, high dye concentrations can be obtained without a demixing, while in low molar mass liquid crystals the solubility of dyes is very limited. In these dye-containing polymers the dyes orient with a high order parameter parallel to the preferred orientation of the mesogenic groups [26]. Different orientations of the dyes along with the mesogens lead to different colours of the polymers if they are viewed with polarized light. The possibility of orienting and reorienting the dye containing LC polymers in electric and magnetic fields [50–52] makes these polymers very interesting as media for optical data storage and as additives for colour displays [26].

◀
Fig. 9.20: Chemical structure, X-ray diagram and structural model for a sanidic nematic phase of a fully aromatic polyamide with ethylene oxide chains [63].

It is found that the structure of the dye containing comonomer modifies the width of the liquid crystalline phase. This is schematically presented in Fig. 9.21. The incorporation of an anthraquinone dye, which is not mesogenic by itself, leads to a narrowing of the liquid crystalline phase, which is lost at high dye contents. Nevertheless, up to 30 % by wt of the dye can be incorporated into the copolymers while still retaining the liquid crystalline phase. For low molar mass liquid crystals the maximal concentration of anthraquinone dyes is limited to a few percent, due to the very low solubility of these dyes.

If, however, dyes of an elongated rod-like structure that resemble mesogens are incorporated into LC polymers, a totally different behaviour is found. This is presented for a tris-azo-dye in Fig. 9.21. The incorporation of this strongly mesogenic dye leads to a broadening of the liquid crystalline phase.

9.2.3.2 Photosensitive Liquid Crystalline Polymers

Photosensitive LC polymers (see Fig. 9.22) can be prepared by the use of mesogens with photocleavable lateral substituents [29], by the incorporation of photochromic dyes into the polymer [28], or by the use of mesogens that undergo photoisomerizations [71]. The endproducts of photoisomerizations or photoreactions of mesogenic molecules will have, in general, a higher or lower tendency towards liquid crystallinity than their precursors. Therefore, it is possible to stabilize or – in the more common case – to destabilize a liquid crystalline phase via a photoreaction.

The photoisomerizations of azobenzene [71] and of 1-iminopyridinium ylides [29] can be used for a *destabilization or destruction of the liquid crystalline phase*. The reversible isomerization of azobenzene is especially interesting in this context because it reversibly modifies the optical properties of azo-group containing polymers upon illumination with light of an appropriate wavelength [71]. The reason for this is that the rod-like trans-configuration of azobenzene is transformed into a non-rod-like, non-mesogenic cis-configuration. These changes are obvious from the UV spectra displayed in Fig. 9.23. They may occur even in the solid glassy state.

This trans-cis transition induces a variation in the anisotropic surroundings of the azobenzene unit and consequently, also of the optical properties. Strong changes are introduced both with respect to the extinction and the refractive index. An important finding is that these changes remain stable in the glassy nematic or smectic state even if the azobenzene experiences a thermal backrelaxation (see Fig. 9.23) into the trans-configuration.

This effect can be exploited for optical application such as holographic storage of information or the holographic manufacturing of optical components such as lenses. Fig. 9.24 gives an impression of the quality of the liquid crystalline films as storage materials. Results obtained on holographically induced cross gratings are shown: the resolution is better than 0.3 μm, corresponding

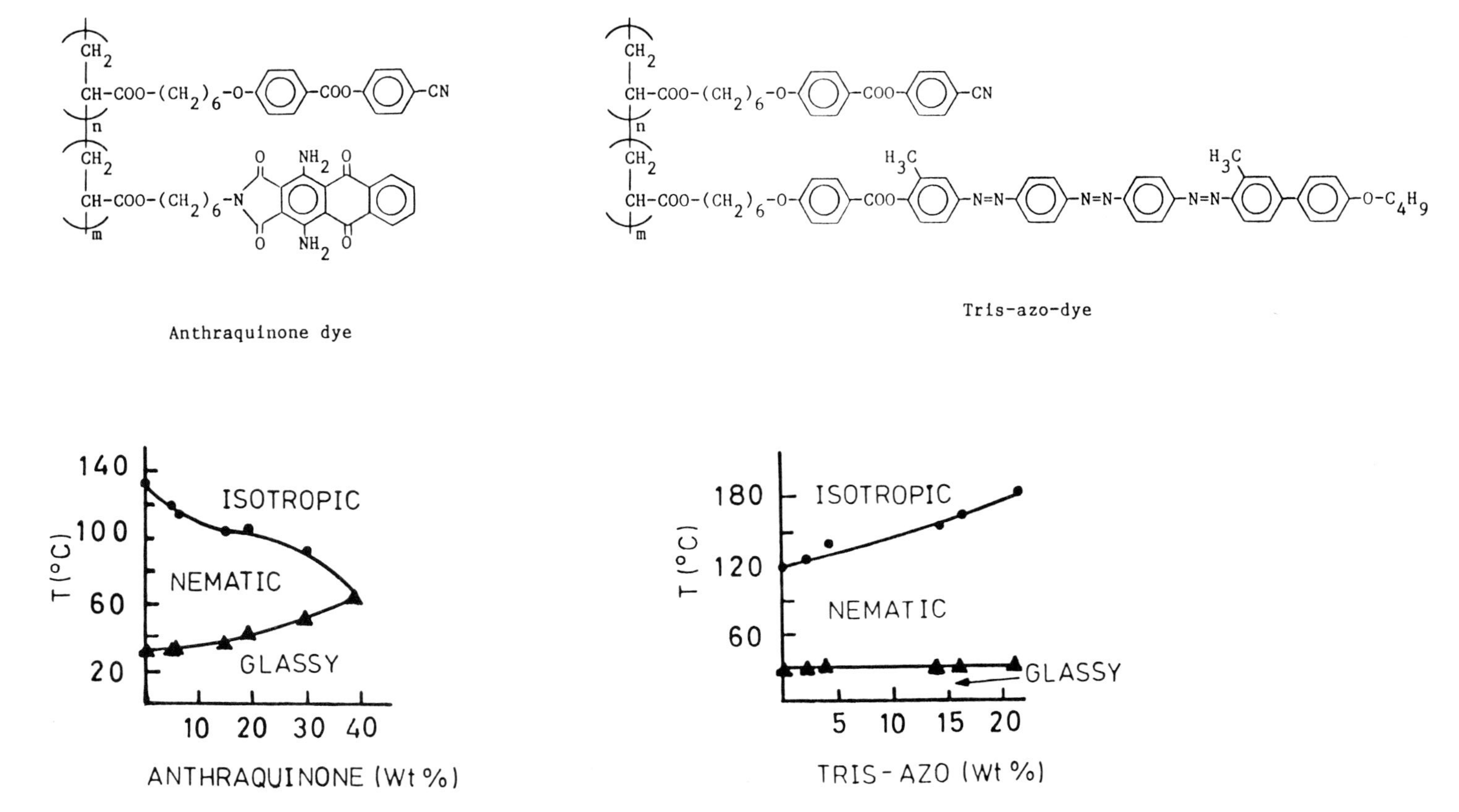

Fig. 9.21: Dependence of the width of the liquid crystalline phase on the dye content for copolymers with anthraquinone dyes or tris-azo-dyes [25].

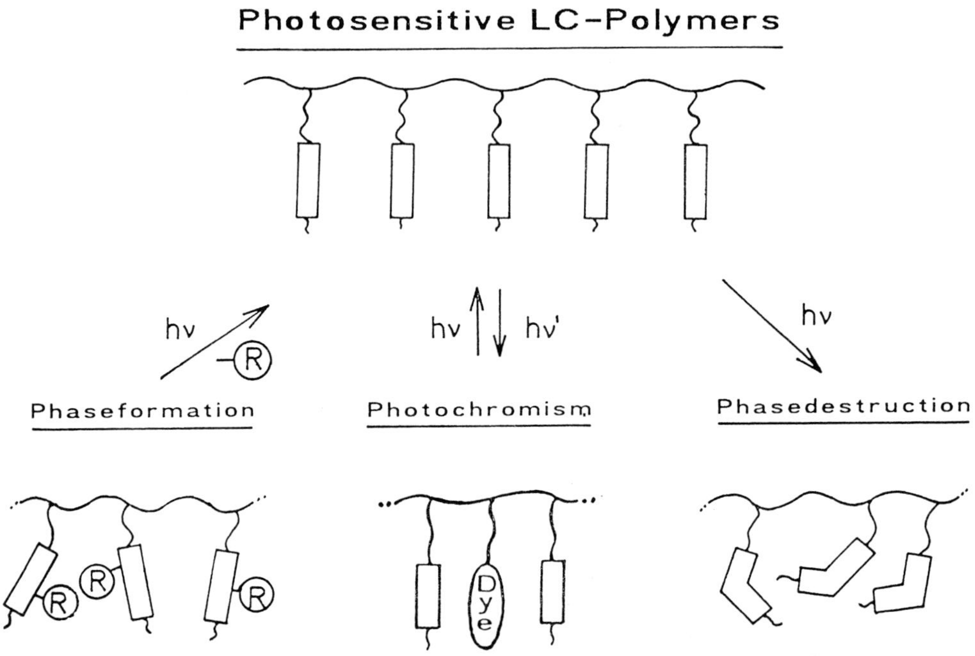

Fig. 9.22: Depending on the kind of photoreaction, the irradiation of photosensitive LC polymers can either lead to a phase formation, to photochromism or to a phase destruction.

to an information density of 1 Gbit/cm^2 and the efficiency, as defined by the ratio of the first order diffraction intensity and of the primary beam intensity, amounts to about 50 %. Recent experiments have shown that transmission and reflection holograms of real objects can be stored reversibly in such liquid crystalline films.

In contrast to the phase destruction of azobenzene containing polymers the *formation of a liquid crystalline phase* can be induced on irradiation. This is observed if mesogenic groups are modified with cleavable lateral substituents (see Fig. 9.25) [29]. The bulky lateral substituents suppress the formation of liquid crystalline phases, but after their cleavage, liquid crystalline phases can be formed. This cleavage can be induced photochemically via the formation of strong protic acids from sulfonium salts, which split off the labile lateral substituents (the TBOC-protective group in Fig. 9.25).

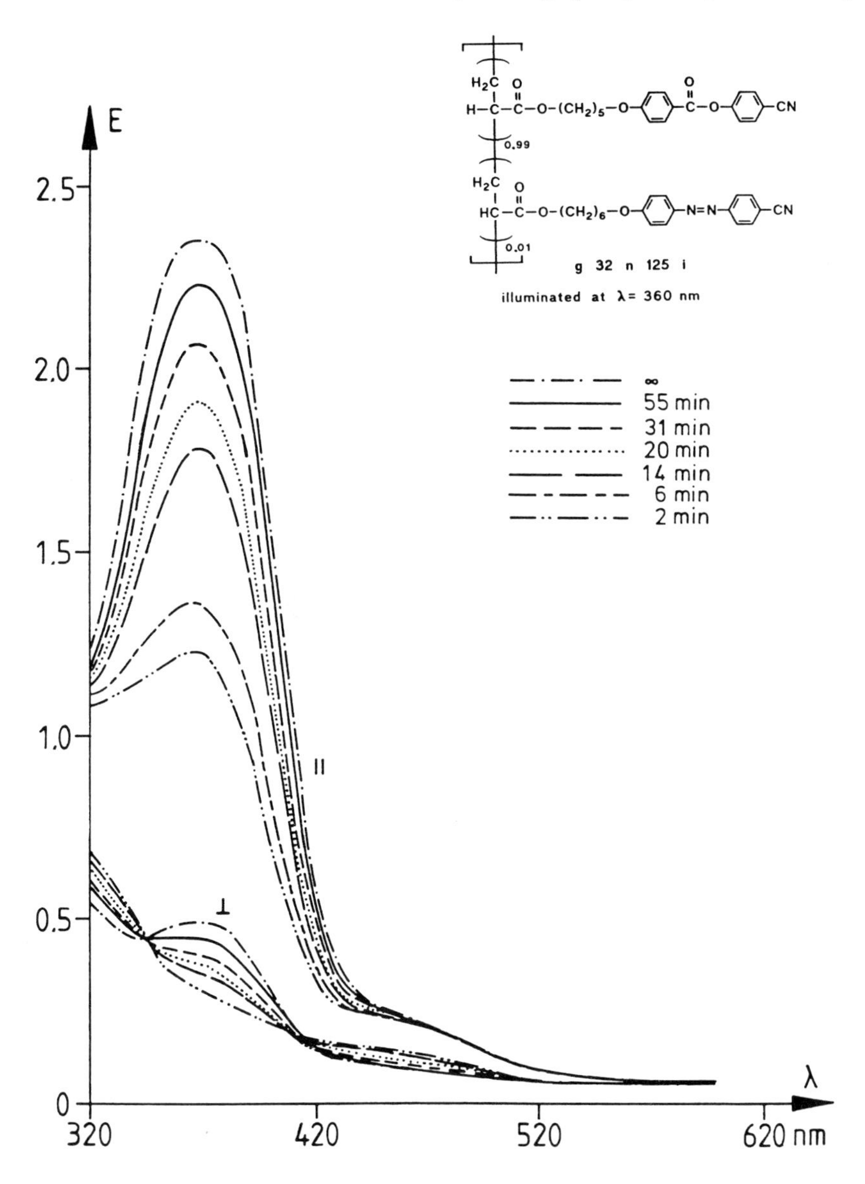

Fig. 9.23: UV spectrum of an azobenzene containing side chain polymer after irradiation measured parallel (||) and perpendicular (⊥) to the director. Parameter is the time after the irradiation.

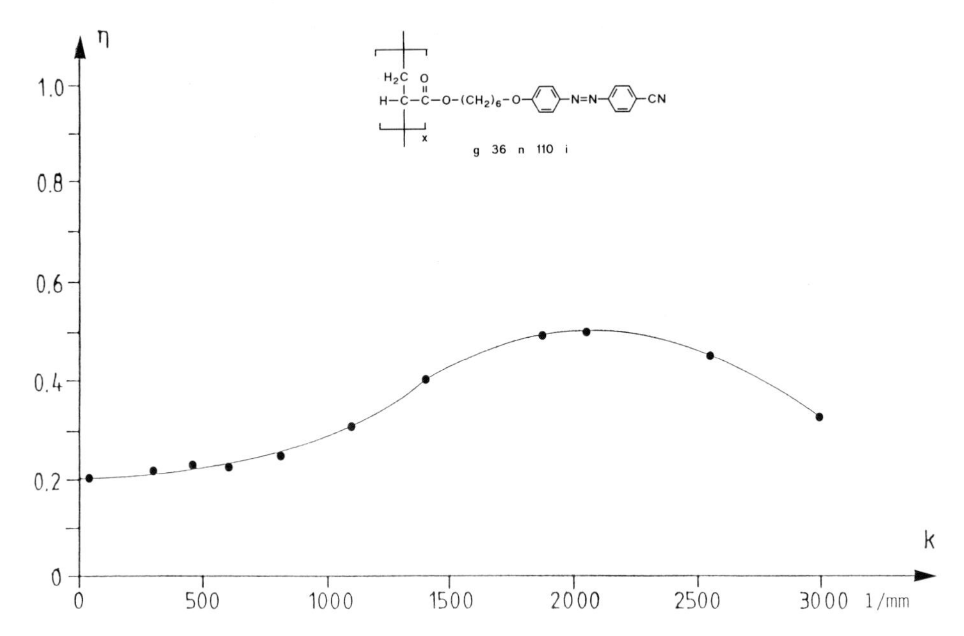

Fig. 9.24: Dependence of the diffraction efficiency on the spatial frequence, as obtained from grating experiments performed on a monodomain of a side chain liquid crystalline polymer. The gratings were obtained holographically.

9.2.3.3 Liquid Crystalline Elastomers

Another type of functionalized LC polymer is the LC elastomer. LC elastomers can be prepared in a one step reaction (see Fig. 9.26a), modifying a functionalized polymer backbone with mesogenic side chains and simultaneously cross-linking [72]. Alternatively, they can be prepared in 2 steps: First, soluble functionalized LC polymers are synthesized which are then cross-linked in a second step [27, 73] (see Fig. 9.26b). We have used this second method to prepare slightly cross-linked side group polymers, main chain polymers and combined main chain/side group polymers (see Fig. 9.27) with elastic properties. These polymers are especially interesting because of their good *mechanical orientability*. A strain of 20–40 % leads to a well oriented sample if it is applied in the liquid crystalline phase [60, 74, 75]. The same strain produces only a small orientation, if applied in the isotropic phase [75].

If liquid crystals are functionalized with chiral groups, then some liquid crystalline phases are transformed into *chiral phases* with a helical superstructure: In this way a nematic phase can be transformed into a *cholesteric phase* which shows *selective reflection* of light, and a smectic C phase can be transformed into a *chiral smectic C* phase* which has *ferro-electric properties* (see Fig. 9.28)

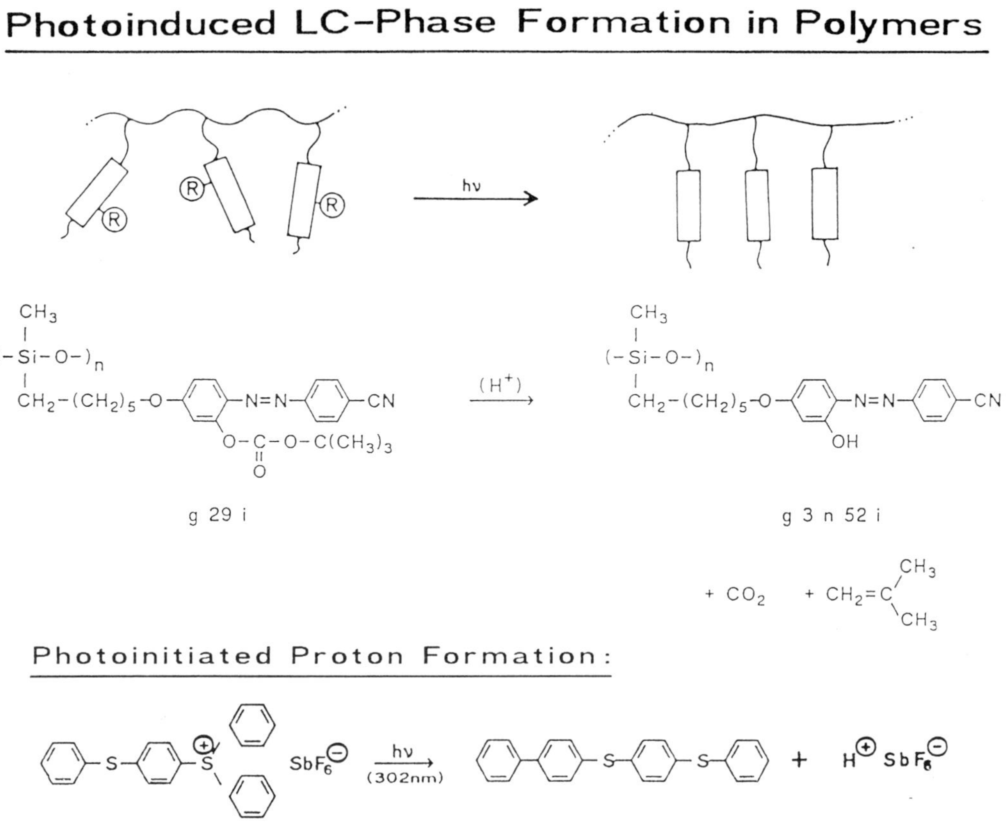

Fig. 9.25: Photoinduced phase formation via a photoinitiated proton formation [29].

[2–4,7]. The properties of these chiral phases are linked to their helical superstructure. They can, therefore, be manipulated by a partial or complete untwisting of this helical superstructure, which is achieved for low molar mass liquid crystals by strong electric or magnetic fields. It seemed, therefore, interesting to investigate if similar effects – an untwisting of the helical superstructure – could be achieved in chiral LC elastomers by mechanical strains. If this is the case, then these chiral elastomers should act like a device that transforms a mechanical signal (the strain) into an optical signal (selective reflection of light for the cholesteric phase) or into an electric signal (spontaneous polarization for the chiral smectic C* phase) [76]. This is schematically illustrated in Fig. 9.28.

In order to prepare *chiral liquid crystalline elastomers*, two functions have to be combined in one polymer: chirality to obtain chiral phases and reactive groups for the cross-linking reaction. Since combined main chain/side group

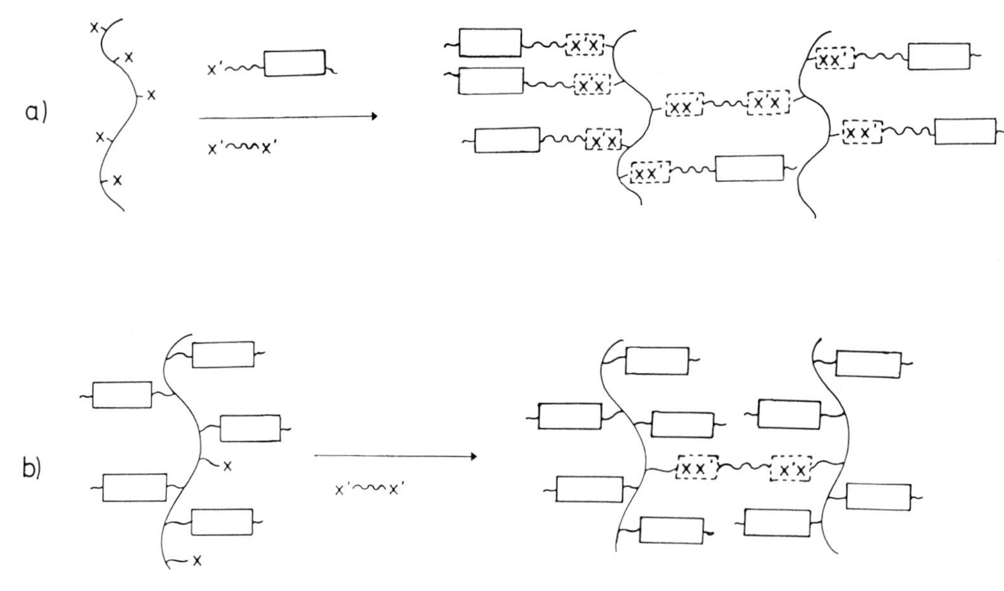

Fig. 9.26: Different ways to prepare LC elastomers in one (a) or two (b) steps [27, 72].

polymers form broad liquid crystalline phases which are often of the smectic C and nematic type, they seem especially interesting in this respect. Therefore, in a first step, noncross-linkable combined main chain/side group polymers with chiral groups have been prepared [30, 31] which show cholesteric and chiral smectic C* phases. The additional functionalization of these polymers with reactive groups can be achieved by the preparation of copolyesters (see Fig. 9.29) in which one half of the mesogenic side groups is functionalized with chiral tails, while the other half is functionalized with olefinic double bonds [76, 77]. These chiral copolyesters can be cross-linked by a hydrosilylation reaction between some of the olefinic double bonds and the Si-H bonds of a α-ω-dihydrooligo(dimethylsiloxane) as described for one example of cross-linked combined polymers (see Fig. 9.27) in Fig. 9.29.

First measurements on these chiral elastomers [76] show that the analogy between electrical (low molar mass liquid crystals) and mechanical fields (LC elastomers) really exists and that it is possible to untwist the helical superstructure by mechanical stretching, as presented in Fig. 9.28.

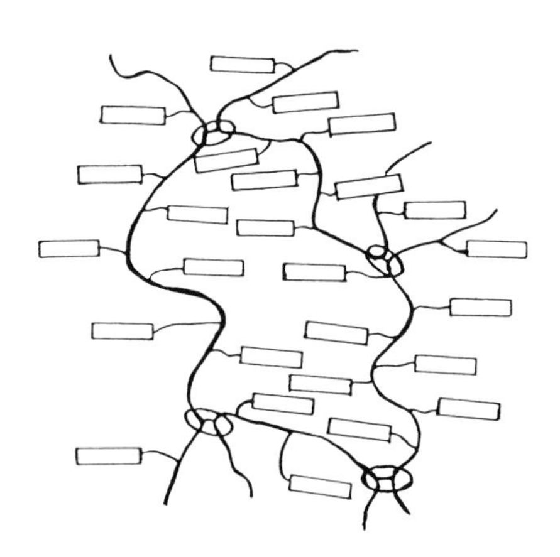

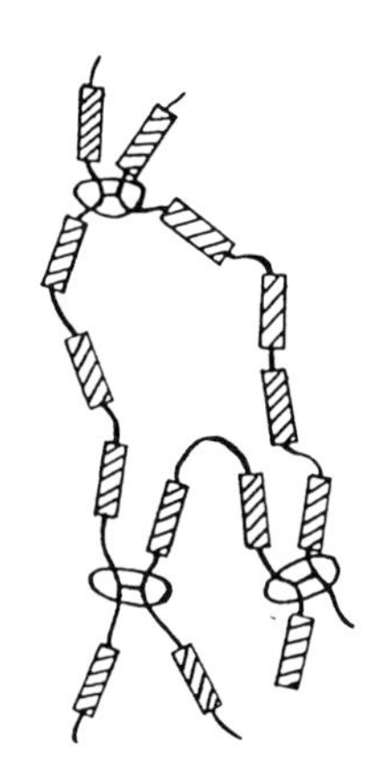

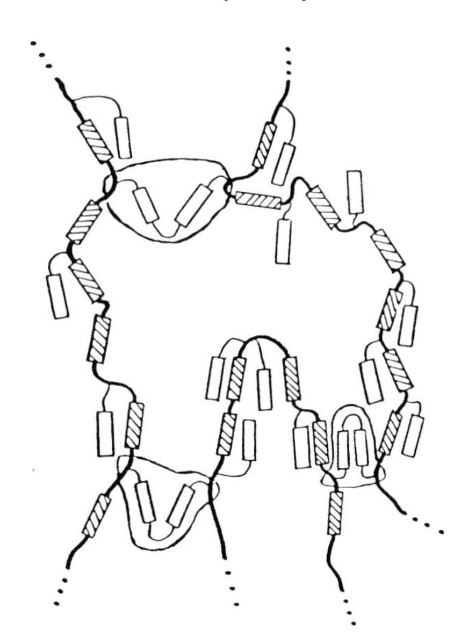

Fig. 9.27: Schematic drawing of networks prepared from side group, main chain and combined main chain/side group polymers [27].

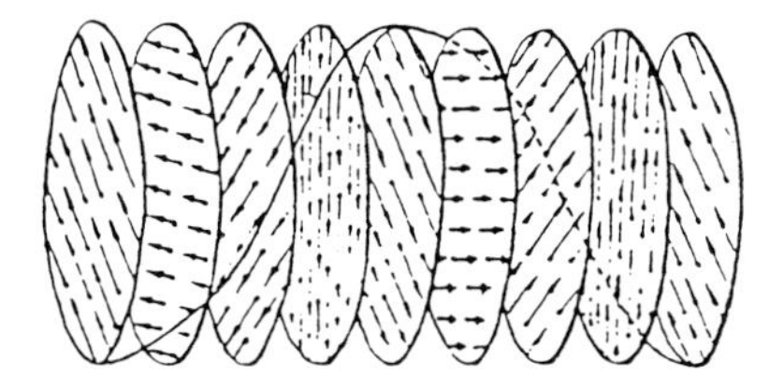

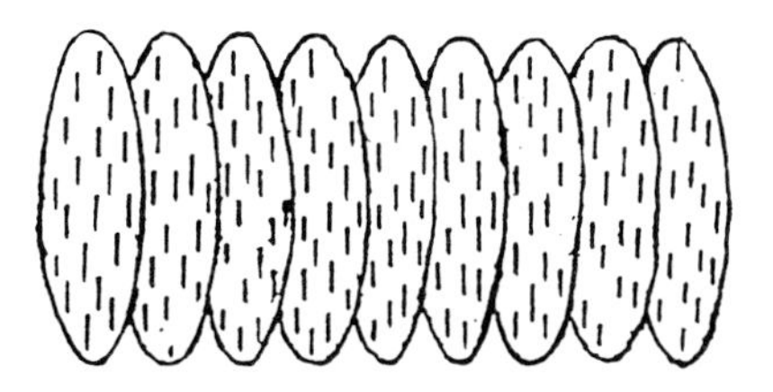

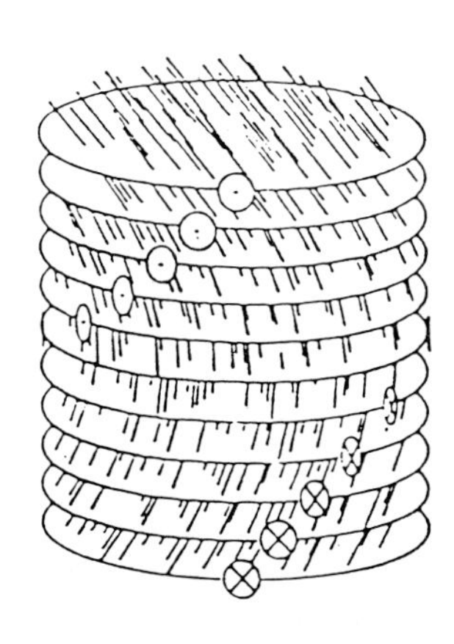

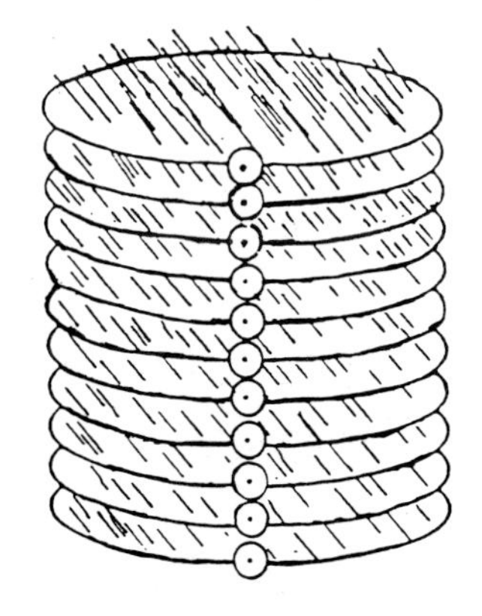

Fig. 9.28: Twisted (a and c) and untwisted (b and d) states of cholesteric and chiral smectic C* phases [76].

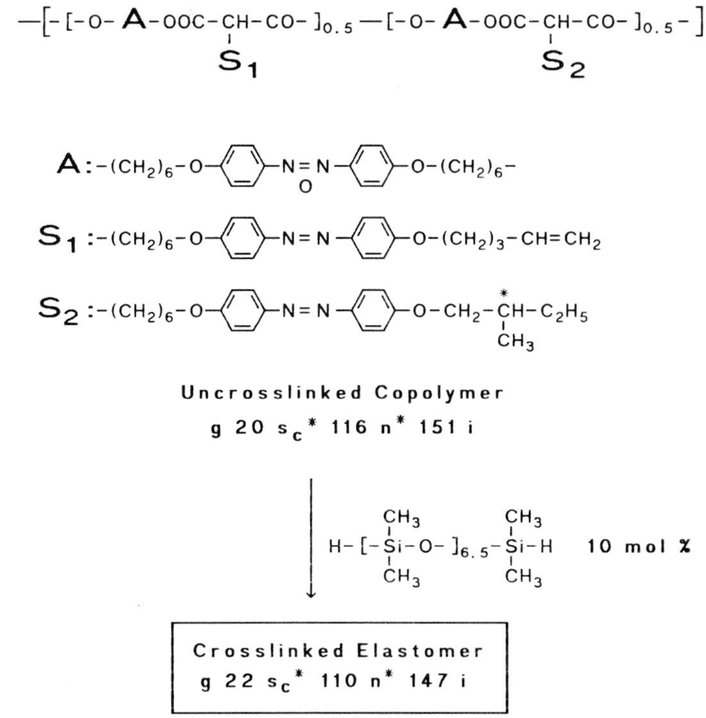

$$-[-[-O-\mathbf{A}-OOC-\underset{S_1}{CH}-CO-]_{0.5}-[-O-\mathbf{A}-OOC-\underset{S_2}{CH}-CO-]_{0.5}-]_z-$$

$\mathbf{A}$: $-(CH_2)_6-O-C_6H_4-N(O)=N-C_6H_4-O-(CH_2)_6-$

S_1: $-(CH_2)_6-O-C_6H_4-N=N-C_6H_4-O-(CH_2)_3-CH=CH_2$

S_2: $-(CH_2)_6-O-C_6H_4-N=N-C_6H_4-O-CH_2-\overset{*}{C}H(CH_3)-C_2H_5$

Uncrosslinked Copolymer
g 20 s_c^* 116 n^* 151 i

↓ $H-[-Si(CH_3)_2-O-]_{6.5}-Si(CH_3)_2-H$ 10 mol %

Crosslinked Elastomer
g 22 s_c^* 110 n^* 147 i

Fig. 9.29: Molecular structure of a chiral, uncross-linked copolymer and its transformation into a cross-linked elastomer (see Fig. 9.27) via a hydrosilylation reaction [76, 77].

9.3 Molecular Architecture and Physical Properties

9.3.1 The Amorphous, Crystalline and Liquid Crystalline State of Matter

Liquid crystals represent a state of matter which combines, in different ways, the properties of a crystal having long range positional and orientational correlations, with those of an amorphous material, having only short range correlations. This gives rise to a multitude of different macroscopic structures which have been extensively documented in the literature [78].

The molecular structure of liquid crystal polymers and the dynamical behaviour of individual molecules and groups of molecules is now being subjected to increasing scrutiny because of their paramount importance in determining the physical properties of the system. Some of the topics in relation to liquid crystal polymers which are gaining increasing importance both in basic research as well as for their potential technological applications are:

a) Non-linear optical properties for application in optical data storage, b) electrical conductivity in precisely ordered structures by controlled charge transfer, c) collective phenomena involving long-range forces which influence the glassy state, d) the physics of critical phenomena, in order to understand and control segregation effects, with potential application for the fabrication of membranes with specific pore sizes and distributions, e) good orientability of main chain systems giving high tensile strength.

Future development in all these fields requires precise knowledge about the molecular structure of these materials and their defects.

Unfortunately nature does not divulge this information easily. The structure of matter as well as its time dependent fluctuations can be measured by various elastic and inelastic scattering techniques such as X-ray, neutron, electron and light scattering as well as by spectroscopic methods such as deuteron nuclear magnetic resonance.

All these experimental methods, as well as interpretation of the results, require specialized knowledge about the crystalline and the amorphous state which has been developed by various groups within the framework of this Sonderforschungsbereich. Only a fraction of the work on amorphous and semi-crystalline polymers which has been published in the specialized literature is indicated here. Thus an enormous research effort has been invested and experience gained in the investigation of the static properties such as polymer chain conformation [79–81], orientational order [82], correlation functions [83], morphology of the semi-crystalline state [84–86], density fluctuation [87] and on dynamic properties of molecules or groups of molecules from light [88] and neutron scattering [89] as well as ^{2}H-NMR-spectroscopy [90]. In the following, therefore, we indicate only a few examples of investigations which have been performed on liquid crystalline systems, specifically those which were introduced in the first part of this chapter.

9.3.2 Fluctuations in Liquid Crystalline Polymers

9.3.2.1 Defining Fluctuations

When discussing the characteristic structures of liquid crystalline phases, one tends to forget that molecular motions such as rotation or translations take place leading to molecular configurations which fluctuate rapidly as a function of time. The fact that the average orientational order parameter $S = \frac{1}{2}(3\cos^2\theta - 1)$, for instance, is zero in the isotropic phase does not imply that fluctuations of the order parameter about its mean value do not take place. This can lead to strong orientational correlations already in the isotropic fluid phase [91–94]. Statistical treatments, based on a consideration of particle distribution functions, particle correlation functions, and correlation parameters connect the fluctuations with thermodynamic, dynamic and structural properties.

It is for this reason that the experimental analysis of such fluctuations yields valuable information on macroscopic properties. Topics of great interest are, for instance – in view of the molecular design of liquid crystalline polymers – an investigation of the difference between the properties of low molar mass and polymeric liquid crystals and the detection of liquid crystalline phase transitions which are hidden below a glass transition (kinetically hidden) or below another liquid crystalline phase transition (thermodynamically hidden). Examples will be given below.

9.3.2.2 Kerr Relaxation Studies on Orientational Fluctuations

Orientational fluctuations, which occur in the isotropic phase near a nematic one, lead to strong electrooptical effects. The application of a static or dynamic electric field causes the induction of birefringence Δn, which depends linearly on the square of the electric field (Fig. 9.30). The Kerr constant K, defined by:

$$\Delta n = \lambda \, K \, E^2$$

is thus a measure of the extent of orientational fluctuations.

Kerr effect studies [91–94] have actually shown that, both in low molar mass and in polymeric liquid crystals, strong orientational order parameter fluctuations occur already in the isotropic phase, and that their magnitude increases with decreasing temperature. The reason for this is that the number of molecules or mesogenic units performing correlated reorientational motions, as reflected in the magnitude of the orientational correlation parameter g_2, increases and tends to diverge as a characteristic temperature T^* is approached. This temperature is located close to, but below, the transition temperature T_{ni} into the nematic phase. These features can be accounted for – as described in detail by Jungnickel and Wendorff (Chapter 13, this Volume) – on the basis of the phenomenological Landau de Gennes theory, which essentially states that the nematic-isotropic transition is a weak first order transition but only a weak one. Pretransitional fluctuations characteristic of a second order phase transition, therefore, must be expected. This can be exploited as follows, for the detection of hidden transitions.

Fig. 9.31 shows the results of Kerr effect studies which were obtained for a side group polymer in which liquid crystalline structures could not be detected: the isotropic phase is frozen in the glassy state at lower temperatures. The Kerr

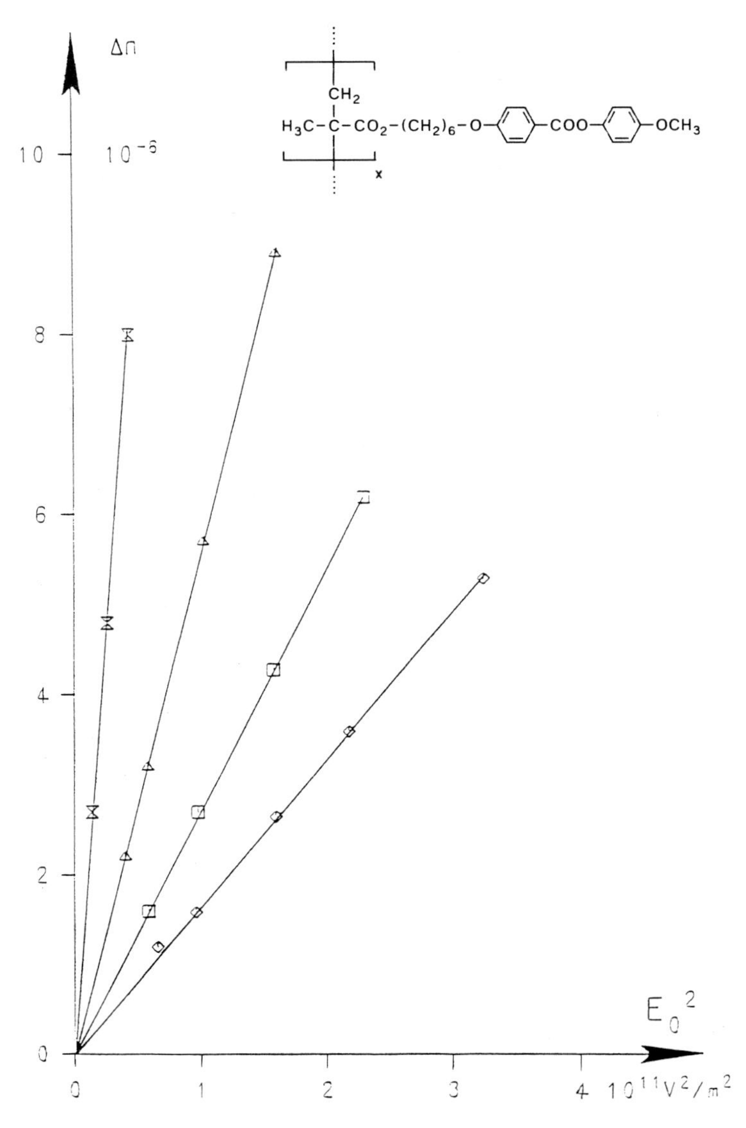

Fig. 9.30: Variation of the electrically induced birefringence with the square of the applied electric field, as obtained for a side group polymer in the isotropic melt neighbouring a nematic phase. Parameter is the temperature.
($_V$ T = 383.15 K, △ T = 383.65 K, □ T = 384.25 K, ◇ T = 388.05 K)

data show, however, that a nematic phase becomes stable with respect to the isotropic phase at temperatures well within the glassy state: the Kerr constant approaches infinity in this temperature range. By adding a low molar mass solvent and allowing it to evaporate, it is subsequently possible to induce the nematic phase, due to the different shifts of the glass and the nematic-isotropic transitions. The general result of the Kerr effect studies is that the absolute magnitude of the correlation parameter, g_2, at a given reduced temperature, T/T_{ni}, as well as the quantity T_{ni}-T are independent of the chain architecture and identical to those found for low molar mass liquid crystals. Thus, the static aspects of orientational fluctuations and the corresponding thermodynamic properties are very similar for low molar mass and polymeric liquid crystals. The dynamic features, on the other hand, which can be deduced from the transient Kerr response, were quite different. The orientational relaxation times are increased by three to five orders of magnitude as one goes from low molar to polymeric liquid crystals, due to the increase of the rotational viscosity (see also Fig. 9.9).

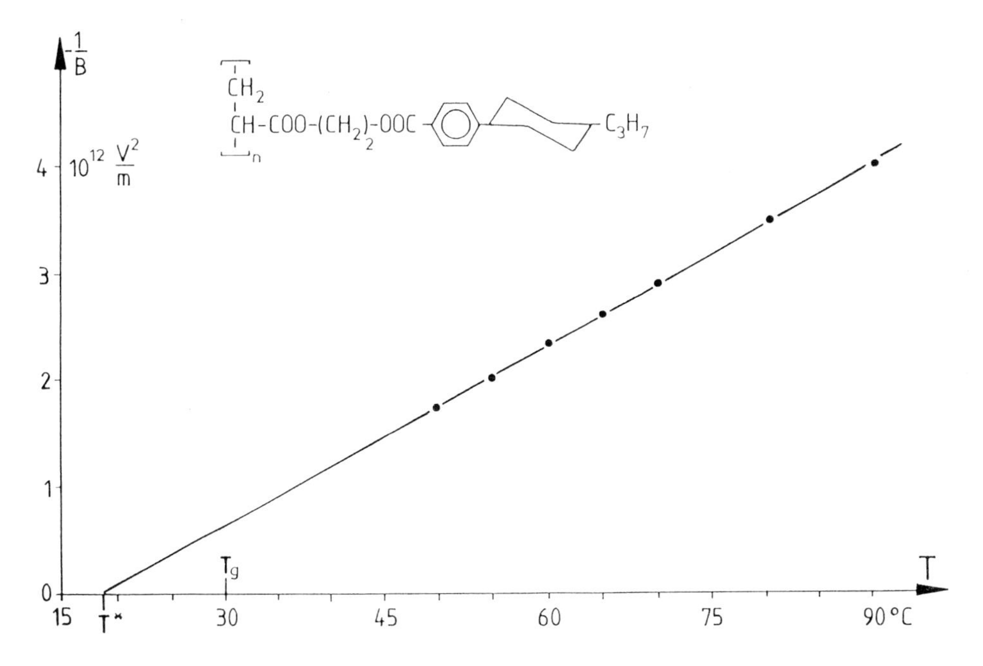

Fig. 9.31: Kerr effect results obtained for a side group polymer displaying a kinetically hidden nematic phase transition.

9.3.2.3 Small Angle X-Ray Scattering Studies on Orientational Correlations in Side Group Polymers

The orientational correlations analyzed by Kerr relaxation studies refer to the mutual orientation of the long axes of pairs of molecules or of mesogenic units. Using absolute small angle X-ray scattering one is able to characterize still another type of orientational correlation in liquid crystalline polymers, namely the correlation between the direction defined by the vector which connects the centers of pairs of molecules and the directions of the long axis of the molecules. There will be a random distribution – on the average – in the isotropic melt, but fluctuations about this mean configuration might occur. Theoretical considerations [95–96] show that such fluctuations lead to variations of the electron density and may thus be detected by X-ray scattering.

The absolute small angle X-ray scattering of fluids is controlled by thermal density fluctuations as long as no orientational correlations of molecular axes

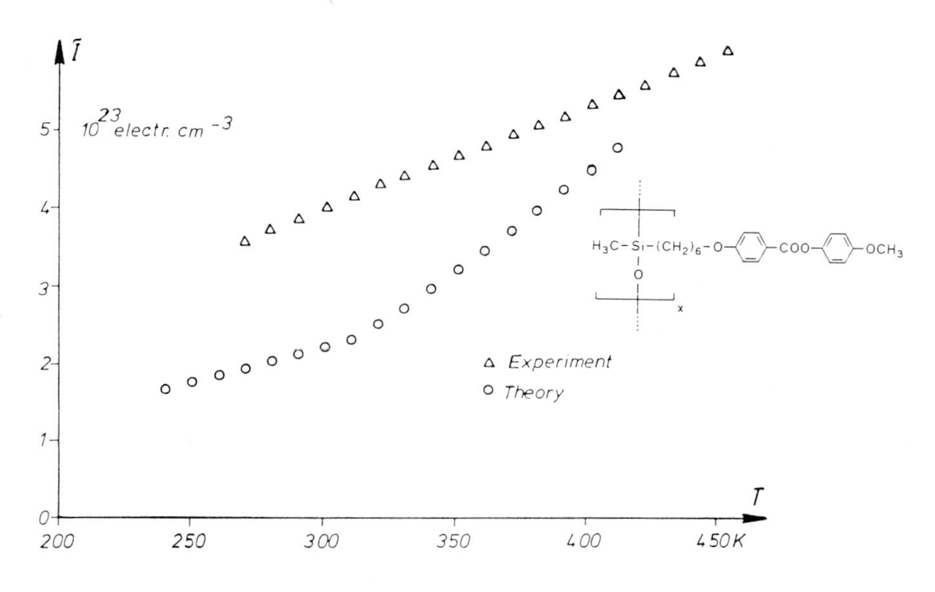

Fig. 9.32: Absolute small angle X-ray scattering intensity as a function of the temperature obtained in the isotropic and liquid crystalline state of a side group liquid crystalline polymer. The data are compared with those, expected to result just from thermal density fluctuations.

occur. Studies on amorphous polymers and low molar mass fluids have already shown this to be true [95–96]. Fig. 9.32 shows, however, that the small angle X-ray scattering of a liquid crystalline system in the isotropic and the nematic state is characterized by the presence of excess scattering, i.e. by an additional scattering which does not result from density fluctuations. The excess scattering increases with decreasing temperature and behaves continuously at the transition into the nematic (and also cholesteric or smectic A) phase. The reason for this is that orientation correlations of the kind described above occur, the magnitude of which again increases with decreasing temperature. Very similar results were obtained for low molar mass liquid crystalline systems, indicative again of the close similarity of polymeric and low molar mass systems as far as static properties are concerned.

These examples have shown that new insights into structural and dynamic features characteristic of liquid crystalline phases are obtained by analyzing fluctuations. It should be mentioned that work is presently in progress concerned with the analysis of smectic A and smectic C fluctuations within the nematic phase using high resolution X-ray scattering.

9.3.3 High Resolution Electronmicroscopy as a Tool for Visualizing Smectic Structures

9.3.3.1 Introduction

In recent years, ultrathin polymeric films have aroused considerable interest as possible materials for use in microelectronics, optical systems, lithography and biotechnology. Specifically liquid crystal polymers in the smectic state can be regarded as possible candidates for various transport processes because of their high orientational order in one direction only. However, while powerful synthesis techniques provide an impressive range of organic compounds based on specific molecular properties, physical performance will depend on structural details involving the precise arrangement of these molecules in an appropriately prepared thin film.

Under these circumstances dark-field electron microscopy and appropriately oriented electron diffraction combined with high resolution techniques was shown to be the most appropriate technique in order to probe structural details.

The liquid crystalline polymer investigated by electron microscopy within the framework of the present project was a main chain/side group polymalonate with azobenzene as mesogen in the main chain and cyanobenzene as mesogenic side group. The chemical structure and molecular geometry is indicated below:

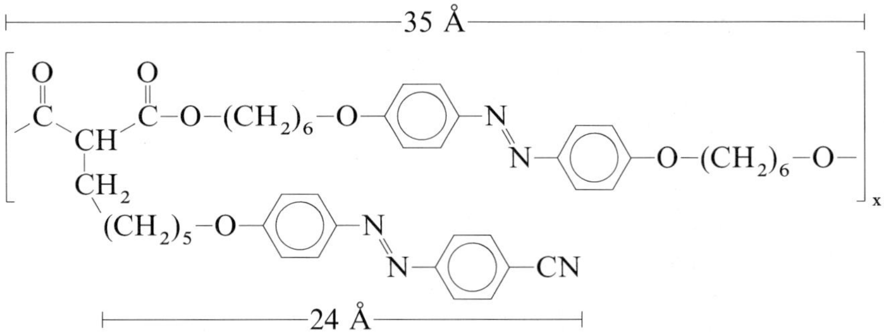

35 Å

O O
C CH C–O–$(CH_2)_6$–O– –N
CH N– –O–$(CH_2)_6$–O–
CH_2 x
$(CH_2)_5$–O– –N
N– –CN

24 Å

Phase transitions are observed at the following temperatures: g 33 k 88 s 195 i. Above the melting temperature, the polymer is in the smectic phase. The transition to the isotropic phase occurs at 195° C. The polymer belongs to a series of polymalonates [23] which indicate how strongly the mesophase behaviour is influenced by the molecular architecture.

9.3.3.2 Sample Preparation

The monomers were synthesized as described previously [23] and polymerized by melt polycondensation. Subsequently thin films were prepared for electronmicroscopic investigation. Several techniques were applied, but the best results were obtained for these samples using solution casting. The preorientation obtained by this procedure was improved by subsequently annealing near the smectic-isotropic transition.

9.3.3.3 Structural Information Obtained by Electron Microscopy

Liquid crystal films which have been pre-oriented and annealed frequently demonstrate dark stripes in the micron range [97–100]. An explanation for the appearance of the stripes is, however, not identical in all cases, since they may be caused by diffraction or scattering contrast. The origin of the stripes observed in the polymalonates (Fig. 9.33) was unambiguously identified as arising from scattering contrast due to thickness variations in the film occurring during annealing. The inference of this statement is that they are not related to orientation of the smectic planes or the molecule.

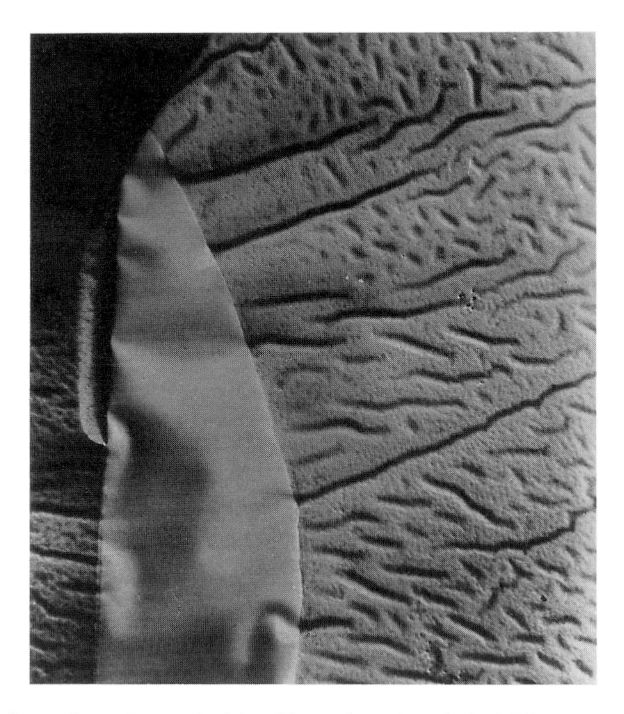

Fig. 9.33: Shadowed replica of thin film of main chain/side group liquid crystal polymer.

In the selected area electron diffraction pattern a small-angle maximum with a second order is observed, plus an oriented amorphous halo perpendicular to it (Fig. 9.34). Such diffraction patterns are typical for smectic liquid crystals and indicate that the mesogens lie in the plane of the film. The measured small angle spacings are:

$$L_1 = 26\text{Å}$$
$$L_2 = 13\text{Å}$$

These spacings correspond to those observed by X-ray small angle scattering [100]. The wide angle halo corresponds to a Van der Waal's spacing of 5.1Å. Careful inspection of the electron diffraction pattern indicates that while the small angle diffraction spot is only slightly arced, the wide angle oriented halo is split. This splitting is not observed in the X-ray scattering pattern. A possible explanation for this is that the volume of material investigated by electron microscopy is in the $(\text{micron})^3$ range whereas the probed volume in X-ray diffraction is in the $(\text{millimeter})^3$ range. Consequently the structural effect observed is averaged over many orientations. The split diffraction halo

Fig. 9.34: Electron diffraction pattern of main chain/side group liquid crystal polymer showing sharp small angle reflection and split wide angle halo.

suggests a structure such as the one indicated schematically in Fig. 9.35. The long spacing of 26Å is less than the value that can be accounted for by a tilt angle of 33°, so that the flexible spacer is clearly not fully extended. Thus selected area electron diffraction is an important tool for highlighting structural details (split amorphous halos) which may well be obliviated by methods probing large volumes and thus averaging over many orientations.

Information about the distribution of the oriented regions giving rise to the small angle maxima is obtained by dark field techniques. In this case, an aperture is placed over the diffraction pattern in the focal plane of the objective lens such that only the electrons scattered into the small angle diffraction maxima are transmitted. These electrons are then used to produce an image. Only the oriented regions now appear bright (Fig. 9.36). It is immediately clear that a very unusual distribution is observed. The oriented areas form parallel regions which are a few Ångstroms thick and several tens of microns long. Their width and frequency depend on annealing time and temperature. Correlation with the diffraction pattern shows that the smectic planes are perpendicular to the long direction (Fig. 9.36). This important dark field observation clearly shows that the oriented regions are not necessarily uniformly distrib-

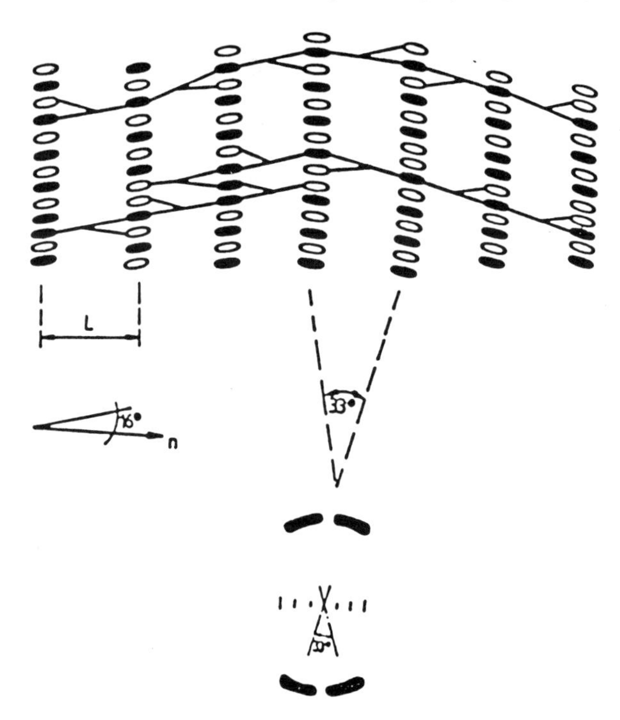

Fig. 9.35: Schematic diagram showing possible model main chain/side group polymalonate.

uted within the film, an erroneous conclusion which may easily be derived from a premature interpretation of the small angle maxima.

In order to apply high resolution techniques to beam sensitive samples such as these, a number of special precautions have to be applied. We have discussed methods of dealing with the most serious problems previously [102]. However, for a correct interpretation it is important to understand that lattice planes are imaged by phase contrast produced by the interference of the scattered electron wave that passes through the objective aperture with the unscattered electron wave.

Both phase and amplitude contrast depend on the electron microscope transfer function. When samples are beam sensitive, the whole transfer function must be calculated in order to pre-set the correct defocussing value for the planes which are to be imaged [102]. These, as we know from the diffraction pattern, have a value of 26Å. Unfortunately the contrast transfer function oscillates rather badly if spatial frequencies in this region of reciprocal space are to be imaged (Fig. 9.37). Furthermore, a considerable amount of high frequency noise is introduced by the amplitude transfer function as well as from the supporting carbon film, leading to a decrease in the signal to noise

Fig. 9.36: Dark field micrograph from main chain/side group liquid crystal polymer with diffraction pattern in correct orientation.

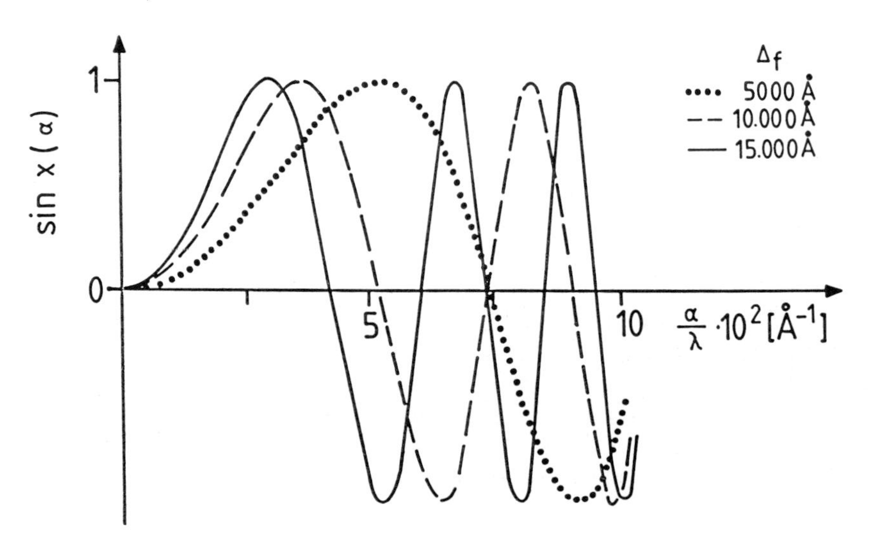

Fig. 9.37: Phase contrast transfer function for the Philips EM 420 at 100 kV for different values of defocus.

ratio (S/N). However, the quality of the micrographs which were obtained is sufficiently good to show a number of important details:

(1) The smectic planes within the elongated regions which were observed in dark field are indeed, perpendicular to the long direction of the oriented regions (Fig. 9.38) as predicted from the diffraction pattern.
(2) The smectic planes in these regions are curved (Fig. 9.38).
(3) After long annealing times the oriented regions increase in size and the smectic planes are shown to undulate (Fig. 9.39).
(4) It is possible to identify edge dislocations in the smectic planes (Fig. 9.39).

The observation and identification of smectic planes and defects [104] in these liquid crystalline polymers is a significant new achievement. All previous discussions on defects in liquid crystals have concentrated on "macroscopic" effects observed by light microscopy [105], but not in microscopic details such as dislocations in the smectic planes.

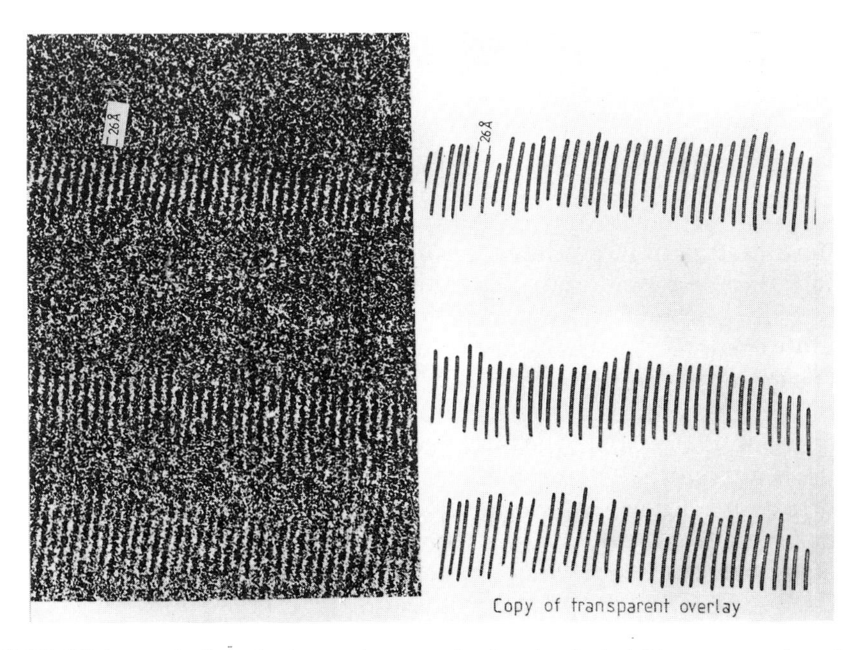

Fig. 9.38: High resolution electron micrograph of main chain/side group polymalonate showing early stages in development of smectic planes.

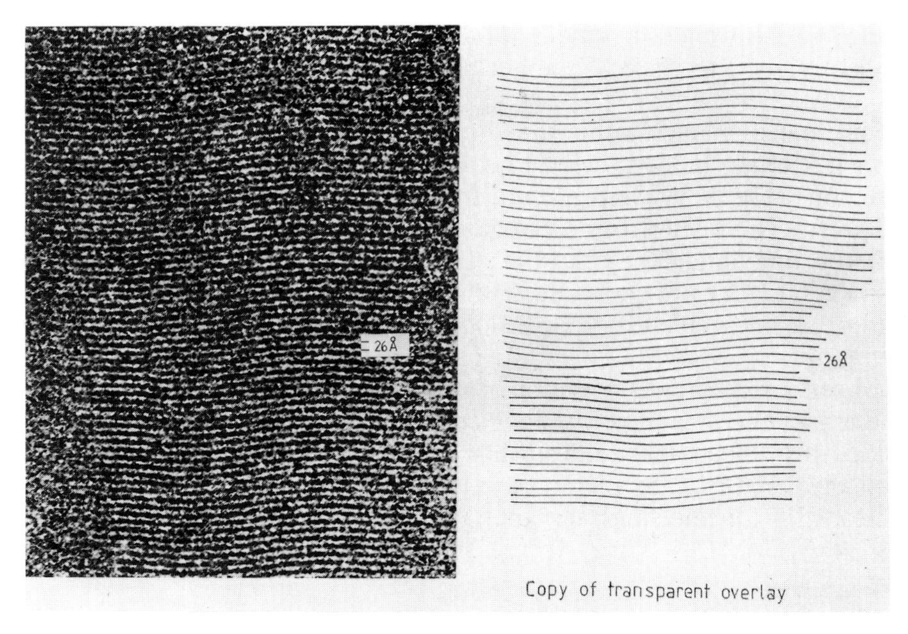

Fig. 9.39: High resolution electron micrograph showing undulating smectic planes and a dislocation.

9.3.3.4 Image Analysis and Processing

The interpretation of lattice images is not straight forward [106] if dynamical effects have to be considered. This is the case if the film thickness is greater than $\xi_g/2$ [107], where ξ is the extinction distance and g indicates the Bragg planes in question. It is necessary to make certain assumptions in calculating ξ_g using an appropriate structure factor, because the sample does not possess three dimensional crystalline order. A rough calculation indicates a value of about 1800Å [102]. The films under investigation are between 50Å and 100Å thick, therefore dynamical effects can be ignored. The observed images thus correspond closely to the true structure, but the effect of the microscope transfer function must be eliminated by deconvolution.

A second stage in image interpretation involves image processing. Clearly it would be desirable to improve the signal/noise (S/N) ratio on our micrographs and to detect defects quickly and easily. Therefore image processing is desirable. Image processing can be achieved by using different techniques [108] such as light diffractometry and digital image processing in real or reciprocal space.

Image reconstruction using a light diffractometer was used in a classical paper by Klug and de Rosier [109] to analyze the tobacco mosaic virus. It involves filtering of individual spatial frequencies in the diffraction plane of an objective lens (Fourier filtering) (Fig. 9.40). Essentially the same principle is applied in digital image processing in reciprocal space, where the image is digitized and amplitudes plus phases are stored in the computer. The Fourier transform is calculated and noise is eliminated by electronic filtering in reciprocal space. A second Fourier transform produces an image with a highly improved S/N ratio. Whichever technique is chosen, great care must be exercized when filtering in reciprocal space because the phase information is lost in this plane. This is of no consequence for periodic images but may cause artefacts in the reconstruction of aperiodic details such as defects. Work is in progress in our laboratories in order to develop reliable methods of image reconstruction by correctly taking account of the amplitudes and phases.

Alternatively, image processing can be performed in real space using modern processing techniques [110–111]. Standard techniques involve low pass or high pass filtering, compass masks or electronic zooming and clipping methods [108]. Such a technique was used to identify and magnify a dislocation in the main chain/side group sample shown in Fig. 9.39 (see Fig. 9.41). However, the technique does not as yet generally produce the high quality of reconstruction which we have obtained by Fourier filtering. Furthermore aperiodic details may easily be obliterated rather than enhanced.

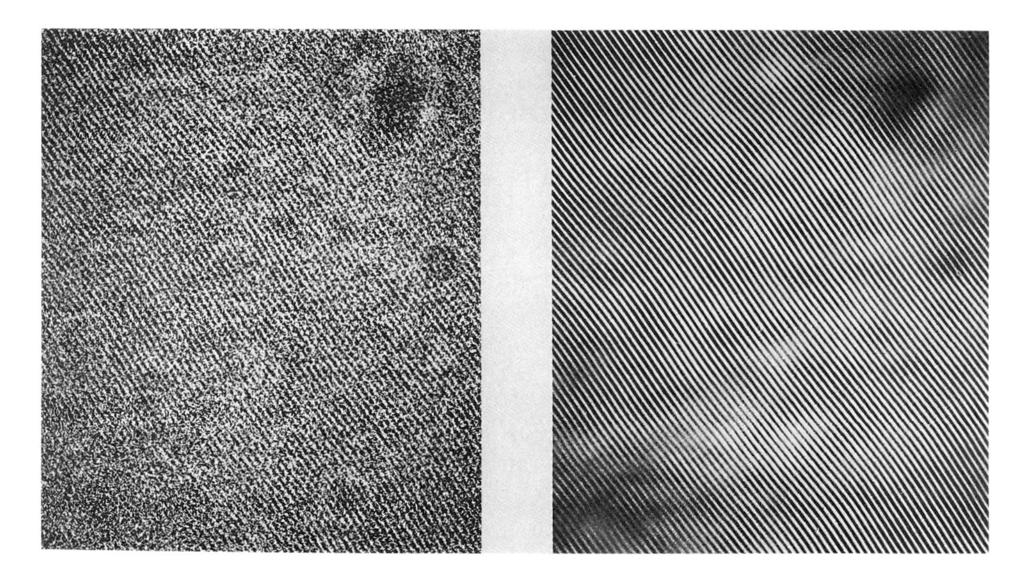

Fig. 9.40: High resolution electron micrograph of smectic planes and image after filtering by optical diffractometry.

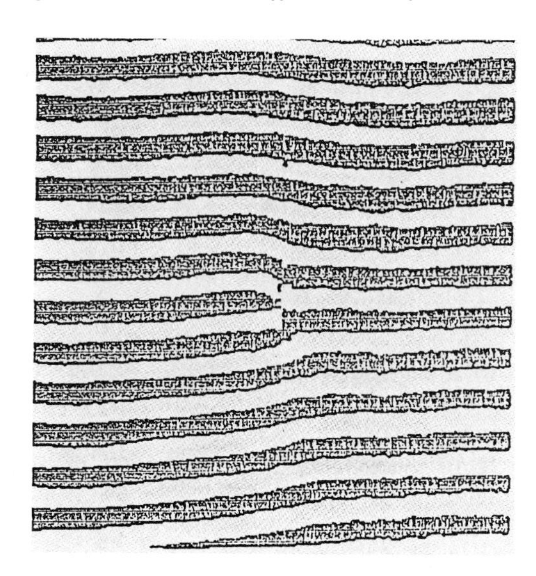

Fig. 9.41: Image of dislocation after electronic zooming and clipping.

9.3.3.5 Conclusion

We have shown that high resolution electron microscopy combined with electron diffraction and dark field methods is a very powerful technique to monitor molecular level information. For the first time images of smectic planes and their defects have been demonstrated. Investigations on other polymer systems are in progress and it is hoped that certain general principles of structural organization applicable to liquid crystalline polymers can be derived. The techniques of image analysis and digital processing are being improved, so that aperiodic images can be reconstructed more reliably.

9.4 References

[1] H. Ringsdorf, B. Schlarb, J. Venzmer: Molecular architecture and function of polymeric oriented systems: models for the study of organization, surface recognition and dynamics of biomembranes, Angew. Chem. Int. Ed. Engl. 27 (1988) 113–158.

[2] P. G. de Gennes: The Physics of Liquid Crystals, the International Series of Monographs on Physics, Clarendon Press, Oxford 1975.

[3] R. Steinsträßer, H. Krüger: Flüssigkristalle. Ullmanns Encykl., Techn. Chem. 4. Auflage, 11 (1976) 657–686.

[4] D. Demus, G. Pelzl, F. Kuschel: Flüssigkristalle in elektrischen Feldern, Z. Chem. 21 (1981) 1–9.
[5] S. Chandrasekhar, B. K. Sadashiva, K. A. Suresh: Liquid crystals of disc-like molecules, Pramana 9 (1977) 471–480; Chem. Abstr. 88 (1978) 30566 y.
[6] D. Destrade, P. Foucher, H. Gasparoux, N. H. Thinh, A. M. Levelut, J. Malthête: Disc-like mesogen polymorphism, Mol. Cryst. Liq. Cryst. 106 (1984) 121–146.
[7] G.W. Gray, J.W. G. Goodby: Smectic Liquid Crystals, Leonhard Hill 1984.
[8] J. L. White: Historical survey of polymer liquid crystals, J. Appl. Polym. Sci. Appl. Polym. Symp. 41 (1985) 3–24.
E.T. Samulski: Macromolecular structure and liquid crystallinity, Faraday Discuss., Chem. Soc. 79 (1985) 7–20.
[9] P.W. Morgan: Synthesis and properties of aromatic and extended chain polyamides, Macromolecules 10 (1977) 1381–1390.
S. L. Kwolek, P.W. Morgan, I. R. Schaefgen, L.W. Gulrich: Synthesis, anisotropic solutions, and fibers of poly(1,4-benzamide), Macromolecules 10 (1977) 1390–1396.
T. I. Bair, P.W. Morgan, F. L. Killian: Poly(1,4-phenyleneterephthalamides). Polymerization and novel liquid-crystalline solutions, Macromolecules 10 (1977) 1396–1400.
M. Panar, L. F. Beste: Structure of poly(1,4-benzamide) solutions, Macromolecules 10 (1977) 1401–1406.
[10] W. J. Jackson, H. F. Kuhfuss: Liquid crystal polymers, I: Preparation and properties of p-hydroxybenzoic acid copolyesters, J. Polym. Sci., Polym. Chem. Ed. 14 (1976) 2043–2058.
W. J. Jackson: Liquid crystal polymers, IV: Liquid crystalline aromatic polyesters, Br. Polym. J. 12 (1980) 154–162.
S. L. Kwolek, P.W. Morgan, J. R. Schaefgen: Liquid crystalline polymers, Encycl. Polym. Sci. 9 (1987) 1–61.
[11] Yu. B. Amerik, J. J. Konstantinov, B. A. Krentsel: Polymerization of p-methacryloxybenzoic acid in mesomorphic and liquid states, J. Polym. Sci. C 23 (1968) 231–238.
L. Strzelecki, L. Liebert: Synthèse et polymérisation de nouveaux monomères mésomorphes, Bull. Soc. Chim. Fr. (1973) 597–602.
E. Perplies, H. Ringsdorf, J. H. Wendorff: Polyreaktionen in orientierten Systemen, 3: Polymerisation ungesättigter Benzylidenaniline mit flüssigkristallinen Eigenschaften, Makromol. Chem. 175 (1974) 553–561.
E. Perplies, H. Ringsdorf, J. H. Wendorff: Polyreaktionen in orientierten Systemen, 5: Polyacryl- und -methacryl-Schiffsche Basen mit flüssigkristallinen Eigenschaften, Ber. Bunsenges. Phys. Chem. 78 (1974) 921–923.
F. Cser: Mesomorphic polymers, J. Physique (Paris) 40 (1979) 459–470.
[12] H. Finkelmann, H. Ringsdorf, J. H. Wendorff: Model considerations and examples of enantiotropic liquid crystalline polymers, Makromol. Chem. 179 (1978) 273–276.
V. P. Shibaev, N. A. Platé, Ya. S. Freidzon: Thermotropic liquid crystalline polymers, I: Cholesterol-containing polymers and copolymers, J. Polym. Sci., Polym. Chem. Ed. 17 (1979) 1655–1670.

[13] A. Blumstein, K. N. Sivaramakrishnan, S. B. Clough, R. Blumstein: Polymeric thermotropic liquid crystals: Polymers with mesogenic elements and flexible spacers in the main chain, Mol. Cryst. Liq. Cryst. 49 (1979) 255–258.
A. Roviello, A. Sirigu: Mesophasic polymers. Copolyalkanoates of 4,4'-dihydroxy α,α'-dimethylbenzolazine, Europ. Polym. J. 15 (1979) 61–67.
[14] M. Engel, B. Hisgen, R. Keller, W. Kreuder, B. Reck, H. Ringsdorf, H.W. Schmidt, P. Tschirner: Synthesis, structure and properties of liquid crystalline polymers, Pure & Appl. Chem. 57 (1985) 1009–1014.
[15] W. Kreuder, H. Ringsdorf: Liquid crystalline polymers with disc-like mesogens, Makromol. Chem., Rapid Commun. 4 (1983) 807–815.
[16] a) W. Kreuder, H. Ringsdorf, P. Tschirner: Liquid crystalline polymers with disc-like mesogens in the main chain, Makromol. Chem., Rapid Commun. 6 (1985) 367–373.
b) O. Herrmann-Schönherr, J. H. Wendorff, W. Kreuder, H. Ringsdorf: Structure of the mesophase of a discotic main-chain polymer, Makromol. Chem., Rapid Commun. 7 (1985) 97–101.
c) G. Wenz: New polymers with disc-shaped mesogenic groups in the main chain, Makromol. Chem., Rapid Commun. 6 (1985) 577–584.
[17] a) F. Hessel, H. Finkelmann: A new class of liquid crystal side chain polymers: mesogenic groups laterally attached to the polymer backbone, Polym. Bull. 14 (1985) 375–378.
b) F. Hessel, H. Finkelmann: Optical biaxiality of nematic LC-side chain polymers with laterally attached mesogenic groups, Polym. Bull. 15 (1986) 349–352.
c) F. Hessel, R.-P. Herr, H. Finkelmann: Synthesis and characterization of biaxial nematic side chain polymers with laterally attached mesogenic groups, Makromol. Chem. 188 (1987) 1597–1611.
d) Q. Zhou, H. Li, X. Feng: Synthesis of liquid-crystalline polyacrylates with laterally substituted mesogens, Macromolecules 20 (1987) 233–234.
[18] S. Berg, V. Krone, H. Ringsdorf: Structural variations of liquid crystalline polymers: cross-shaped and laterally linked mesogens in main chain and side group polymers, Makromol. Chem., Rapid Commun. 7 (1986) 381–388.
[19] a) O. Herrmann-Schönherr, J. H. Wendorff, H. Ringsdorf, P. Tschirner: Structure of an aromatic polyamide with disc-like mesogens in the main chain, Makromol. Chem., Rapid Commun. 7 (1986) 791–796.
b) H. Ringsdorf, P. Tschirner, O. Herrmann-Schönherr, J. H. Wendorff: Synthesis, structure, and phase behaviour of liquid-crystalline rigid-rod polyesters and polyamides with disc-like mesogens in the main chain, Makromol. Chem. 188 (1987) 1431–1445.
[20] M. Ballauff, G. F. Schmidt: Rigid rod polymers with flexible side chains, 2: Observation of a novel type of layered mesophase, Makromol. Chem., Rapid Commun. 8 (1987) 93–97.
[21] W. Kreuder, H. Ringsdorf, O. Herrmann-Schönherr, J. H. Wendorff: The "Wheel of Mainz" as a liquid crystal?, Angew. Chem. Int. Ed. Engl. 26 (1987) 1249–1252; Angew. Chem. 99 (1987) 1300–1303.

[22] M. Ballauff: Phase equilibria in rodlike systems with flexible side chains, Macromolecules 19 (1986) 1366–1374.
M. Ballauff: Rigid-rod polymers having flexible side chains, 1: Thermotropic poly(1,4-phenylene-2,5-dialkoxyterephthalate)s, Makromol. Chem., Rapid Commun. 7 (1986) 407–414.
[23] B. Reck, H. Ringsdorf: Combined liquid crystalline polymers: mesogens in the main chain and as side groups, Makromol. Chem., Rapid Commun. 6 (1985) 291–299.
[24] B. Reck, H. Ringsdorf: Combined liquid crystalline polymers: rigid rod and semiflexible main chain polyesters with a lateral mesogenic group, Makromol. Chem., Rapid Commun. 7 (1986) 389–396.
[25] H. Ringsdorf, H.W. Schmidt, H. Eilingsfeld, K.-H. Etzbach: Synthesis and characterization of liquid-crystalline copolymers with dichroic dyes and mesogens as side groups, Makromol. Chem. 188 (1987) 1355–1366.
[26] H. Ringsdorf, H.W. Schmidt, G. Baur, R. Kiefer, F. Windscheid: Orientational ordering of dyes in the glassy state of liquid-crystalline side group polymers, Liq. Cryst. 1 (1986) 319–325.
[27] R. Zentel, G. Reckert: Liquid crystalline elastomers based on liquid crystalline side group, main chain and combined polymers, Makromol. Chem. 187 (1986) 1915–1926.
[28] I. Cabrera, V. Krongauz, H. Ringsdorf: Photo- and thermochromic liquid crystal polysiloxancs, Angcw. Chem. Int. Ed. Engl. 26 (1987) 1178–1180.
[29] M. Engel: Photoreaktionen in flüssigkristallinen Polymeren. Diss. Univ. Mainz 1988.
[30] S. Bualek, R. Zentel: Combined liquid-crystalline polymers with chiral phases; 1: 2-octanol as chiral end of the mesogens, Makromol. Chem. 189 (1988) 979–804.
[31] H. Kapitza, R. Zentel: Combined liquid crystalline polymers with chiral phases, 2: Lateral substituents, Makromol. Chem. 189 (1988) 1793–1807.
[32] H. Finkelmann, M. Happ, M. Portugall, H. Ringsdorf: Liquid crystalline polymers with biphenyl-moieties as mesogenic group, Makromol. Chem. 179 (1978) 2541–2544.
[33] H. Ringsdorf, A. Schneller: Liquid crystalline side chain polymers with low glass transition temperatures, Makromol. Chem., Rapid Commun. 3 (1982) 557–562.
[34] B. Hisgen, W. Kreuder, H. Ringsdorf: New liquid crystalline polymers: phenylcyclohexane derivatives and „en bloc" systems. 13. Arbeitstagung Flüssigkristalle, 23.03.–25.03.1983, Freiburg, FR-Germany.
[35] V. Krone, H. Ringsdorf: Liquid-crystalline monomers, dimers and side group polymers containing phenylpyrimidine mesogens, Liq. Cryst. 2 (1987) 411–422.
[36] M. Portugall, H. Ringsdorf, R. Zentel: Synthesis and phase behaviour of liquid crystalline polyacrylates, Makromol. Chem. 183 (1982) 2311–2321.
[37] H. Ringsdorf, A. Schneller: Synthesis, structure and properties of liquid crystalline polymers, Br. Polym. J. 13 (1981) 43–46.
[38] R. Zentel, H. Ringsdorf: Synthesis and phase behaviour of liquid crystalline polymers from chloroacrylates and methacrylates, Makromol. Chem., Rapid Commun. 5 (1984) 393–398.

[39] H. Finkelmann: Flüssigkristalline Polymere, Angew. Chem. 99 (1987) 840–848; Angew. Chem. Int. Ed. Engl. 26 (1987) 816.
A. Ciferri, W. R. Krigbaum, R. B. Meyer (eds.): Polymer Liquid Crystals, Academic Press, New York 1982.
M. Gordon, N. A. Platé (eds.): Liquid Crystal Polymers, Vol. I–III (Adv. Polym. Sci. 59/60–61) (1984).
A. Blumstein (ed.): Polymer Liquid Crystals, Plenum Press, New York 1985.
[40] R. Zentel: Polymers with side-chain mesogenic units, in Comprehensive Polymer Science, Vol. 5, Pergamon Press, 1989, 723.
[41] R. Eidenschink, D. Erdmann, J. Krause, L. Pohl: Substitutierte Cyclohexane – eine neue Klasse flüssigkristalliner Verbindungen, Angew. Chem. 89 (1977) 103–106.
[42] J. Brandrup, E. M. Immergut: Polymer Handbook, 2nd. ed., John Wiley & Sons, New York 1975.
[43] H. Finkelmann, J. Koldehoff, H. Ringsdorf: Synthesis and characterization of liquid crystalline polymers with cholesteric phases, Angew. Chem. Int. Ed. Engl. 17 (1978) 935–936.
[44] P. Zugenmaier, J. Mügge: X-ray investigations on polymethylsiloxanes with mesogenic side chains in the crystalline and smectic phase, 1: poly [5-[4-(4-methoxyphenoxycarbonyl)phenoxy]pentenylmethylsiloxane), Makromol. Chem., Rapid Commun. 5 (1984) 11–19.
[45] R. Zentel, G. Strobl: Structures of liquid crystalline side group polymers oriented by drawing, Makromol. Chem. 185 (1984) 2669–2674.
[46] S. G. Kostromin, V.V. Sinitzyn, R.V. Talroze, V. P. Shibaev, N. A. Platé: Thermotropic liquid crystalline polymers, 12: Smectic "C" phase in liquid crystalline polyacrylates with CN-containing mesogenic groups, Makromol. Chem., Rapid Commun. 3 (1982) 809–814.
N. A. Platé, R.V. Talroze, Ya. S. Freidzon, V. P. Shibaev: Polymeric liquid crystals – problems and trends, Polym. J. 19 (1987) 135–145.
[47] P. Davidson, P. Keller, A. M. Levelut: Molecular organization in side chain liquid crystalline polymers, J. Physique (Paris) 46 (1985) 939–946.
[48] a) C. Boeffel, B. Hisgen, U. Pschorn, H. Ringsdorf, H.W. Spiess: Structure and dynamics of liquid crystalline polymers from deuteron NMR, Israel J. Chem. 23 (1983) 388–394.
b) C. Boeffel, H.W. Spiess, B. Hisgen, H. Ringsdorf, H. Ohm, R. G. Kirste: Molecular order of spacer and main chain in polymeric side group liquid crystals, Makromol. Chem., Rapid Commun. 7 (1986) 777–783.
c) U. Pschorn, H.W. Spiess, B. Hisgen, H. Ringsdorf: Deuteron NMR study of molecular order and motion of the mesogenic side groups in liquid-crystalline polymers, Makromol. Chem. 187 (1986) 2711–2723.
d) R.W. Lenz: Balancing mesogenic and non-mesogenic groups in the design of thermotropic polyesters. Faraday Discuss. Chem. Soc. 79 (1985) 21–32; and the remarks of: R. Zentel, E.T. Samulski, P. J. Flory on page 89–90.
[49] K.-H. Wassmer, E. Ohmes, M. Portugall, H. Ringsdorf, G. Kothe: Molecular order and dynamics of liquid-crystal side-chain polymers: an electron spin resonance study employing rigid nitroxide spin probes, J. Am. Chem. Soc. 107 (1985) 1511–1519.
[50] H. Ringsdorf, R. Zentel: Liquid crystalline side chain polymers and their behaviour in the electric field, Makromol. Chem. 183 (1982) 1245–1256.

[51] S. G. Kostromin, R.V. Talroze, V. P. Shibaev, N. A. Platé: Thermotropic liquid crystalline polymers, 11: Influence of molar mass of the liquid crystalline polymer on mesophase properties, Makromol. Chem., Rapid Commun. 3 (1982) 803–808.
[52] H. J. Coles, R. Simon: Investigations of smectic polysiloxanes for high contrast dyed smectic polymer storage effect, Mol. Cryst. Liq. Cryst. Lett. 3 (1986) 37–42.
[53] H. Kresse, R.V. Talroze: Dielectric investigations on a comb-like polymer with a liquid crystal phase, Makromol. Chem., Rapid Commun. 2 (1981) 369–374.
H. Kresse, E. Tennstedt, R. Zentel: Dielectric investigations on liquid crystalline side-group polymers, Makromol. Chem., Rapid Commun. 6 (1985) 261–265.
[54] R. Zentel, G. Strobl, H. Ringsdorf: Dielectric relaxation of liquid crystalline polyacrylates and polymethacrylates, Macromolecules 18 (1985) 960–965.
[55] W. Haase, H. Pranoto, F. J. Bormuth: Dielectric properties of some side chain liquid crystalline polymers, Ber. Bunsenges. Phys. Chem. 89 (1985) 1229–1234.
H. Pranoto, F. J. Bormuth, W. Haase, U. Kiechle, H. Finkelmann: Dielectric properties of a liquid crystalline siloxane copolymer, Makromol. Chem. 187 (1986) 2453–2460.
[56] W. Heinrich, B. Stoll: Dielectric relaxation in two liquid crystalline polyacrylates under high hydrostatic pressure, Colloid Polym. Sci. 263 (1985) 895–898.
[57] G. S. Attard, J. J. Moura-Ramos, G. Williams: Molecular dynamics of a smectic liquid crystalline side chain polymer. The dielectric properties of aligned and nonaligned side chain polymer. The range of frequency and temperature, J. Polym. Sci., Polym. Phys. Ed. 25 (1987) 1099–1111.
[58] C. Aguilera, J. Bartulin, B. Hisgen, H. Ringsdorf: Liquid crystalline main chain polymers with highly flexible siloxane spacers, Makromol. Chem. 184 (1983) 253–262.
[59] S. Diele, St. Oelsner, F. Kuschel, B. Hisgen, H. Ringsdorf, R. Zentel: X-ray investigations of liquid crystalline homo- and copolysiloxanes with paired mesogens, Makromol. Chem. 188 (1987) 1993–2000.
St. Westphal, S. Diele, A. Mädicke, F. Kuschel, U. Scheim, K. Rühlmann, B. Hisgen, H. Ringsdorf: Microphase separation in thermotropic liquid crystalline polysiloxanes with paired mesogens, Makromol. Chem., Rapid Commun. 9 (1988) 489–493.
[60] R. Zentel, G. F. Schmidt, J. Meyer, M. Benalia: X-ray investigations of linear and cross-linked liquid crystalline main chain and combined polymers, Liq. Cryst. 2 (1987) 651–664.
[61] B.W. Endres, J. H. Wendorff, B. Reck, H. Ringsdorf: Dielectric properties of a combined main chain/side chain liquid crystalline polymer, Makromol. Chem. 188 (1987) 1501–1509.
[62] a) C. Destrade, N. H. Tinh, H. Gasparoux, J. Malthête, A. M. Levelut: Disc-like mesogens: A classification, Mol. Cryst. Liq. Cryst. 71 (1981) 111–135.
b) S. Chandrasekhar: Liquid crystals of disc-like molecules, Phil. Trans. R. Soc. Lond. A. 309 (1983) 93–103.
c) J. Billard: Discotic mesophases, a review, in W. Helfrich, G. Heppke (eds.): Liquid Crystals of One- and Two-dimensional Order. Berlin 1980, 383–395.
d) J. Billard, J. C. Dubois: Discotic mesophase, a complementary review, in A. C. Griffin, J. F. Johnson (eds.): Liquid Crystals and Ordered Fluids, Vol. 4.
[63] M. Ebert, O. Herrmann-Schönherr, J. H. Wendorff, H. Ringsdorf, P. Tschirner: Evidence for a biaxial nematic phase in sanidic aromatic polyamides, Makromol. Chem., Rapid Commun. 9 (1988) 445–451.

[64] a) W. Kreuder: Diskotische Polymere aus funktionalisierten Derivaten des Triphenylens, Diss. Univ. Mainz 1986.
b) O. Karthaus: Kombinierte flüssigkristalline Polymere mit calamitischen und diskotischen Mesogenen, Diplom-Arbeit Univ. Mainz 1988.
c) H. Ringsdorf, R. Wüstefeld, E. Zerta, M. Ebert, J.H. Wendorff, Angew. Chem. 101 (1989) 934–938, Angew. Chem. Int. Ed. Engl. 28 (1989) 914–918.
d) B. Hüber, W. Kranig, H.W. Spiess, W. Kreuder, H. Ringsdorf, H. Zimmermann, Adv. Mater. 2 (1990) 36–40.
[65] a) C. Sirlin, L. Bosio, J. Simon: Spinal Columnar Liquid crystals, polymeric octasubstituted μ-oxo-(phthalocyaninato)tin (IV), J. Chem. Soc., Chem. Commun. (1987) 379–380.
b) A. Beck, M. Hanack, H. Lehmann: Syntheses of liquid crystalline phthalocyanines, Synthesis 8 (1987) 703–705.
[66] B. Hüser, T. Pakula, H.W. Spiess: Macroscopic ordering of liquid crystalline polymers with discotic polymers, Macromolecules 22 (1989) 1960–1963.
[67] a) B. Hüser, H.W. Spiess: Macroscopic alignment of discotic liquid-crystalline polymers in a magnetic field, Makromol. Chem., Rapid Commun. 9 (1988) 337–343.
b) B. Hüser: Untersuchung der molekularen Ordnung von flüssigkristallinen Polymeren mit diskotischen Mesogenen. Diss. Univ. Mainz 1987.
[68] a) C. Piechocki, J. Simon, A. Skoulios, D. Guillon, P. Weber: Discotic mesophases obtained from substituted metallophthalocyanines. Toward liquid crystalline one-dimensional conductors, J. Am. Chem. Soc. 104 (1982) 5245–5247.
b) L.Y. Chiang, J.P. Stokes, C.R. Safinya, A.N. Bloch: Charge transfer salts of highly oriented fibres of discotic liquid crystal triphenylenes, Mol. Cryst. 125 (1985) 279–288.
c) J. van Keulen, T.W. Warmerdam, R.J.M. Nolte, W. Drenth: Electrical conductivity in hexaalkoxytriphenylenes, Recl. Trav. Chim. Pays-Bas 106 (1987) 534–536.
d) Z. Witkiewicz, I. Szulc, R. Dabrowski: Disc-like liquid crystalline stationary phases from the triphenylene derivatives group, J. Chromatogr. 315 (1984) 145–159.
[69] a) O. Herrmann-Schönherr: Strukturuntersuchungen an neuen Mesophasen, Diss. TH Darmstadt, Deutsches Kunststoff-Institut 1987.
b) M. Ebert, O. Herrmann-Schönherr, J.H. Wendorff, H. Ringsdorf, P. Tschirner: Sanidics: A new class of mesophases displayed by highly substituted rigid-rod polyesters and polyamides, Liq. Cryst. 1 (1990) 63–79.
[70] a) S. Chandrasekhar, B.R. Ratna, B.K. Sadashiva, N. Raja: Biaxial nematic phase in a low molecular weight thermotropic system, The 12th International Liquid Crystal Conference, Freiburg 1988, Abstract SY 09.
b) S. Chandrasekhar, B.K. Sadashiva, B.S. Srikanta: Paramagnetic nematic liquid crystals, Mol. Cryst. Liq. Cryst. 151 (1987) 93–107.
c) J. Malthête, L. Liebert, A.M. Levelut, Y. Galerne: Physique de la matière condensée – Nématique biaxe thermotrope, C.R. Acad. Sci. Paris 303 (1986) 1073–1076.

[71] a) M. Eich, J. H. Wendorff, B. Reck, H. Ringsdorf: Reversible digital and holographic optical storage in polymeric liquid crystals, Makromol. Chem., Rapid Commun. 8 (1987) 59–63.
b) M. Eich, J. H. Wendorff, H. Ringsdorf, H.W. Schmidt: Non-linear optical self-diffraction in a mesogenic side chain polymer, Makromol. Chem. 186 (1985) 2639–2647.
c) M. Eich, B. Reck, H. Ringsdorf, J. H. Wendorff: Reversible digital and holographic optical storage in polymeric liquid crystals (PLC), Proc. SPIE Mol. Polym. Optoelectr. Mat. 862 (1986) 93–96.
d) M. Eich, J. H. Wendorff: Erasable holograms in polymeric liquid crystals, Makromol. Chem., Rapid Commun. 8 (1987) 467–471.
e) M. Eich, J. H. Wendorff: Reversible optical information storage in liquid crystalline polymers, in Polymers for advanced technologies, Weinheim 1988.

[72] H. Finkelmann, H.-J. Kock, G. Rehage: Investigations on liquid crystalline polysiloxanes 3, Liquid crystalline elastomers – a new type of liquid crystalline material, Makromol. Chem., Rapid Commun. 2 (1981) 317–322.
H. Finkelmann, H.-J. Kock, W. Gleim, G. Rehage: Investigations on liquid crystalline polysiloxanes 5, Orientation of lc-elastomers by mechanical forces, Makromol. Chem., Rapid Commun 5 (1984) 287–293.
W. Gleim, H. Finkelmann: Thermoelastic and photoelastic properties of crosslinked liquid crystalline side chain polymers, Makromol. Chem. 188 (1987) 1489–1500.

[73] S. Bualek, R. Zentel: Crosslinkable liquid crystalline combined main chain/side group polymers with low transition temperatures, Makromol. Chem. 189 (1988) 791–796.

[74] R. Zentel, M. Benalia: Stress induced orientation in lightly crosslinked liquid crystalline side group polymers, Makromol. Chem. 188 (1987) 665–674.

[75] S. Bualek, H. Kapitza, J. Meyer, G. F. Schmidt, R. Zentel: Orientability of crosslinked and of chiral liquid crystalline polymers, Mol. Cryst. Liq. Cryst. 155 (1988) 47–56.

[76] R. Zentel: Untwisting of the helical superstructure in the cholesteric and chiral smectic C* phases of crosslinked LC-polymers by strain, Liq. Cryst. 3 (1988) 531–536.

[77] R. Zentel, G. Reckert, B. Reck: New liquid crystalline polymers with chiral phases, Liq. Cryst. 2 (1987) 83–89.
R. Zentel: Interrelation between the orientation of the polymer chains and the mesogenic groups in crosslinked liquid crystalline polymers, Progr. Coll. Polym. Sci. 75 (1987) 239–242.

[78] G. Gray: Polymer liquid crystals, in A. Ciferri, W. R. Krigbaum, R. B. Meyer (eds.), Academic Press, New York 1982.

[79] R. G. Kirste, W. A. Kruse, K. Ibel: Determination of the conformation of polymers in the amorphous solid state and in concentrated solution by neutron diffraction, Polymer 16 (1975) 120–124.

[80] M. Stamm, E.W. Fischer, M. Dettenmaier, P. Convert: Chain conformation in the crystalline state by means of neutron scattering methods, Trans. Faraday Soc. 68 (1979) 263–278.

[81] J. Kugler, E.W. Fischer, M. Puscher, C. D. Eisenbach: Small angle neutron scattering studies of poly (ethylene oxid) in the melt, Makromol. Chem. 184 (1983) 2325–2334.

[82] E.W. Fischer, M. Dettenmaier: Non-crystall. Sol. 31 (1978) 181.
[83] I. G. Voigt-Martin, J. Wendorff: Amorphous polymers, Mark encyclopedia of polymer science and engineering, Vol. I, 2nd ed., John Wiley, 1985, 789–842.
[84] U. Kalepky, E.W. Fischer, P. Herchenröder: Characterization of semicrystalline random copolymers by small angle neutron scattering, J. Poly. Sci. (Phys.) 17 (1979) 2117.
[85] G. Strobl, M. Schneider, I. G. Voigt-Martin: Model of partial crystallization and melting derived from small angle X-ray scattering and electron microscopic studies on low-density polyethylene, J. Poly. Sci. 18 (1980) 1361–1381.
[86] I. G. Voigt-Martin: Use of transmission electron microscopy to obtain quantitative information about polymers, Adv. Poly. Sci. 67 (1985) 196–218.
[87] J. H. Wendorff, E.W. Fischer: Thermal density fluctuations in amorphous polymers as revealed by small angle X-ray diffraction, Koll. Z.u. Z. Polym. 251 (1973) 876–883.
[88] C. H. Wang, E.W. Fischer: Density fluctuations, dynamic light scattering, longitudinal compliance, and stress modules in a viscoelastic medium, J. Chem. Phys. 82 (1985) 632–638.
[89] B. Ewen, D. Richter, B. Lehnen: Segmental diffusion of polymer molecules in solution as studied by means of quasi-elastic neutron scattering, Macromolecules 13 (1980) 876–880.
[90] C. Boeffel, H.W. Spiess, B. Hisgen, H. Ringsdorf, H. Ohm, R. G. Kirste: Molecular order of spacer and main chain in polymeric side-group liquid crystals, Makromol. Chem., Rapid Commun. 7 (1986) 777–783.
[91] M. Eich, K. Ullrich, J. H. Wendorff: Investigations on pretransitional phenomena of the isotropic-nematic phase transition of mesogenic materials by means of electrically induced birefingence, Progr. Colloid Polym. Sci. 69 (1984) 94–99.
[92] M. Eich, K. Ullrich, J. H. Wendorff: Pretransitional phenomena in the isotropic melt of a mesogenic side chain polymer, Polymer 25 (1985) 1271–1276.
[93] K. H. Ullrich, J. H. Wendorff: Orientation correlations in the isotropic state of low molecular weight and polymeric fluids, Mol. Cryst. Liq. Cryst. 313 (1985) 361–375.
[94] D. Jungbauer, J. H. Wendorff, W. Kreuder, B. Reck, C. Urban, H. Ringsdorf: Ordered and disordered glasses: A comparison of thermodynamic and dynamical properties, Makromol. Chem. 189 (1988) 1345–1351.
[95] W. Kopp, J. H. Wendorff: Analysis of orientation fluctuation in fluids by small angle X-ray scattering, Colloid Polym. Sci. 260 (1982) 1071–1078.
[96] J. H. Wendorff: Studies on the orientational order of polymeric liquid crystals by means of electric birefringence and small angle X-ray scattering, Makromol. Chem., Suppl. 6, (1984) 41–45.
[97] A. M. Donald, A. H. Windle: Electron microscopy of banded structures in oriented thermotropic polymers, J. Mat. Sci. 18 (1983) 1143–1150.
[98] A. M. Donald, A. H. Windle: Transformation of banded structures on annealing thin films of thermotropic liquid crystalline polymers, J. Mat. Sci. 19 (1984) 2085–2097.
[99] I. G. Voigt-Martin, H. Durst: Structure analysis of side chain liquid crystal polymer films by means of electron microscopy, Liq. Cryst. 2 (1987) 585–600.
[100] B. Reck: Synthesen und Struktur-Eigenschafts-Beziehungen von flüssigkristallinen Polyestern. Diss. Univ. Mainz 1988.

[101] H. Erichson, A. Klug: Measurement and compensation of defocusing and aberrations by Fourier processing of electron micrographs, Phil. Trans. Roy. Soc. London B 261 (1971) 105–118.
[102] I. G. Voigt-Martin, H. Durst: Direct observation of smectic layers in side chain liquid crystal polymer films, Liq. Cryst. 2 (1987) 601–610.
[103] I. G. Voigt-Martin, H. Durst, B. Reck, H. Ringsdorf: Structure analysis of a combined main-chain/side-group liquid crystalline polymer by electron microscopy, Macromolecules 21 (1988) 1620–1626.
[104] I. G. Voigt-Martin, H. Durst: High resolution images of defects in liquid crystalline polymers in the smectic and crystalline phases. Macromolecules 22 (1989) 168–173.
[105] D. Demus, L. Richter: Textures of Liquid Crystals, Verlag Chemie, Weinheim 1978.
[106] J. M. Cowley: Diffraction Physics, North Holland 1975.
[107] P. B. Hirsch, A. Howie, M. J. Whelan: Electron Microscopy of Thin Crystals, Butterworth, London 1965.
[108] I. G. Voigt-Martin, H. Durst, H. Krug: High resolution images of defects in liquid crystalline polymers: Image analysis using light diffractometer and computer techniques, Macromolecules 22 (1989) 595–600.
[109] A. Klug, D. J. de Rosier: Optical filtering of electron micrographs: reconstruction of one-sided images, Nature 212 (1966) 29–32.
[110] N. K. Pratt (ed.): Digital Image Processing, J. Wiley & Sons, New York 1978.
[111] F. M. Wahl: Digitale Bildsignalverarbeitung, Springer, Berlin 1974.

10 Thermodynamics and Rheology of Polymer Solutions

Bernhard A. Wolf*

Before starting this report on the work performed within the frame of the Sonderforschungsbereich, a few comments concerning the motivation for the choice of research topics and the criteria for the selection of results for their presentation in this article appear appropriate.

The primacy of the dissolved state of polymers for their synthesis and for many of their applications is beyond doubt. Extensive research, pioneered by G.V. Schulz, was therefore carried out in the field of physical chemistry of polymer solutions from the very beginning. Despite the voluminous work already done, the present knowledge is far from satisfactory in many fields. This is particularly true as soon as the variables of state exceed the customary ranges or further polymers are investigated (i), as one tries to describe new experimental information by means of existing theories (ii), and as one is not interested in single limited aspects but in the mutual influence of different phenomena (iii). In order to improve this situation and to supply some basic physico-chemical information for other projects of the present programme, corresponding research was performed.

Some examples of the work done in the past years, which could be of more general interest, were chosen for this report so that they illustrate the above three items. Section 10.1 deals with the extension of the normal pressure- and temperature ranges. In section 10.2 it is exemplified how new experimental material calls for new theoretical approaches. And section 10.3, finally, illustrates how thermodynamic and rheological properties of polymer solutions interrelate.

* Institut für Physikalische Chemie der Universität Mainz, Jakob-Welder-Weg, D-6500 Mainz

B. A. Wolf

10.1 Pressure, the Neglected Variable

In spite of the important role of elevated hydrostatic pressures in the synthesis of polymers (particularly from gaseous monomers) and in their practical application (for instance as viscosity-index improvers in motor oils), only very few research groups deal with that variable world-wide [1–5]. In addition to the need of mastering the effects of pressure on the thermodynamic and hydrodynamic properties of polymer solutions, the main theoretical incentives for the present kind of studies lie in the possibility to separate the influences of temperature into the contributions which are due to changes in thermal motion and changes in free volume.

10.1.1 Thermodynamics

In this field two questions are obvious: How do the boundaries of complete miscibility of the polymer with low molecular weight solvents depend on pressure, and how does the solvent quality change with that variable within the homogeneous region?

10.1.1.1 Pressure Dependence of Polymer Solubility

All results presented in this section refer to systems which exhibit a demixing into two *liquid* phases. In order to monitor phase separation, two methods have turned out to be particularly useful: One rests on the turbidity which normally develops as a second phase is segregated. The other procedure utilizes the fact that the viscosity of the two-phase system lies considerably below that of the corresponding homogeneous system in the case of polymer solutions. In other words, demixing is detected from the viscosity summits.

The apparatus [6] used for most of the turbidimetric measurements can be operated from −70 to +500° C and up to 4000 bar. Fig. 10.1 shows one of the early experimental results demonstrating a very general case of separation behaviour: At low temperatures and atmospheric pressure the solution demixes for enthalpic reasons (occurrence of an upper critical solution temperature, UCST). At high temperatures and under the equilibrium vapour pressure of the solvent, the solution demixes for entropic reasons (occurrence of a lower critical solution temperature, LCST). In the present example the homogeneous region is generally increased by pressure, but more so in the case of the LCST. This result is only typical as far as the high temperature phase separation is concerned. In all cases studied so far, the LCSTs are raised by p. With the low temperature demixing, on the other hand, the situation is more complicated. Depending on the particular system or even on the chain length of

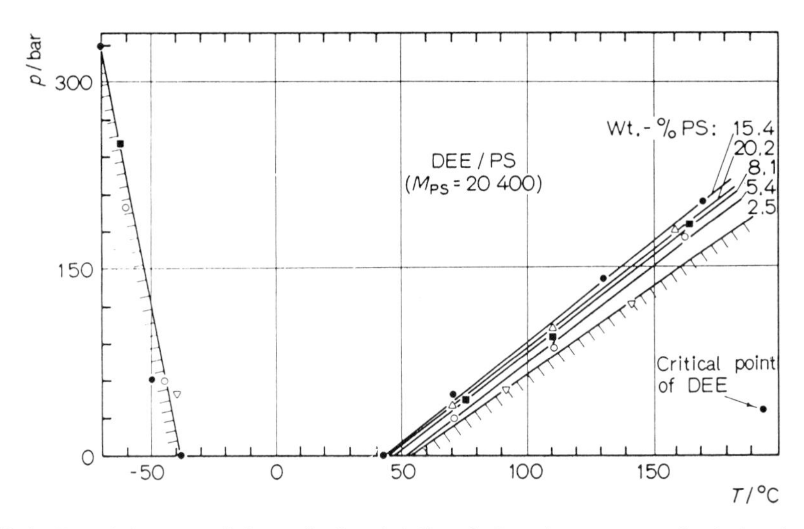

Fig. 10.1: Demixing conditions (spinodal lines) for the system diethyl ether/poly-(styrene) for the indicated polymer concentrations [6]. The unstable states lie below the individual lines.

the polymer, the solvent quality can be either deteriorated or improved. The effect of p may even change sign, i.e. the UCSTs can exhibit extrema as demonstrated by Fig. 10.2; the data [7] shown in this graph have been obtained viscometrically (cf. 10.1.2.2).

For a quantitative theoretical calculation of the influences of pressure on polymer solubility it is indispensable to use an approach which considers non-zero excess volumes. Because of the relative simplicity of the calculations and the fact that the quality of the results is comparable with that of more com-

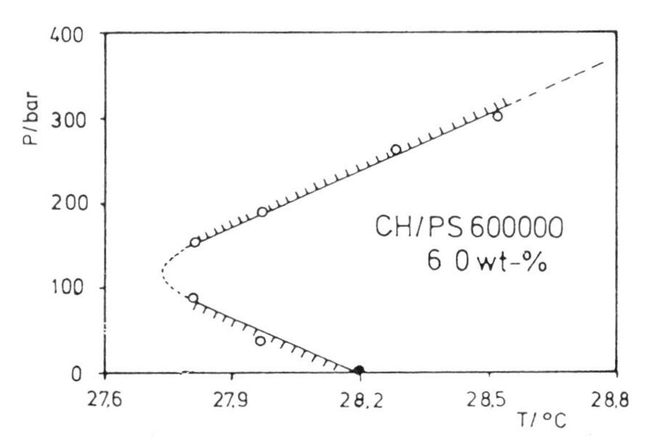

Fig. 10.2: Demixing curve of a solution of 6 wt % of poly(styrene) in cyclohexane [7]. The pressure-temperature region of the two phase system is indicated by the hatching, the filled circle gives the cloud point of the solution under atmospheric conditions.

plicated attempts, the combination [8] of Flory's equation of state with the solubility parameter theory was used to discuss the extensive experimental material obtained for solutions of chemically very different types of polymers in various single or mixed solvents. With the LCSTs the agreement between theory and experiment is reasonable, but not even the sign of the effect can be predicted for UCSTs. A simple (and theoretically intelligible) rule turned out to be more reliable: Whenever $(T_c-T_{mp})/T_{mp}$, the relative distance of the UCST from the melting point T_{mp} of the pure solvent, exceeds ca. 0.25 K/K, the solubility is enhanced by pressure, if it falls below ca. 0.2, it is reduced [7].

10.1.1.2 Pressure Dependence of Interaction Parameters

To assess the pressure influence on the thermodynamic quality of a given solvent also within the homogeneous region, light scattering measurements were performed [9] to obtain the second osmotic virial coefficient A_2, from which χ, the Flory-Huggins interaction parameter at infinite dilution (cf. 10.2.2.1), can be calculated by standard procedure. An example of the results, which depend markedly on the system under investigation, is given in Fig. 10.3. Since A_2 and the solvent quality vary in the same direction, these results shown in the above figure mean that the application of pressure makes the solvent less favourable

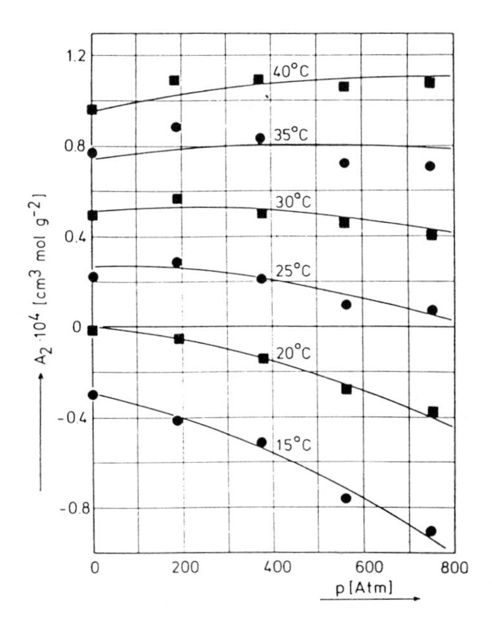

Fig. 10.3: Pressure dependence of the second osmotic virial coefficient A_2 for the system *trans*-decalin/poly(styrene) (M_w = 100 000) at the indicated temperatures; the drawn lines were calculated by means of the excluded volume theory [9].

at low temperatures (where it is already marginal), but more favourable at sufficiently high temperatures (where it is already moderately good). Since the volumes of mixing are obviously minute, the data can be well reproduced theoretically on the basis of the excluded volume theory. Phase separation experiments [10] with the same system are in accord with the light scattering results: The UCSTs increase with pressure, i.e. the solvents becomes poorer.

10.1.2 Viscometry

In order to study the pressure influences on the flow behaviour of polymer solutions, two different approaches and types of apparatus are necessary, depending on the concentration regime of interest. Within the range of pair interactions between the macromolecules, η, the viscosity of the solution, is still comparable with η_s, that of the pure solvent, and the decisive quantities are the intrinsic viscosity $[\eta](p)$ and Huggins coefficient $k_H(p)$. At moderate polymer concentrations (e.g. $c_2 \approx 150$ g/L) $\eta \approx 10^2 – 10^3\ \eta_s$, and one is primarily interested in η (p) itself.

10.1.2.1 Intrinsic Viscosities

A rolling ball viscometer, which can be operated up to 4000 bar, was constructed [11] for the measurements with dilute solutions. From the rolling times (at different angles of inclination to check Newtonian behaviour) the viscosities were calculated and the data evaluated according to

$$(\eta - \eta_s)/c_2\eta_s = [\eta] + k_H[\eta]^2 c_2 \tag{1}$$

Since $[\eta]$ is proportional to the hydrodynamic volume of isolated polymer coils, the present measurements give direct access to changes in dimension caused by temperature or pressure. An example of $[\eta](p,T)$ is shown in Fig. 10.4. For atmospheric pressure a one-to-one correspondence between $[\eta]$ and A_2 is discernible; $[\eta]$ assumes a maximum when the solution becomes athermal. In the case shown, this maximum is shifted by +6K/100bar on the application of pressure. From all experimental data obtained so far it can be concluded that p influences the intrinsic viscosity only via the changes in the thermodynamic quality of the solvent. With k_H the situation is different, since the pair interaction turns out to depend on the packing density of the solution even if $[\eta]$ is kept constant.

A recent study on the effects of free volume [12] has demonstrated that maxima in $[\eta](T)$ observed under isobaric conditions at low pressures do not show up as the volume is kept constant at its starting value at the lowest temperature. In this case the change in the sign of the heat of dilution at constant

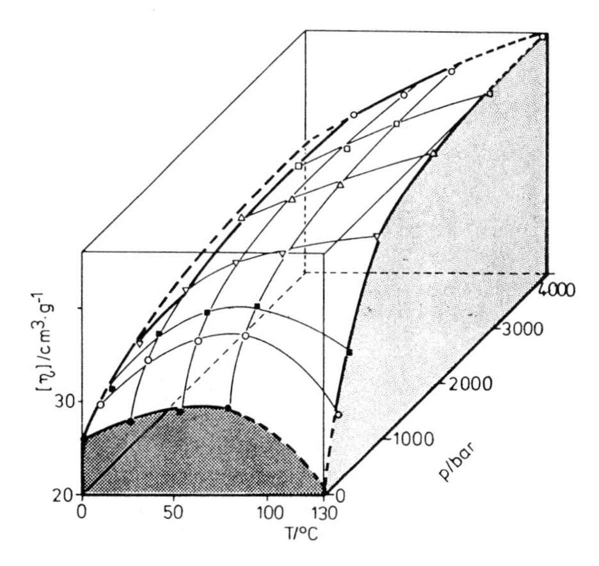

Fig. 10.4: Temperature and pressure dependence of the intrinsic viscosity [η] for the system *tert.*-butyl acetate/poly(styrene) (M_w = 110 000) [11].

pressure is obviously caused by the expansion of the system. If, on the other hand, the isothermal and the isochoric [η](T) are compared at higher initial pressures, maxima are resulting for both cases. This finding indicates that inversions of the T-influence on solvent quality need not be due to free volume effects in all cases.

10.1.2.2 Activation Volumes

For moderately concentrated polymer solutions the increase of η caused by the application of pressure is normally measured in terms of activation volumes $V^{\neq}$, defined by

$$\left(\frac{\partial \ln \eta}{\partial p}\right)_T = \frac{V^{\neq}}{RT} \tag{2}$$

Most experiments were carried out with a modified commercial (Haake) rotational viscometer [13], which can be operated up to 1000 bar. Meanwhile a new apparatus has been constructed with a three-fold higher pressure range and a considerably stronger magnetic clutch allowing, for the first time, to investigate the p-dependence of non-Newtonian effects.

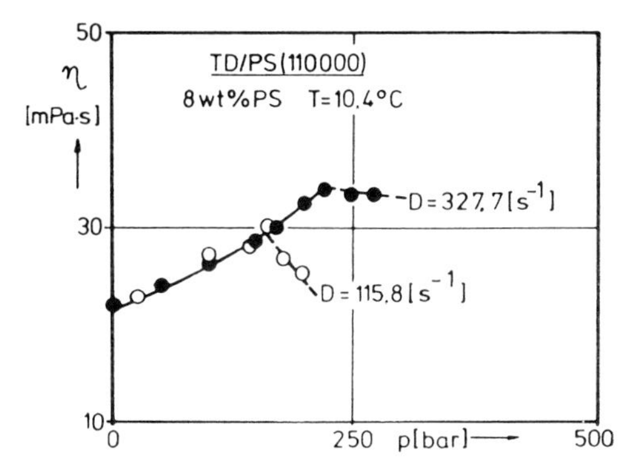

Fig. 10.5: Pressure dependence of the viscosity η in the region of demixing at various shear rates [13]; the onset of phase separation is indicated by a reduction of η upon an increase of p.

Numerous systems studied so far have shown that the viscosity of moderately concentrated polymer solutions is typically increased by a factor of ca. 3–5 as p goes up from 1 to 1000 bar. The actual value of this factor is obviously already substantially fixed by the choice of the solvent, which still dominates the pressure effects at the polymer concentrations of interest. The effects of molecular architecture and thermodynamic interactions (cf. 10.3.1.1) are superimposed and of less importance. In some cases log η varies linearly with p, so that $V^{\neq}$ does not depend on pressure, normally this quantity declines as p becomes larger, but there are also examples known where it increases. Typically the volumes of activation range from 15–50 cm^3/mol.

Fig. 10.5 shown as an example for the present type of experiments, demonstrates two things: The approximately exponential increase of η with pressure, and the fact that the discontinuity, indicating the entrance into the two-phase region, depends on shear rate. Section 10.3.2 will deal with the last mentioned effect in more detail.

10.2 On the Role of Chain Length

As monomers are tied together to form chain molecules, their entropy of mixing with a second compound is reduced drastically (since they can no longer spread statistically over the entire system) whereas the corresponding enthalpy of mixing remains essentially unchanged. This fact bears a number of implications, two of which are the topic of the following sections.

10.2.1 Development of Incompatibility

In the extreme case where both components of a mixture are high molecular weight chain molecules, complete miscibility is – for the reason described above – uncommon. Furthermore, the conditions of homogeneity are hard to measure in a reliable manner in those exceptional cases in which limited miscibility is encountered, because such systems are enormously viscous. It was therefore attempted to obtain some insight into the development of incompatibility of polymers by studying the reduction of the homogeneous region upon a step-wise increase of the degrees of polymerization N, of two chemically different oligomers which are still liquid at room temperature.

10.2.1.1 Experiments with Oligomers

Three systems have been studied in great detail [14, 15]: Oligo(isobutene)/oligo(propylene glycol) (OIB/OPG), oligo(styrene)/oligo(ethylene glycol) (OS/OEG) and oligo(dimethylsiloxane)/oligo(propylene glycol) (ODMS/OPG) with N_{OIB} 2–7, N_{OPG} 7–60, N_{OS} 1–4, N_{OEG} 4–9 and N_{DMS} 2–5. Up to 20 representatives of each system (comprised of well fractionated and characterized components) were studied with respect to the phase separation conditions and their variation with pressure. Demixing takes exclusively place upon cooling (UCSTs). The application of pressure favours phase separation in the case of all representatives of the systems OIB/OPG and OS/OEG; for ODMS/OPG the opposite is true, with the exception of the shortest chain OPGs where one observes pressures of optimum miscibility, i. e. p becomes unfavourable beyond a certain minimum value.

As the chains of the non-polar components become longer at constant N of the polar ones, the two-phase region extends as expected. If, on the other hand, the degree of oligomerization of the non-polar component is kept constant and that of the glycol raised, one observes pronounced minima in the corresponding UCSTs. The qualitative explanation is obvious: The monomeric glycols are immiscible with all non-polar oligomers for enthalpic reasons (their favourable hydrogen bonds would have to be opened upon mixing). As the molecular weight of the glycols go up, the effect of the end-groups becomes less important, but the normal incompatibility develops, so that higher molecular weight oligoglycols become immiscible again. The next section deals with the question, to what extent a quantitative theoretical description of the present results is possible.

10.2.1.2 Solubility Parameter Theory

For the calculation of the dependence of the critical demixing temperatures T_c on the degrees of oligomerization N, a simple modification [15] of the solubility

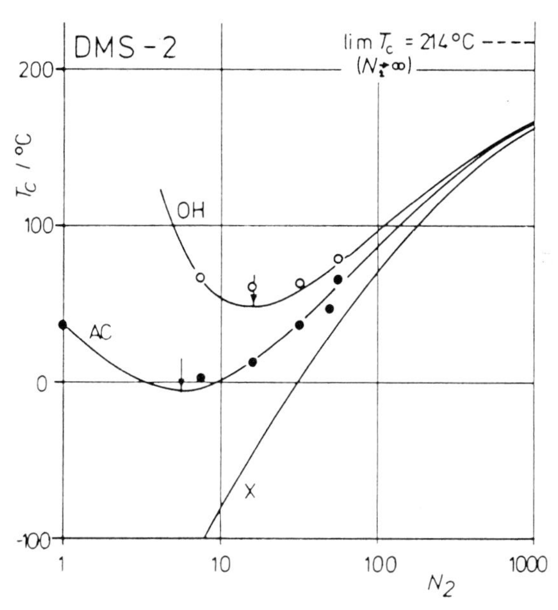

Fig. 10.6: Example for the calculated and the measured dependences of the upper critical solution temperatures T_c on the number of monomeric units N_2 of the propylene glycol component (logarithmic scale) in the system oligo(dimethyl siloxane)/-oligo(propylene glycol) at a constant degree of oligomerization (N = 2) of the siloxane component [15]. Three kinds of end groups are considered: The original OH-groups, acetyl groups (marked AC in the graph) and the hypothetical case that the solubility parameters and volumes of the end groups are identical with that of the middle groups (marked X). The degrees N_2 of optimum miscibility are indicated by arrows and the calculated limiting value of T_c for infinite N_2 is also given.

parameter theory has proven particularly successful, as exemplified in Fig. 10.6. One starts from the usual equation

$$T_c = \frac{2\,(\delta_i - \delta_2)^2}{R\,(V_i^{-0.5} + V_2^{-0.5})^2} \tag{3}$$

in which δ_i stands for the solubility parameters, and V_i for the molar volumes of the components. Use is then made of an experimentally well confirmed theoretical approach [16], according to which the δ values of chain molecules should be built up from the solubility parameters of middle groups, δ_m, and of end groups, δ_e, in the following manner:

$$\delta_1 = \delta_m\left(1 - \frac{V_e}{V_i}\right) + \delta_e \frac{V_e}{V_i} \; . \tag{4}$$

V_e and V_m, the volumes of the end-groups and of middle-groups, resp., are accessible [15] from experimental data concerning $V_i(N)$ by means of

$$V_i = (N-1)V_m + V_e. \tag{5}$$

The δ_m and δ_e values still required to calculate δ_1 from Equ. (4) can be obtained from tabulated molar attraction constants. By means of the thus accessible solubility parameters of the individual oligomers, the critical temperatures can be easily calculated for any combination of chain lengths on the basis of Equ. (3).

All dependences $T_c(N)$ measured for the different representatives of a given system are reproduced theoretically with surprisingly high accuracy using just one set of parameters. It is also noteworthy how well (cf.Fig.10.6) the influence of chemically differing end-groups is taken into account by the present simple approach.

10.2.2 Molecular Weight Dependence of Pair Interactions

As already stated in the introduction of section 10.2, the monomeric units of chain molecules cannot spread out uniformly over the entire volume. This feature becomes of particular importance at low polymer concentrations (i.e. in the region of pair interaction), where the volume fraction of segments within the equivalent sphere of a polymer coil remains comparatively high, whereas it is nil outside. The consequences of this situation can be seen especially clear-cut in the molecular weight dependence of the second osmotic virial coefficient A_2 (or equivalently in the Flory-Huggins interaction parameter χ), and in k_H, the corresponding hydrodynamic pair interaction parameter.

10.2.2.1 Theoretical Concept

The many published approaches to comprehend the uneven distribution of polymer segments over the entire solution have become increasingly more complicated as they try to cover new experimental material. Notwithstanding that complexity, some features (like the increase of A_2 with polymer molecular weight M in sufficiently exothermal solutions) still lie beyond their scope. For this reason an attempt was made to start from an entirely new and simple point of view.

A_2 (or χ) and k_H measure the effect associated with the addition of solvent, i.e. the thermodynamic or hydrodynamic consequence of opening intersegmental contacts by the insertion of solvent molecules. The new approach [17, 18] is therefore based on the fact, that (with the exception of theta conditions) the dimensions of dissolved polymer coils change upon dilution. Consequently, in addition to contacts between polymer segments belonging to different molecules, *intra*molecular intersegmental contacts are also opened as solvent is added. It is now postulated that the effect for intermolecular contacts normally differs from that of intramolecular ones. This assumption rests on the fact that segments belonging to different molecules are separable without restriction, whereas that belonging to the same molecule only to a certain extent given by their mutual distance on the polymer chain.

The calculations based on these assumptions, lead to Equ. (6)

$$\varepsilon = \frac{\varepsilon^{)(} + \varepsilon^{()}}{2} + \frac{\varepsilon^{)(} - \varepsilon^{()}}{2} \frac{K_\theta}{K} M^{-(a-0.5)} \qquad (6)$$

in which ε is a general parameter that measures the effect associated with the opening of a contact between a pair of segments by adding solvent; $\varepsilon^{)(}$ and $\varepsilon^{()}$ specify the corresponding effects for an inter- and intramolecular contact, respectively. K and a are the pre-exponential factor and the exponent of the Kuhn-Mark-Houwink (KMH) equation relating the intrinsic viscosity $[\eta]$ and M; K_θ, like $a = 0.5$, signifies theta conditions, for which only intermolecular effects exist. Equ. (6), if correct, should describe all measured molecular weight dependences of pair interaction constants ε by means of only two parameters, characterizing the inter- and the intramolecular effect, respectively.

10.2.2.2 Experimental Results and Discussion

The evaluation of published material and of own measurements, particularly performed to check the validity of the new concept, has meanwhile yielded enough data for a number of polymers to cover a wide range of thermodynamic conditions. Fig. 10.7 shows how the magnitude of inter- and of intramolecular effects varies with solvent quality (measured in terms of the KMH-exponent a) in the case of the Flory-Huggins interaction parameter χ; Fig. 10.8 gives the corresponding diagram for the Huggins coefficient k_H, i.e. the hydrodynamic pair interaction. Postponing the explanation of the shifts observed for the two polymers under investigation, the results for χ and k_H are discussed separately in the following sections.

As can be seen from Fig. 10.7 for the case of the *thermodynamic pair interactions*, $\chi^{)(}$ and $\chi^{()}$ assume the identical value of 0.5 for Θ-conditions ($a = 0.5$). As the separation of intermolecular contacts becomes thermodynamically more favourable (i.e. as $\chi^{)(}$ decreases) $\chi^{()}$ ascends. This finding can be rationalized by bearing in mind that the spatial constraints, which hinder the segments of a polymer molecule from distribution uniformly throughout the solution, become more effective as the tendency to surround the segments by solvent molecules becomes larger, i.e. as $\chi^{)(}$ becomes smaller. If the solvent power is

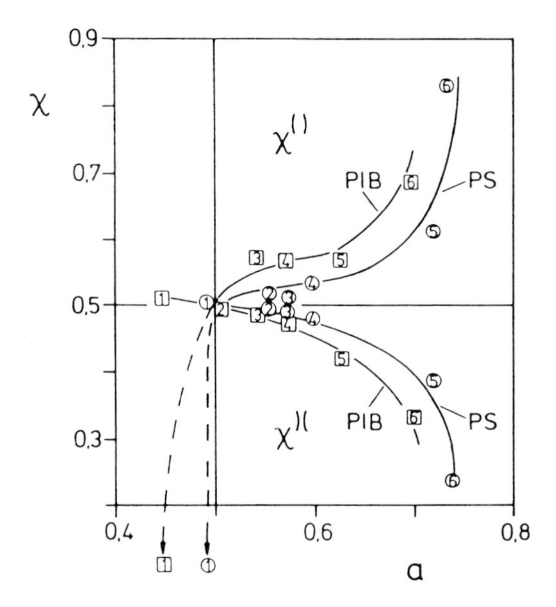

Fig. 10.7: Variation of the *inter*molecular and of the *intra*molecular Flory-Huggins interaction parameters at infinite dilution, $\chi^{)(}$ and $\chi^{()}$, with the thermodynamic quality of the solvent, as measured by the Kuhn-Mark-Houwink exponent a for poly(styrene) (PS) and poly(isobutene) (PIB), each number represents a different solvent and/or different temperature [18].

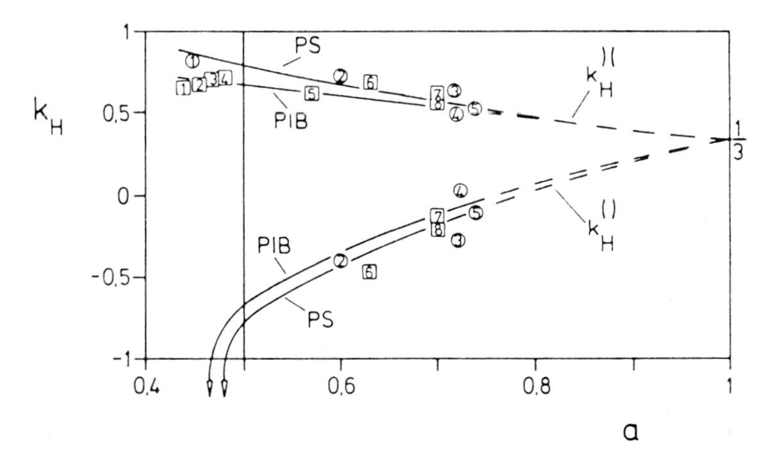

Fig. 10.8: Variation of the *inter*molecular and of the *intra*molecular Huggins coefficients, $k_H^{)(}$ and $k_H^{()}$, with the thermodynamic quality of the solvent as measured by the Kuhn-Mark-Houwink exponent a for poly(styrene) (PS) and poly(isobutene) (PIB); each number represents a different solvent and/or different temperature [18].

not improved, but deteriorated starting from Θ-conditions, $\chi^{)(}$ exceeds the Θ-value only slightly, while $\chi^{()}$ decreases strongly, assuming even negative values. This observation can be explained by the reasonable assumption that $\chi^{()}$ no longer measures the effect associated with the opening of only one intersegmental contact by insertion of a solvent molecule, but the effect associated with the rupture of a whole cluster (cf. coil-collapse phenomena).

In the case of the *hydrodynamic pair interactions* the situation is completely different, as one can see from Fig. 10.8. First of all, negative $k_H^{()}$ values are already observed with good solvents. Furthermore, $k_H^{)(}$ and $k_H^{()}$ approach the same value of ca. 1/3 at a = 1 (instead of 0.5) and separate as the solvent power decreases. To explain this findings, the following can be said:

For near-theta conditions we are dealing with non-draining coils; it should consequently make a big difference, whether an intersegmental contact opened upon dilution belongs to the intra- or to the intermolecular type. In the latter case its opening contributes to the achievement of the infinitely dilute state, i. e. leads to a reduction of $(\eta-\eta_s)/c_2\eta_s$ ($k_H^{)(} > 0$), since the coils can move more independently of each other. If, however, an intramolecular contact is opened, this only signifies that the hydrodynamically active volume per solute molecule increases, which is equivalent to a rise in $(\eta-\eta_s)/c_2\eta_s$ and therefore to negative $k_H^{()}$ values.

As the solvent power improves, the flow mechanics changes, until complete draining should be finally achieved at a = 1. Under these conditions it is no longer possible to distinguish between inter- and intramolecular contacts, since each polymer segment contributes individually towards the increase of viscosity so that $k_H^{)(}$ and $k_H^{()}$ should become identical.

Concerning the parallel shift of the relations shown in Figs. 10.7 and 10.8 for polystyrene and for polyisobutene, only tentative explanations can be offered [18] at the moment. In the thermodynamic case the a-value corresponding to a given thermodynamic quality of the solvent (measured by $\chi^{()}$) should depend on the special molecular architecture of the polymer. In the hydrodynamic case, this effect could be superimposed by an unequal flexibility of the polymers which leads to different shapes of the polymer coils and consequently to somewhat modified k_H values.

B. A. Wolf

10.3 Interrelation of Thermodynamic and Rheological Properties

The following two sections deal with the mutual influence of mixing tendency and flow behaviour. In paragraph 10.3.1 the equilibrium thermodynamic properties constitute the independent variables and it is asked, how they can be felt in the response of a system towards laminar shear. In paragraph 10.3.2, the statement of the problem is reversed. This time the shear rate $\dot{\gamma}$ is fixed and the question reads: How does the phase separation behaviour of a polymer solution change as it flows? At the first glance it appears as if these questions cannot be answered separately because of the reciprocal nature of the influences. On a closer inspection of the situation it turns out, however, that the changes in the thermodynamic interaction parameters caused by laminar flow are generally only minute, and that it is only because of the extreme sensitivity of the phase separation conditions against even the tiniest variations of χ that they become measurable.

10.3.1 Flow under Different Thermodynamic Conditions

For the flow resistance of any solution, the ease of the relative motion of the components with respect to their neighbours plays an important role. The more the molecules stick together, the higher η_o, the zero-shear viscosity, will become under else identical conditions. In the case of moderately concentrated polymer solutions η_o is primarily determined by the high molecular weight component. A deterioration of the solvent quality, leading to an interactional preference of intersegmental contacts between the polymer, will consequently increase the extent of common motion. This effect is in the following called pull-along effect. In its different manifestations it is the topic of section 10.3.1.1. What can happen if the thermodynamic interactions become very special is demonstrated in the section 10.3.1.2, which deals with the phenomenon of thermoreversible gelation.

10.3.1.1 The Pull-along Effect

In the case of the zero-shear viscosity, the most obvious manifestation of the pull-along effect can be observed as one approaches LCSTs. When the solutions of polymers in thermodynamically sufficiently good solvents are heated, their η_o decreases approx. exponentially. Not so, however, in the present case where the interaction between the polymer molecules becomes so favourable as compared with the solute/solvent interaction that a second phase is segregated upon heating. Under these circumstances the pull-along effect can become so dominant that it leads to an inversion [19] of the normal temperature dependence of η_o.

Naturally the pull-along effect is also visible in the activation energies and volumes of the viscous flow. As already mentioned in the section dealing with $V^{\neq}$, the dependence of these activation parameters on the variables of state can be split up into three contributions [20]: The basis is laid by the choice of the solvent. The mere presence of the polymer in the absence of special energetic interactions with the solvent (athermal mixing conditions) contributes a certain increment on which the effects of thermodynamic interaction are built-up. Like η_o itself, $V^{\neq}$ (T) can pass a minimum [21], in this case the effect is due to a change from endo- to exothermal conditions.

In a very pronounced manner the pull-along effect also manifests itself in the non-Newtonian flow behaviour of polymer solutions. Under good solvent conditions, Graessley's master curve [22] provides an excellent fit up to high shear rates, whereas in the vicinity of Θ-conditions the extent of shear-thinning (measured by the slope in the double-logarithmic plots) is markedly reduced. The reason for this behaviour lies in the occurrence of considerable frictional contributions which do not depend on $\dot{\gamma}$ and which are again a consequence of the increased number of intersegmental contacts and of the pull along-effect associated with it.

In an even more drastic manner this effect manifests itself [23] in the characteristic time of mechanical response of the chain molecules to flow, i. e. in the viscometric relaxation time τ_o (the inverse of which is a rough measure for the $\dot{\gamma}$-values at which shear thinning sets in). As demonstrated in Fig. 10.9, τ_o is found to be on the order of the Rouse relaxation time τ_R ($\approx 6\eta_o M/(\pi^2 c_2 RT)$) in good solvents. In poor solvents, particularly close to demixing, τ_o tends to much higher values, as a result of the reduced chain mobility caused by the pull-along effect. Under certain conditions this reduction in chain mobility (which is not fully visible in the reduced representation of Fig. 10.9, since τ_R results already much larger for poor solvents than for good ones) can become so pronounced that the macromolecules are unable to escape even the comparatively small drag associated with laminar shear so that they brake. This phenomenon, which was termed thermodynamically induced shear degradation, was studied with a number of polymers.

10.3.1.2 Thermoreversible Gelation

In the course of a study concerning the solution properties of various poly(n-alkyl methacrylate)s it was observed that the butyl polymer behaves in a completely different manner as compared with the corresponding methyl or decyl compound, or with other ordinary polymers. The most striking feature is that

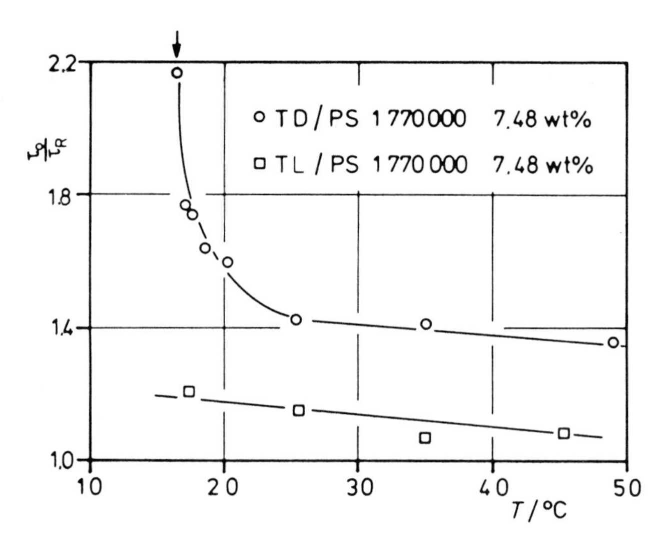

Fig. 10.9: Ratio of the viscometric relaxation time τ_o and the Rouse relaxation time τ_R for solutions of polystyrene (PS) of identical concentration in a good solvent, toluene(TL), and in a Θ-solvent, *trans*-decalin (TD), as function of temperature [23]. The arrow indicates the demixing temperature of the solution in TD.

with poly(n-butyl methacrylate) (PBMA) thermoreversible gelation [24–27] accompanies phase separation. At high polymer concentrations gelation sets in upon cooling prior to the segregation of a second phase, at low c_2 the opposite is true. In order to rationalize the very special behaviour of PBMA, detailed thermodynamic and rheological measurements were performed and the results compared with that of ordinary systems [28].

According to that analysis the gelation in the vicinity of phase separation can be attributed to extraordinarily large enthalpies of mixing which exceed that of normal systems by almost one order of magnitude. These specific thermodynamic conditions render the pull-along effect so powerful that the opening of intersegmental contacts by thermal motion becomes very difficult. If the heat of gelation per mol of crosslinks (obtained from the evaluation of the concentration dependence of the gelation temperatures according to Ferry and Eldridge [29]) which amounts to −37 kJ, is compared with the value of −6 kJ, measured for the heat of formation of 1 mol of intersegmental contacts out of 2 mol of solvent/segment contacts, one can estimate that a single crosslink may consist of ca. 6 segments only. Within a conceivable range of gelation mechanisms extending from the formation of crystallites on one side to the formation of very stable bond-like structures between individual polymer segments on the other, the present system is obviously situated very close to the latter case.

The explanation for that fact lies in the possibility of *dissolved* PBMA to build up locally highly ordered and energetically very favourable contacts between the n-butyl side groups of its monomeric units.

10.3.2 Phase Separation of Flowing Polymer Solutions

For ternary systems, composed of two incompatible polymers dissolved in a common solvent, it was already reported more than thirty years ago [30] that the homogeneous region can be extended up to 10 K by shear. With polymer solutions in a single solvent, similar observations have been made (cf. Fig. 10.5 of section 10.1.2.2) under certain conditions. The effects of shear dissolution is, however, normally much less pronounced and in some cases the flowing systems demix *earlier* than the quiescent ones. In order to work out the criteria which decide whether a certain solution shows shear dissolution or shear demixing behaviour, a systematic study was performed. The experimental results are presented in the next section; a theoretical discussion is given in paragraph 10.3.2.2.

10.3.2.1 Experimental Results

Demixing data of flowing systems have been essentially obtained by turbidimetric and viscometric methods for solutions of polystyrene of different molecular weights in Θ-solvents, mainly *trans*-decalin. To measure the transmittance of sheared polymer solutions, an apparatus was constructed [31] which can be operated up to ca. 2000 s^{-1}. The viscometric determination was performed on various commercial instruments.

The results demonstrate that the effect of shear rate depends primarily on the molecular weight M of the polymer, but also on c_2, its concentration, and on the applied shear rate $\dot{\gamma}$. Irrespective of c_2 and $\dot{\gamma}$ no measurable influences are observed for M less than ca. 100 000. As the molecular weight increases, there follows a region which is dominated by shear dissolution. At still higher M values shear demixing becomes more common, particularly for larger c_2. In some special cases (like the one shown in Fig. 10.11) it is possible to see an inversion of the shear effects from shear dissolution at low $\dot{\gamma}$ values to shear demixing at high.

10.3.2.2 Theory and Discussion

The present treatment [32] rests on the introduction of a generalized molar Gibbs energy of mixing for the sheared system, as the sum of G_z, the Gibbs energy at zero shear, and E_s, the energy the solution can store during stationary flow:

$$G_{\dot{\gamma}} = G_z + E_s \qquad (7)$$

G_z is obtained by means of standard thermodynamic equilibrium measurements and E_s can be calculated from viscometric data according to

$$E_s = V_1(1+(N-1)x_2)\eta\tau_o\dot{\gamma}^2(\eta/\eta_o)|\eta\dot{\gamma}|^{-2d^*}. \qquad (8)$$

V_1 is the molar volume of the solvent, N the degree of polymerization of the solute, x_2 its mole fraction, η the viscosity of the solution at shear rate $\dot{\gamma}$ and $d^* = -\partial(\ln\eta)/\partial(\ln\dot{\gamma})$ was introduced to be able to include non-linear viscoelastic behaviour in the present treatment. The stored energy increases proportionally with the viscometric relaxation time τ_o (cf. 10.3.1).

Whether the generalized Gibbs energy of Equ. (7) is really applicable for theoretical calculations along the normal thermodynamic lines, i.e. whether the phase separation of flowing polymer solutions can be considered as a near-equilibrium process, depends on the relative magnitude of G_z and E_s. For the polymer solutions and shear rates of present interest it turns out that $E_s < 10^{-3}G_z$, phase separation conditions can therefore be obtained from the concentration dependence of $G_{\dot{\gamma}}$ as normal.

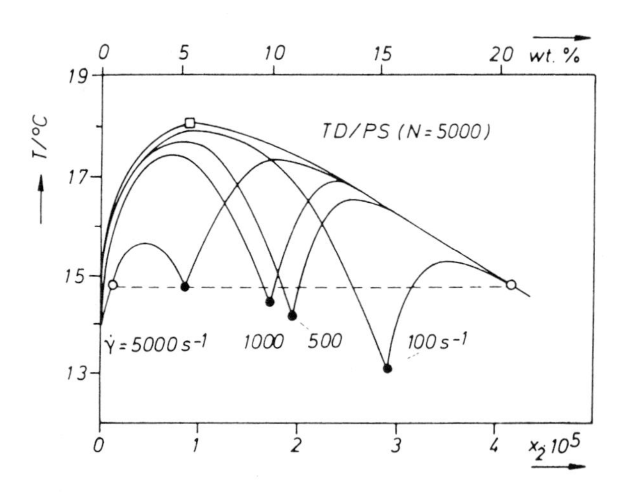

Fig. 10.10: Demixing temperatures of the system *trans*-decalin/poly(styrene) as a function of the mol fraction x_2 of the polymer at the indicated constant shear rates $\dot{\gamma}$, calculated on the basis of Equs.(7) and (8) for a degree of polymerization N of 5000 [32]. The solid circles represent the eulytic points; the dashed line shows, as an example, the coexistence of three phases at $\dot{\gamma} = 5000\ s^{-1}$.

Without any quantitative theoretical calculations the following statements can already be made on the basis of the Equs. (7) and (8): As long as the solutions behave Newtonian, ($d^* = 0, \eta = \eta_o$ in Equ. 8), and only linear viscoelastic effects have to be considered, E_s increases steadily as c_2 becomes larger. This means that the stored energy, when added to the zero-shear Gibbs energy of mixing can only undo a hump in $G_z(x_2)$ indicating phase separation of the unsheared solution, but never produce one if it is absent in the above function. In other words: Under Newtonian flow conditions only the phenomenon of shear dissolution should be observable. The situation changes, however, as non-linear viscoelasticity sets in and the energy stored for constant $\dot{\gamma}$ at higher c_2 can become less than that at lower, due to disentanglement processes. Under these circumstances $E_s(x_2)$ passes a maximum and, when added to a smooth function $G_z(x_2)$, can produce a hump in $G_{\dot{\gamma}}(x_2)$, i.e. lead to shear demixing.

Quantitative calculations performed for Newtonian conditions on the basis of the present concept [32] have yielded the phase diagrams presented in Fig. 10.10. The most striking feature of these results is the occurrence of bimodal demixing curves for the sheared system due to the particularly large areas of shear induced homogenization in the vicinity of the thermodynamic critical composition of the system, which reflect the extreme sensitivity of G_z against any disturbance in this region where it depends almost linearly on x_2. Furthermore, a new phenomenon turns up, namely a shear-induced coexistence of three liquid phases at the temperature at which the two neighbouring branches of the bimodal demixing curve cross. This intersection, which represents the largest extension of the homogeneous state for given $\dot{\gamma}$, is called eulytic point by analogy with the eutectic point.

For non-linear viscoelastic conditions the quantitative theoretical calculations [33] yield phase diagrams in which the demixing curves for constant $\dot{\gamma}$ are raised with respect to the zero-shear curve so that they even surpass them partially or in total. The exact shape of the curves varies strongly with the molecular weight of the polymer; the position of the maximum in $E_s(x_2)$ as compared with the thermodynamic critical composition of the system turns out to be decisive. By means of the data for a solution which shows shear dissolution at low $\dot{\gamma}$ values but shear demixing at high, Fig. 10.11 demonstrates how well the phase separation conditions of flowing solutions can be predicted on the basis of the present concept, if the equilibrium thermodynamic data and the flow behaviour of the system are known with sufficient accuracy.

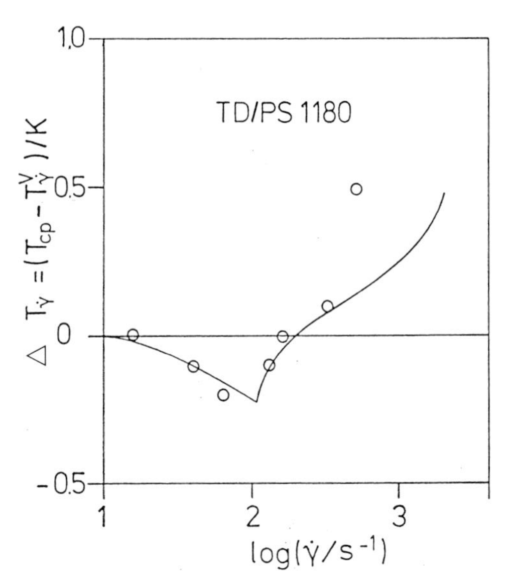

Fig. 10.11: Comparison of measured and calculated dependences of demixing temperatures on shear rate $\dot\gamma$ for the example of a 6.15 wt.% solution of poly(styrene) (M_w = 1180000) in *trans*-decaline [33]; T_{cp} is the cloud point temperature at zero shear and $T\dot\gamma^{v}$ the demixing temperature determined viscometrically at different $\dot\gamma$ values.

10.4 Outlook

As normal, more new questions have turned up than could be answered in the course of the reported investigations. To name just one point for each section: In the context of the pressure investigations it would be particularly interesting to study the influences of p also on the non-linear viscoelastic effects. With section 10.2, it would be tempting to describe the entire concentration dependence of the Flory-Huggins interaction parameter χ in terms of $\chi^{)(}$ and $\chi^{()}$ by making use of the fact that the mixing ratio of these two contributions changes with composition. In the case of section 10.3, finally, the inclusion of polymer mixtures into the study of shear influences on phase separation would be of great practical importance for the processing of polymer blends. Furthermore, an extension of the work so far primarily performed with dilute and moderately concentrated polymer solutions to systems almost entirely consisting of the high molecular weight component would be highly desirable to complete the picture and further the development of generally valid descriptions of concentration influences.

10.5 References

[1] L. Zeman, J. Biroš, G. Delmas, D. Patterson: Pressure effect in polymer solutions – phase equilibria I, J. Phys. Chem. 76 (1975) 1206–1213.
[2] E. Schröder, K.-F. Arndt: Löslichkeitsverhalten von Makromolekülen in komprimierten Gasen I und II, Faserforschung und Textiltechnik 27 (1976) 135–146.
[3] S. Saeki, N. Kuwahara, K. Hamano, Y. Kenmochi, T. Yamaguchi: Pressure dependence of upper critical solution temperatures in polymer solutions, Macromolecules 19 (1986) 2353–2356.
[4] G. Luft, N. S. Subramanian: Phase behaviour of mixtures of ethylene, methyl acrylate and copolymers under high pressure, I & EC Research 26 (1987) 750–753.
[5] F. Kiepen, W. Borchard: Pressure-pulse-induced critical scattering of oligostyrene in n-pentane, Macromolecules 21 (1988) 1784–1790.
[6] B. A. Wolf, G. Blaum: Pressure influence on true cosolvency, Makromol. Chem. 177 (1976) 1073–1088.
[7] B. A. Wolf, H. Geerissen: Pressure dependence of the demixing of polymer solutions determined by viscometry, Colloid & Polym. Sci. 259 (1981) 1214–1220.
[8] J. Biroš, L. Zeman, D. Patterson: Prediction of the χ parameter by solubility parameter and corresponding states theory, Macromolecules 4 (1971) 30–35.
[9] M. D. Lechner, G.V. Schulz, B. A. Wolf: Thermodynamics of polymer solutions as functions of pressure and temperature, J. Coll. Interface Sci. 30 (1972) 462–468.
[10] B. A. Wolf, R. Jend: Über die Möglichkeiten zur Bestimmung von Mischungsenthalpien und -volumina aus der M-Abhängigkeit der kritischen Entmischungstemperaturen, Makromol. Chem. 178 (1977) 1811–1822.
[11] J. R. Schmidt, B. A. Wolf: Pressure dependence of intrinsic viscosities and Huggins constants for polystyrene in *tert*-butyl acetate, Macromolecules 15 (1982) 1192–1195.
[12] N. Schott, B. Will, B. A. Wolf: Thermodynamics and high-pressure viscosity of dilute solutions of poly(decyl methacrylate) and how the free volume influences them, Makromol. Chem. 189 (1988) 2067–2075.
[13] B. A. Wolf, R. Jend: Pressure and temperature dependence of the viscosity of polymer solutions in the region of phase separation, Macromolecules 12 (1979) 732–737.
[14] B. A. Wolf, G. Blaum: Dependence of oligomer-oligomer incompatibility on chain length and pressure, J. Polym. Sci., Symposia 61 (1977) 251–270.
[15] B. A. Wolf, W. Schuch: Oligomer-oligomer incompatibility 3. End-group effects, Makromol. Chem. 162 (1981) 1901–1818.
[16] B. A. Wolf: An extrapolation method for the determination of solubility parameters of polymers demonstrated for polyethylene, Makromol. Chem. 178 (1977) 1869–1871.
[17] B. A. Wolf: Second osmotic virial coefficient revisited. 2. Build-up from contributions of inter- and intramolecular contacts between polymer segments, Macromolecules 18 (1985) 2474–2478.
[18] F. Gundert, B. A. Wolf: On the molecular weight dependence of the thermodynamic and of the hydrodynamic pair interaction between chain molecules. IV Second virial coefficients revisited, J. Chem. Phys. 87 (1987) 6156–6165.

[19] J. K. Rigler, B. A. Wolf, J.W. Breitenbach: Die Viskosität von Polymerlösungen in der Nähe von oberen und unteren Entmischungstemperaturen, Angew. Makromol. Chem. 57 (1977) 15–27.
[20] H. Geerißen, J. Roos, B. A. Wolf: Continuous fractionation and solution properties of PVC, 5. Pressure dependence of the viscosity – Influence of solvent, Makromol. Chem. 186 (1985) 787–799.
[21] J. R. Schmidt, B. A. Wolf: The pressure dependence of the viscosity of polymer solutions and how it reflects the thermodynamic conditions, Makromol. Chem. 180 (1976) 521–571.
[22] W.W. Graessley: The entanglement concept in polymer rheology, Adv. Pol. Sci. 16 (1974) 1–179.
[23] M. Ballauff, H. Krämer, B. A. Wolf: Rheological studies of moderately concentrated polystyrene solutions in the vicinity of the Θ-temperature, II Shear-rate dependence for different thermodynamic conditions, J. Pol. Sci., Phys. Ed. 21 (1983) 1271–1226.
[24] A. Takahashi, M. Sakai, T. Kato: Melting temperature and thermally reversible gels, VI. Effect of branching on the sol-gel transition of polyethylene gels, Polymer J. Japan 12 (1980) 335–341.
[25] R. F. Boyer, E. Baer, A. Hiltner: Concerning gelation effects in atactic polystyrene solutions, Macromolecules 18 (1985) 427–434.
[26] J.Y. S. Gan, J. François, J.-M. Guenet: Enhanced low-angle scattering from moderately concentrated solutions of atactic polystyrene and its relation to physical gelation, Macromolecules 19 (1986) 173–178.
[27] R. C. Domszy, R. Alamo, C. O. Edwards, L. Mandelkern: Thermoreversible gelation and crystallization of homopolymers and copolymers, Macromolecules 19 (1986) 310–325.
[28] L. M. Jelich, S. P. Nunes, E. Paul, H. E. Jeberien, B. A. Wolf: On the cooccurrence of demixing and thermoreversible gelation of polymer solutions 1–3, Macromolecules 20 (1987) 1943–1975.
[29] J. E. Eldridge, J. D. Ferry: Studies of the cross-linking process in gelatin gels III. Dependence of melting point on concentration and molecular weight, J. Phys. Chem. 58 (1954) 992–995.
[30] A. Silberberg, W. Kuhn: Size and shape of droplets of demixing polymer solutions in a field of flow, J. Pol. Sci. 13 (1954) 21–42.
[31] H. Krämer-Lucas, H. Schenck, B. A. Wolf: Influence of shear on the demixing of polymer solutions 1, Apparatus and experimental results, Makromol. Chem. 189 (1988) 1613–1625.
[32] B. A. Wolf: Thermodynamic theory of flowing polymer solutions and its application to phase separation, Macromolecules 17 (1984) 615–618.
[33] H. Krämer-Lucas, H. Schenck, B. A. Wolf: Influence of shear rate on the demixing of polymer solutions 2, Stored energy and theoretical calculations, Makromol. Chem. 189 (1988) 1627–1634.

11 Polymer Diffusion as Studied by Holographic Grating Techniques

Hans Sillescu*

11.1 Introduction

Diffusion of macromolecular chains in bulk polymers was first investigated by Bueche [1] and interpreted along with the early ideas of entanglement whereby the frictional force upon a molecule is proportional to some "effective" chain length L^* with $L^* \sim L$ for short chains and $L^* \sim L^{3.5}$ for long entangled chains. The diffusion coefficient D was assumed inversely proportional to L^*, and Bueche's first experiments [1, 2] with ^{14}C labeled tracer chains seemed to support this view. However, later experiments in molten polyethyleneoxide and polydimethylsiloxane using the NMR pulsed field gradient technique [3] yielded $D \sim L^{-1.8}$, and no change in slope was found in the regime of the viscosity crossover from short to long chain behaviour.

The invention of the reptation model by Edwards and de Gennes [4–6], assuming that entangled chains move predominantly in a snake like fashion, stimulated the development of further experimental techniques for measuring very slow diffusion in polymer systems, in particular, IR-scanning [7], neutron scattering [8], forward recoil scattering (FRES) of α-particles [9], and the analysis of holographic gratings by forced Rayleigh scattering (FRS) [10–12]. A large number of polymer diffusion coefficients has been measured in recent years mostly by application of the FRES and FRS techniques which are complementary in many respects since they use different kinds of tracers and sample preparation thus covering a wide field of polymer diffusion problems. A recent review covering also interdiffusion in polymer blends is given in Ref. [11]. The present article concentrates on the FRS technique which was first applied to polymer solutions by Hervet et al. [12]. Since we were more interested in the dynamics of bulk polymers our main effort was in extending the range to very slow diffusion. In the following, we describe the physical

* Institut für Physikalische Chemie der Universität Mainz, Jakob-Welder-Weg, D-6500 Mainz

principles of the technique also adressing the labeling of polymers with photoreactive dyes. In the main part, applications to polymer-polymer diffusion and dye probe diffusion at the glass transition are reviewed and compared with other experiments for studying translational and rotational diffusion in polymer systems.

11.2 Physical Principles and Experimental Aspects

11.2.1 Hologram Formation

A holographic grating can be formed in a light-sensitive material by recording the interference pattern of two coherent plane wave fields of equal intensity (Fig. 11.1). For a simple one-photon process of the photoreactive dye the concentration profile after short reaction times can be formulated as (Fig. 11.1C)

$$C(x) = 1/2\,(C_{max} + C_{min}) + 1/2\,(C_{max} - C_{min})\cos(2\pi x/d) \tag{1}$$

$$d = \frac{\lambda}{2\sin(\theta/2)} \tag{2}$$

C_{max} and C_{min} correspond to the minima and maxima, respectively, of the light intensity. The grating distance d is related with the wave length λ and the intersecting angle θ of the incoming light beams. It should be noted that inside the sample λ and $\sin(\theta/2)$ are replaced by $\lambda_i = \lambda/n$ and $\sin(\theta_i/2) = \sin(\theta/2)/n$. Whereas Equ. (2) is independent of the index of refraction n the situation is different for the formation of a "reflection hologram" [13]. Here, the two light beams come from above and below the sample and cross at the angle $\alpha_i = \pi - \theta_i$ (Fig. 11.1B). The grating distance becomes

$$d_r = \frac{\lambda_i}{2\sin(\alpha_i/2)} = \frac{\lambda}{2\,[n^2 - \sin^2(\theta/2)]^{1/2}} \tag{3}$$

since $n\cos(\alpha_i/2) = \sin(\theta/2)$. The minimum distance $d_r\,(\alpha = \pi) = \lambda/2\,n$ is smaller than the minimum $d\,(\theta = \pi) = \lambda/2$ of Equ. (2). Thus for $\lambda = 360$ nm and $n = 1.5$ a minimum grid spacing of 120 nm is achievable.

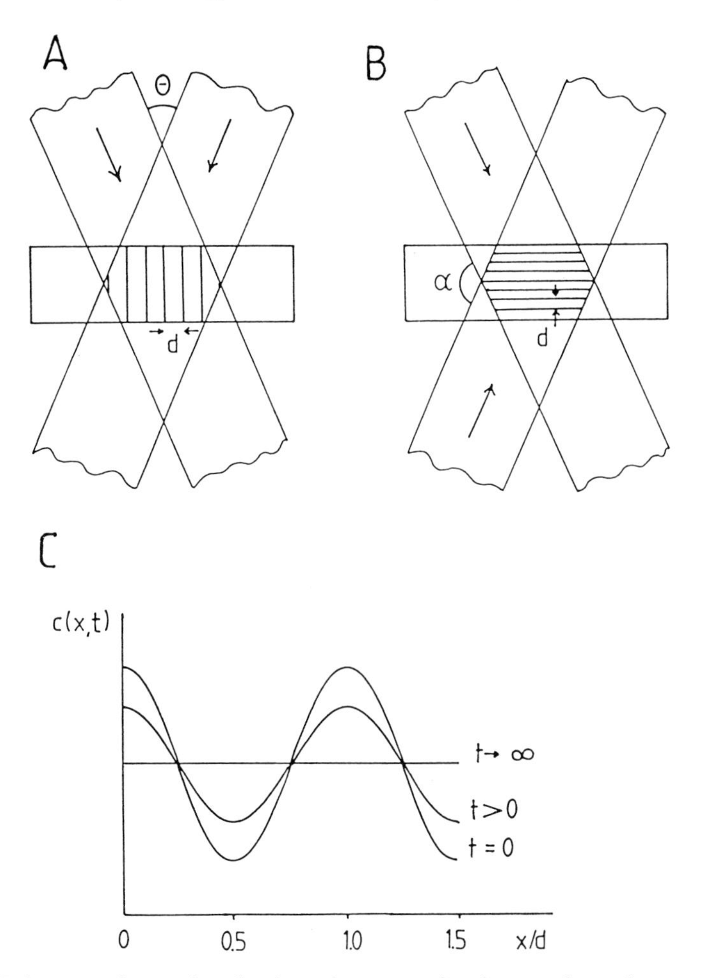

Fig. 11.1: Hologram formation by interference of coherent laser beams. A: Holographic grating with grid spacing d; B: Reflection hologram; C: Diffusive decay of concentration profile (see text).

11.2.2 Hologram Reading by Forced Rayleigh Scattering

Elastic light scattering from the optical heterogeneity of light induced gratings is traditionally termed "forced Rayleigh scattering" [13]. As the sample thickness is much larger than the grating distance (thick holograms) the main scattering intensity is detected at the Bragg angle given by Equ.(2) where λ is now the wave length of the "reading" light beam. In the experimental setup shown in Fig. 11.2, the wave lengths of writing and reading beams are identical, but the latter is reduced in intensity by 3–4 orders of magnitude in order to prevent further photoreaction. In diffusion experiments, the concentration of the photoreactive dye label is usually kept as low as possible. The maximum deviations Δa and Δn from the average absorption constant a and index of refraction n, respectively, are thus also rather small, and the scattering intensity at the Bragg angle is given by an approximation of the Kogelnick equation [13], namely

$$I = I_o\, D^2 e^{-aD}\, (\pi^2 \Delta n^2/\lambda^2 + \Delta a^2/16) \quad (4)$$

D is the sample thickness, and I_o the intensity of the incoming light beam. Maximum efficiencies $\eta = I/I_o$ can be achieved for pure phase holograms ($a = \Delta a = 0$) where the reading wavelength is outside the absorption spectrum of the dye label. However, we have also obtained sufficient efficiencies by keeping the reading and writing wave lengths identical resulting in some mixture of phase and amplitude (absorption) holograms.

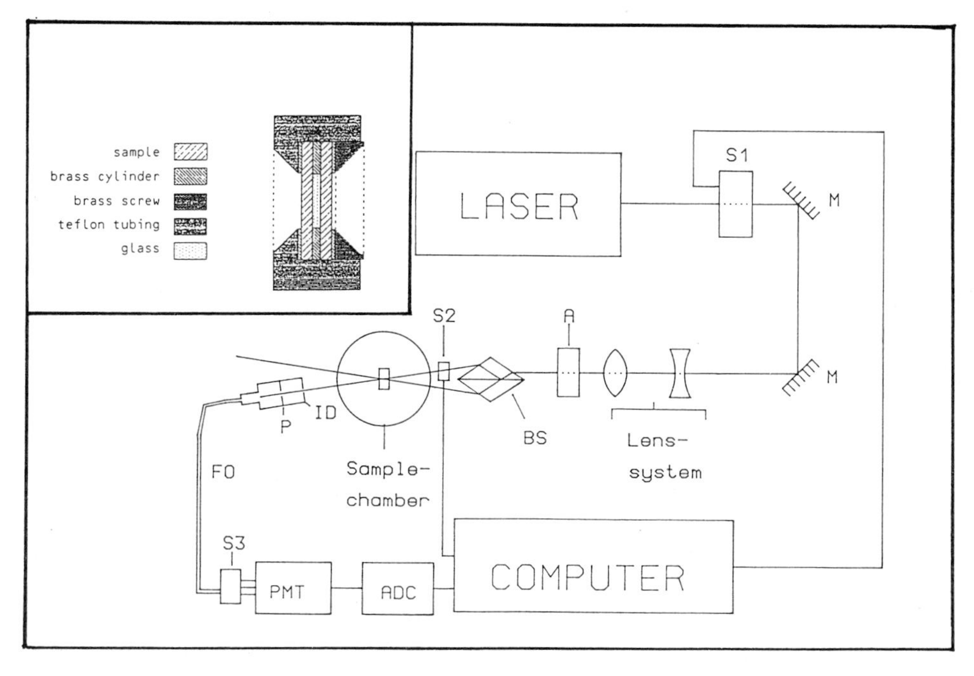

Fig. 11.2: Experimental setup. S 1–S 3: Shutters. A: Attenuator, M: Mirrors, BS: Beam splitter, P: Pinhole, PMT: Photomultiplier, ID: Iris diaphragm, FO: Fiber optics, ADC: Analogue digital converter, Inset: Sample holder.

11.2.3 Analysis of Diffusion Experiments

Since Δ n and Δ a are proportional to the concentration change $\Delta C = C_{max} - C_{min}$ defined in Equ. (1) the FRS intensity is proportional to ΔC^2. In simple cases (see below) ΔC is an exponential decay function resulting in an exponential FRS decay

$$I(t) = [A \exp(-t/\tau) + B]^2 + C \tag{5}$$

The parameters B and C originate from impurities inside and spurious light sources outside the sample contributing to the coherent and incoherent background, respectively.

In diffusion studies, the label concentration is a time function following Fick's law

$$\frac{\partial}{\partial t} C(x,t) = D \frac{\partial^2}{\partial x^2} C(x,t) \tag{6}$$

The initial condition of Equ. (1) leads to the solution

$$C(x,t) = 1/2\,(C_{max} + C_{min}) + 1/2\,(C_{max} - C_{min})\cos(2\pi x/d)\exp(-t/\tau) \tag{7}$$

$$\tau^{-1} = 4\pi^2 D/d^2 \tag{8}$$

In this case, Equ. (5) applies, and D can be determined from the FRS decay by evaluating τ as a fit parameter. The grating distance d is obtained from Equs.(2) or (3) with the intersecting angle θ and wave length λ of the writing beams.

The analysis of FRS decay functions can be complicated if C (x, t) changes with time because of chemical reactions. In this case, Equ. (5) may still apply, however, with a decay constant given by

$$\tau^{-1} = 4\pi^2 D/d^2 + \tau_{chem}^{-1} \tag{9}$$

D and τ_{chem} can be determined separately by analysing the FRS decay for different values of the grating distance d.

In hologram formation, the concentration profile C(x) shown in Fig. 11.1C is accompanied by a corresponding concentration profile $C_p(x)$ of the photoproduct shifted in phase by half the grating distance: $C_p(x) \sim C(x-d/2)$. If the photoreactive dye serves as label covalently bound to a polymer molecule it should have little influence upon the polymer diffusion coefficient. In this case one has to choose photoreactions where the photoproduct also should have little influence and Equ. (5) can be applied. However, for diffusion of small dye molecules (see 11.3.3) the diffusion coefficient D_p of the product may differ from D resulting in a more complex FRS decay function

$$I(t) = [A \exp(-t/\tau) + A_p \exp(-t/\tau_p) + B]^2 + C \tag{10}$$

where τ_p is related with D_p through Equ. (8) or Equ. (9). Since A and A_p can differ in sign I (t) may decay to $B^2 + C$, rise again and come back to $B^2 + C$ in the limit $t \gg \tau, \tau_p$. In this case, D and D_p can be determined separately [14] from a fit of Equ. (10).

Finally, it should be noted that holographic gratings different from Equ. (1) can be obtained for longer reaction times [13]. The periodic concentration profile can then be expanded in a Fourier series with components proportional to $\cos(2\pi \upsilon x/d)$ and $\upsilon = 1, 2, 3, \ldots$ The resultant FRS intensity is given by higher order Bragg peaks at $\upsilon > 1$. Whereas the latter can be analyzed as described above the higher order peaks correspond to smaller effective grid spacings d/2, d/3, ... In principle, this should be advantageous for determining extremely small D values, however, the low intensity of higher order peaks prevents their use in diffusion studies.

In Fig. 11.2 the experimental setup is shown which is presently used in our laboratory for diffusion in bulk polymer systems. A 4 W argon ion laser operating in single mode at 458 or 488 nm is used for hologram formation with typical bleaching times below 100 ms. The crystal beam splitter (BS) shown in Fig. 11.2 is used for small intersecting angles $\theta < 10°$. For larger angles, the beam is split by a semi-reflecting mirror, and the two coherent beams are guided by further mirrors to the sample. The attenuator A and shutter S 2 are in operation during the detection period if the writing and reading beam wave lengths are equal. If a He-Ne-laser is used for detection, provisions for precision adjustment of the reading angle are necessary. Further technical details are given in Refs. [15] and [16].

11.2.4 Photo-Labeling and Sample Preparation

Good labels for FRS studies have to fulfill a number of requirements which sometimes must be met simultaneously: The light-sensitivity should be sufficient at one of the laser lines (360–514 nm for an Ar^+-laser) and result in large Δn or a values (see Equ. 4) at the reading wave length. In our experiments, we have typically about one dye molecule per 10^5 carbon atoms. Secondary reactions after bleaching should not alter the dynamics of the polymer system which is investigated, e.g., by cross-linking. For reversible photochromic labels the thermal back reaction should not be faster than the diffusion process to be studied (see Equ. 9).

Most of our diffusion experiments with labeled polystyrene (PS) molecules were done with the same label, namely, an ester of 2-nitro-4-carboxy-4'-(dimethylamino)stilbene

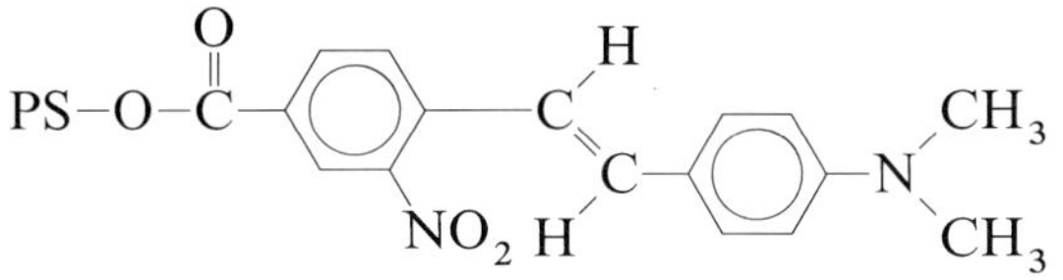

The label was introduced by reacting the Cs-salt of the dye acid with a chloromethylend- or side-group at the PS molecule. The photoreaction at 458 or 488 nm occurs essentially in two steps where the first photochemical step leads to a colourless intermediate which reacts thermally to the blue isatogene:

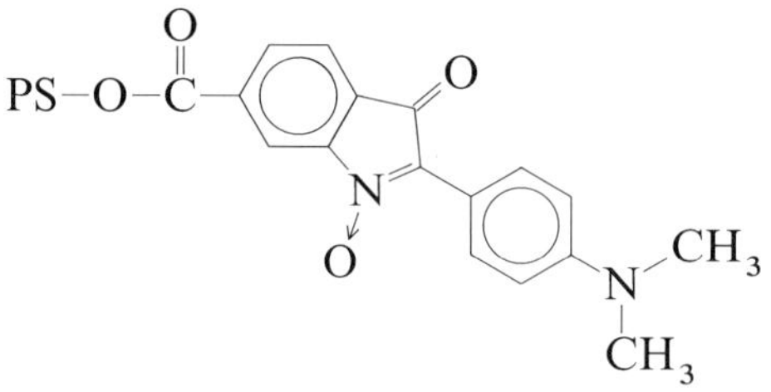

The thermal step is very rapid above ~ 150 °C, and very slow below ~ 50 °C. In between, it may interfere with diffusion which requires the complex analysis through (Equ. 9). We have also prepared the corresponding tolane derivative which differs from the stilbene by having a triple instead of a double bond in the center. Here, the photoreaction leads in one step to the isatogene. However, the tolane system is much more sensitive to light, and requires extensive darkening procedures during sample preparation. Therefore, we prefer working with the stilbene label when possible. A collection of further labels in present use is listed in Ref. [16].

The samples of labeled bulk polymer systems are usually prepared by mixing solutions of the labeled and unlabeled species in a volatile solvent, e.g., benzene or dioxane, and freeze drying the mixture. The resulting powder is pressed into pellets of 8.5 mm diameter and 0.3–0.8 mm thickness using a modified IR pellet press. The pellet is placed into the sample holder shown in the inset of Fig. 11.2 and annealed at the diffusion temperature until any flow processes have relaxed. Since the laser spot size used in hologram formation is chosen below 1 mm, up to ~ 30 diffusion coefficients can be determined in one sample thus allowing for varying the temperature or the intersecting angle θ of the light beams.

11.3 Applications

11.3.1 Diffusion of Polymer Chains by Reptation and other Transport Mechanisms

Translational diffusion of polymer chains depends critically upon the molecular weights M and M' of the diffusant and the surrounding matrix, respectively. For short diffusant chains, the Rouse model predicts [6].

$$D_{ROUSE} = \frac{k_B\, TM_o}{\varsigma_o\, M} \tag{11}$$

where M_o is the monomer molecular weight and ς_o the monomer friction coefficient which depends upon the properties of the matrix. For long entangled diffusant chains, the reptation model as devised by Doi and Edwards [6] predicts curvilinear diffusion along the chain contour with a (one dimensional) "tube" diffusion coefficient that is identified with D_{ROUSE}. The tube of length L is modeled as a random walk with L = N a where the step length a (also the tube diameter) is identified with the r.m.s. distance between entanglements. A detailed analysis yields for the diffusion coefficient of reptating chains [17]:

$$D_{REP} = \frac{D_{ROUSE}}{3\, N} = \frac{4}{15}\,\frac{M_e}{M}\, D_{ROUSE} = \frac{D_o}{M^2} \tag{12}$$

where M_e is the molecular weight between entanglements. In Equ. (12), one assumes that the matrix is fully entangled which implies a fixed tube. Matrix chain diffusion contributes to the diffusion of the M chains by a "constraint release" term

$$D_{CR} = k_{CR} M'^{-\alpha} D_{ROUSE} \tag{13}$$

where k_{CR} depends upon the "number of constraints per monomer" [17] and $\alpha \leq 3$ depending of the model assumed for matrix chain motion (see the discussion in Ref. [18]).

The first experiments confirming the M^{-2} power law, Equ. (12), were published by Klein [7] for chain diffusion in molten polyethylene. Further support for the reptation model was provided by measuring diffusion coefficients of chains in networks with a distance between cross-links much smaller than the entanglement spacing [19]. Since almost no difference was found for diffusion in networks and uncross-linked entangled matrices we concluded that any "lateral" transport mechanisms are negligible in the latter. However, it was

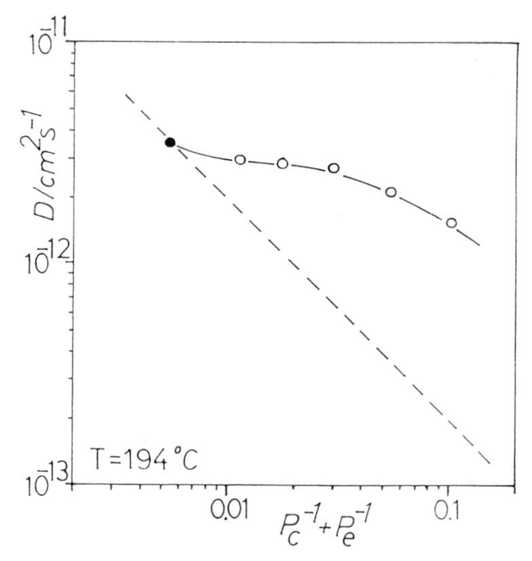

Fig. 11.3: Diffusion coefficients of linear polystyrene (M_w = 103000, T = 194 °C) in microgel matrices and a linear reference (full symbol). P_e and P_c are the average number of monomers between entanglements and cross-links, respectively. The dashed line represents the expectation calculated from the tube model.

rather surprising that Equ. (12) was a good approximation even in dense networks where we expected that M_e should be replaced by the molecular weight between cross-links, $M_c < M_e$ (see Fig. 11.3, [20]). Our findings were confirmed for different network topologies [21]. Only, after we minimized the number of cyclic structures in networks prepared by suspension copolymerization, chain diffusion became sizably slower than in the uncross-linked matrix [22].

The crossover between short and long chain behaviour is not really explained by Equs. (11) and (12) since one should expect that no "tube" can be defined unless $N \gg 1$. Nevertheless, Equ. (12) was found applicable to polystyrene chain diffusion in fully entangled matrices ($M' \gg M$) down to $M \approx M_e = 18000$ or $N \approx 1$, [23, 24]. However, the D values around M = 7000 shown in the upper left corner of Fig. 11.4A are clearly below the prediction of Equ. (12). Kramer [25] has obtained a quantitative fit of the crossover regime by assuming the rule $D = (D^{-1}_{ROUSE} + D_{REP}^{-1})^{-1}$ for combining Equs. (11) and (12). If the matrix molecular weight M' is also varied the friction constant ζ_o should also depend upon M' which is partly accounted for by the constraint release contribution Equ. (13). We have interpreted our polystyrene data in the crossover regime by assuming that Equ. (12) applies, however, with M_e replaced by

$$M_e' = M_e(1 + k\, M^{\gamma}/M'^{\gamma'}) \tag{14}$$

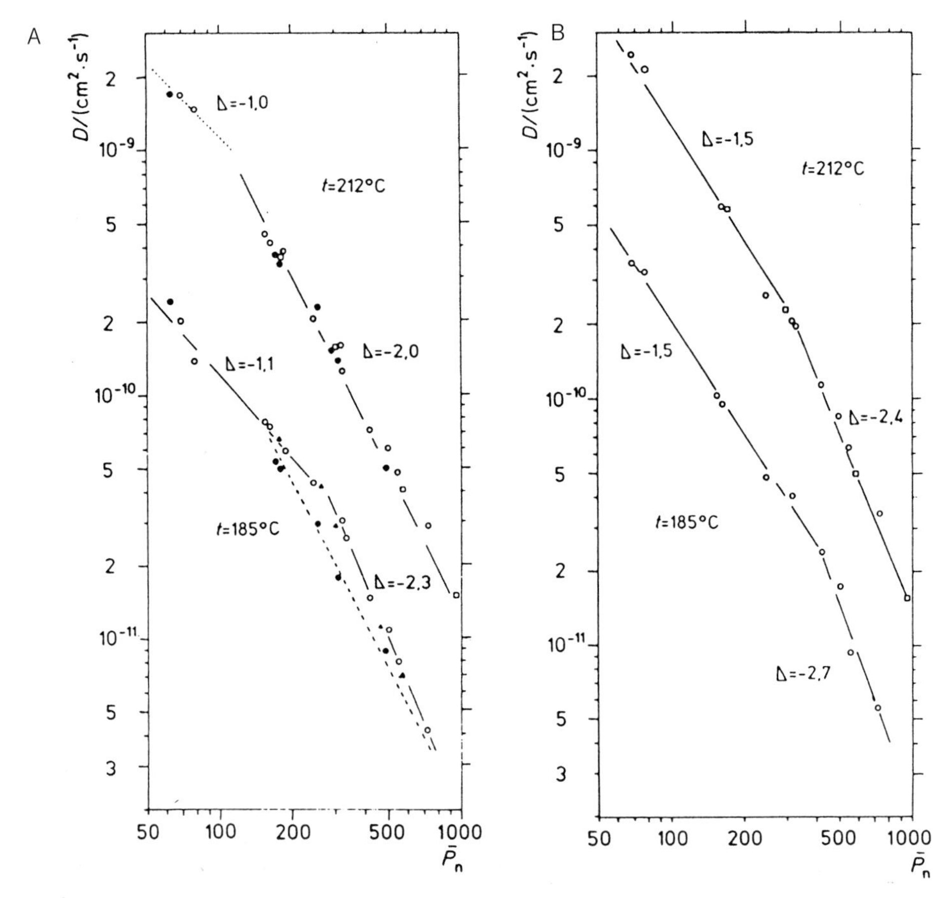

Fig. 11.4: Diffusion coefficients of polystyrene (PS) in different PS matrices versus the number average degree of polymerization $\overline{P}_n$ of the labeled PS (Δ:slope). (A) Matrix of large molecular weight (M_w = 111 000). Open symbols: PS labeled at one end. Full symbols: PS labeled at both ends. (B) Matrix with same molecular weight as labeled PS.

where the fit parameters were determined as $k = 1.5 \times 10^4, \gamma = 0.6, \gamma' = 1.6$ [23, 24]. Equ. (14) is similar to the assumption $D = D_{REP} + D_{CR}$ with $\gamma = 1$ and $\gamma' = \alpha \leq 3$ given by Equ. (13). It should be noted that Equs. (12–14) predict D $\sim M^{-\alpha}$ with $\alpha > 2$ for the case of equal diffusant and matrix molecular weight (M' = M, self-diffusion). This is clearly seen in Fig. 11.4B where $\alpha = 2.4$ was found and the crossover is shifted to larger M in accordance with Equ. (14). The D values of Fig. 11.4B have been corrected for end-group free volume of the matrix which causes faster diffusion of short chains thus explaining why no crossover was seen in the early experiments [3, 11].

The results shown in Fig. 11.4 for the lower temperature of 185° C exhibit a label influence which is not yet fully understood [24]. However, we should point out that these effects are relatively small, and our results are in good agreement with those of other techniques reporting overall agreement with free-volume theory [26]. We should also mention a recent computer simulation of polymer chain dynamics where the crossover between Rouse and reptation behaviour was found in good agreement with our experiments [27].

11.3.2 Diffusion of Polymer Rings, Microgels, and Stars

If the reptation model is applied to ring shaped macromolecules no center of mass transport is expected. Experiments with polystyrene rings, however, produced ring diffusion coefficients much faster than predicted by theory [28–30]. The diffusion PS microgels behaves similarly, and is only slowed down to a large extent for high intramolecular cross-linking and large molecular weight [29]. Clearly, cooperative processes play a more important role for the diffusion of medium size non-linear macromolecules than is expected from the tube model.

A star-branched molecule can only diffuse by reptation if the arms are retracted to the branch point and threaded into a neighbouring mesh of the entanglement network thus building up a new tube. As predicted by de Gennes [31] the diffusion coefficient depends exponentially on the length of the arms. This has been confirmed by experiments in polyethylene and polystyrene [11] and is clearly demonstrated in Fig. 11.5A where a semi-logarithmic plot of PS data in microgel networks is shown [18]. Previous theoretical treatments requiring simultaneous retraction of all but two of the arms for center of mass transport [11] are not confirmed by experiment, since the D values for three-, four-, and eight-arm stars fall on the same line. It should further be noted that results for dye labeled three-arm stars determined by the FRS technique are in harmony with the others in Fig. 11.5A and B determined by the FRES technique [9]. Finally, we point out that the D values of the two smallest 3-arm stars shown in Fig. 11.5A are not much smaller than those of PS chains of equal molecular weight [29]. This emphasizes the importance of non-reptative transport in the regime where the "tube" of the Doi-Edwards model is not well defined ($N \lesssim 4$, see Equ. 12).

In Fig. 11.5B, diffusion coefficients of PS stars in matrices of linear PS and PS networks are shown. The upturn in the regime of shorter matrix chains is well described by a constraint release model that corresponds to Equ. (13). Clearly, the network matrix behaves like a fully entangled matrix of very long uncross-linked chains. It is very remarkable that qualitatively the same behaviour has been found in experiments where diffusion coefficients of linear

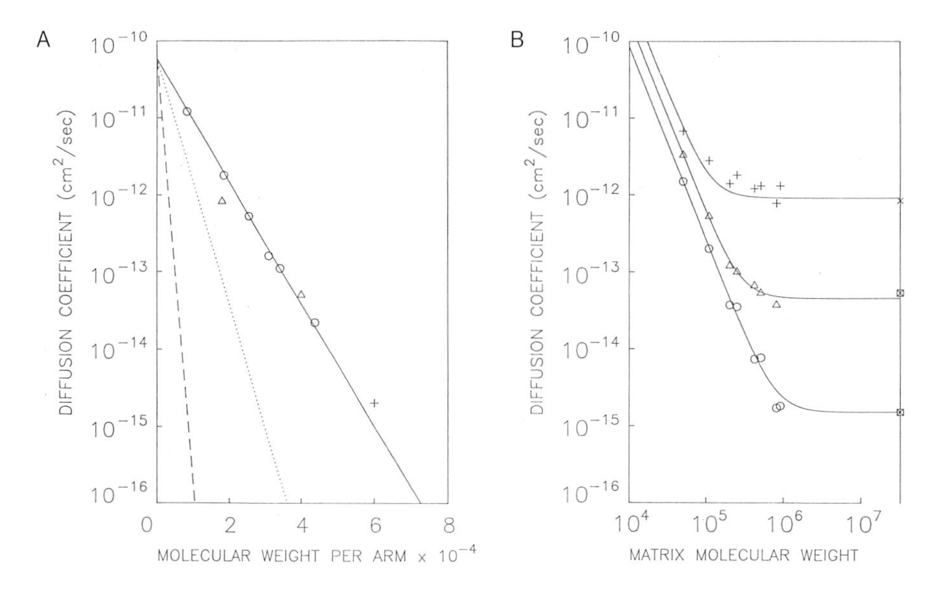

Fig. 11.5: Diffusion coefficients of star-branched polystyrene (T = 178 °C). (A) Diffusion in microgel matrix; Circles: three-arm stars; triangles: four-arm stars; cross: eight-arm star; dotted and dashed lines: theoretical predictions for four- and eight-arm stars, respectively [18]. (B) Diffusion in linear matrices; crosses: four-arm star with molecular weight per arm $M_a = 18\,000$; triangles: four-arm star with $M_a = 40\,000$; circles: eight-arm star with $M_a = 60\,000$.

chains [26], rings [30], and microgels [29] were measured as a function of the matrix molecular weight in linear matrices. The behaviour of diffusant linear chains is not surprising and in harmony with the reptation model. However, rings and microgels are interpenetrated by the matrix chains as is known from neutron scattering. Therefore, constraint release of matrix chains should slow down dramatically the diffusion of rings and microgels in the limit of large chain length, but this is clearly *not* seen in the experiment. We find that microgels move in the entanglement network as in a true network without feeling the center of mass motion of the interpenetrating matrix chains which can be much slower than that of the diffusing microgels [29]. Apparently, matrix chain ends and loops can penetrate in and out of the cyclic structures on a time scale which is much faster than center of mass diffusion of both, the diffusant and the matrix molecules.

11.3.3 Diffusion of Monomer Dyes at the Glass Transition of Amorphous Polymers

Previous diffusion studies of small probe molecules in polymers, using various permeation and sorption techniques, were limited to relatively high diffusion coefficients $D \gtrsim 10^{-10}$ $cm^2\ s^{-1}$ [32]. The FRS technique has extended the dynamical range to below 10^{-16} $cm^2\ s^{-1}$ thus providing a tool for investigating the diffusion of larger probe molecules in the vicinity of the glass transition [14–16, 33–35]. In Fig. 11.6, we give an example of probe diffusion in different polymer matrices above and below the glass transition temperature T_g of the matrix determined by DSC [15, 35]. The probe molecule, tetrahydrothiophene-indigo (TTI)

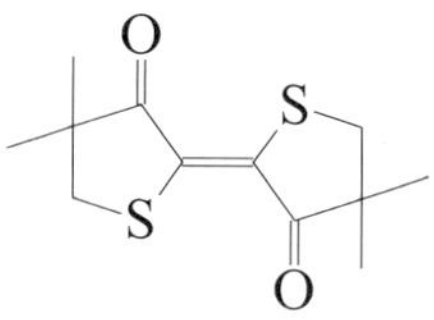

undergoes trans-cis-isomerization on irradiation at $\lambda = 488$ nm (Ar^+-laser). The thermal backreaction is sufficiently slow at the temperatures of our investigation. Since the absorption spectrum of the photoproduct (cis-TTI) is shifted to shorter wave lengths, we predominantly observe the diffusion of trans-TTI at 488 nm. The D values at T_g vary between 10^{-11} and 10^{-14} $cm^2\ s^{-1}$ depending upon the internal mobility of the matrix. The main chain motion of polycarbonate, which has been investigated by dielectric and NMR-techniques [36], also accounts for the fast probe diffusion. It is remarkable that the modification of polystyrene by adding 10 % tricresylphosphate or by reducing the chain length to a degree of polymerization of $\overline{P}_n = 20$ leads to the same "master curve" if D is plotted versus $T\text{-}T_g$ though T_g is considerably reduced (cf. Table 11.1). On the other hand, the "internal" plasticization by binding an ethyl group to each phenyl ring of PS increases the probe diffusivity, in particular, at temperatures below T_g. The full lines in Fig. 11.6 were obtained by a fit to the equation [37, 38].

$$\log\left[\frac{D(T)}{D(T_g)}\right] = \frac{(\xi\alpha_f B/2.303\ f_g)\ (T-T_g)}{f_g/\alpha_f + T-T_g} \tag{15}$$

where f_g is the fractional free volumne at T_g and α_f is the free volume expansion coefficient at $T > T_g$. Following the proposal of Vrentas and Duda [38] we have taken T_g, f_g/B and α_f/B from results of dynamic mechanical experiments of the polymers [35, 37]. Thus $D(T_g)$ and ξ are the only fit parameters used for fitting the data of Fig. 11.6 by Equ. (15). Apparently, ξ characterizes the coupling of probe diffusion to the α-process of the matrix glass transition. If we assume

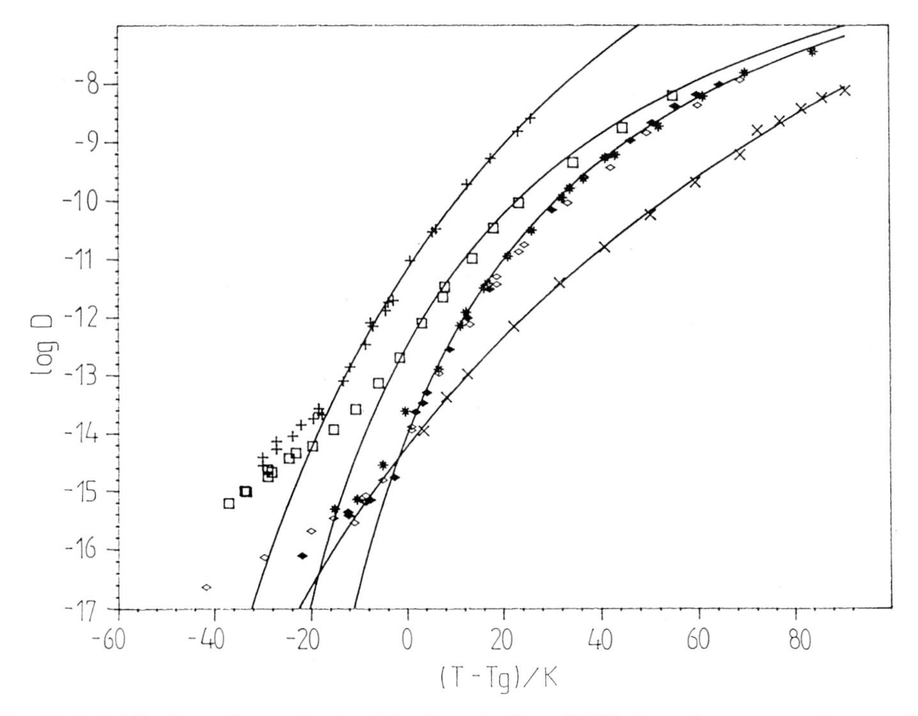

Fig. 11.6: Diffusion of tetrahydrothiopheneindigo (TTI) in polymer matrices: poly(bisphenol A carbonate) (+), poly(p-ethyl-styrene) (□), polystyrene (◆), polystyrene, plasticized with 10 % tricresylphosphate (◇), oligostyrene ($\bar{P}_n$ = 20) (⋆), polyethylmethacrylate (×).

Table 11.1: Free volume parameters for probe diffusion in polymers at $T \geq T_g$.

Polymer	Probe	T_g/°C	f_g/B	$\frac{\alpha_f/B}{10^{-4}\,K^{-1}}$	ξ
PS	TTI	100	0.032	5.6	0.84
PS($\bar{P}_n = 20$)	TTI	66	0.032	5.6	0.80
PS/TCP[a)]	TTI	69	0.028	4.9	0.74
PES[b)]	TI	82	0.032	4.7	0.72
PEMA[c)]	TTI	69	0.028	3.1	0.81
PC[d)]	TTI	150	0.027	4.8	0.50
CPC	ONS-N[e)]	150	0.027	4.8	0.60
PC	ONS-B	150	0.027	4.8	0.59
PC	ONS-A	150	0.027	4.8	0.77

[a)] PS with 10 % tricresylphosphate, [b)] polyethylstyrene, [c)] polyethylmethacrylate, [d)] poly(bisphenol A carbonate), [e)] see legend of Fig. 11.7.

full coupling [$D_\alpha(T) = D(T)$] at some high temperature $T-T_g = 100$ K we can estimate $D_\alpha(T_g)$ from Equ. (15) by letting $\xi = 1$. For example, this yields $D_\alpha(T_g) = 2 \times 10^{-17}$ cm^2 s^{-1} for a probe with "full coupling" to the α-process of PS. This is not far from the value estimated for TTI diffusion in an o-terphenyl matrix at the glass transition [39]. In Fig. 11.7, diffusion coefficients of different probe molecules in polycarbonate are shown. In addition to TTI, derivatives of the o-nitrostilbene (ONS) dye (see 11.2.4) have been investigated. The largest probe molecule has a D value of 3.8×10^{-14} cm^2 s^{-1} at T_g = 150° C similar to that of TTI in PS at its T_g. The values of the size parameter ξ listed in Table 11.1 are in qualitative agreement with the expectations from the molecular volumes and shapes of the probe molecules (see however [35]).

The D values below T_g show three remarkable features. First, there are pronounced annealing effects. For example, the D values of TTI in polycarbonate shown at the lowest temperature in Fig. 11.7 have been obtained from successive measurements where the lowest value corresponds to the longest annealing time. Since the slow FRS decay (see Equ. 5) requires minimum measurement times of 1 day for $D \lesssim 10^{-16}$ cm^2 s^{-1} the analysis of diffusion and

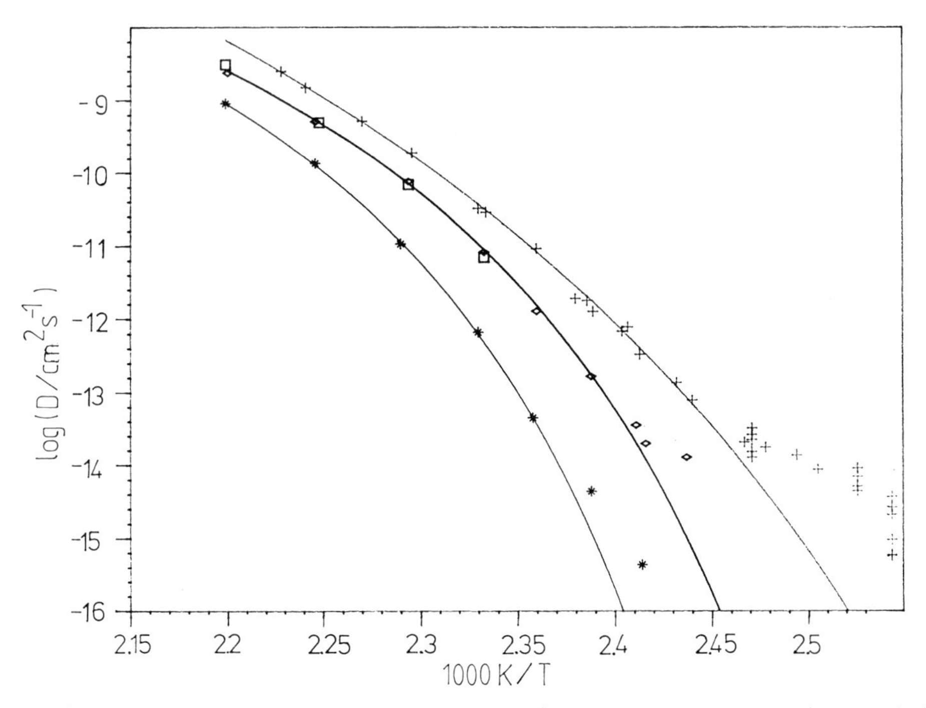

Fig. 11.7: Diffusion of different probes in poly(bisphenol A carbonate): TTI (+), ONS-N (□), ONS-B (◇), ONS-A (⋆). ONS is the o-nitrostilbene derivative described in Section 11.2.4 where in ONS-N the carboxyl group is replaced by $-C{\equiv}N$, in ONS-B by $CO-O-CH_2-C_6H_5$, and in ONS-A by $CO-O-CH_2-C_6H_4-N{=}N-C_6H_5$.

physical ageing below T_g becomes very difficult and time consuming [15]. Nevertheless, the upturn of the diffusion curves obtained below T_g for "reasonable" annealing times demonstrates that probe diffusion is decoupled from the α-process in polymer glasses. Although, this behaviour can be described within the free volume picture by fitting with a reduced $\alpha_f(\text{glass}) = \lambda\ \alpha_f$ [38], other models assuming thermally activated hopping with certain distributions of barrier heights [40] are probably more appropriate for describing probe diffusion below T_g. Finally, the full curves in Figs. 11.6 and 11.7 obtained by fitting Equ. (15) to experiments above T_g provide a good fit also in some temperature range below T_g. The different time scales of the DSC and diffusion experiments can only account for the behaviour in PS where the upturn starts ~ 5 K below T_g. However, in polycarbonate the D values of TTI (Fig. 11.6) suggest a "T_g" which is about 20 K below the DSC value. This effect can hardly be understood from a simple free volume picture [37, 38] but should be related with the main chain mobility in polycarbonate. It should be very interesting to find out how probe Diffusion is influenced by the suppression of main chain mobility in mixtures with additives [36].

Future more extensive diffusion studies in plasticized polymer glasses will certainly increase our understanding of the molecular basis of dynamical properties which determine the usefulness of these materials in practical applications.

11.4 References

[1] F. Bueche: Viscosity, self-diffusion, and allied effects in solid polymers, J. Chem. Phys. 20 (1952) 1959–1964.

[2] F. Bueche: Diffusion of polystyrene in polystyrene: Effect of matrix molecular weight, J. Chem. Phys. 48 (1968) 1410–1411.

[3] J. E. Tanner, K. J. Liu, J. E. Anderson: Proton magnetic resonance self-diffusion studies of polyethyleneoxide and polydimethylsiloxane solutions, Macromolecules 4 (1971) 586–588.

[4] S. F. Edwards: Statistical mechanics with topological constraints I, Proc. Phys. Soc. 91 (1967) 513–519.

[5] P. G. de Gennes: Reptation of a polymer chain in the presence of fixed obstacles, J. Chem. Phys. 55 (1971) 572–579.

[6] M. Doi, S. F. Edwards: The Theory of Polymer Dynamics, Clarendon Press, Oxford 1986.

[7] J. Klein: Evidence for reptation in an entangled polymer melt, Nature (London) 271 (1978) 143–145.

[8] C. R. Bartels, W.W. Graessley, B. Crist: Measurement of self-diffusion coefficient in polymer melts by a small angle neutron scattering method, J. Polym. Sci., Lett. 21 (1983) 495–499.

[9] P. F. Green, P. J. Mills, Ch. J. Palmstrom, J.W. Mayer, E. J. Kramer: Limits of reptation in polymer melts, Phys. Rev. Lett. 53 (1984) 2145–2148.
[10] H. Hervet, W. Urbach, F. Rondelez: Mass diffusion measurements in liquid crystals by a novel optical method, J. Chem. Phys. 68 (1978) 2725–2729.
[11] K. Binder, H. Sillescu: Diffusion, polymer-polymer, in Encyclopedia of Polym. Sci. and Eng. Suppl. Vol., 2nd ed., John Wiley, New York 1989.
[12] H. Hervet, L. Leger, F. Rondelez: Self-diffusion in polymer solutions: a test for scaling and reptation, Phys. Rev. Lett. 42 (1979) 1681–1684.
[13] H. J. Eichler, P. Günther, D.W. Pohl: Laser-Induced Dynamic Gratings, Springer, Berlin 1986.
[14] J. Zhang, C. H. Wang, D. Ehlich: Investigation of the mass diffusion of camphorquinone in amorphous poly (methylmethacrylate) and poly (tert-butylmethacrylate) hosts by the induced holographic grating relaxation technique, Macromolecules 19 (1986) 1390–1394.
[15] D. Ehlich: Diss. Univ. Mainz 1988.
[16] H. Sillescu, D. Ehlich: Application of holographic grating techniques to the study of diffusion processes in polymers, in J. P. Fouassier, J. F. Rabek (eds.): Application of Lasers in Polymer Science and Technology, CRC Press, Boca Raton, Florida, 1990, p. 211–226.
[17] W.W. Graessley: Viscoelastic properties of entangled flexible polymers, Faraday Symposia Chem. Soc. 18 (1983) 7–27.
[18] K. R. Shull, E. J. Kramer, G. Hadziioannou, M. Antonietti, H. Sillescu: Diffusion of macromolecular stars in linear, microgel, and network matrices, Macromolecules 21 (1988) 2578–2580.
[19] M. Antonietti, H. Sillescu: Self-diffusion of polystyrene chains in networks, Macromolecules 18 (1985) 1162–1166.
[20] M. Antonietti, H. Sillescu, M. Schmidt, H. Schuch: Solution properties and dynamic bulk behaviour of intramolecular cross-linked polystyrene, Macromolecules 21 (1988) 736–742.
[21] M. Antonietti, K. J. Fölsch, H. Sillescu, T. Pakula: Micronetworks by endlinking of polystyrene II: dynamic mechanical behaviour and diffusion experiments in the bulk, Macromolecules.
[22] K. J. Fölsch: Diss. Univ. Mainz 1988.
[23] M. Antonietti, J. Coutandin, H. Sillescu: Diffusion of linear polystyrene molecules in matrices of different molecular weights, Macromolecules 19 (1986) 793–798.
[24] M. Antonietti, K. J. Fölsch, H. Sillescu: Critical chain lengths in polystyrene bulk diffusion, Macromol. Chem. 88 (1987) 2317–2324.
[25] E. J. Kramer, private communication.
[26] P. F. Green, E. J. Kramer: Matrix effects on the diffusion of long polymer chains, Macromolecules 19 (1986) 1108–1114.
[27] K. Kremer, G. S. Grest, I. Carmesin: Crossover from Rouse to reptation: a molecular dynamics simulation, Phys. Rev. Lett.
[28] M. Antonietti, J. Coutandin, R. Grütter, H. Sillescu: Diffusion of labeled macromolecules in molten polystyrenes studies by a holographic grating technique, Macromolecules 17 (1984) 798–802.
[29] M. Antonietti, H. Sillescu: Diffusion of intramolecular cross-linked and three-arm-star branched polystyrene molecules in different matrices, Macromolecules 19 (1986) 798–803.

[30] P. J. Mills, J.W. Mayer, E. J. Kramer, G. Hadziioannou, P. Lutz, C. Strazielle, P. Rempp, A. J. Kovacs: Diffusion of polymer rings in linear polymer matrices, Macromolecules 20 (1987) 513–518.
[31] P. G. de Gennes, P. G. Reptation of stars, J. Phys. 36 (1975) 1199–1203.
[32] J. Crank, G. S. Park (eds.): Diffusion in Polymers, Academic Press, London 1968.
[33] J. Coutandin, D. Ehlich, H. Sillescu, C. H. Wang: Diffusion of dye molecules in polymers above and below the glass transition temperature studied by the holographic grating technique, Macromolecules 18 (1985) 587–589.
[34] J. Zhang, C. H. Wang: Effect of plasticization on mass diffusion of camphorquinone in polystyrene, Macromolecules 21 (1988) 1811–1813, and references therein.
[35] D. Ehlich, H. Sillescu: Tracer diffusion at the glass transition, Macromolecules 23 (1990) 1600–1610.
[36] E.W. Fischer, G. P. Hellmann, H.W. Spiess, F. J. Hörth, U. Ecarius, M. Wehrle: Mechanical properties, molecular motions, and density fluctuations in polymer additive mixtures, Makromol. Chem. Suppl. 12 (1985) 189–214.
[37] J. D. Ferry: Viscoelastic Properties of Polymers, 3rd ed., Wiley, New York 1980.
[38] J. S. Vrentas, J. L. Duda: A free volume interpretation of the influence of the glass transition on diffusion in amorphous polymers, J. Appl. Polym. Sci. 22 (1978) 2325–2339.
[39] M. Lohfink: Diplomarbeit Univ. Mainz 1988.
[40] R. Richert: Merocyanine-Spiropyran photochemical transformation in polymers, probing effects of random matrices, Macromolecules 21 (1988) 923–929, and references therein.

12 Deuteron NMR. A New Tool for Investigating Order and Dynamics in Polymers

Hans Wolfgang Spiess*, Hans Sillescu**

12.1 Introduction

^{2}H-NMR in solids was governed for a long time by broad line techniques where typically only nuclear quadrupole coupling constants could be determined because of poor signal to noise ratios [1]. The signal intensity was considerably improved by application of the solid echo Fourier transform technique first applied to deuterated liquid crystals where the quadrupole splittings are largely reduced through fast anisotropic molecular motion [2]. Undistorted line shapes of ^{2}H spectra in solids were first obtained in the high field of a super-conducting magnet by combining the solid echo with the spin alignment (or stimulated) echo following a three pulse sequence [3, 4]. This achievement opened up the way to applications in polymers where line shape analysis in oriented systems allows, in principle, for determining the full orientation distribution [5]. The study of slow and ultra-slow motions in solid polymers became possible through investigation of their influence upon the shape of solid echo [6] and spin alignment spectra [4]. Here, one can obtain important information upon the type of molecular reorientation on a time scale ranging to very slow motions with correlation times of the order of many seconds. A particular advantage of ^{2}H spectroscopy is its selectivity in partially deuterated systems which was already utilized when investigating ^{1}H- and ^{2}H spin-lattice relaxation times in molten polystyrene [7]. The facilities for preparing and characterizing deuterated polymer systems in the Sonderforschungsbereich 41 were extremely helpful for exploring the potential of ^{2}H-NMR techniques, and stimulating their further development.

* Max-Planck-Institut für Polymerforschung, Postfach 3148, D-6500 Mainz
** Institut für Physikalische Chemie der Universität Mainz, Jakob-Welder-Weg, D-6500 Mainz

In the following review we at first describe in some detail the physics of the ^{2}H-NMR technique in order to help non-specialists appreciating its potential for application in polymer science. The second part provides some illustrative examples showing ^{2}H spectroscopy at work in some important polymer systems. New developments in 2 D-spectroscopy and some comparisons with other solid state NMR techniques in particular ^{13}C-Magic Angle Spinning (MAS) NMR are also included.

12.2 Techniques

12.2.1 Dynamics

12.2.1.1 Slow Motion and Solid Echo Spectra

In a rigid solid, the coupling of the deuteron quadrupole moment with the electric field gradient tensor (FGT) gives rise to a symmetric splitting $2\omega_Q$ of the ^{2}H Larmor frequency where [8]

$$\omega_Q = \delta\,(3\cos^2\theta - 1 - \eta\,\sin^2\theta\,\cos 2\,\Phi),$$

$$\delta = \frac{3e^2Qq}{8\hbar} \tag{1}$$

where $e^2qQ/\hbar$ is the quadrupole coupling constant, θ and Φ are the polar angles of the external magnetic field in the principal axes systems of the ^{2}H FGT. In most polymers, the FGT is axially symmetric ($\eta = 0$) around the C-^{2}H bond which then is at the angle of θ with respect to the external field. The weighted superposition of the doublets defined by Equ. (1) yields a powder line shape which has the form of the Pake spectrum shown in Fig. 12.1a if $\eta = 0$ and the C-^{2}H bonds are distributed at random in an isotropic system, say, a polymer glass. In oriented polymers, the line shape can be rather different, as shown in the example of Fig. 12.1b, and yields important information on the orientational distribution (see Section 12.2.2). Likewise rapid molecular motion leads to a partially averaged quadrupole coupling reflected in yet other line shapes as shown in Fig. 12.1c for the Kink motion of polymer chains [9]. The full width ($\sim$ 250 kHz) of the powder spectrum is given by $3e^2Qq/2\hbar$. Since this constant can always be measured at low temperatures, and varies

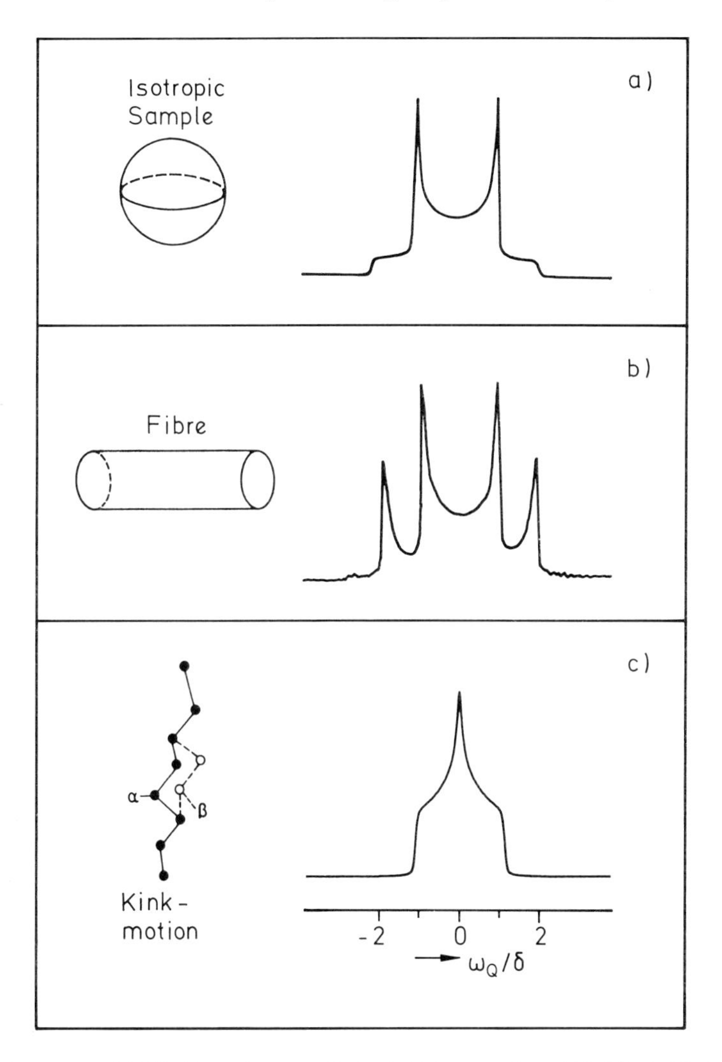

Fig. 12.1: Deuteron NMR line shapes of a) rigid isotropic solid, b) rigid drawn fibre, c) isotropic polymer with flexible chain.

little with temperature, Equ. (1) provides a measure of the orientation of the FGT and its evolution in time which is denoted below as

$$\omega_i = \omega_Q(t_i) = \omega_Q\,[\Omega(t_i)] = \omega_Q[\theta(t_i),\, \Phi(t_i)] \tag{2}$$

t_i being one of the time spacings defined in Fig. 12.2.

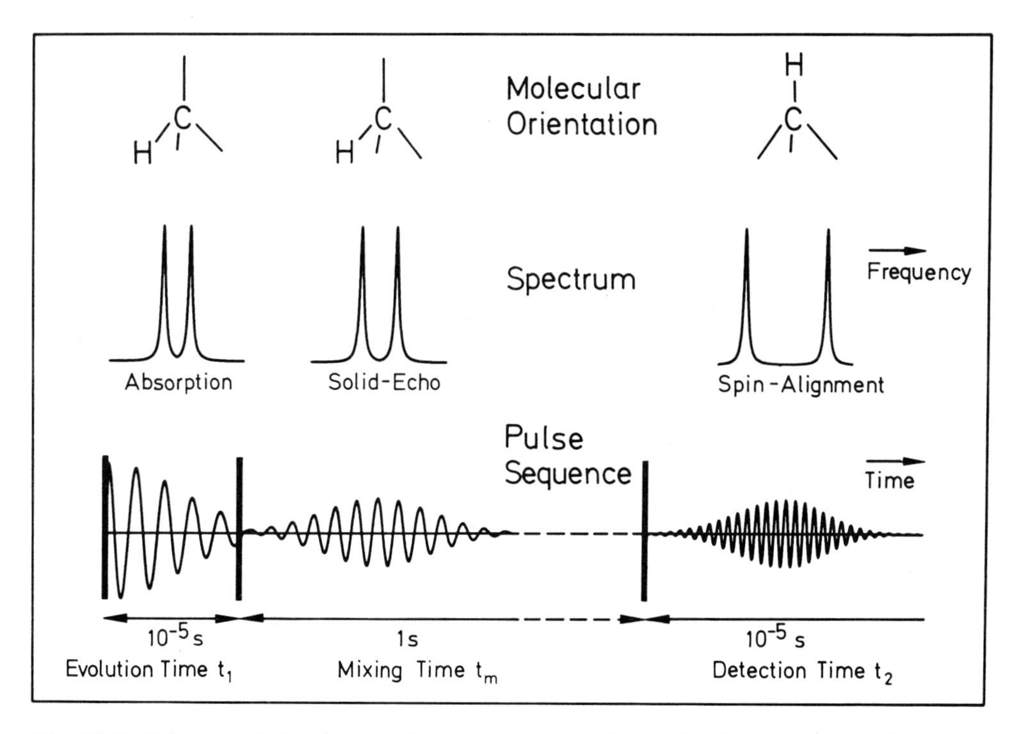

Fig. 12.2: Scheme of the three-pulse sequence used in pulsed deuteron NMR.

Molecular reorientation is given by a stochastic process Ω (t) which can be described by models of different inherent complexity. Numerical procedures can be most easily applied if Ω (t) is approximated by a Markoff process defined by a rate equation [8].

$$\frac{\partial}{\partial t} \underset{\sim}{P} = \underset{\sim}{\Pi} \, \underset{\sim}{P} \tag{3}$$

In the well known Debye model of rotational diffusion

$$\underset{\sim}{\Pi} = D \, \nabla_{\Omega}^2 \tag{4}$$

defines the rotational diffusion constant D, and the Laplacian in spherical coordinates, ∇_{Ω}^2, operates upon $\underset{\sim}{P} = P(\Omega_0 | \Omega, t)$, the probability density that the orientation is Ω at time t if it was Ω_0 at $t = 0$. However, the form of Equ. (3) allows for defining more specialized rotational models by specifying the rate matrix $\underset{\sim}{\Pi}$. In particular, models for rotational jumps between few well defined sites have been applied in previous ^{2}H NMR applications [4, 6, 9]. A thorough analysis of molecular reorientation utilizing the full symmetries in relation with 2 D exchange spectra is contained in Ref. [10]. It should be noted that $\underset{\sim}{P}$ contains more information about the reorientation process than the usual rotational correlation functions which are given for the Debye rotational diffusion process by

$$G_l\,(t) = \langle P_l\,(\cos\theta)\rangle = \exp\,[-l(l+1)D\,t] \tag{5}$$

where P_l is the Legendre polynomial of order l (in NMR experiments l = 2). The pointed brackets denote the average over the random process (see Equs. (9) and (10) below).

The NMR line shape can be described in the whole range from the rigid solid powder spectrum to the motionally narrowed Lorentzian by the Fourier transform (FT) of the free induction decay (FID):

$$S_o\,(t) = \left\langle \exp\,i\int_0^t \omega_Q\,(t')dt' \right\rangle \tag{6}$$

The average can be evaluated numerically if the integral is approximated by a discrete summation, and $\Omega(t)$ follows a discrete approximation of the rate equation, Equ. (3). These "slow motion" line shapes have been analyzed for NMR- and ESR-spectra in many systems applying theories originating from the treatment in Chapter X of Abragam's book [8]. However, the possibilities of measuring slow motion spectra in solids are limited by "dead time" effects due to pulse length, recovery times etc. which prohibit the detection of the FID at short times. In order to circumvent these problems, the solid echo technique has been devised where two 90° pulses separated by a time delay t_1 and shifted

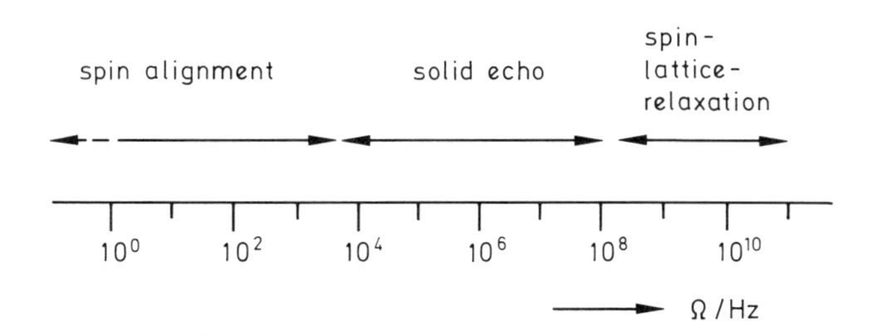

Fig. 12.3: Timescales of pulsed deuteron NMR employing the pulse sequence of Fig. 12.2.

in phase by 90° are applied giving rise to an echo at $2\,t_1$ (see Fig. 12.2). The solid echo spectrum is the FT of (see Ref. [6])

$$S_1(t, t_1) = \left\langle \exp\left[i \int_0^{t_1} \omega_Q(t')dt' - i \int_{t_1}^{t} \omega_Q(t')dt' \right] \right\rangle \tag{7}$$

which is similar to the FID, Equ. (6), but depends critically upon the pulse spacing t_1 if there is molecular motion with correlation times in the range of

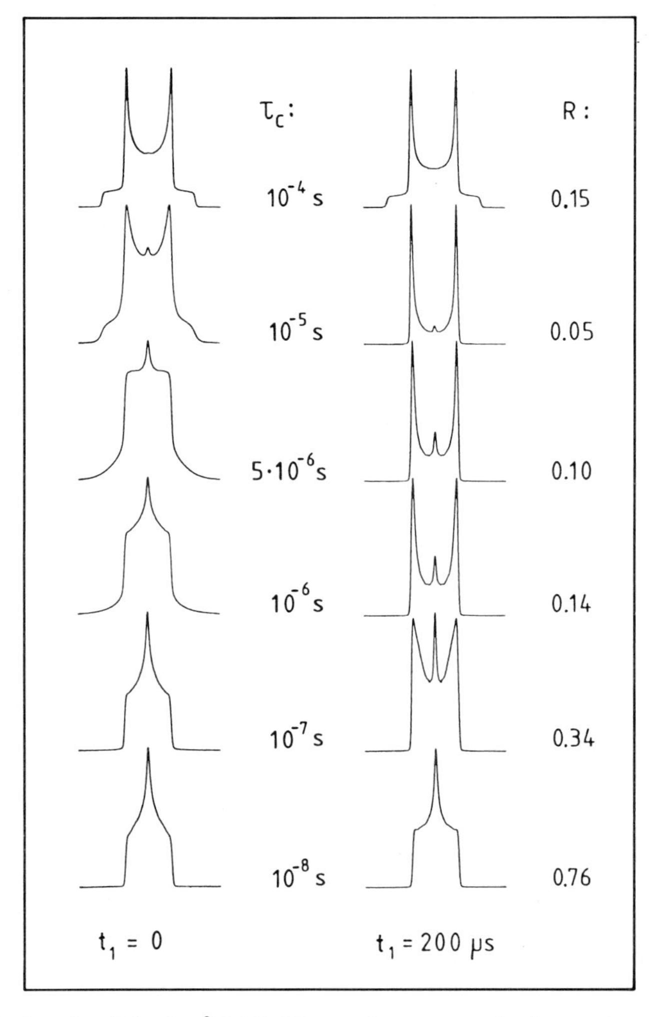

Fig. 12.4: Calculated solid echo ^{2}H-NMR-powder-spectra for jumps between two sites related by the tetrahedral angel for different correlation times τ_c. The reduction factor R gives the ratio of the spectra for $t_1 = 200\ \mu s$ and $t_1 = 0$.

$10^{-8} - 10^{-4}$ s (see Fig. 12.3). Furthermore, the total signal intensity changes considerably as a function of t_1 and can be utilized for analyzing the dynamics. In Fig. 12.4, calculated line shapes [6] are shown for the example of jumps between two sites related by the tetrahedral angle (e.g. trans-gauche isomerization) for $t_1 = 0$ and $t_1 = 200\ \mu s$, respectively. Of particular interest is the case of reorientation by small angular steps where line shape changes can be generated for large $t_1 \gg \omega_Q^{-1}$ if $\tau \lesssim t_1$ for the time τ between two steps. In Debye rotational diffusion (see Equ. 5), we have $\tau \ll \tau_c = (6\ D)^{-1}$ and line shape effects are observable for $\tau_c \lesssim 100$ ms if $t_1 \gtrsim 50\ \mu s$ [11].

A transverse relaxation time T_2^* can be defined by the decay of the solid echo amplitude recorded as a function of the pulse spacing t_1. In the solid, $(T_2^*)^{-1}$ is given by the average half width of a single crystal line. Therefore T_2^* is longer by about two orders of magnitude than the FID decay time, which is given by the inverse of the full width of the powder spectrum.

12.2.1.2 Spin-Alignment

The three pulse sequence shown in Fig. 12.2 gives rise to a spin-alignment or stimulated echo which has a shape and intensity depending upon both pulse spacings t_1 and t_m where the former is typically in the μs-range ($t_1 \lesssim T_2^*$) whereas the "mixing time" t_m is only limited by spin-lattice relaxation ($t_m \lesssim T_1$). Thus, "ultra-slow" motions with correlation times up to the time scale of minutes become observable through spin-alignment techniques. The spin-alignment echo signal can be formulated as [4, 12]

$$S_2\ (t_1, t_2, t_m) = \langle \exp\left[-\, i\omega_Q(O)t_1\right] \exp\left[i\omega_Q(t_m)t_2\right]\rangle \tag{8}$$

where the FT with respect to the detection time t_2 is the spin-alignment spectrum. If a further FT is computed with respect to t_1 one obtains the two-dimensional spectra discussed further below. Note that the integrals contained in Equs. (6, 7) are absent in Equ. (8) since we assume that $\omega_Q\ (O)$ is approximately constant over the evolution time t_1 whereas $\omega_Q\ (t_m)$ is constant over t_2. Thus $t_m (\gg t_1, t_2)$ is the only time left for characterizing molecular reorientation, and the average becomes simply

$$S_2\,(t_1, t_2, t_m) = \int d\Omega_1 \int d\Omega_2\, W(\Omega_1, \Omega_2, t_m) \exp\,(-i\omega_1 t_1) \exp\,(i\omega_2 t_2) \tag{9}$$

where $\omega_i = \omega_Q\,(\Omega_i)$ (see Equ. 2) and

$$W\,(\Omega_1, \Omega_2, t_m) = W\,(\Omega_1)\,P\,(\Omega_1 | \Omega_2, t_m) \tag{10}$$

is the joint probability density related with the probability density $W(\Omega_1)$ to find a C-^{2}H bond in the orientation Ω_1, and $P(\Omega_1|\Omega_2, t_m)$ is the conditional probability discussed below Equ. (4).

It should now have become apparent how the spin-alignment technique can be put to work in order to probe particular models of molecular reorientation: The model is defined through the rate matrix $\underset{\sim}{\Pi}$. Equ. (3) is integrated resulting in $P(\Omega_1|\Omega_2, t_m)$ which is placed in Equ. (9) yielding the theoretical spin-alignment shape. The comparison (after FT) with the experimental shape provides information on whether a particular model is compatible with experiment. (It should be noted that the numerical procedures are more complex and require consideration of the full symmetries inherent in the formalism [10]).

The potential of the method can best be appreciated by a comparison with quasi-elastic incoherent neutron scattering. Here, the intermediate scattering function is

$$I(\underset{\sim}{q}, t) = \langle \exp(-i\, \underset{\sim}{r_o} \cdot \underset{\sim}{q}) \exp(i\, \underset{\sim}{r} \cdot \underset{\sim}{q}) \rangle \tag{11}$$

where r_o and r denote the positions of the scattering particle at $t = 0$ and t, respectively, $\underset{\sim}{q}$ is the wave vector, and the average is over the van Hove function. Comparison of Equ. (11) and Equ. (8) reveals an interesting analogy [12]. For the special case $t_2 = t_1$, the pulse spacing t_1 of the NMR experiment corresponds to $\underset{\sim}{q}$ (or the scattering angle) in neutron scattering. The orientation $\Omega(t_m)$ in NMR corresponds to the position $\underset{\sim}{r}(t)$ in neutron scattering. Of course, the time scales are vastly different, $t \lesssim 10^{-8}$s, and $10^{-3}\text{s} \lesssim t_m \lesssim 10^3$s for neutron scattering and NMR, respectively. Furthermore, the NMR technique is more sensitive, since one varies the two independent variables t_1 and t_2 rather than one scattering angle, and, by setting the r.f. phase between the first two pulses of the sequence shown in Fig. 12.2 one can measure the two averages $< \sin \omega_1 t_1 \sin \omega_2 t_2 >$ and $< \cos \omega_1 t_1 \cos \omega_2 t_2 >$ contained in Equ. (9) where the former is particular sensitive to details of the reorientation process. Finally, in the limit $\omega_i t_i \ll 1$ one obtains $< \sin \omega_1 t_1 \sin \omega_2 t_2 > = t_1 t_2 < \omega_1 \omega_2 >$ which is proportional to the rotational correlation function G_2 defined in Equ. (5).

12.2.1.3 Spin Lattice Relaxation

The ^{2}H spin lattice relaxation time can be formulated as [8]

$$\frac{1}{T_1} = \frac{3}{40} \left[\frac{e^2Qq}{\hbar}\right]^2 \left[J_1(\omega_o) + 4\, J_2(2\omega_o)\right] \tag{12}$$

where ω_o is the Larmor frequency, and the spectral density is given for the example of Debye rotational diffusion (see Equ. 5) by

$$J_1(\omega) = J_2(\omega) = \mathrm{re}\int_0^\infty G_2(t)e^{i\omega t}dt = \frac{\tau_c}{1+\omega^2\tau_c^2};$$

$$\text{with } \tau_c^{-1} = 6\,D. \tag{13}$$

T_1 measurements of ^1H, ^2H, and ^{13}C have been extensively applied to investigating the local mobility of polymer chains in solutions and melts [7, 13–15]. Though the time scale is usually limited to fast motions with correlation times below 10^{-8}s (see Fig. 12.3) one can extend the range up to $\sim 10^{-3}$s by measuring rotating frame relaxation times $T_{1\varrho}$ [13, 16] or low field relaxation times T_1 that correspond to Larmor frequencies in the kHz regime [14]. However, it is apparent from Equ. (13) that spin-lattice relaxation can only provide information about the correlation function $G_2(t)$ which is the second moment of the probability function $P(\Omega_o|\Omega, t)$ (see Equs. 3–5). Thus, spin-alignment and solid echo experiments yield more detailed information upon the mechanisms of molecular reorientation.

In solids, the spectral densities $J_1(\omega_o)$ and $J_2(2\omega_o)$ depend upon the orientation of the time averaged quadrupole coupling tensor (e.g. the axis of methyl group rotation) in the laboratory system. In practice the resulting orientation dependence of T_1 is rather weak since the effects of J_1 and J_2 partly cancel in Equ. (12). The situation is different for the spin-lattice relaxation time T_{1Q} of the spin alignment which is given by [4].

$$\frac{1}{T_{1Q}} = \frac{9}{40}\left[\frac{e^2Qq}{\hbar}\right]^2 J_1(\omega_o) \tag{14}$$

The orientation dependence of $J_1(\omega_o)$ can produce considerable changes in the shape of spin-alignment spectra in solids [17] which should not be confused with the direct reorientation influence described by Equs. (8–10) where T_{1Q} has been assumed to be very long compared with the correlation times of spin-alignment decay. However, the time scale of molecular motion can extend over many decades in solid polymers, and the analysis of ^2H experiments should become extremely complex unless the different motional processes are sufficiently decoupled in order to allow for assuming distributions of correlation times.

12.2.1.4 Distribution of Correlation Times

The rate equation, Equ. (3), appears to be quite general for modeling various types of molecular reorientation. Nevertheless, it is not sufficient for describing the complex motion in the vicinity of the glass transition of amorphous polymers or other glass forming liquids. Here, the time evolution of many relaxation experiments (e.g. mechanical or dielectric relaxation) can be fitted to a stretched exponential, $\exp[-(t/\tau_o)^\beta]$, which is the Laplace transform of the Kohlrausch-Williams-Watts distribution of correlation times [18]. This corresponds to replacing Equ. (5) by a linear superposition of exponentials, say (for $l = 2$),

$$G_2(t) = \int p(x) e^{-t/\tau_x} dx = \sum_\alpha p_\alpha e^{-t/\tau_\alpha}. \tag{15}$$

where x can be chosen as $\ln \tau_x$ in order to account for broad correlation time distributions. Similarly, one can define a linear superposition of probabilities $\underset{\sim}{P}$ each being the solution of a corresponding rate equation. One can also account for coupling between these rate equations by formulating a set of coupled equations of the form of Equ. (3). Thus, the formalism is sufficiently flexible in order to account for increased complexity of molecular reorientation. However, one should realize that ^2H spectroscopy always detects the reorientation of some well defined molecular vector, and yields no direct information on cooperative motion. Indirect information is nevertheless possible because of the different time scales seen by the different ^2H experiments. An example will be discussed below (Section 12.3.1.2) where a correlation time distribution was determined in solid polycarbonate by fitting a superposition of solid echo spectra each one corresponding to a single correlation time within the distribution [19]. A subsequent investigation of spin-lattice relaxation revealed a corresponding distribution of T_1 on a much longer time scale. This proves that the deuterons belonging to a particular correlation time τ_c as determined by the solid echo experiment remain in this “motional state” on the time scale of the T_1 experiment longer by several orders of magnitude. Any rapid exchange indicative of faster cooperative motion would result in exponential spin-lattice relaxation given by one average T_1 rather than a T_1 distribution. In conclusion, ^2H spectroscopy can reveal “motional” heterogeneities which often have an obvious relation with spatial heterogeneity.

12.2.2 Order

12.2.2.1 Orientational Distribution

The molecular order in partially ordered polymers, e.g. drawn fibres or oriented frozen liquid crystalline polymers is described by the orientational distribution function $P(\alpha, \beta, \gamma)$. It represents the probability density to find a molecule oriented within a range between α and $\alpha + d\alpha$, β and $\beta + d\beta$, γ and $\gamma + d\gamma$ with respect to the coordinate system describing the macroscopic order. Thus one has to define a *molecular frame* and a *director frame* (cf. Fig. 12.5). The

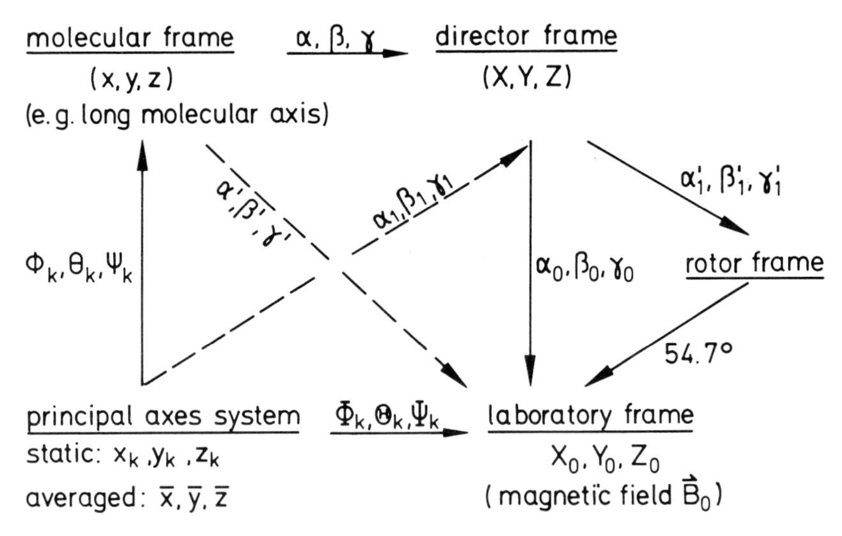

Fig. 12.5: Relation between different coordinate systems in partially ordered solids. The transformations between the different frames are described by the sets of Euler as indicated.

^{2}H-NMR line shape, however, primarily depends on the orientation of the *principal axes systems* of the FGT with respect to the magnetic field, specifying the *laboratory frame.*

In most cases of interest, e.g. in drawn amorphous polymers or frozen liquid crystalline polymers the orientational distribution function has a full width at half height exceeding 20°. It can then be expanded in terms of Wigner rotation matrices $D^{(l)}_{kk'}(\alpha, \beta, \gamma)$:

$$P(\alpha, \beta, \gamma) = \sum_{l=o}^{\infty} \sum_{k=-l}^{l} \sum_{k'=-l}^{l} P_{lkk'} D^{(l)}_{kk'}(\alpha, \beta, \gamma). \tag{16}$$

The expansion coeficients $P_{lkk'}$ are the moments of the orientational distribution. The analysis of NMR line shapes for this general case has been described in detail in Ref. [5]. If we confine ourselves to symmetric systems, straightforward ways to analyze NMR-spectra can be applied [20, 21].

i. Expansion in terms of spherical functions.
ii. Expansion in terms of planar distributions.
iii. Expansion in terms of conical distributions.

The latter two are particularly useful for axially symmetric coupling tensors (e.g. ^{2}H in the rigid limit) in highly oriented systems. For moderately ordered polymers, the expansion i., however, not only is most convenient, it also provides the moments of the orientational distribution function for easy comparison with the results of other techniques characterizing the molecular order through the determination of second and – maybe – fourth moments. Moreover, it also provides a convenient way of analyzing 2D-^{13}C-MAS-NMR-spectra of partially ordered polymers described below.

12.2.2.2 Expansion in Terms of Spherical Functions

If the molecules are uniformly distributed around the director and a molecular frame can be chosen such, that the principal axes of the field gradient tensor are also uniformly distributed around the unique axis of the frame, we deal with systems of *transverse isotropy.* In this case the orientational distribution function can be expanded in terms of Legendre polynomials [20, 21], representing the simplest form of spherical functions

$$P(\beta) = \sum_{l=o}^{\infty} P_{loo} P_l(\cos\beta) \tag{17a}$$

$$= \sum_{l=o}^{\infty} \frac{2l+1}{8\pi^2} < P_l(\cos\beta) > P_l(\cos\beta) \tag{17b}$$

$$= \sum_{l=o}^{\infty} \frac{2l+1}{8\pi^2} < P_l > P_l(\cos\beta) \tag{17c}$$

where P_{loo} are the moments of the orientational distribution functions and the expansion coefficients $<P_l(\cos\beta)> = <P_l>$ are the order parameters for a given molecular direction. The maximum of the orientational distribution function as defined in Equ. (17) can be interpreted as the most probable angle between the director and the molecular axis. In absence of motion the NMR lineshape can easily be calculated using this expansion. The total spectrum is then obtained by a superposition of subspectra $F_l(\omega)$ (see Fig. 12.6) weighted by the order parameters $<P_l>$ with even l [5, 20, 21]:

$$F(\omega) = \sum_{l=0,2,4..}^{\infty} (2l+1) <P_l> F_l(\omega)\ P_l(\cos\beta_o)$$

$$F(\omega) = F_o(\omega) \sum_{l=0,2,4..}^{\infty} (2l+1) <P_l> P_l(\chi(\omega))\ P_l(\cos\beta_o) \tag{18}$$

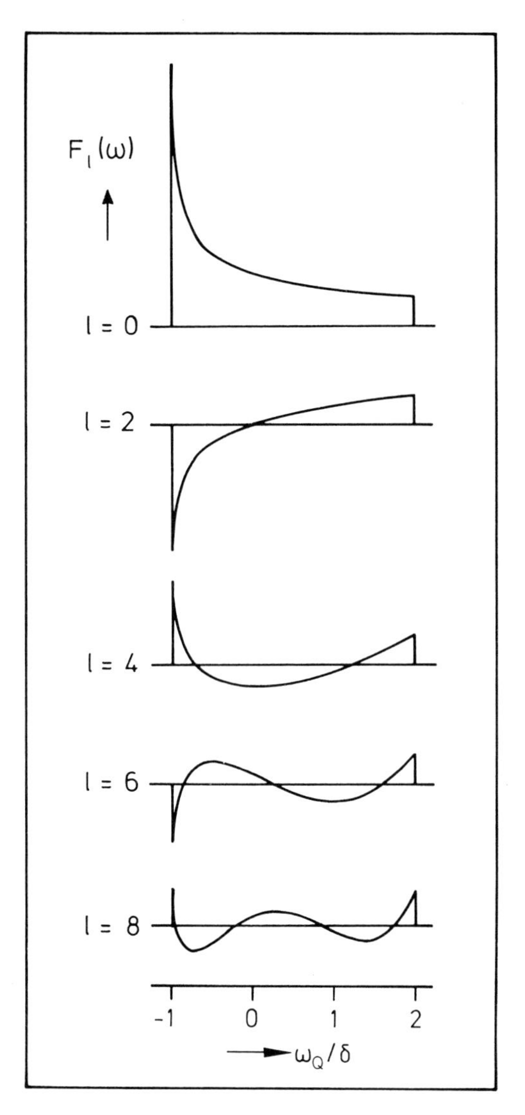

Fig. 12.6: Subspectra $F_l(\omega)$ for an axially symmetric coupling tensor for composing NMR line shapes of partially ordered polymers according to Equ. (18).

β_o is the angle between the director and the magnetic field and $F_o(\omega)$ represents the lineshape of an isotropic sample, centered at ω_o:

$$F_o(\omega) = \frac{1}{6\delta}\frac{1}{\chi(\omega)}, \text{ with}$$

$$\chi(\omega) = \frac{1}{\sqrt{3}}\sqrt{\frac{\omega}{\delta}+1}, \quad -1 < \frac{\omega}{\delta} \leq 2 \tag{19}$$

From Equ. (18) it is clear, that the subspectra $F_l(\omega)$ are obtained by simply weighting $F_o(\omega)$ by the corresponding Legendre polynomial of order l. $F(\omega)$ describes the lineshape of a single NMR transition only. The total ^{2}H-NMR-spectrum consists of a superposition of $F(\omega)$ and its mirror image with respect to the magnetic field. Therefore the order parameters $<P_l>$ obtained from the experiment are refered to the unique axis of the field gradient, typically along the C-H bond direction. This order parameter can easily be related to a molecular order parameter, if the conformation of that molecular site is known:

$$<P_l>_{CH} = <P_l>_{mol}\, P_l(\cos\theta) \tag{20}$$

where θ is the angle between the C-H bond and the molecular axis. This simple relation is only valid, if there exists one defined conformation. In the presence of different conformations the mean value $<P(\cos\theta)>$ has to be considered. By measuring the NMR line shape as a function of the angle β_o (rotation pattern) the different order parameters can be determined.

12.2.2.3 ^{13}C-MAS-NMR

^{2}H-NMR is a useful method for studying molecular order in various systems, but it requires isotopic labelling, which is an expensive and laborious step. The ^{13}C chemical shifts are large enough, however, to discriminate between the signals from carbons in different sites in ^{13}C-high resolution NMR in solids [22]. This is achieved by applying high power proton decoupling and magic angle spinning (MAS). Proton decoupling removes the dipolar interaction with the protons. The resulting powder patterns due to anisotropic chemical shifts overlap, however, so that the different sites cannot be resolved. By magic angle spinning the resolution can further be increased (see Fig. 12.7). For spinning rates smaller than the frequency shifts due to anisotropic chemical shifts (1 to 4 kHz compared to 5 to 15 kHz) the powder pattern is split into sidebands, retaining the information about the anisotropic interaction and, therefore, about the molecular orientation. If the sample is partially ordered, the MAS-NMR-spectrum depends on the rotor position at the beginning of the

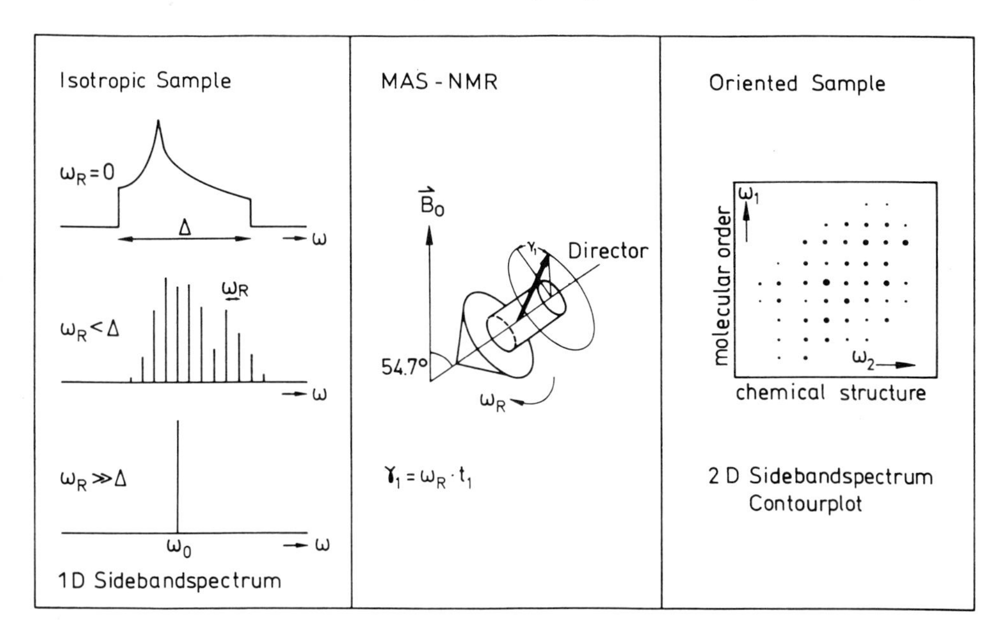

Fig. 12.7: Principle of 2D-MAS-NMR of partially ordered polymers.

experiment. This leads to a periodic modulation of phase and amplitude of individual sidebands. Through a second Fourier transform over the rotor phase this modulation is translated into a second frequency dimension [23] as a measure of the molecular order.

The analysis of such 2D-MAS-NMR-spectra closely follows the approach i. outlined above. The spectrum $F(\omega_1,\omega_2)$ is again obtained by a weighted superposition of subspectra in analogy to Equs. (17) and (18), where the subspectra, however, now themselves are 2D-sideband spectra [24]. For systems with transverse isotropy we thus have:

$$F(\omega_1,\omega_2) = \sum_{l=0}^{\infty} \frac{2l+1}{8\pi^2} \langle P_l \rangle F_l(\omega_1,\omega_2) \tag{21}$$

where the subspectra $F_l(\omega_1,\omega_2)$ depend on the shielding tensor, the spinning speed and the orientation of the sample in the rotor. Representative examples are plotted in Fig. 12.8. Due to the orthogonality of the Legendre polynomials a subspectrum $F_l(\omega_1,\omega_2)$ contains only sidebands up to order l. Thus the higher order sidebands primarily measure the higher moments of the orientational distribution function. The method also can be expanded to systems lacking axial symmetry [24].

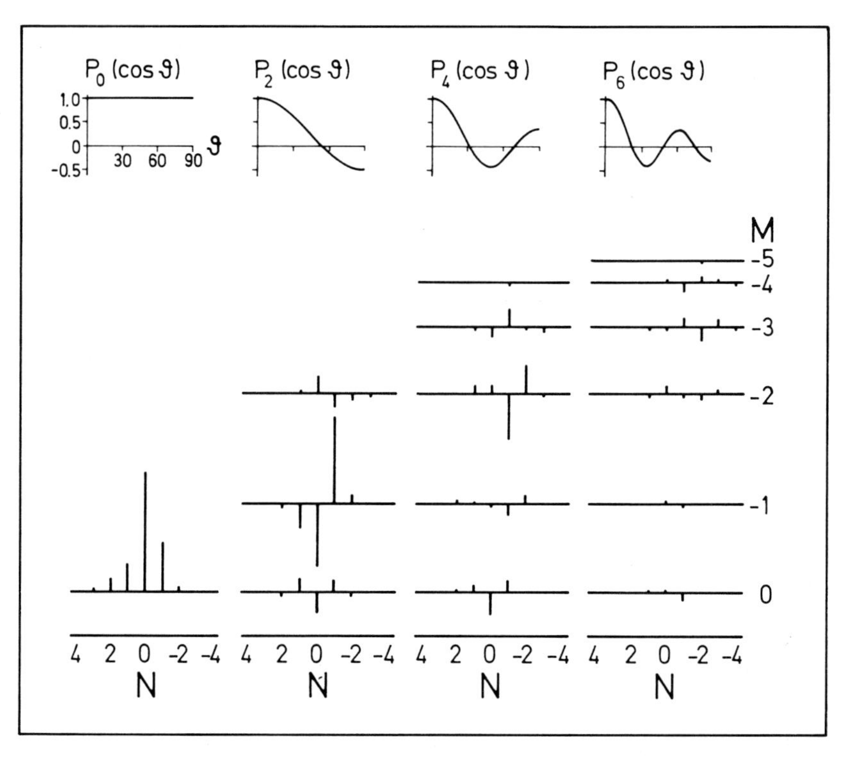

Fig. 12.8: Subspectra F_l (ω_1, ω_2) for an axially symmetric coupling tensor for composing 2D-MAS-spectra of partially ordered polymers according to Equ. (21), for details see Ref. [24].

The angular accuracy with which the orientational distribution can be specified by 2D-MAS-NMR is inherently lower than that of ^{2}H-NMR on selectively deuterated static samples, because the number of sidebands is limited. By reducing the spinning speed the number of sidebands can be increased, but it is generally difficult to detect higher than eight moments, corresponding to an angular resolution of approximately ± 10°. This, however, is highly sufficient for most amorphous systems.

12.3 Experimental Examples

12.3.1 Dynamics

12.3.1.1 Chain Motion in Linear Polyethylene

The techniques described in section 12.2.1 were first applied to study chain dynamics in perdeuterated linear polyethylene, isothermally crystallized from the melt [25, 26]. Representative ^{2}H-NMR-spectra are plotted as a function of temperature in Fig. 12.9a. They clearly show the presence of two regions with grossly different chain mobility: the deuterons in the rigid crystalline regions give rise to a Pake spectrum spanning the full width of 250 kHz (cf. Fig. 12.1) all the way up to the melting point. In addition, however, a broad central component is observed at room temperature, which narrows considerably with increasing temperature. This part of the spectrum naturally is attributed to the deuterons in the mobile non-crystalline regions. The two components can quantitatively be separated due to their different spin lattice relaxation times T_1. The crystallinity decreases at elevated temperatures [26] in accord with analysis of Raman and X-ray data (cf. Chapter 16).

The line shape changes displayed in Fig. 12.9a are ascribed to rapid conformational fluctuations where the number of conformations accessible for a given segment increases steadily with increasing temperature. The data could be analyzed quantitatively within a simple model [27, 28] of highly constraint localized chain motions involving flexible units of finite length between topological constraints, Fig. 12.9b. The length of the flexible unit increases from 3–5 carbon-carbon bonds at low temperatures to more than 10 bonds close to the melting point [26] (Fig. 12.9c). The presence of long lived topological constraints on a time scale of at least 50 ms was proved by the spin-alignment technique, which can detect ultraslow changes in the quadrupole coupling which would result from a finite lifetime of these constraints to the motion [25, 26]. From analysis of the spin lattice relaxation rate, Equ. (12), on the other hand, we can deduce the correlation time for a single conformational change within a flexible unit. At temperatures close to the melting point, it is below 10^{-9} s, thus shorter by more than 7 orders of magnitude. In the melt, relatively long lived constraints to the chain mobility can still be detected by ^{2}H-NMR. They do have, however, a limited lifetime below 1 ms [29].

This first example demonstrates already that by exploiting the different techniques of ^{2}H-NMR we not only can monitor molecular motions in polymers over an extraordinary wide range of time scales but that this information is crucial for the derivation of a meaningful picture of polymer dynamics.

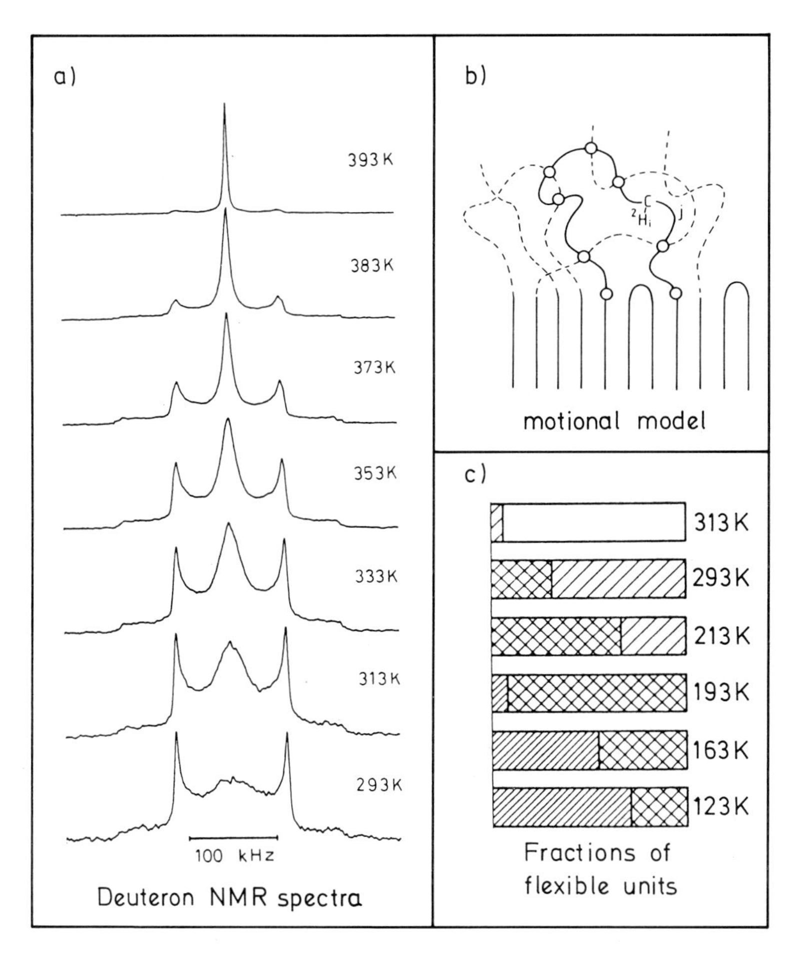

Fig. 12.9: Chain motion in linear polyethylene, a) ^{2}H-NMR-spectra, b) sketch of selected chain in the non-crystalline regions, c) fraction of flexible units between topological constraints as determined from the analysis of the ^{2}H-NMR-spectra: rigid, 3 bonds, 5 bonds, 7 bonds

12.3.1.2 Molecular Motion and Mechanical Properties of Polycarbonate

The close connection between local motions in glassy polymers and their mechanical properties was demonstrated in polycarbonate (PC). This material exhibits a pronounced low temperature β-relaxation as displayed in Fig. 12.10, but the nature of the corresponding motional mechanism was not clear. By ^{2}H-NMR we were able to show that the phenylene groups exhibit major mobility, consisting in 180° flips augmented by substantial small angle fluctuations about the same axis reaching an rms amplitude of ± 35° [28]. The detailed analysis of the line shapes yields the distribution of correlation times for this process, i.e. a distribution of jump rates (Fig. 12.10 [30]). The close connection of the phenylene motion and the mechanical relaxation is not only demonstrated by the fact that both processes have similar activation energies [30, 31]. It also shows up in mixtures of PC with low molar mass additives. These not only act as plasticizers in lowering the glass temperature but also change the mechanical properties drastically suppressing the mechanical relaxation. As a consequence the material becomes brittle, which severely limits the application of such polymers [31]. As shown in Fig. 12.10 the ^{2}H-NMR-spectrum in the mixture is dominated by the rigid Pake type spectrum contrary to that of PC itself which is dominated by the motionally narrowed one. Analysis of these data shows that the additives not only lead to an increase in the activation energy of the flip motion, they also cause a severe broadening of the distribution of jump rates. It should be clear from Fig. 12.10 that a close connection exists between the mechanical relaxation measured macroscopically (top of Fig. 12.10) and the jump rates of the phenylene rings determined on a molecular level by the use of ^{2}H-NMR (bottom of Fig. 12.10).

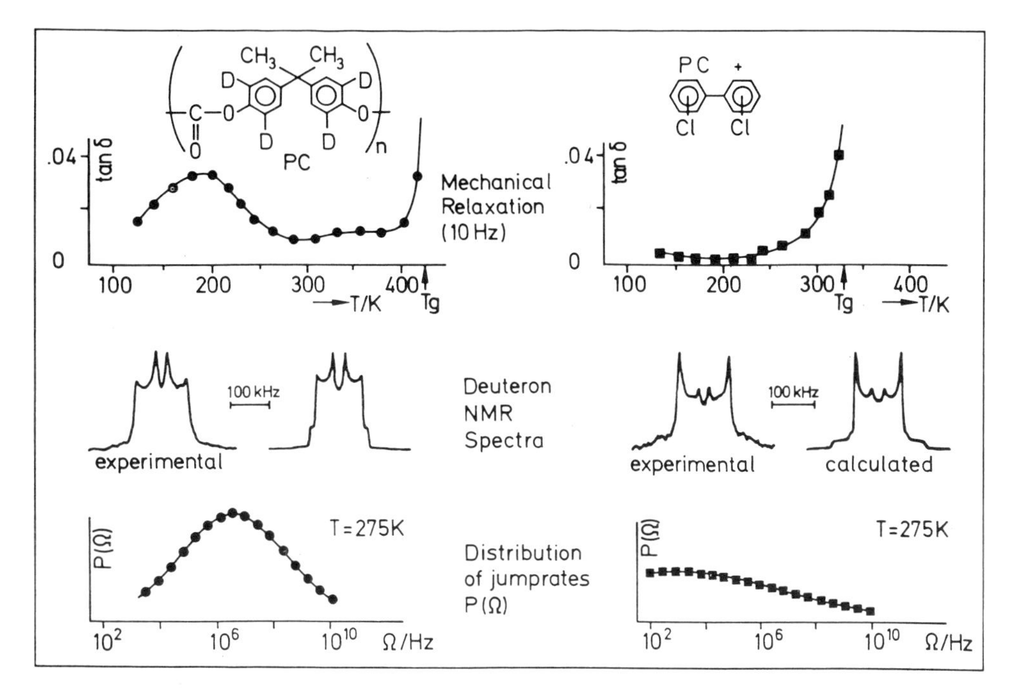

Fig. 12.10: Molecular motion and mechanical relaxation in polycarbonate.

12.3.1.3 Molecular Motion in Interfacial Regions of Segmented Polyurethanes

The selectivity of ^{2}H-NMR also offers a means to study the molecular motion on internal surfaces. In segmented blockcopolymers of rigid hard-segments and flexible soft-segments phase separation occurs in microdomains which are rich in one or the other component. The mechanical properties of such polymers depend on the degree of this phase separation and the mobility in the interfacial regions because the hard-segment domains act as cross-links in these elastomers. Examples of considerable scientific and practical interest are segmented polyurethanes, where the hard-segment domains are stabilized by hydrogen bonding. Since the microphase separation is not complete it can be anticipated, e.g., that hard-segments with different mobility exist. Those in the hard-segment domains should be more rigid than those immersed in the soft-segment phase. This has indeed been observed [32]. Thus the fraction of immobilized hard segments at elevated temperatures can be related to the degree of phase-separation. Even more interesting is the question concerning the possibly different mobilities of the hard segments in the interior and at the surface of hard-segment domains, respectively.

In order to tackle this problem ^{2}H-NMR studies were performed [33] on model polyurethanes involving methyldiisocyanate (MDI) as hard segment, butanediol as chain extender and poly(oxytetramethylen) as soft segment, monodispers in the hard-segment length [34]. The results on a system with 3 MDI units in the hard-segment block are summarized in Fig. 12.11. At low temperatures (193 K) a rigid solid spectrum is observed for samples selectively deuterated either at the phenylene rings of the outer MDI or of the inner MDI, respectively. At 323 K, on the other hand, the spectrum for the sample deuterated at the outer MDI is dominated by a narrow central line, indicating high mobility. The broad component due to MDI with low mobility is hardly visible. It is much stronger, however, in the spectrum of the sample deuterated at the inner MDI. The plot of the intensity of the rigid component for the inner MDT as a function of temperature demonstrates that the phase separation in these systems with short, but well defined hard segments is high. At least 60 % of the hard segments are in domains as revealed by the rigid fraction of the inner MDI at elevated temperatures. The much lower immobile fraction of the outer MDI (below 30 % at the highest temperatures studied) indicates that the hard segments are not regularly packed but are staggered, such that on the average about one of the 3 MDI is in a mobile interfacial environment. The narrow ^{2}H-NMR-spectra observed, indicate that despite of being fixed to the hard-segment domain through one butanediol unit only the local rotational mobility of these MDI in the interfacial region is comparable with those immersed in the soft-segment phase.

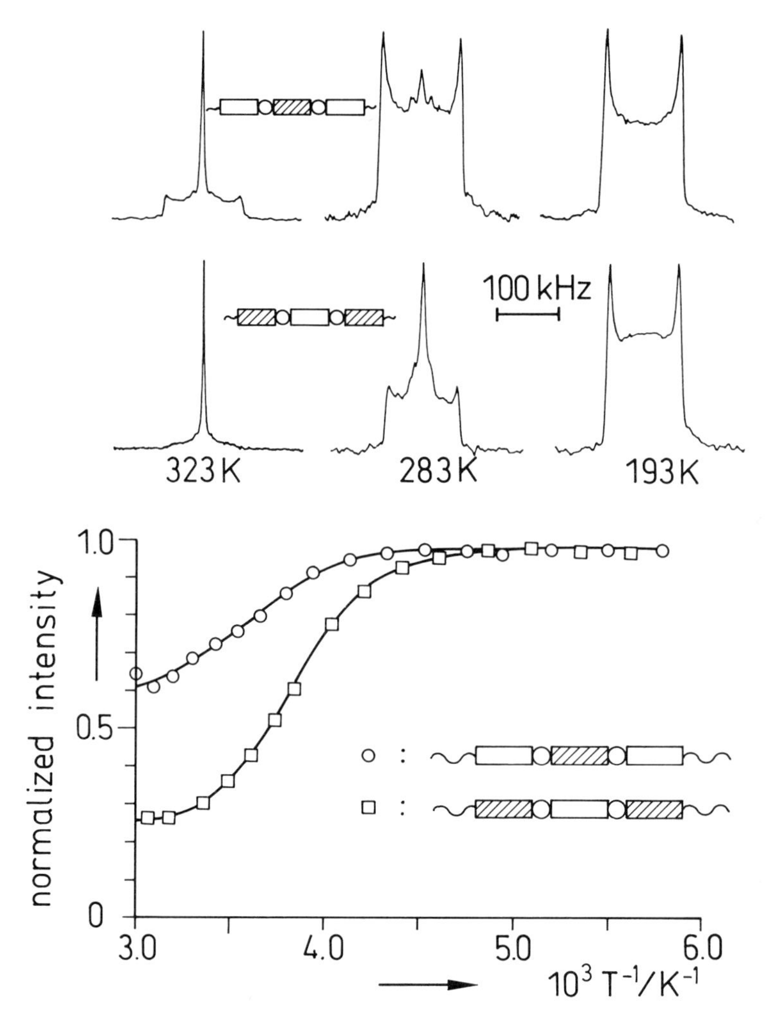

Fig. 12.11: Molecular motion of hardsegments in segmented polyurethanes; top: ^{2}H-NMR-spectra, bottom: normalized ^{2}H-NMR signal intensity of the rigid component. The labelled MDI units are indicated by hatched rectangles.

12.3.1.4 Motion of Mesogens in Liquid Crystalline Polymers

^{2}H-NMR was also used to study the novel polymeric materials prepared by Ringsdorf and coworkers within the Sonderforschungsbereich 41. Here we describe our investigation of the local motions involving the mesogenic groups in side group liquid crystalline polymers (cf. Chapter 9). These systems at elevated temperatures form mesophases similar to low molar mass liquid crystals. On cooling the low temperature mesophase is frozen at the glass transition into a partially ordered glassy state. Collective dynamic processes like the translational motion of the polymer backbone is frozen in. Local motions, however, involving the phenylene groups of the mesogens and restricted conformational fluctuations within the spacer are detected in the glassy state also.

The phenylene motion may depend on the mesophase structure, the flexibility of the polymer backbone and the spacer length. Therefore polymethacrylates, polyacrylates and polysiloxanes all with phenylbenzoate as the mesogenic group (cf. Fig. 12.12) were studied [35–37]. They exhibit nematic, smectic A and smectic C phases as well as significantly different glass temperatures T_g. All systems were selectively deuterated at the pending phenylene ring.

In all systems the phenylene groups were found to exhibit 180° flips as discussed above for polycarbonate. An activation diagram collecting the results of ^{2}H-NMR line shape analysis as a function of temperature is given in Fig. 12.12. The mean activation energies for the different systems are in a rather narrow range between 42 and 48 kJ/mol. The width of the distribution of jump rates (between 2.2 and 2.6 decades) is indicated by the two extra solid lines in order to illustrate that differences in the mean jump rate between different systems are smaller than the variation within one system. The phenylene flip motion thus is a thermally activated process which will also take place at higher temperatures in the liquid crystalline phases as suggested long

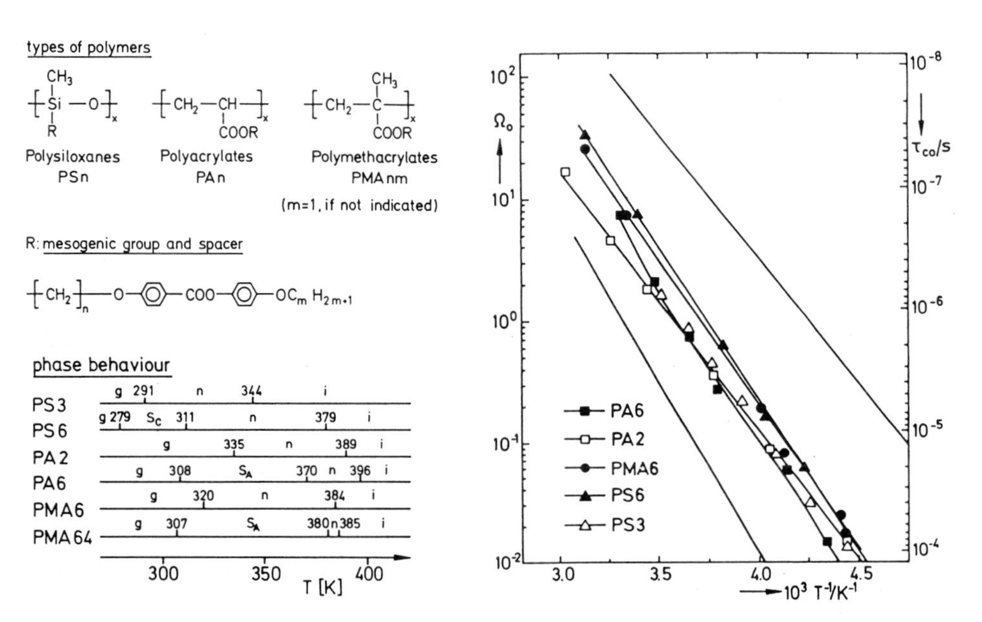

Fig. 12.12: Liquid crystalline side group polymers; left: molecular structure and phase behaviour of the systems studied, right: temperature dependence of the mean jump rate for the 180° flip motion of the pending phenylene group in the mesogenic group.

ago for low molar mass systems. There the flip motion is augmented by diffusive rotational motions about the long axis of the mesogen, resulting in rotations about arbitrary angles. These two motions can thus be separated in polymers by cooling into the glassy state, where the diffusive motion is frozen, whereas the local motion persists.

The 180° flip motion of phenylene groups is a common process observed in numerous amorphous polymers [9, 28, 30, 32]. Moreover, the temperature dependence as well as the distribution of jump rates is of comparable magnitude. It may be surprising that the phenylene motion in the highly ordered frozen liquid crystalline phases is similar to that in amorphous polymers. Our results show, however, that the overall packing differences between amorphous systems and the different mesophases are too small compared with the packing differences within each phase itself in order to cause vastly different motional behaviour. It also means that it is difficult to suppress the mobility in the frozen ordered phases altogether, which is desired, if these liquid crystalline systems are to be used as storage devices.

12.3.1.5 Mobility of Lipid Chains in Polymer Model Membranes

Model membranes composed of simple lipid analogues easily organize themselves to form liposomes, hexagonal or lamellar phases. The stability of such systems can substantially be increased by introducing a polymerizable group and subsequent polymerization (cf. Chapter 9). Several concepts, where the polymerization is achieved either in the hydrophobic or in the hydrophylic region, respectively, have been realized. The stabilization, however, is always accompanied by restrictions on the mobility of the membrane, which limits its applicability. ^{2}H-NMR of selectively deuterated systems is ideally suited to study changes in mobility that occur as a consequence of the polymerization. Since membrane fluidity is known to be an essential feature of biological membranes such studies were only carried out on polymer membranes which retain the main phase transition from the lamellar gel phase into the mobile liquid crystalline phase.

As an example, Fig. 12.13 shows the scheme of polymerization and ^{2}H-NMR-spectra in the liquid crystalline phase for a model membrane polymerized in the hydrophylic region [38, 39]. In its monomeric form such a membrane has a considerable number of degrees of motional freedom: rotation about the long axis, fluctuation of the long axis itself, translational motion within the smectic layer and local conformational changes. They lead to the narrow ^{2}H-NMR-spectra plotted in Fig. 12.13 at the left. For a qualitative discussion it suffices to note that the *width* of the spectrum is a mesure of the number of motional degrees of freedom that are activated, the spectrum becoming narrower when this number is increased. Thus in the monomer the local mobility as probed by ^{2}H-NMR is alike for the lipid chains, the head group and the polymerizable group, although differences in the correlation times are detected [39].

Polymerization is expected to suppress the translational motion, it also hinders the local dynamics, however, introducing a pronounced motional gradient within each lipid moiety. As shown in Fig. 12.13 the ^{2}H-NMR-spectra for deuterons adjacent to the polymer backbone exhibit broad, solid-like line shapes, whereas those in the middle of the lipid chain remain narrow. This scheme of polymerization thus results in polymer membranes with highly mobile lipid chains but largely immobilized hydrophylic regions. The mobillity of the head group which is low in the system with a short spacer of only 3 methylene units [38, 39] (cf. the broad spectrum in Fig. 12.13) can substantially be increased by extending the spacer [40].

By use of a lipid analogue carrying a polymerizable group at the end of only one of its two alkyl chains, membranes retaining even higher mobility in the hydrophobic regions can be generated. As shown by ^{2}H-NMR and other spectroscopic techniques typical membrane constituents, i.e. cholesterol, gramicidin, and ubiquinone can be incorporated and the mixtures show a behaviour similar to that of natural model membranes, despite a lack of lateral diffusion as a consequence of the polymerization [40].

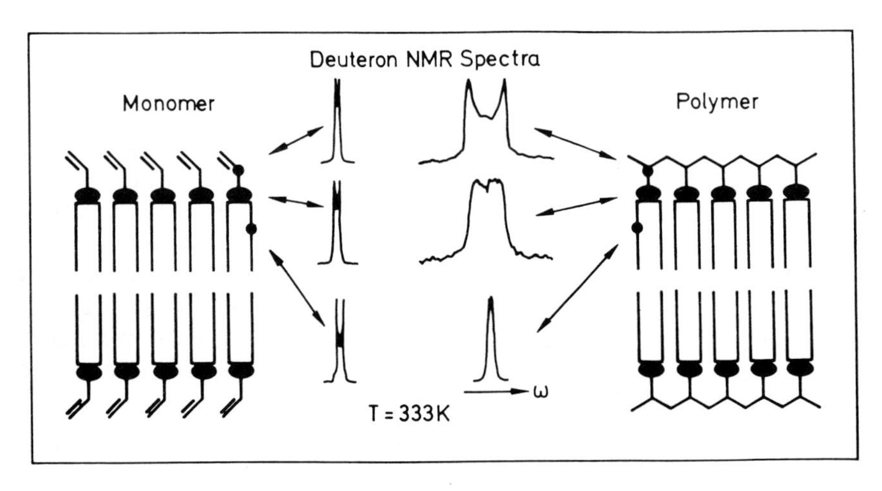

Fig. 12.13: Restriction of mobility in a model membrane.
● : hydrophylic head group
⸝ : polymerizable group
⊓ : hydrophobic alkyl chains
⌒⌒ : polymer chain

12.3.2 Order

12.3.2.1 Molecular Order in Liquid Crystalline Polymers

The new ^{2}H-NMR techniques for studying the orientational distributions were first applied to ordered polyethylene samples [3, 9, 20, 21]. This demonstrated that largely different degrees of molecular order could be handled in crystalline as well as amorphous environments. Thus ^{2}H-NMR was found to be an attractive tool for determining the different degrees of alignment of the mesogenic groups, the spacer and the polymer chain in liquid crystalline side group polymers, expected from the spacer model (cf. Chapter 9), for a recent review see Ref. [37].

Angular dependent ^{2}H-NMR-spectra of the smectic polyacrylate (PA6, Fig. 12.12) selectively deuterated at the CH_2 group of the spacer adjacent to the polymer backbone are shown in Fig. 12.14 [41]. The sample was aligned in its nematic phase in the high magnetic field (7.4 T) of the NMR-spectrometer, then slowly cooled into the smectic phase and further down into the glassy state. The fitted line shapes were calculated employing the expansion of the orientational distribution function as described above. The results of this analysis together with those obtained in a similar way for the mesogenic group [35, 36] and the polymer chain [41] are collected in Fig. 12.15. The order parameters $<P_2>$ are 0.88, 0.52 and 0.25 for the mesogenic group, the spacer and the polymer chain, respectively. Thus a pronounced order gradient is observed as predicted by the spacer model. Analysis of 2D-^{13}C-MAS-spectra on the same polymer [42] shows that the order parameter is alike for the different methylene groups within the spacer, presumably due to conformational disorder involving all 6 carbons of the spacer. Moreover, the order parameter of the acrylic carboxylcarbon next to the polymer chain is very similar to that of the spacer carbons. The decrease in order from the spacer to the polymer chain thus can be localized to the carbon-carbon bond connecting the carboxylcarbon to the polymer backbone.

In the polyacrylate the ^{2}H-NMR data suggest that even in this frozen smectic systems the polymer chain is only slightly distorted from a random coil conformation. In the corresponding methacrylate with the same phase behaviour we find a completely different orientational distribution of the polymer chain [43] consistent with a marked elongation of the chain perpendicular to the mesogenic groups, presumably because the chain is largely confined between the smectic layers. While the ^{2}H-NMR data on the methacrylate are consistent with those of small angle neutron scattering [44], our results on the acrylates have been disputed recently [45] on the basis of neutron scattering data.

The chain orientation also strongly depends on the mesophase structures studied. In nematics the main chain typically is only slightly distorted as detected both by ^{2}H-NMR [41] and neutron scattering [46, 44]. The delicate

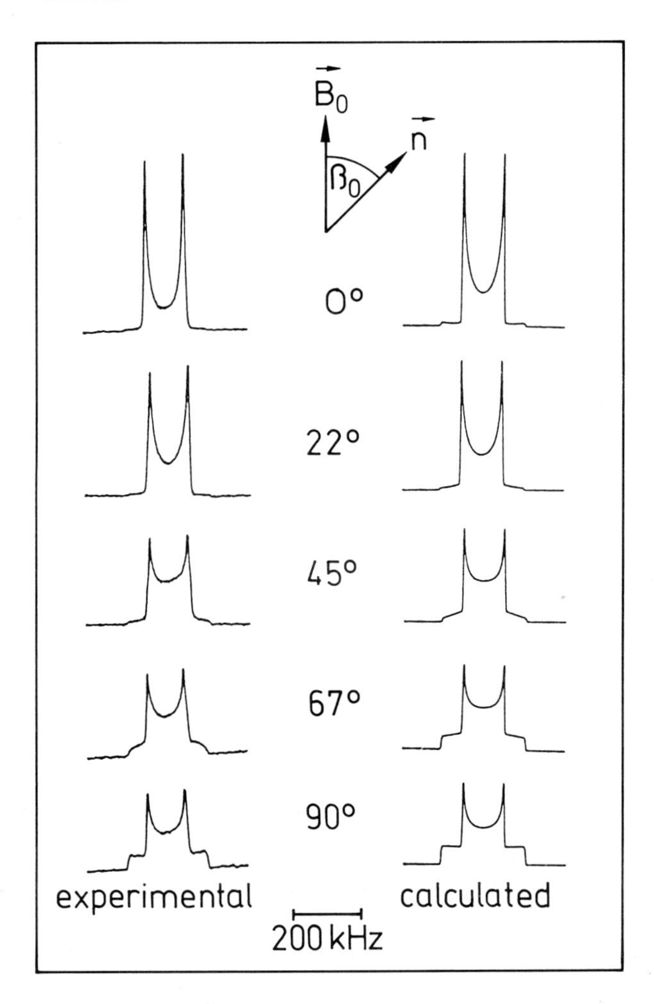

Fig. 12.14: Angular dependent ^{2}H-NMR-spectra of the frozen smectic polymer PA6, selectively labelled at the spacer.

coupling between the highly ordered mesogenic groups and the polymer backbone also manifests itself in the observation that in strained elastomers the mesogenic groups can be oriented parallel or perpendicular, respectively, to the polymer backbone in polyacrylates and polymethacrylates [47] (cf. also Chapter 9).

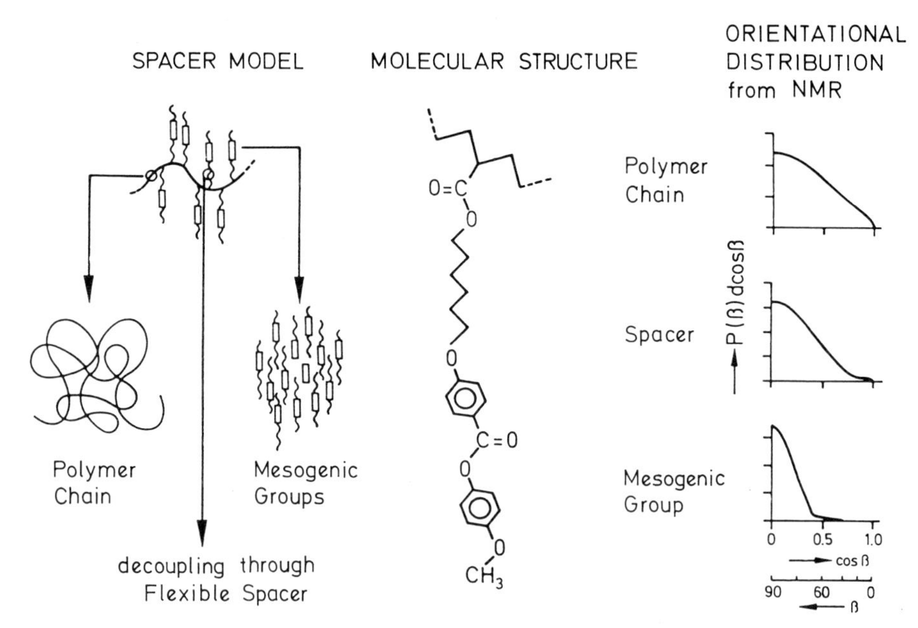

Fig. 12.15: Molecular order in a liquid crystalline side group polymer as determined by NMR.

12.3.2.2 Drawn Fibres of Poly(ethyleneterephthalate) (PET)

As an experimental example of the use of 2D-^{13}C-MAS-NMR for the determination of molecular order let us consider drawn fibres of PET [24]. Fig. 12.16 displays slices of the 2D-spectrum, where ω_2 corresponds to the frequency in a conventional NMR-spectrum and therefore, displays the differences in chemical shift of the different carbons in the monomer unit. In ω_1 a total of 16 slices only is recorded in the experiment [24], carrying all the information about molecular order. Although the signals from all carbons are resolved in the spectrum of Fig. 12.16 the resolution can further be improved by varying the cross polarization time or by gated decoupling to discriminate the carbons involved in C-H bonds from the "unprotonated" carbons [42].

The analysis of these spectra again yields the orientational distribution function. Since the carbon chemical shift tensor typically is not axially symmetric, it depends on two angles rather than one as for ^{2}H-NMR where the direction of the C-H bond with respect to B_o can be specified by a single angle (cf. Fig. 12.5). The contour plot of the orientational distribution function may directly be interpretated in terms of the molecular structure as shown in Fig. 12.16 for the "protonated" aromatic carbon. The maxima at about 35° and 85° are consistent with the 24°-tilt of the aromatic ring of the monomer unit away from the draw direction in the crystalline regions of PET as determined by X-rays.

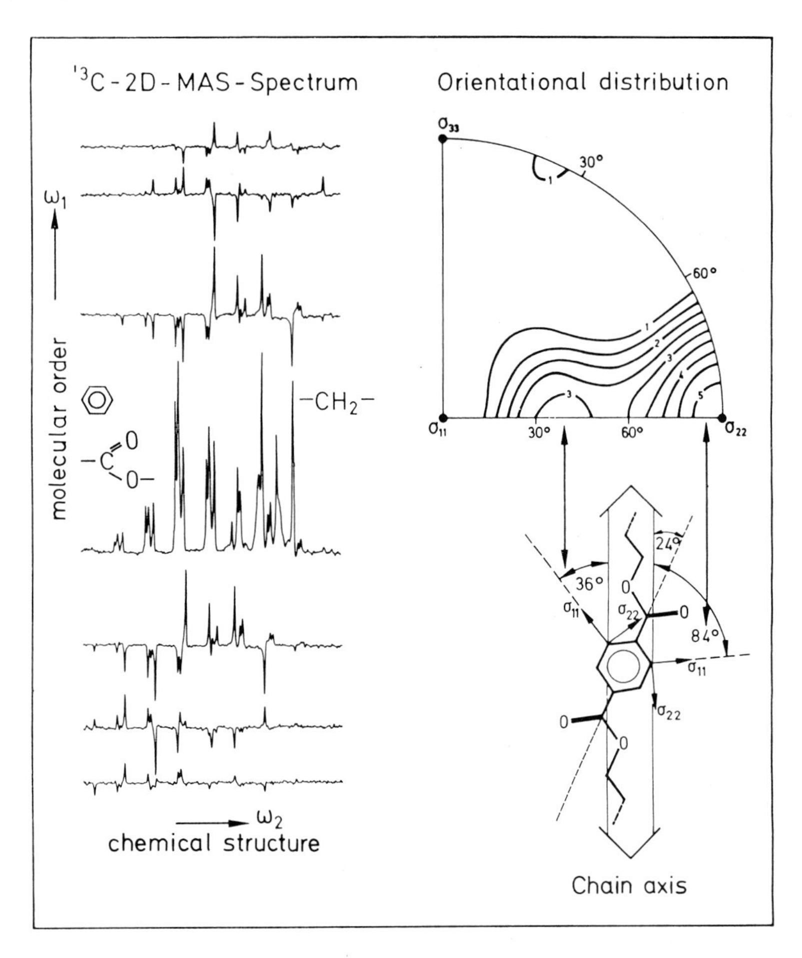

Fig. 12.16: ^{13}C-2D-MAS-spectrum, orientational distribution and molecular alignment in a PET-fibre.

The NMR-method is not restricted to crystalline systems and in a single experiment yields structural information in amorphous solids also, based on various carbon positions with an accuracy of about ± 10°. Thus both, the conformation and the alignment of macromolecules can be determined. This method should, therefore, be widely applicable to both polymer science and biophysics.

12.3.3 Overview of ^{2}H-NMR in Polymers

This review is based on the authors' work, mainly obtained within the program of the Sonderforschungsbereich 41. It is rewarding that pulsed ^{2}H-NMR meanwhile has been widely applied to study the molecular structure and dynamics in a large variety of polymeric systems. It is outside the scope of this chapter to give a comprehensive account of all these applications. A few selected examples, however, are collected here in order to put the results described above into perspective.

As far as synthetic polymers are concerned, L.W. Jelinski and coworkers (for a review see Ref. [32]) applied the solid echo technique to study molecular dynamics in a number of carefully selected systems, e.g. chain dynamics in poly(butylene terephthalate), segmented copolyesters and polyurethanes, phenylene ring flips in poly(butylene terephthalate), poly(arylene ether sulfone) and the interaction of water with epoxy resins.

The advantages of ^{2}H-NMR have also been realized in other industrial research laboratories. P. M. Henrichs of Kodak concentrated on polycarbonate [48] and A. D. English (DuPont) performed a detailed study of the chain dynamics in nylons [49]. G. Kothe and coworkers have performed detailed studies of conformation, alignment and molecular dynamics in semiflexible main chain liquid crystalline polymers. Moreover, they have developed an elaborate computer program based on the stochastic Liouville equation for handling solid echo line shapes for complex motional behaviour [50].

^{2}H-NMR is also highly suited to study the chain statistics in liquid crystalline phases [51, 52] and strained elastomers [53], where the information is mainly obtained from the strength of the quadrupole splitting observed.

In biophysics the application of ^{2}H-NMR at first concentrated on the study of lipid chain motion in model membranes [54–57]. More recently it has also been used to study the chain folding in nucleic acids [58, 59].

The number of applications increases steadily. Thus ten years after the publication of the first Fourier transform ^{2}H-NMR-spectrum covering the full width of 250 kHz [3] pulsed ^{2}H-NMR of solid materials has become what is usually called a "standard technique".

12.4 New Developments: Two-dimensional ^{2}H-NMR

Despite of the usefulness of the ^{2}H-NMR techniques described above the information about the molecular dynamics that can be deduced from the analysis of the line shapes observed is limited because it is based on model assumptions. Different motional mechanisms can be discriminated much better, if all the information contained in the 3-pulse exchange experiment (cf. Fig. 12.2) is exploited. Through systematic variation of the pulse spacing t_1 and two subsequent Fourier transforms with respect to both, t_1 and t_2 a two-dimensional NMR-spectrum [60] can be generated [61]. Its two frequency axes ω_1 and ω_2 describe the NMR-frequencies and, therefore, the molecular orientations at the beginning, and the end of the mixing time t_m, respectively.

If during both the evolution time t_1 and the detection time t_2 the molecules have fixed orientation but rotate during the much longer mixing time about a well defined angle, such that the C-D bond directions – or the C_3 axes of rapidly rotating methyl groups – rotate about the angle ϑ, elliptical ridges occur in the 2D exchanges spectrum defined by the frequency coordinates [61]:

$$\omega_1 = \frac{\delta}{2} [1 + 3 \cos 2 \theta] \tag{22a}$$

$$\omega_2 = \frac{\delta}{2} [1 + 3 \cos (2 \theta \pm 2 \vartheta)] \tag{22b}$$

The rotational angle ϑ is thus directly projected into the 2D-NMR-spectrum, the geometry of the dynamic process can be read off the spectrum without the need of interfacing a model. The first experimental example [61] was provided through a simple crystalline solid, dimethylsulfone (DMS) (cf. Fig. 12.17a). Here the two site exchange results in the elliptical ridges for half of the molecules, and a Pake doublet along the diagonal for the other half.

More complex motions will lead to a distribution of rotational angles after the mixing time. A comprehensive treatment of the 2D exchange experiment has recently been given [10]. The central element of this description is the identification of the 2D pure absorption mode spectrum with a two-time distribution function. Each 2D-spectrum is an image of a special state of the motional process detected. The whole three-dimensional data set, where the mixing time t_m defines the third dimension, represents a direct image of the motional process.

As an example, Fig 12.17b shows a ^{2}H-2D-NMR-spectrum of polystyrene selectively deuterated at the polymer chain at a temperature above the glass transition [62]. Here the diffuse reorientation by small angles of part of the chains leads to the spectral intensity near the diagonal, whereas the large angle reorientation of other chain segments during the same mixing time is reflected in the ridges parallel to the frequency axes and the broad featureless spectral intensity covering the whole 2D-plane.

Thus rotational motions in well defined crystalline environments and in amorphous systems lead to vastly different 2D-spectra and can easily be distinguished. 2D-NMR of static samples is not restricted to ^{2}H-NMR. It has meanwhile also been applied to ^{13}C-NMR of both isotropic and drawn poly-

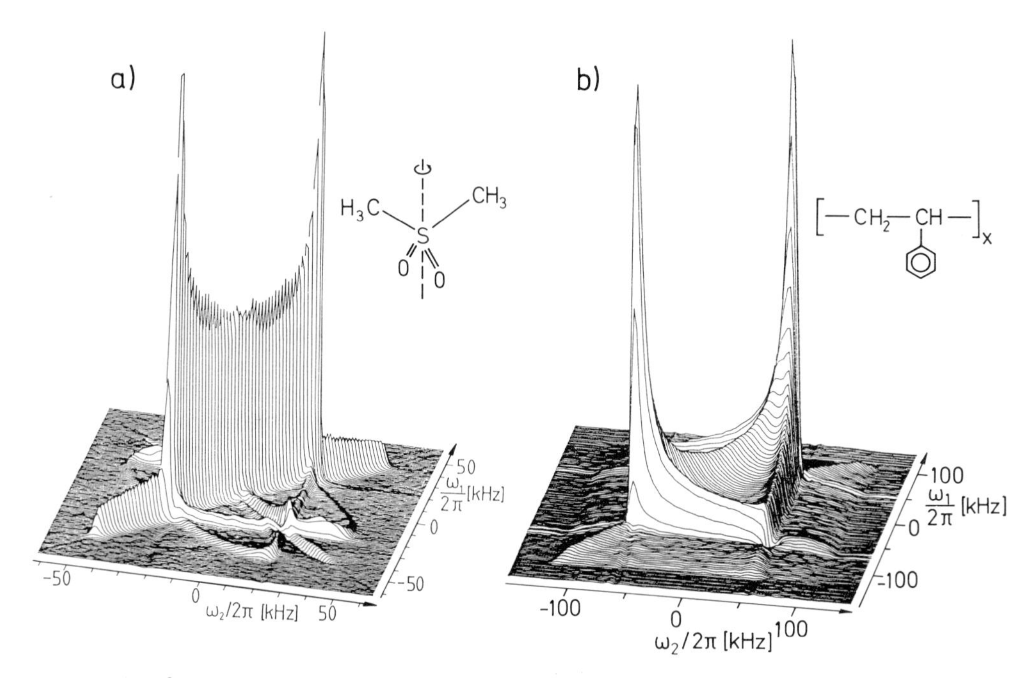

Fig. 12.17: ^{2}H-2D-exchange spectra; a) dimethylsulfone, b) polystyrene deuterated at the polymer chain.

(oxymethylene) [63], where the reorientation of the helical polymer chain leads to pronounced exchange patterns. The 2D experiments described here have to be distinguished from 2D-spectra based on the solid echo technique [64, 65], in which the second dimension reflects the angular dependent transverse relaxation time and the correspondingly different widths of the lines comprising the powder lineshape.

Two-dimensional NMR of solids is a field of rapid development [66]. It not only opens up new prospects for the characterization of order and mobility. As demonstrated recently [67] it also offers a means to correlate these two important molecular parameters which are believed to be crucial for the understanding of the macroscopic behaviour of solid polymers.

12.5 References

[1] R. G. Barnes: Deuteron quadrupole coupling tensors in solids, Adv. Nucl. Quadrupole Resonance 1 (1974) 335–355.
[2] J. H. Davis, K. R. Jeffrey, M. Bloom, M. I. Valic, T. P. Higgs: Quadrupolar echo deuteron magnetic resonance spectroscopy in ordered hydrocarbon chains, Chem. Phys. Lett. 42 (1976) 390–395.
[3] R. Hentschel, H.W. Spiess: Deuterium Fourier Transform NMR in solids und solid polymers, J. Magn. Reson. 35 (1979) 157–162.
[4] H.W.: Spiess: Deuteron spin alignment: A probe for studying ultraslow motions in solids and solid polymers, J. Chem. Phys. 72 (1980) 6755–6762.
[5] R. Hentschel, J. Schlitter, H. Sillescu, H.W. Spiess: Orientational distributions in partially ordered solids as determined from NMR and ESR line shapes, J. Chem. Phys. 68 (1978), 56–66.
[6] H.W. Spiess, H. Sillescu: Solid echoes in the slow-motion region, J. Magn. Res. 42 (1981), 381–389.
[7] B. Willenberg, H. Sillescu: ^{1}H NMR and ^{2}D NMR relaxation and wide line measurements in partially deuterated polystyrenes, Makromol. Chem. 178 (1977), 2401–2412.
[8] A. Abragam: The Principles of Nuclear Magnetism, Clarendon Press, Oxford 1961.
[9] H.W. Spiess: Deuteron NMR – a new tool for studying chain mobility and orientation in polymers, Adv. Polym. Sci. 66 (1985) 23–58.
[10] S. Wefing and H.W. Spiess: Two-dimensional exchange NMR of powder samples I. Two-time distribution functions, J. Chem. Phys. 89 (1988) 1219–1233.
[11] S. Wefing: Ph. D. Thesis, Mainz 1988.
[12] F. Fujara, S. Wefing, H.W. Spiess: Dynamics of molecular reorientations: Analogies between quasielastic neutron scattering and deuteron spin alignment, J. Chem. Phys. 84 (1986) 4579–4584.
[13] P. Lindner, E. Rössler, H. Sillescu: NMR study of molecular motion in solid and molten polystyrene: application of the deuteron dilution technique, Makromol. Chem. 182 (1981) 3653–3669.
[14] R. Kimmich, G. Schnur, M. Köpf: The tube concept of macromolecular liquids in the light of NMR experiments. Progress in NMR Spectroscopy 20 (1988) 385–421.
[15] R. A. Komoroski, (ed.): High Resolution NMR Spectroscopy of Synthetic Polymers in Bulk, VCH Publ., Deerfield Beach 1986.
[16] Chapter 3 in ref. 15.
[17] S. Wefing, S. Jurga, H.W. Spiess: Orientation dependent spin-lattice relaxation of deuteron spin alignment, in K. A. Müller, R. Kind, J. Roos (eds.): Proc. XXIInd Congress Ampere, Zürich 1984, 375–376.
[18] A. Blumen, J. Klafter, G. Zumofen: Models for reaction dynamics in glasses, in I. Zschoffe (ed.): Optical Spectroscopy of Glasses, Reidel Publ, Dordrecht 1986.
[19] C. Schmidt, K. J. Kuhn, H.W. Spiess: Distribution of correlation times in glassy polymers from pulsed deuteron NMR, Progr. Colloid & Polymer Sci. 71 (1985) 71–76.
[20] R. Hentschel, H. Sillescu, H.W. Spiess: Orientational distribution of polymer chains studied by ^{2}H-NMR line shapes, Polymer 22 (1981) 1516–1521.

[21] H.W. Spiess: NMR in oriented polymers, in Developments in Oriented Polymers – 1, ed. by I. M. Ward. Appl. Science Publ. London (1982) 47–48.
[22] M. Mehring: High Resolution NMR in Solids, Springer, Berlin 1983.
[23] G. S. Harbison, H.W. Spiess: Two-dimensional magic-angle-spinning NMR of partially ordered systems. Chem. Phys. Letters 124 (1986) 128–134.
[24] G. S. Harbison, V.-D. Vogt, H.W. Spiess: Structure and order in partially oriented solids: Characterization by 2D-magic-angle-spinning NMR, J. Chem. Phys. 86 (1987). 1206–1218.
[25] D. Hentschel, H. Sillescu, H.W. Spiess: Chain motion in the amorphous regions of polyethylene as revealed by deuteron magnetic resonance, Macromolecules 14 (1981) 1605–1607.
[26] D. Hentschel, H. Sillescu, H.W. Spiess: Deuteron NMR study of chain motion in solid polyethylene, Polymer 25 (1984) 1078–1086.
[27] K. Rosenke, H. Sillescu, H.W. Spiess: Chain motion in amorphous regions of polyethylene: interpretion of deuteron NMR lineshapes, Polymer 21 (1980) 757–763.
[28] H.W. Spiess: Molecular dynamics of solid polymers as revealed by deuteron NMR, Colloid & Polymer Sci. 261 (1983) 193–209.
[29] J. Collignon, H. Sillescu, H.W. Spiess: Pseudo-solid echoes of proton and deuteron NMR in polyethylene melts, Colloid & Polymer Sci. 259 (1981) 220–226.
[30] M. Wehrle, G. P. Hellmann, H.W. Spiess: Phenylene motion in polycarbonate and polycarbonate/additive mixtures, Colloid & Polymer Sci. 265 (1987) 815–822.
[31] E.W. Fischer, G. P. Hellmann, H.W. Spiess, F. J. Hörth, U. Ecarius, M. Wehrle: Mechanical properties, molecular motions and density fluctuations in polymer additive mixtures, Makromol. Chem. Suppl. 12 (1985) 189–214.
[32] L.W. Jelinski: Deuterium NMR of Solid Polymers. Chapter 10 of ref. [15].
[33] A. Olinger: Molekulare Dynamik und Phasenseparation in Polyurethan-Elastomeren, Ph. D. Thesis, 1987.
[34] C. D. Eisenbach, M. Baumgartner, C. Günter: Polyurethane elastomers with monodisperse segments and their model precursors: Synthesis and properties, in J. Lal, J. E. Mark (eds.): Advances in Elastomers and Rubber Elasticity, Plenum (1987) 51–87.
[35] H. Geib, B. Hisgen, U. Pschorn, H. Ringsdorf, H.W. Spiess: Deuteron NMR study of molecular order and motion in a liquid crystalline polymer, J. Am. Chem. Soc. 104 (1982) 917–919.
[36] U. Pschorn, H.W. Spiess, B. Hisgen, H. Ringsdorf: Deuteron NMR study of molecular order and motion of the mesogenic side group in liquid-crystalline polymers, Makromol. Chem. 187 (1986) 2711–2723.
[37] C. Boeffel, H.W. Spiess: NMR methods for studying molecular order and motion in liquid crystalline side group polymers, in C. B. McArdle (ed.): Side Chain Liquid Crystal Polymers, Blackie Publ. Group, Glasgow 1988, Chapter 8, 224–259.
[38] R. Ebelhäuser, H.W. Spiess: Deuteron NMR study of molecular mobility in a polymer model membrane, Ber. Bunsenges. Phys. Chem. 89 (1985) 1208–1214.
[39] R. Ebelhäuser, H.W. Spiess: Deuteron NMR study of molecular motion and phase behaviour in a polymer model membrane, Makromol. Chem. 188 (1987) 2935–2949.
[40] T. Fahmy, J. Wesser, H.W. Spiess: Structure and molecular dynamics of highly mobile polymer membranes, Angew. Makromol. Chem. 166/167 (1989) 39–56.

[41] C. Boeffel, H.W. Spiess, B. Hisgen, H. Ringsdorf, M. Ohm, R. Kirste: Molecular order of spacer and main chain in polymeric side group liquid crystals, Makromol. Chem. Rapid Commun. 7 (1986) 777–783.
[42] B. Blümich, C. Boeffel, G. S. Harbison, Y. Yang, H.W. Spiess: Two-dimensional MAS NMR: New prospects for the investigation of partially ordered polymers, Ber. Bunsenges. Phys. Chem. 91 (1987) 1100–1103.
[43] C. Boeffel, H.W. Spiess: Highly ordered main chain in a liquid crystalline side-group polymer, Macromolecules 21 (1988) 1626–1629.
[44] P. Keller, B. Carvalho, J. Cotton, M. Lambert, F. Moussa, G. Pepy: Side chain mesomorphic polymers: studies of labelled backbones by neutron scattering, J. Physique Lett. 46 (1985) 1065–1071.
[45] L. Noirez, J. P. Cotton, F. Hardouin, P. Keller, F. Moussa, G. Pepy, C. Strazielle: Backbone confirmation of a mesogenic side-chain polyacrylate, Macromolecules 21 (1988) 2889–2891.
[46] R. Kirste, H. Ohm: The conformation of liquid crystalline polymers as revealed by neutron scattering, Makromol. Chem. Rapid Commun. 6 (1985) 179–185.
[47] R. Zentel, M. Benalia: Stress induced orientation in lightly crosslinked liquid-crystalline side-group polymers, Makromol. Chem. 188 (1987) 665–674.
[48] P. M. Henrichs, H. R. Luss: Ring dynamics in a crystalline analogue of bisphenol A polycarbonate, Macromolecules 21 (1988) 860–862.
[49] H. Miura, A. D. English: Segmental dynamics in Nylon 66, Macromolecules 21 (1988) 1543–1544, ibid. 22 (1990) 2153–2182.
[50] K. Müller, P. Meier, G. Kothe: Multipulse dynamic NMR of liquid crystal polymers, Progress in NMR Spectroscopy 17 (1985) 211–239.
[51] E.T. Samulski, M. M. Gauthier, R. B. Blumstein, A. Blumstein: Alkyl chain order in a linear polymeric liquid crystal, Macromolecules 17 (1984) 479–483.
[52] S. Brückner, J. D. Scott, D.Y. Yoon, A. C. Griffin: NMR study of nematic order of semiflexible thermotropic polymers, Macromolecules 18 (1985) 2709–2713.
[53] W. Gronski, R. Stadler, M. Maldaner Jacobi: Evidence of nonaffine and inhomogeneous deformation of network chains in strained rubber-elastic networks by deuterium magnetic resonance, Macromolecules 17 (1984) 741–748.
[54] J. H. Davis: Deuterium magnetic resonance study of the gel and liquid crystalline phases of dipalmytoyl phosphatidylcholine, Biophys. J. 27 (1979) 339–358.
[55] J. H. Davis, M. Bloom, K.W. Butler, I. C. P. Smith: The temperature dependence of molecular order and the influence of cholesterol in Acholeplasma Laidlawii membranes, Biochim. Biophys. Acta 597 (1980) 477–491.
[56] T. H. Huang, R. P. Skarjune, R. J. Wittebort, R. G. Griffin, E. Oldfield: Restricted rotational isomerization in polymethylene chains, J. Amer. Chem. Sci. 102 (1980) 7377–7379.
[57] A. Blume, D. M. Rice, R. J. Wittebort, R. G. Griffin: Molecular dynamics and conformation in the gel and liquid crystalline phases of phosphatidylethanolamine bilayers, Biochemistry 24 (1982) 6220–6230.
[58] R. Brandes, R. R. Vold, R. L. Vold, D. R. Kearns: Effects of hydration on purine motion in solid DNA, Biochemistry 25 (1986) 7744–7751.
[59] A. Kintanar, W. C. Huang, D. C. Schindele, D. E. Wemmer, G. Drobny: Dynamics of bases in hydrated d [$(CGCGAATTCGCG)_2$], Biochemistry 28 (1989) 282–293.
[60] R. R. Ernst, G. Bodenhausen, A. Wokaun: Principles of Nuclear Magnetic Resonance in one and two Dimensions, Clarendon Press. Oxford 1987.

[61] C. Schmidt, S. Wefing, B. Blümich, H.W. Spiess: Dynamics of molecular reorientations: Direct determination of rotational angles from two-dimensional NMR of powders, Chem. Phys. Letters 130 (1986) 84–90.
[62] S. Wefing, S. Kaufmann, H.W. Spiess: Two-dimensional exchange NMR of powder samples, II. The dynamic evolution of two-time distribution functions, J. Chem. Phys. 89 (1988) 1234–1244, III., ibid. 93 (1990) 197–214.
[63] A. Hagemeyer, K. Schmidt-Rohr, H.W. Spiess: Two-dimensional exchange NMR in static and rotating samples, Adv. Mag. Res. 13 (1989) 85–130.
[64] L. Müller, S. I. Chan: Two-dimensional deuterium NMR of lipid membranes, J. Chem. Phys. 78 (1983) 4341–4348.
[65] K. Müller, A. Schleicher, G. Kothe: Two-dimensional NMR relaxation spectroscopy of liquid crystal polymers, Mol. Cryst. Liq. Cryst. 153 (1987) 117–131.
[66] B. Blümich, H.W. Spiess: Two-dimensional solid state NMR spectroscopy: New possibilities for the investigation of structure and dynamics of solid polymers, Angew. Chem. 100 (1988) 1716–1734; Int. Ed. Engl. 27 (1988) 1655–1672.
[67] Y. Yang, A. Hagemeyer, B. Blümich, H.W. Spiess: 2D Magic angle spinning NMR spectroscopy: Correlation between molecular order and dynamics, Chem. Phys. Letters 150 (1988) 1–5.

13 The Kerr Effect as Applied to the Investigation of Polymers

Bernd-J. Jungnickel and Joachim H. Wendorff*

13.1 Introduction

In the year 1871, the Scotch physicist John Kerr observed that usual glass although positioned between crossed polarizers became transparent when subjected to a strong electric field [1]. He also found that the effect rised and – when switching off the field – decayed in a time scale of typically some seconds. Obviously, electric fields can cause the occurrence of birefringence in an otherwise isotropic sample. Soon after his first publication of the effect, Kerr also reported this birefringence to increase with the second power of the electric field strength [2]. This widely valid rule, the proportionality constant, as well as the effect of the "electric birefringence" itself were soon named according to their discoverer.

Langevin [3] and Born [4] showed that the Kerr effect is caused by an electric field induced orientation of permanent molecular dipoles, or that of induced molecular dipoles originating from the polarizability, in the material. The magnitude of the effect should consequently allow conclusion on these molecular electrical parameters. The efficient permanent moment and the polarizability, moreover, reflect the conformation of large molecules since they result from a tensor addition of the contributions of all subunits [5]. That is why the effect is so interesting for polymer science [6, 7]: It allows the determination of the chain conformation as a function of external parameters such as temperature or concentration in solution, or of material parameters like molecular weight. This, in turn, allows conclusions on other physical or physico-chemical properties like interactions between subunits or monomers, chain flexibility, orientation correlations between chains and chain parts, on solvent-solute interactions and so on. Conclusions on phase transition properties are also possible.

* Deutsches Kunststoff-Institut, Schloßgartenstraße 6, D-6100 Darmstadt

So far, only the "static Kerr effect" has been considered. The temporal development of birefringence after a sudden application of the electric field ("dynamic Kerr effect"), however, is an important additional source of information. It allows insight into the type and the rate of molecular reorientation processes. This is important for the description of the dynamics of polymer molecules as a whole, of their subunits, or of their aggregation.

The aim of the present article is twofold. First of all, it is intended to give a short survey on the theoretical and experimental fundamentals of the Kerr effect, and on the kind of information on a sample material which can be drawn basically from the effect. Next, we want to report selected results which were obtained in the last years at the Deutsches Kunststoff-Institut, Darmstadt. These results are used not only to demonstrate which kind of information is available on the polymeric systems studied but also to show which specific information on topical problems of polymer physics can be obtained by Kerr effect measurements, how such measurements can be extended to new types of sample materials, particularly glassy-amorphous and liquid-crystalline polymers, and how they complete and support results of other kinds of experiments, in particular those which are used in other partial projects in the Sonderforschungsbereich.

13.2 Fundamentals

13.2.1 Experimental Setup and Procedure

The experimental arrangement, basically, has to enable the measurement of the electric field induced birefringence as a function of time and temperature. The usual set-up is shown in Fig. 13.1 [8]. It is composed of three main parts. The optical part consists of a light source, a polarizer that is oriented at an angle of 45° with respect to the electrical field, a quarter wave retarder, the slow axis of which is oriented perpendicularly to the polarizer, and an analyzer which is rotated by a known fixed angle α away from the position of extinction. The sign of this rotation is, suitably, chosen such that a positive birefringence causes a decrease of transmitted light intensity. A quartz optical compensator that is positioned immediately behind the sample case is necessary in order to compensate intrinsic birefringences in the sample which can be caused by frozen-in stresses or slight chain orientations. The electrical part consists of a high voltage pulse generator and a signal detection array. A retardation of the pulses should be possible. The components of the signal detection array are a fast photo diode, a noise reducting band filter, an amplifier, a storage oscilloscope or a transient recorder (which in turn are triggered by the high voltage pulses), and a recorder. Last but not least, there is the sample unit consisting of

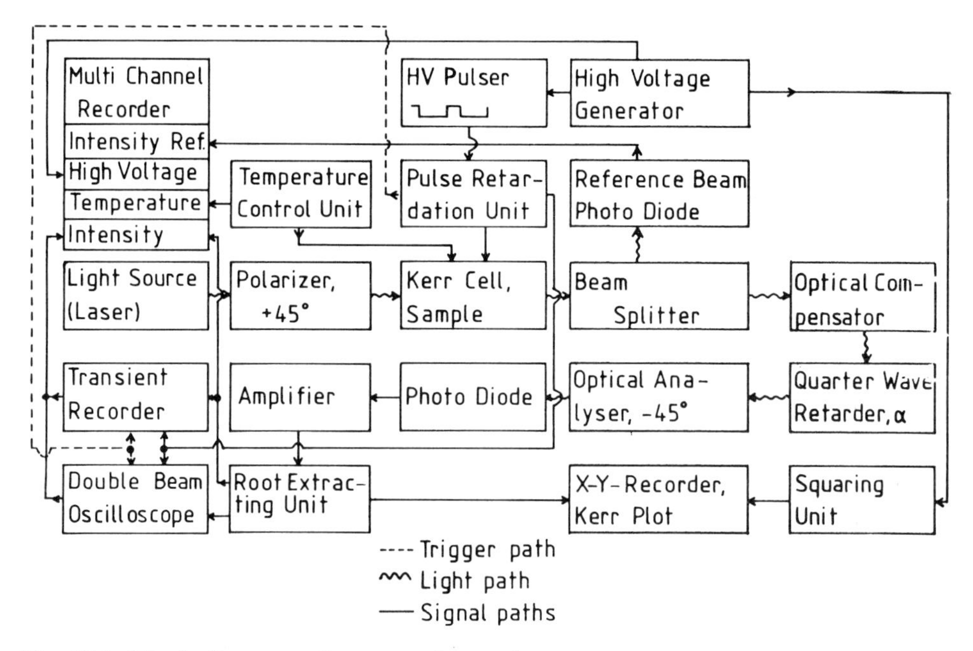

Fig. 13.1: Block diagram of an experimental set-up.

the heatable sample cell ("Kerr cell"), and a temperature control unit. Our Kerr cell for solid samples consists of a precisely adjustable aluminium plate on earth potential on which the sample is placed. A specific construction allows to put a ceramic coated steel electrode pressurefree on the sample such that, nevertheless, there is no air gap. Maxwell stress which is caused by the attractive force between the electrodes – due to the electric field – can then not influence the sample. It could cause deformation induced chain orientations and, therefore, additional birefringence. The whole arrangement is embedded in a well isolated heatable aluminium chamber. The characteristics of our set-up are collected in Table 13.1.

From the light intensities with (I_E) and without electric field (I_0), the birefringence can be calculated according to

$$\Delta\ n = (\lambda/(\pi L))\ ((\arcsin(\sin\alpha(I_E/I_0)^{1/2})) - \alpha) \tag{1a}$$

(L = optical path length, λ = light wave length). Since usually $\alpha \ll 1$, we have approximately

$$\Delta\ n = (\alpha\lambda)/(\pi L)\ ((I_E/I_0)^{1/2}-1), \tag{1b}$$

Table 13.1: Characteristics of the experimental set-up.

Light source	He-Ne laser (λ = 632.8 nm, 5 mW polarized, beam diameter < 1 mm)
High voltage source	
maximal voltage	10 kV
pulse rise time (including contributions of the detection array)	500 ns
pulse duration	10 μs
pulse decay time (including contributions of the detection array)	500 ns
pulse repetition frequency	1 kHz
pulse retardation time t	4 μs < t < 1 ms
Sample parameters	
length L	10 cm
thickness	2 mm
width	3 cm
electrode length	8 cm
Temperature T	
range	−150 °C < T < 160 °C
stability Δ T	<0.1 °C
Sensitivity Δ n	$< 10^{-10}$

and the birefringence is proportional to the square root from the light intensity,

$$\Delta n \sim I_E^{1/2}. \qquad (1c)$$

13.2.2 Basic Theoretical Relations

The electric birefringence is caused by a preferred orientation of the molecules or parts of them which try to minimize the energy of their permanent or induced electric dipoles in the external electric field. This birefringence can be calculated as follows [9–11]. The refractive index n of a material is a function of the macroscopic expectation value of the polarizability at optical frequencies $\boldsymbol{\alpha}^o$ of its molecules:

$$n_i = f(<\alpha_{ii}^o>). \qquad (2)$$

We have by definition

$$<\alpha_{ii}^{o}> = \varepsilon_o (Nc)^{-1} \chi_{ii}^{o} \tag{3}$$

where ε_o is the permittivity of the vacuum, N is the number density of the particles, $\boldsymbol{\chi}^o$ is the dielectric susceptibility at optical frequencies, and $c = E_i/E_o$ is the ratio between the internal and the externally applied electric field; c in its turn can depend on the susceptibilities in the high frequency and in the static limit as well. The values of the tensors $\boldsymbol{\alpha}$ and $\boldsymbol{\chi}$, here, have to be taken in the sample-fixed major axes system. For the birefringence, then, we have

$$\Delta n \overset{!}{=} n_3 - n_1 = dn/d<\alpha^o> <\Delta\alpha^o> \tag{4a}$$

$$= dn/d\chi^o (d<\alpha^o>/d\chi^o)^{-1} <\alpha_{33}^{o} - \alpha_{11}^{o}> = z<\Delta\alpha^o> \tag{4b}$$

with

$$z = (n^2 + 2)^2 N/(18 n \varepsilon_o) \tag{5}$$

if the "Lorentz field approximation", $c = (\chi^o + 3)/3$, and the Maxwell formula, $n^2 = \chi^o + 1$, are introduced. From statistical thermodynamics it is further known that

$$<\Delta\alpha^o> = \int \Delta\alpha^o \exp(-\beta H) d\mathbf{x} \,/ \int \exp(-\beta H) d\mathbf{x} \tag{6}$$

where

$$H = \sum_{i} \mu_i E_i + \sum_{i,j} \alpha_{ij}^{s} E_i E_j \tag{7}$$

is the energy of the permanent ($\boldsymbol{\mu}$) and induced dipoles in the electric field, $\mathbf{x}$ is the set of the (relevant) degrees of freedom of the molecules (here: the orientation angles; cf. Fig. 13.2), the superscript "s" indicates the static limit of the polarizability, and $\beta = 1/kT$. The integration, considering that $\boldsymbol{\mu}$ and $\boldsymbol{\alpha}^s$ (Eq. 7) are usually expressed in a molecule-fixed coordinate system whereas H and $<\Delta\alpha^o>$ (Eq. 6) must be expressed in a sample-fixed one, can only be performed by power series expansion of the integrand and gives

$$<\Delta\alpha^o> = (\beta c E_o)^2 <R>/30 + \text{terms of higher powers of } E_o \tag{8}$$

with

$$<R> = \sum_{i,j} (3<\mu_i \alpha_{ij}^{o} \mu_j> - <\mu_i \mu_i \alpha_{jj}^{o}> + (3\beta^{-1}<\alpha_{ij}^{o}\alpha_{ji}^{s}>) -$$

$$-\beta^{-1} < \sum_{i} \alpha_{ii}^{o} \sum_{i} \alpha_{ii}^{s}> \tag{9}$$

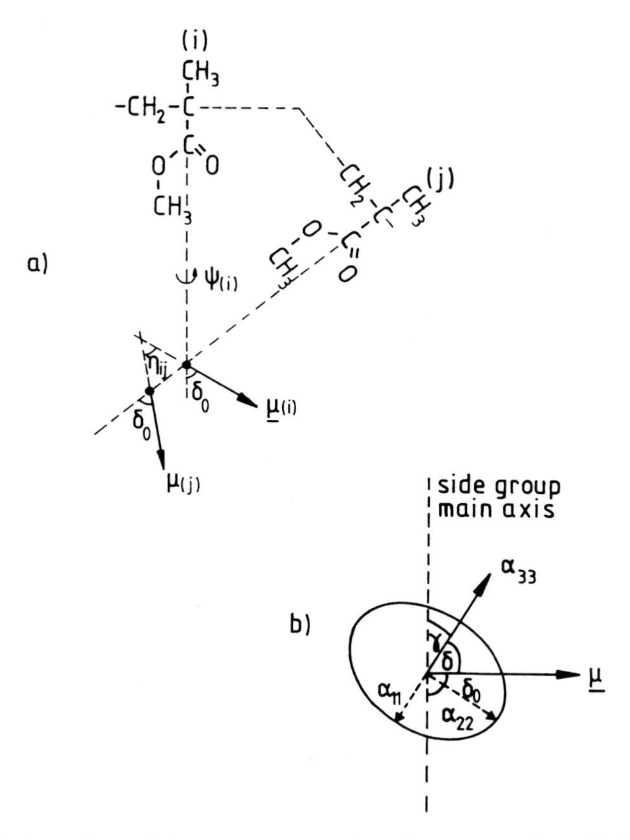

Fig. 13.2: a) Orientation of the permanent electric moment in a PMMA monomer unit, and relative orientation of two units; b) definition of the angles occuring in the equations.

where the **μ** and **α** are the molecular quantities, and the angular brackets design the conformation average over all molecules. Combination of all equations easily gives an expression for the Kerr constant:

$$K = z\langle R\rangle(\beta c)^2/30. \tag{10}$$

We have

$$\langle R\rangle = 2\ \Delta\alpha^{o}\ (f\mu^2 + \beta^{-1}\ \Delta\alpha^{s}) \tag{11}$$

if the polarizability tensors at optical frequencies and in the static limit have equal orientations and are both axially symmetric around the "3"-axis.

$$f \overset{!}{=} P_2\ (\delta) \overset{!}{=} 1/2(3\cos^2\delta - 1), \tag{12}$$

and δ is the angle between the molecular dipole and the "3"-direction of the polarizability (Fig. 13.2).

Let us now assume that the particles cannot move independently in the external field since their orientation is correlated. This correlation is treated in a simplified approach on the basis of the orientation dependent part of the pair correlation function $P_{ij}(\mathbf{\Omega}_i(0),\mathbf{\Omega}_j(t))$ [12]. P_{ij} is the probability that particle "j" has the orientation $\mathbf{\Omega}_j$ at the time "t" if particle "i" had the orientation $\mathbf{\Omega}_i$ at the time zero, and can be expanded in the following manner [12]:

$$P_{ij} = \sum_{l,m,n} g^{ij}_{lmn}(t)\Phi_{lmn}(\eta_{ij}(t)) \tag{13}$$

where the Φ are generalized spherical harmonics, the "g" are orientation correlation parameters, and $\eta_{ij}(t)$ is the orientation difference between $\Omega_i(0)$ and $\Omega_j(t)$ where the two particles are again assumed to be rotationally symmetric. Consideration of the pair orientation correlations modifies Eq. (11) in the following manner [9, 11, 13]:

$$<R> = 2\,\Delta\alpha^o\,(f\mu^2(2(g_1-1) + g_2) + \beta^{-1}g_2\Delta\alpha^s). \tag{14}$$

In this equation, g_1 is the "Kirkwood factor" that is well known from the theoretical description of dielectric properties [14]. It yields information on the orientation correlation of directed entities (vectors) like dipoles:

$$g_1 = 1 + \sum_i <\cos\eta_{ij}>_j,\ i \neq j,\ t = 0 \tag{15}$$

(cf. Fig. 13.2). We have $g_1 < 1$ if the dipoles are preferentially antiparallel and $g_1 > 1$ with preferentially parallel orientation. If there are no such correlations, we have $g_1 = 1$. The correlation parameter g_2 is defined by

$$g_2 = 1 + 1/2 \sum_i <3\cos^2\eta_{ij} - 1>_j,\ i \neq j,\ t = 0 \tag{16}$$

Spatial orientation correlations of molecular major axes which are invariant with respect to a reversion of their direction can be characterized by means of g_2. This concerns the major axes of the polarizability ellipsoid. g_2 assumes as g_1 a value of $g_2 = 1$ if no orientation correlations of the described kind are present. It should be pointed out that the g_i reflect the orientation correlations in the field free state since they are taken at $t = 0$.

In order to describe the reorientational kinetics, the time dependence of the pair orientation correlation function has to be taken into account. Let us assume for the moment that there are no orientation correlations between the particles. Then, the theory shows that usually [15]:

$$\Delta n_r(t) \sim c_1 g_1'(t) + c_2 g_2'(t), \tag{17}$$

and

$$\Delta n_d(t) \sim c_3 g_2'(t) \tag{18}$$

where

$$g_1'(t) = 1 + \langle \cos\eta_{ii}(t) \rangle \tag{19}$$

and

$$g_2'(t) = 1 + 1/2 \langle 3\cos^2\eta_{ii}(t) - 1 \rangle \tag{20}$$

(cf. Eqs. 15 and 16). $\Delta n_r(t)$ and $\Delta n_d(t)$ are the transient functions of the birefringence after a sudden switching on and switching off, respectively, of the electric field. The time dependent correlation functions g', on the one hand, can therefore be determined by dynamic Kerr effect measurements, or by complementary Kerr and dielectric studies. They can, on the other hand, be calculated on the basis of different reorientation models. For the "rotation diffusion model", the molecular rearrangement is assumed to proceed by infinitesimal angular steps. The calculation shows that $g_n' = \exp(-n(n+1)Dt)$ where D is the rotation diffusion coefficient. The decay of the birefringence, therefore, follows an exponential law with the time constant $\tau = (6D)^{-1}$ whereas the time constant of dielectric relaxation equals $(2D)^{-1}$. The transient of the rise curve is not simply exponential and depends on the ratio of the permanent to the induced dipoles. For the "fluctuation diffusion model", the molecular reorientation steps are assumed to be possible only after local density fluctuations, supplying the necessary free volume in the vicinity of the reorienting particle. The reorientation proceeds then via large angular steps. Rise and decay transient curves are symmetrical in this case, and $g_1' = g_2'$. Therefore, electric birefringence and dielectric relaxation obey the same transient function.

13.2.3 Information Content of the Effect

It follows from the equations discussed above that by Kerr effect measurements, direct information can be obtained on the molecular polarizability and on the strength of molecular dipoles, on their mutual orientation and on orientation correlations. Of greater importance, however, is the evaluation of the dependence of these quantities on preparational or experimental conditions. They can depend on the temperature, on the concentration if in solution, and, for polymers, on the molecular weight. The polarizability and the dipole of a large molecule are available from a tensor addition of the contributions of all subunits, in particular monomers [5]. The Kerr constant, therefore, will reflect the conformation of the macromolecule in terms of flexibility parameters like persistence length, and the dependence of them on the experimental and material boundary conditions mentioned above. It is, however, also possible to evaluate the equations by treating the monomer unit as a "particle" instead of the whole macromolecule. Then, the conformation of the macromolecule is characterized by orientation correlation parameters. Both these procedures are equivalent. The temperature dependence of all these quantities will allow conclusions on thermodynamic properties, on the phase type and the phase transition behaviour of the sample material as will be shown in more detail later. Sometimes, the final thermodynamical equilibrium which the ensemble of molecules has to assume for the validity of the above equations cannot be reached. Then, also conclusions on the type and the activation temperature of the constrained motions can be drawn as will be shown in the next paragraph.

13.3 Glassy-Amorphous Polymers

Almost all work on the Kerr effect of macromolecules as reported up to now concerns solutions whereas only few publications treat solid polymers [16]. This is due to major preparational difficulties which make a quantitative evaluation of the effect almost impossible. These difficulties are mainly caused by the fact that solid samples almost always exhibit an intrinsic birefringence due to inherent stresses or weak textures. This birefringence can amount to about 10^{-6} even when the material has been carefully homogenized, this value being negligible normally even for optical applications. It is, however, sometimes by orders of magnitude stronger then the electrically induced birefringence. The quantitative detection of the Kerr effect in polymer solids, nevertheless, is an important task since it should allow interesting and valuable insight into structural and dynamical properties of such materials, and support or complete information from comparable measuring techniques like (inelastic) light scattering or dielectric measurements. The experimental difficulties can be overcome as will be reported later provided the materials used are at least optically transmittent and homogeneous. Some glassy-amorphous polymers fullfill these requirements. Poly (methyl methacrylate), PMMA, is a particularly suitable sample material: its optical properties are excellent, its dielectric and light scattering behaviour are already well investigated, and the possible molecular relaxation modes are known and are in a well accessable temperature and time range. The occurrence of several distinct relaxation processes, however, can complicate the electro-optical behaviour. Therefore, some corresponding theoretical considerations are necessary which will be outlined here before presenting the experimental results.

13.3.1 Theoretical Considerations

Most molecules, particularly polymers, are flexible or consist of separate chemical groups with individual mobilities. Each of these dynamical modes are characterized by a temperature below which it will be frozen in. The thermodynamic equilibrium that was supposed to exist when deriving Eqs. (8–11) and (14), however, can only be obtained if all modes are thawed. Only an equilibrium under constraint will be established if some modes are still frozen in, this leadig to deviations between the predictions on the birefringence based on the equations introduced earlier and the one found experimentally. In PMMA, particularly, there is the side chain rotation that is initiated already below 0 °C ("β-relaxation"), whereas above the glass transition temperature T_g of (105...110)°C the rotation axis itself can reorient too ("α-relaxation"). At room temperature, therefore, the permanent dipoles which are located in the side chain can orient in an external electric field only to a limited extent. Rising the temperature above T_g, then, allows complete relaxation and occurrence of the total Kerr response. It is the aim of the present paragraph to derive equations for the limited relaxation strengths belonging to the reorientation modes which can occur at the actual temperature.

Naturally, a complete "freeze in" is impossible. Here, frozen in just means that the corresponding relaxation time is long in comparison to that of the thawed motions which can be observed. Rising the temperature shifts the former relaxation time in the experimentally accessible range. The above considerations in the temperature domain can such be translated into the time domain. Both procedures are equally entitled from the point of view of the time-temperature equivalence principle.

We denote those parts of the reorientational degrees of freedom which are still frozen in by $\mathbf{x}_2$ whereas $\mathbf{x}_1$ denotes the ones which can elapse and allow establishment of a restrained equilibrium when the energetical boundary conditions (here: switching on or off the electric field) are suddenly changed. It is furthermore assumed that $\mathbf{x}_1$ and $\mathbf{x}_2$ have no common components, and that $\mathbf{x} = (\mathbf{x}_1, \mathbf{x}_2)$. For the restrained equilibrium of the thermodynamic variable quantity Q after the sudden application of the distortion "E", starting from the complete equilibrium, Eq. (6) has then to be replaced by [9, 11]

$$<Q(+E)>_r \stackrel{!}{=} <Q>_{r+} = (|\mathbf{x}_1|/|\mathbf{x}|)$$

$$\int((\int Q\exp(-\beta H(E))d\mathbf{x}_1)/(\int \exp(-\beta H(E))d\mathbf{x}_1))d\mathbf{x}_2, \quad (21)$$

and for the restrained equilibrium after sudden removal of the distortion, again starting from the complete equilibrium that belongs to H(E)

$$<Q(-E)> \stackrel{!}{=} <Q>_{r-} = (1/|\mathbf{x}_1|)$$

$$\int((\int Q d\mathbf{x}_1 \int \exp(-\beta H(E)) d\mathbf{x}_1)/(\int \exp(-\beta H(E)) d\mathbf{x})) d\mathbf{x}_2 \tag{22}$$

(cf. Fig. 13.3). Solving these equations with respect to the motion of the PMMA permanent electric moment with β-relaxation (orientational degree of freedom: angle ψ; cf. Fig. 13.2) we get in the field-on case

$$\langle R\rangle_{r+} = 3\,\mu^2\,\Delta\alpha^{o}((\cos\delta - \cos\delta_o \cos\gamma)^2 - (\sin^2\delta_o \sin^2\gamma)/2), \tag{23}$$

and in the field-off case

$$\langle R\rangle_{r-} = \mu^2\,\Delta\alpha^{o} P_2(\delta_o) P_2(\delta). \tag{24}$$

The relations (23) and (24) complete Eq. (11) which describes the total equilibrium. They apply not only to the β-relaxation strength of PMMA in the Kerr experiment but are valid for all molecules (also those of low molecular weight)

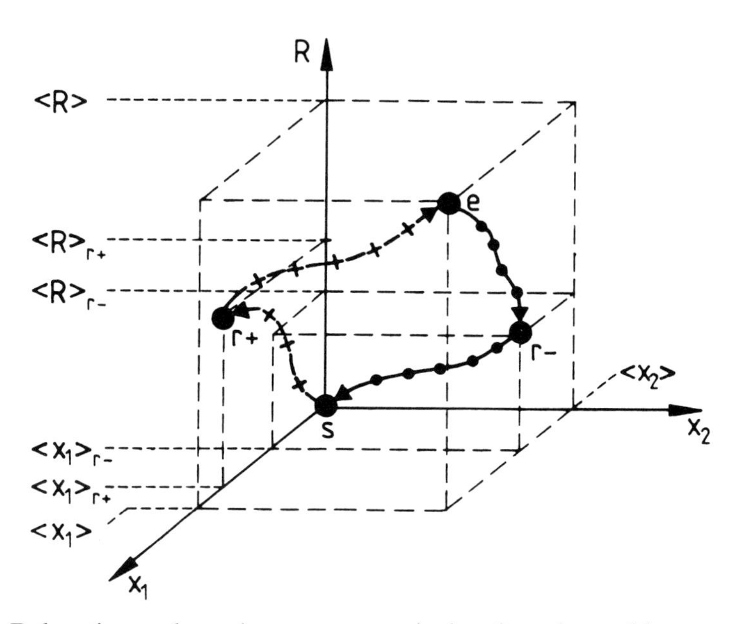

Fig. 13.3: Relaxation paths and stages, respectively, of an observable quantity R after application (+ + + +) and removal (oooo) of a distortion; x_1: relaxing degree of freedom at the initial observation temperature; x_2: remaining degree of freedom relaxing only at a sufficiently elevated temperature; $\langle R\rangle_{r+}$: expectation value of R after sudden application of a distortion in the restrained equilibrium caused by the sole relaxation of x_1; $\langle R\rangle$: complete equilibrium after relaxation of both degrees of freedom $\mathbf{x} = (x_1, x_2)$; $\langle R\rangle_{r-}$: expectation value of R after sudden removal of the distortion in the restrained equilibrium after the sole relaxation of x_1. The point "s" must not necessarily be the origin of the coordinate system. The signs of $\langle R\rangle$, $\langle R\rangle_{r+}$, and $\langle R\rangle_{r-}$ may differ. For the points "s", "r+" "e", and "r−"; cf. also Fig. 13.2.

which obey the presuppositions introduced above and possess comparable rotational degrees of freedom as PMMA. Depending on the values of the set of parameters (δ, δ_o, γ) (cf. Fig. 13.2), a variety of different ratios between the relaxation strengths $<R>_{r+}$, $<R>$, and $<R>_{r-}$ can be observed, some of which exhibiting different signs for the several relaxation steps (Fig. 13.4). The parameter set ($\pi/2$, $\pi/2$, $\pi/2$) causes a relaxation behaviour according to Fig.

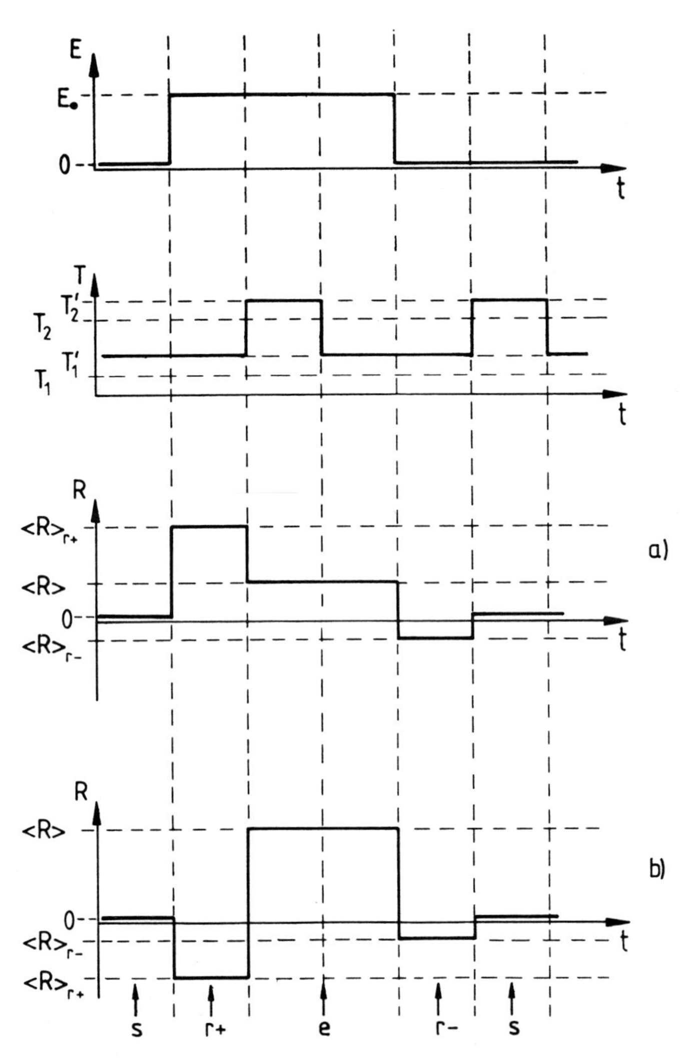

Fig. 13.4: Kerr effect relaxation patterns for different particle conformations (a, b; cf. text) after sudden changes of electric field and temperature; schematically. T_1: Thawing temperature of x_1; T_2: Thawing temperature of x_2; T_1': initial measuring temperature; T_2': elevated measuring temperature. Cf. also Fig. 13.3.

13.4a whereas Fig. 13.4b corresponds to the set $(\pi/2, \pi/2, \pi/4)$. With $(0, 0, 0)$, the first relaxation does not occur. In the first mentioned case, the permanent dipole and the symmetry axis of the optical polarizability are oriented perpendicularly to the axis, rotation around which causes the first relaxation step, and include a right angle. In the second example, these two physical quantities include again a right angle but are symmetrically oriented with respect to the rotation axis. In the third case, finally, the orientation of the permanent dipole and the symmetry axis of the polarizability coincide both with the rotation axis.

Consideration of pair orientation correlations leads to

$$<R>_{r+} = 2\mu^2\,\Delta\alpha^o(1 - P_2(\delta_o))^2/3, \quad (25)$$

and

$$<R>_{r-} = 2\mu^2\,\Delta\alpha^o((P_2(\delta_o))^2 + 2P_2(\delta_o)(g_1 - 1) + g_2 - 1), \quad (26)$$

respectively, these equations belonging to Eq. (14). The angle δ has here simplifyingly been put zero.

These theoretical considerations are presented in more detail in Ref. [11].

13.3.2 Experimental Investigations and Results

Experiments were performed with PMMA GS 218 (Röhm GmbH, Table 13.2). This material is specifically designed for optical applications. The polymer was annealed at 120 °C for 24 h and subsequently slowly cooled down for several days. The remaining birefringence of about 10^{-6} could be optically compensated.

The Kerr measurements [8, 17] revealed at room temperature three distinct relaxations (Figs. 13.5 and 13.6): a relaxation (I) that was built up faster then 0.5 μs, a second one (II) that rised in the time domain of ms, and a third one (III), the rise time of which amounted to several hours. In some samples, relaxations (I) and (II) could not be separated since relaxation (II) became so fast that it moved into the time scale of relaxation (I), and some samples did

Table 13.2: PMMA sample characteristics.

Grade	PMMA GS 218 (Röhm GmbH, Germany), atactic
Molecular weight $\overline{M}_w$	4×10^6
Refractive index n_{20}	1.491
Dielectric loss angle $\tan\delta$ ($f = 50$ Hz)	0.06
Absorption coefficient $\varkappa$ ($\lambda = 514.5$ nm)	$< 1.7 \times 10^{-4}$ mm^{-1}

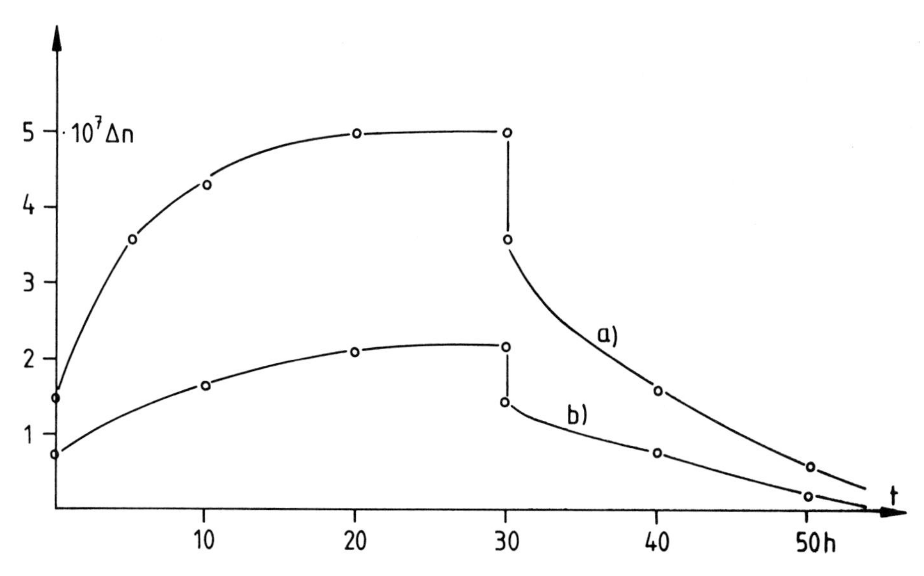

Fig. 13.5: Birefringence transient curves. Room temperature. a) $E_o = 9.4\ MVm^{-1}$, b) $E_o = 6.3\ MVm^{-1}$.

not exhibit relaxation three in the time scale which was experimentally accessable. All relaxations obeyed the Kerr law at all applied fields and within the total temperature range used (cf. Table 13.1). Fig. 13.7 shows the temperature dependence of the several relaxations. Whereas the relaxation times of relaxations (I) and (II) could not be measured or resolved, respectively, those of relaxation (III) shortened drastically into the time domain of minutes when exceeding the glass temperature. It is remarkable that the relative contributions of the several relaxations differed for rise and decay, and changed with temperature (Table 13.3).

Table 13.3: Relative contributions K of the several relaxation steps at different temperatures. Total Kerr constant normalized to unity; r = rise, d = decay.

T/°C/	K_{Ir}	$K_{I+II,r}$	K_{IIr}	K_{IIIr}	K_{Id}	$K_{I+II,d}$	K_{IId}	K_{IIId}
20	0.18		0.074	0.74	0.18		0.074	0.74
130		0.20		0.80		0.53		0.47
142		0.14		0.86		0.57		0.43

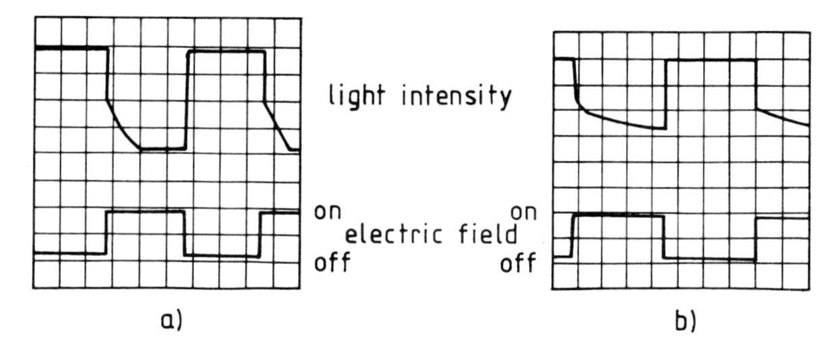

Fig. 13.6: Oscillographic light intensity transient curves. $E_o = 9.4\,\mathrm{MVm^{-1}}$. Room temperature. Grid scale: a) 1.07 s, b) 0.027 s. Light intensity scales not comparable. $\alpha = 0$ (cf. Eq. 1).

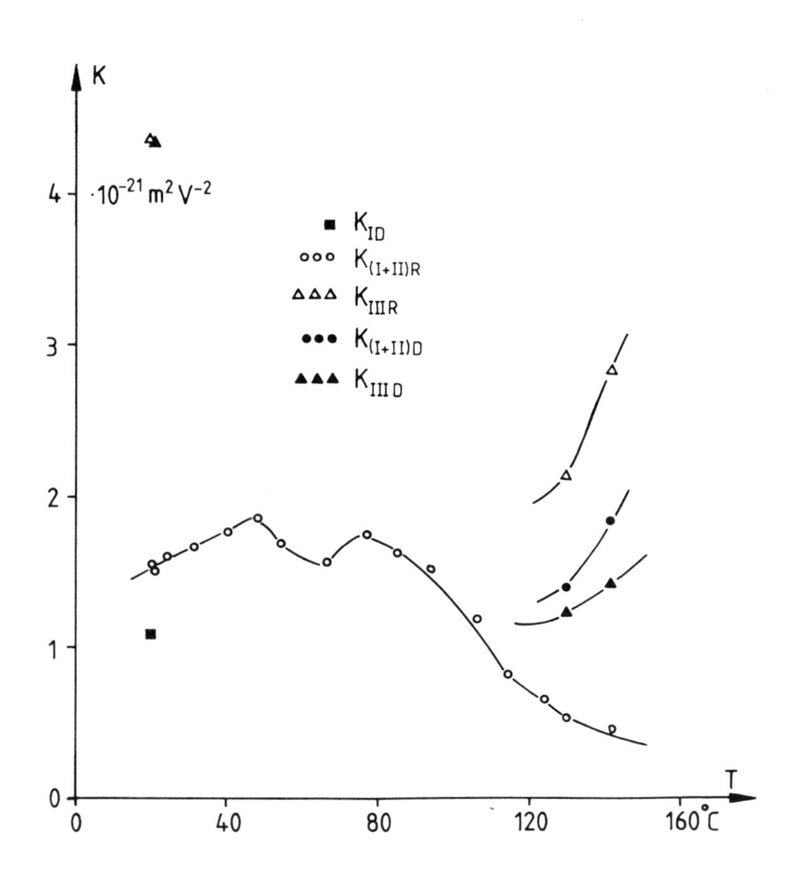

Fig. 13.7: Contributions of the several relaxation steps (cf. text) to the whole Kerr constant in dependence on the temperature.

13.3.3 Discussion

From the scales of the different relaxations it must be concluded that relaxation (I) is due to the so-called γ-relaxation caused by local rearrangement of methyl and ethyl groups and is, therefore, connected only with the reorientation of the polarizability of the PMMA monomer unit. This relaxation causes a dielectric damping maximum at about 40 MHz, corresponding to a rise time of 5 ns [18]. It should be mentioned that a pure rotation of the methyl groups would not change the orientation of the polarizability, and that, therefore, other dynamical modes or an electric field induced change in the potential distribution must be responsible for this relaxation. Relaxation (II) is caused by the β-relaxation (side chain motion, in particular rotation around the linking bond between main chain and side group), and relaxation (III) is linked with the α-relaxation which sets in when larger sections of the chain become mobile. The last two motional modes, therefore, should allow also a reorientation of the permanent moments.

It can be estimated from Eq. (11) that the ratio of the contributions of the permanent moments and of the polarizability to the Kerr effect amounts to $\mu^2/(kT\ \Delta\alpha^s)$, that is, about 2500 if both are fully efficient [8]. According to our measurements, however, this ratio amounts only to about one. The dipolar contribution, therefore, must be diminuished by deleterious g_i-values. It can be estimated from dielectric measurements [19] that g_1 increases from 0.3 at room temperature to 0.7 in the vicinity of the glass temperature. This indicates a preferentially pairwise antiparallel orientation correlation of the dipoles that deteriorates when approaching the glass transition, or changes to a more perpendicular orientation. From Eq. (14), then, it turns out that in the temperature range considered, g_2 varies between 1.4 and 0.7, this being consistent with the conclusions from the g_1-values. In any case, there is only a short range orientation correlation. A more detailed quantitative analysis of the g_i-values is not possible because of the uncertain ratios between the contributions of the several relaxations at the different temperatures and some unknown geometric parameters.

Another interesting and important observation is that the relaxation strength ratios characteristic of the various types of relaxation processes differ for the rise and decay transients, and that they depend on temperature (Table 13.3). This, however, is consistent with the calculations in paragraph 13.3.1, and with Eqs. (23–26). It is a consequence of the non-linear relation between electric field strength and birefringence which leads to an overrepresentation of the contribution of the initial relaxations to the total response. The temperature dependence of the strength ratios can be due either to a temperature dependent geometry of the monomer unit or to the temperature dependence of the orientation correlation parameters. In the light of the foregoing considerations, the latter is assumed to be true. A quantitative analysis, however, is again impossible from the same reasons as above.

The origin of some experimental observations is still unclear. This concerns the non-occurence of relaxation (III) in some samples, and the strong variation of the relaxation times of relaxation (II). Different states of aging of the samples may be one reason for this effect, unknown experimental artefacts another one.

13.4 Liquid Crystalline Polymers

13.4.1 Introduction: Pretransitional Effects

Studies on the Kerr effect have turned out to be very useful for the characterization of structural and dynamical properties of liquid crystalline systems, including liquid crystalline polymers. Most of the studies published so far were concerned with the isotropic phase neighbouring a nematic or cholesteric phase [20, 21]. It is a particular feature of the isotropic phase in this case that it displays pretransitional effects connected with fluctuations of the orientational order [20–22]. The occurrence of such fluctuations has been attributed to the fact that the isotropic-nematic (or cholesteric) phase transition – although being first order – is extremely weak. It therefore displays features which are characteristic of a second order phase transition, among them pretransitional fluctuations of order parameters.

De Gennes [22] was able to account for the pretransitional orientation fluctuations in nematic systems in terms of a phenomenological Landau treatment as far as static – i.e. thermodynamical and structural – properties are concerned. The dynamical properties of such fluctuations, on the other hand, were analyzed on the basis of a hydrodynamic theory [21, 22]. Thus, by evaluating the static and dynamical properties of these orientational fluctuations, one is able to obtain detailed information on thermodynamical, structural, and hydrodynamic properties of systems displaying liquid crystalline phases.

Now, Kerr effect studies have been found to be a very powerful method for investigating such fluctuations. The studies performed by us on liquid crystalline polymers [23–31] which will be discussed in the following had the aim to characterize selected static and dynamical properties of the fluctuations, to find out in which way polymer and low molar mass liquid crystals differ in this respect, and to establish the dependence of these properties on the chain architecture. Some examples will be given below. Prior to the discussion of experimental results a short review of pertinent features of the theoretical treatment of the Kerr effect in liquid crystalline systems will be given.

13.4.2 Kerr Effect and Order Parameter Fluctuations

13.4.2.1 Static Properties

The phenomenological approach used by de Gennes to represent the static behaviour of the pretransitional fluctuations in the isotropic phase [22] consisted in the most simplest approach in expending the free energy density F(T, S) in the vicinity of this transition as a function of the scalar nematic order parameter S:

$$F(T,S) = F_o\,(T) + (A(T)/2)\,S^2 - (b(T)/3)S^3 + (c(T)/4)S^4 \qquad (27)$$

where S is the nematic order parameter:

$$S = (3\,\cos^2\theta - 1)/2, \qquad (28)$$

and θ is the angle between director and molecular long axis. The coefficient b of the third order term is nonzero for symmetry reasons, and the coefficient of the quadratic term a(T) is expressed in the spirit of the Landau treatment as

$$a(T) = a_o\,(T - T^*)^v \qquad (29)$$

T* is the hypothetical temperature at which the isotropic phase becomes instabil with respect to the nematic one. The exponent v is taken to be one. The mean value of the order parameter S is zero in the isotropic state and the prediction is that the mean square value of the order parameter fluctuations behaves as

$$(\overline{\Delta S^2})_{S=0} \sim (\partial^2 G/\partial S^2) \sim (a_o(T - T^*))^{-1} \qquad (30)$$

i.e. it diverges as the characteristic temperature T* is approached.

The extension of this theory to the case that an electric field interacts with the isotropic phase and thus couples to the orientational fluctuations leads to the prediction, that the mean value of the order parameter S is no longer zero in the isotropic phase and varies with the temperature as follows:

$$S_E = (\varepsilon_o \Delta\varepsilon_s\, E_o^2)/(3\,a_o\,(T - T^*)) \qquad (31)$$

where $\Delta\varepsilon_s$ is the anisotropy of the dielectric constant and E_o the electric field.

This equation can be used to derive the following expression for the Kerr constant within the isotropic phase [21, 25], where Δn_s is the anisotropy of the refractive index:

$$K = (\Delta n_s \; \Delta\varepsilon_s \; \varepsilon_o)/(3 \; a_o \; (T - T^*)) \tag{32}$$

The Kerr constant is thus predicted to diverge as the characteristic temperature T* is approached. The reason is that the correlation length of the order parameter fluctuations approaches infinity with decreasing temperature. This becomes apparent if one compares the expression for the Kerr constant given above with the one introduced earlier (Eq. 14) on the basis of a statistical approach. The orientation correlation parameter g_2 characterizing the spatial extension of the fluctuations may be expressed in this case (assuming g_1 to be one) as follows:

$$g_2 = (45 \; N \; kT)/\left((\overline{n^2} + 2)^2 \; a_o \; (T - T^*)\right) \tag{33}$$

N is the number of units per unit volume. The parameter depends on the coefficient a_o of the expansion and on the inverse of the distance between the actual temperature T and the characteristic temperature T*. g_2 is thus expected to increase strongly as the temperature is decreased. The absolute magnitude of the Kerr constant provides thus information on the anisotropy of dielectric and optical properties as well as on thermodynamic and structural properties.

13.4.2.2 Dynamical Properties

The theoretical treatment of the dynamical behaviour of the orientational order parameter fluctuations and of the dynamics of the Kerr effect in the isotropic phase of liquid crystals will be reviewed only very briefly. De Gennes used a hydrodynamic approach [20–22] which connects the Kerr relaxation time with the viscosity of the isotropic phase and with the thermodynamic driving force as follows:

$$\tau(T) = 3 \; \eta \; (T)/(2 \; a_o \; (T - T^*)) \tag{34}$$

The basic assumptions are that the time dependent fluctuations of the order parameter are much faster than the spatial fluctuations of the velocity field [20–22].

So, the prediction is that the Kerr relaxation time diverges at the same temperature T* as the static Kerr constant. Further predictions of the hydrodynamic treatment are that the rise time (presence of applied field) and decay time (absence of electric field) are identical and that the Kerr relaxation obeys a simple exponential behaviour. So the dynamic Kerr data can be analyzed to yield the viscosity but also the characteristic temperature T*.

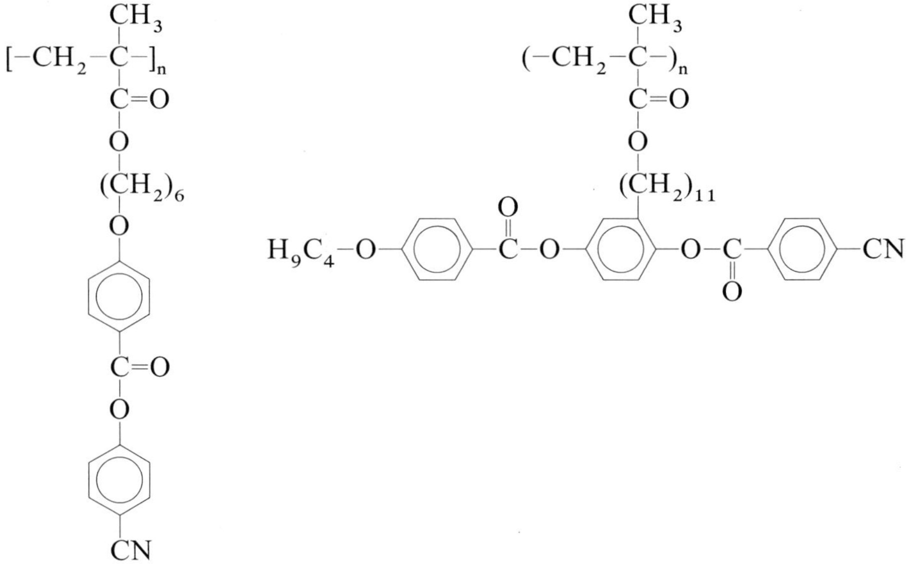

Fig. 13.8: Chemical structure of the liquid crystalline side chain polymers studied; a) longitudinal and b) lateral configuration of the mesogenic group.

13.4.3 Kerr Effect Studies in Liquid Crystalline Side Chain Polymers

In the following selected results obtained by Kerr effect studies on liquid crystalline polymers will be reported and discussed with respect to their implications. The first example is concerned with normal liquid crystalline side chain polymers where the mesogenic side group is attached longitudinally to the flexible spacer and the second example with the particular case of a lateral attachment (Fig. 13.8). To start with, a comparison between static and dynamical features of the orientation fluctuations between those of low molar mass and of a normal liquid crystalline side chain polymers will be drawn.

Fig. 13.9a displays the variation of the static Kerr constant with the temperature within the isotropic phase as obtained for a normal side chain liquid crystalline polymer displaying a lower temperature nematic phase. Fig. 13.9b shows the corresponding variation of the Kerr relaxation time. It is evident that in this case both the static Kerr constant as well as the Kerr relaxation time

Fig. 13.9: Variation of the reduced static Kerr constant $B = K/\lambda$ (a) and of the Kerr relaxation time τ (b) with the temperature for the polymer shown in Fig. 13.8a. A WLF correction was made in order to account for the temperature dependence of the viscosity. ▶

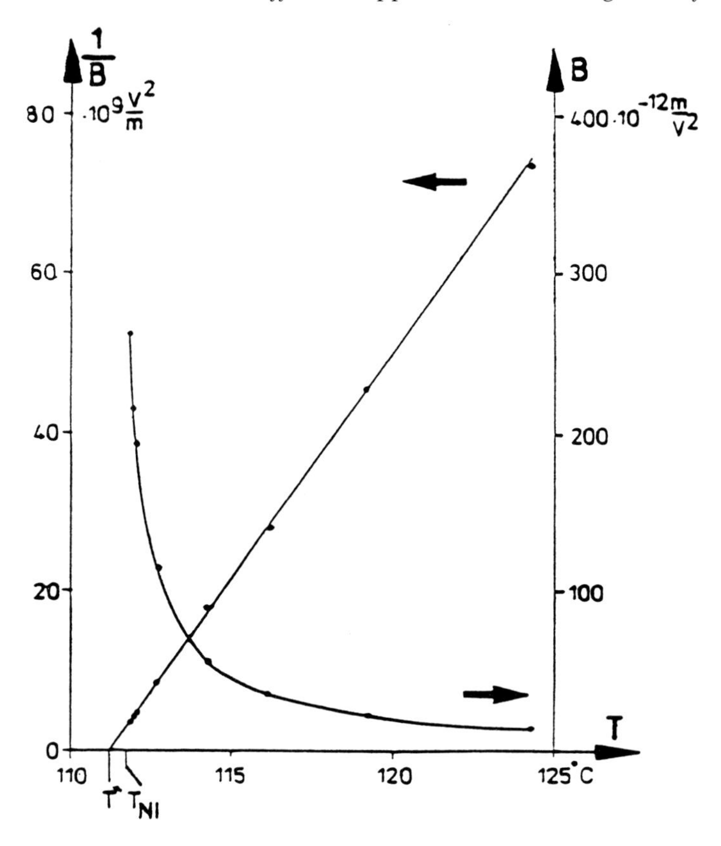
1/B
B
80
60
40
20
0
$\cdot 10^9 \frac{V^2}{m}$
$400 \cdot 10^{-12} \frac{m}{V^2}$
300
200
100
T
110
115
120
125°C
T*
T_{NI}

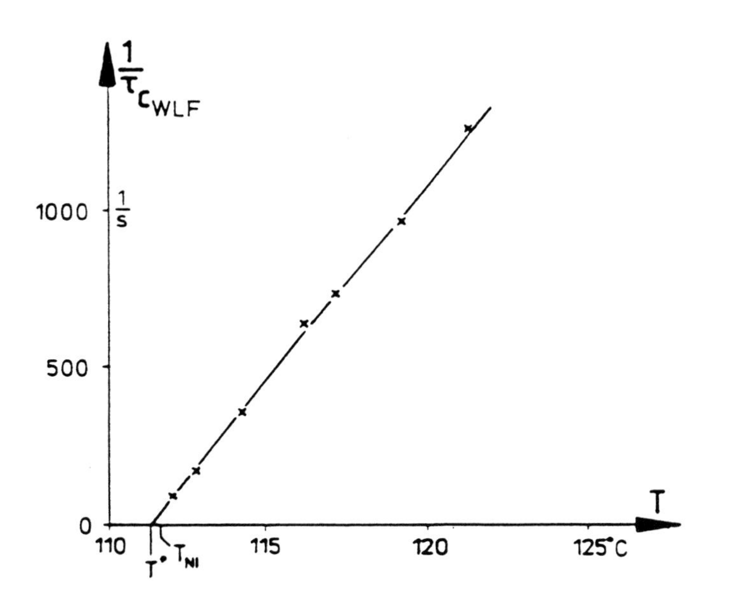
$\frac{1}{\tau_{c_{WLF}}}$
1000
$\frac{1}{s}$
500
0
T
110
115
120
125°C
T*
T_{NI}

diverge at a common temperature T* which is about 1 K below the first order phase transition T_{ni}. The critical exponent amounts to 1. The variation of the correlation parameter g_2 with the temperature is displayed in Fig. 13.10. It is of the order of 250 at a temperature located 1 K above the characteristic temperature T*.

A comparison with the results reported by us and by others on Kerr effect studies on low molar mass systems [21, 23–31] reveals that both the thermodynamical properties characteristic of the phase transition – i.e. magnitude of a_o and of $\Delta T = (T_{ni} - T^*)$ – as well as the structural properties – i.e. the magnitude of g_2 – agree closely. This was found to be the case not only for normal side chain polymers but also for liquid crystalline polymers possessing a more complex chain architecture such as for polymers with laterally attached mesogenic groups (Fig. 13.8b) or for polymers containing mesogenic groups both as side groups and in the chain backbone. So the general conclusion is that the transition from a low molecular to a polymer system does not influence the character

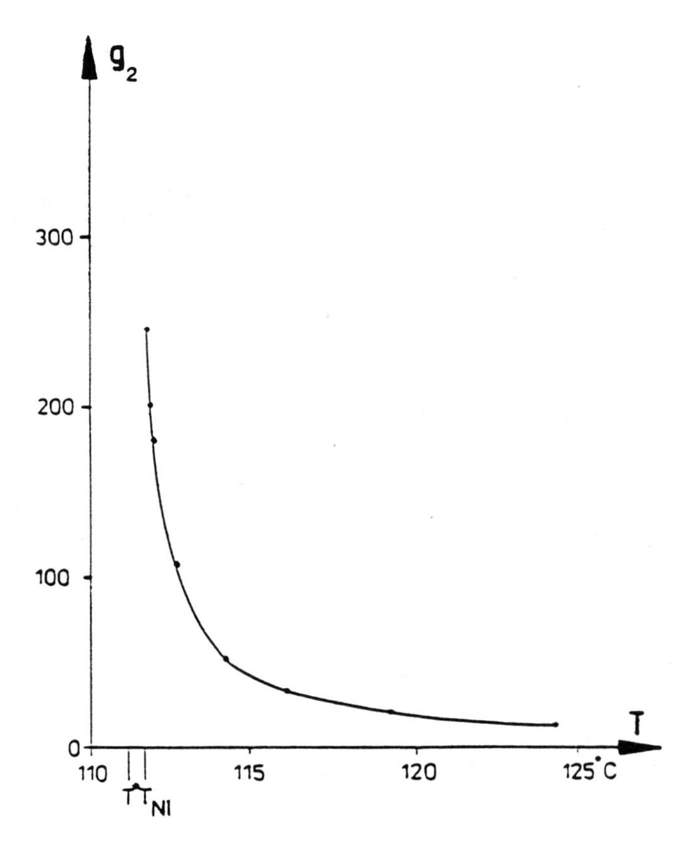

Fig. 13.10: Variation of the correlation parameter g_2 with the temperature for the polymer shown in Fig. 13.8a.

of the nematic isotropic phase transition nor the static properties of the orientational fluctuations significantly.

The dynamic response of the normal liquid crystalline side chain polymers, on the other hand, differs greatly from the one observed for low molar mass liquid crystals as far as the time scale is concerned. A single exponential decay and rise curve were found in agreement with the predictions of the de Gennes approach allowing for a simple determination of the relaxation time. The rise and the decay time were found to be equal, again in agreement with the theoretical predictions. The relaxation times of side chain polymers are typically several orders of magnitude larger than the ones found for low molar mass liquid crystals, due to their much larger viscosity. Investigations of the dependence of the Kerr relaxation time on the average chain length, and on the nature of the chain backbone (for instance acrylate chain backbone vs methacrylate chain backbone) have revealed that the increase of the viscosity from low molar mass to polymer systems does not only originate from the increase of the chain length but often also from the closeness of the glass transition to the nematic isotropic phase transition temperature.

This is often apparent also from the dependence of the Kerr relaxation time on the temperature. A frequent finding is that the plot of the inverse of the relaxation time versus the temperature does not lead to a straight line, approaching zero at the characteristic temperature. The reason for the occurrence of a curvature in this plot is, however, not a critical exponent deviating from one but a strong dependence of the viscosity on the temperature due to the neighbourhood of the glass transition temperature. A WLF-correction applied to the viscosity causes the plot of $1/\tau$ versus T to display a straight line, extrapolating to T^*. So, the general conclusion is that the dynamical properties of the orientational fluctuations of low molar mass and polymer systems agree on a qualitative level whereas the absolute magnitude of the Kerr relaxation times differs by many order of magnitudes.

A very surprising dynamic behaviour was observed for liquid crystalline side chain polymers with laterally attached mesogenic groups (Fig. 13.8b). Fig. 13.11 displays the time dependence of the induced birefringence after a sudden switching on and switching off of the electric field, respectively. It is apparent that the response is characterized by the presence of a fast and a slow relaxation process and secondly that the saturation values of the two processes differ. The saturation values of the two dynamical components were found to depend linearly on the square of the electric field, obeying thus the Kerr law, and to diverge as the common characteristic temperature T^* is approached, obeying thus the Landau-de Gennes predictions [21, 22].

An important observation is that the ratio of the magnitudes of the fast and the slow responses found in the presence of the field is not identical to the one observed for the field free case but differ strongly:

	rise	decay	
K_{fast} / K_{total}	0.37	0.83	(35)

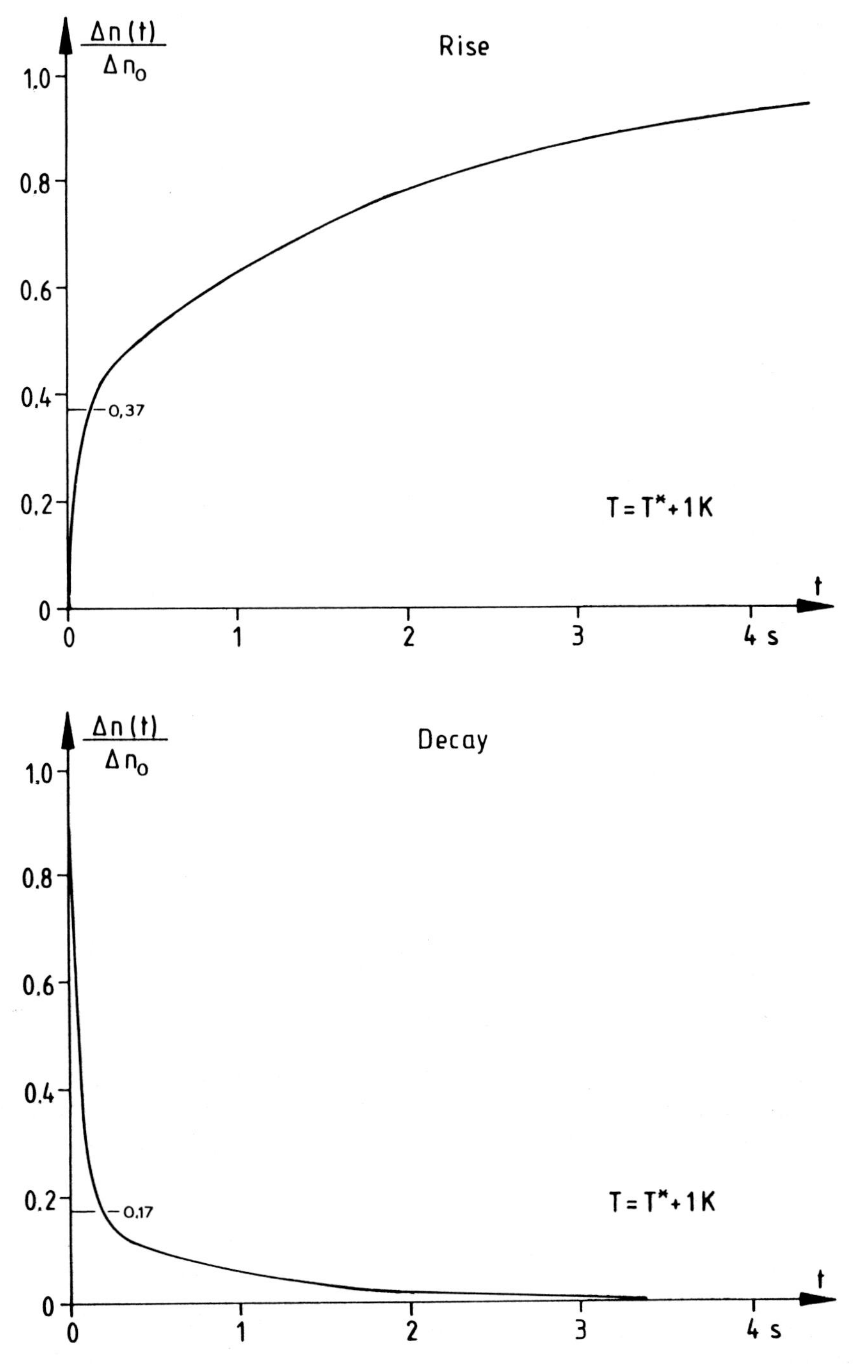

Fig. 13.11: Kerr rise (a) and decay curve (b) for the side chain polymer with laterally attached mesogenic group (Fig. 13.8b).

Furthermore the experiments reveal that the time scales both of the fast and of the slow response are not identical for the case of the rise and the decay. They differ by a factor of about 2:

$$\tau_{\text{rise, fast}} \simeq \tau_{\text{decay, fast}} \tag{36}$$

$$\tau_{\text{rise, slow}} \simeq 2\,\tau_{\text{decay, slow}} \tag{37}$$

Nevertheless all four relaxation times diverge at the temperature T*. So it is apparent that the electro-optical response discussed here is, indeed, very complex.

The properties reported so far as well as results obtained on the dielectric behaviour of these polymers seem to indicate that the dynamics of the Kerr effect and thus the dynamical properties of the orientational fluctuations are strongly influenced by restrictions imposed on the reorientation of the mesogenic units of the polymer. They are not able to reorient as freely as in the case of low molar mass or normal side chain. The results suggest that the dynamical response of the laterally substituted side chain polymers corresponds to the case which was treated theoretically and described above (Section 13.3.1). Constraints apparently exist, which have to arise from the lateral coupling of the mesogenic units to the chain backbone. The constraints force the molecular long axis to reorient only along given pathes, leading to a quasi-equilibrium state (fast response) whereas the relaxation towards the true equilibrium state occurs in a second step (slow response).

We were actually able [31] to account for all the features described above on the basis of this theoretical treatment and to derive in this way the following information on the local configuration of the side chain polymer in terms of the set of parameters introduced in part 13.3.1:

(i) the angle between the rotation axis (spacer axis) and the direction of the dipole moment of the mesogenic unit: 69 deg rad,
(ii) the angle between the principal axis of the optical polarizability and the rotational axis: 69 deg rad,
(iii) the angle between the directions defined by the dipole moment and the principle axis of the optical polarization, as obtained for the projection on a plane perpendicular to the rotation axis: 34 deg rad.

These values agree very well with the ones obtained from calculations on the dipolar configuration and from space filling models. Variations of the dipole moment lead to changes in the dynamical Kerr response again in agreement with the theoretical treatment [31].

The examples discussed above have demonstrated that Kerr effect studies provide an effective means for the characterization of thermodynamical, structural, conformational and dynamical properties of liquid crystalline systems.

13.5 References

[1] J. Kerr: A new relation between electricity and light: Dielectrified media birefringent, Phil. Mag. Ser. 4, 50 (1875) 337–348.
[2] J. Kerr: Measurement and law in electro-optics, Phil. Mag. Ser. 5, 9 (1880) 157–174.
[3] P. Langevin: Sur les biréfringence électronique et magnetique, Compt. Rend. 151 (1910) 475–478.
[4] M. Born: Elektronentheorie des natürlichen optischen Drehungsvermögens isotroper und anisotroper Flüssigkeiten, Ann. Phys. 55 (1918) 177–240.
[5] P. J. Flory: Statistical Mechanics of Chain Molecules, Wiley & Sons, New York 1969.
[6] S. Krause (ed.): Molecular Electro-Optics, Plenum Press, New York – London 1981.
[7] B. R. Jennings (ed.): Electro-Optics and Dielectrics of Macromolecules and Colloids, Plenum Press, New York – London 1979.
[8] K. Ullrich, B.-J. Jungnickel: Determination of angular correlations in atactic PMMA by measurement of the temperature dependent Kerr effect, Eur. Polym. J. 21 (1985) 991–997.
[9] U. Gallenkamp: Quantitative Auswertung relativer Relaxationsstärken bei Anlegen einer äußeren Störung, Master Thesis, TH Darmstadt 1986.
[10] K. Nagai, T. Ishikawa: Internal rotation and Kerr effect in polymer molecules, J. Chem. Phys. 43 (1965) 4508–4515.
[11] U. Gallenkamp, B.-J. Jungnickel: Thermodynamic expectation values in restrained equilibria, and the dielectric and electro-optical relaxation in PMMA, Ber. Bunsenges. Phys. Chem. 93 (1989) 585–593.
[12] H. Versmold: Neue experimentelle Möglichkeiten für das Studium der molekularen Struktur und Dynamik von Flüssigkeiten, Ber. Bunsenges. Phys. Chem. 85 (1981) 979–992.
[13] S. Kielich: A statistical theory of the Kerr effect in multi-component systems, Mol. Phys. 6 (1963) 49–59.
[14] N. G. McCrum, B. E. Read, G. Williams: Anelastic and Dielectric Effects in Polymeric Solids, Wiley & Sons, London – New York – Sidney 1967.
[15] R. H. Cole: Correlation function theory for Kerr effect relaxation of axially symmetric polar molecules, J. Chem. Phys. 86 (1982) 4700–4704.
[16] D. E. Cooper, T. C. Cheng, K. S. Kim, K. Kantak: Kerr-type electro-optic effect in solid dielectrics, IEEE Trans. Electr. Ins. 15 (1980) 294–300.
[17] B.-J. Jungnickel: Kerr effect relaxation measurements on glassy-amorphous poly (methyl methacrylate), Polymer 22 (1981) 720–725.
[18] M. Soliman: Untersuchungen zum Kerr-Effekt in glasig-amorphen Polymeren, Master Thesis, TH Darmstadt 1988.
[19] P. Hedwig: Dielectric Spectroscopy of Polymers, Adam Hilger, Bristol 1977.
[20] L. M. Blinov: Electro-Optical and Magneto-Optical Properties of Liquid Crystals, Wiley Interscience, New York 1983.
[21] D. A. Dunmur: Electro-optical properties of liquid crystals, in S. Krause (ed.): Molecular Electro-Optics, Plenum Press, New York 1981.
[22] P. G. de Gennes: The Physics of Liquid Crystals, Clarendon Press, New York 1974.

[23] M. Eich, K. Ullrich, J. H. Wendorff, H. Ringsdorf: Pretransitional phenomena in the isotropic melt of mesogenic side chain polymers, Polymer 25 (1984) 1271.
[24] J. H. Wendorff: Studies on the orientational order of polymeric liquid crystals by means of electric birefringence and small angle X-ray scattering, Makromol. Chem. Suppl. 6 (1984) 41–46.
[25] K. Ullrich: Charakterisierung von Orientierungskorrelationen in nichtkristallisierten molekularen Systemen mittels der elektrisch induzierten Doppelbrechung. Ph. D. Thesis, TH Darmstadt 1984.
[26] K. Ullrich, J. H. Wendorff: Orientation correlations in the isotropic state of low molecular weight and polymeric fluids, Mol. Cryst. Liq. Cryst. 313 (1985) 361–385.
[27] M. Eich, K. Ullrich, J. H. Wendorff: Investigations on pretransitional phenomena of the isotropic-nematic phase transition of mesogenic materials by means of electrically induced birefringence, Progr. Colloid Polymer Sci. 69 (1984) 94–99.
[28] D. Jungbauer, J. H. Wendorff, W. Kreuder, B. Reck, C. Urban, H. Ringsdorf: Ordered and disordered glasses: A comparison of thermodynamic and dynamical properties, Makromol. Chem. 189 (1988) 1345–1351.
[29] D. Jungbauer, J. H. Wendorff: Electro-optical studies on a monotropic nematic-smectic A transition, Makromol. Chem. Rapid Commun. 9 (1988) 165–169.
[30] B. Endres: Struktur und dynamische Eigenschaften von kombinierten Polymeren Haupt/Seitenketten-Flüssigkristalle, Ph. D. Thesis, TH Darmstadt 1988.
[31] H. Hirschmann: Elektro-optische Eigenschaften flüssig-kristalliner Polymerer: Seitenkettenpolymere mit longitudinaler und lateraler Ankopplung der Mesogenen, Masters Thesis, TH Darmstadt 1988.

14 Concentration Fluctuations and Kinetics of Phase Separation in Polymer Blends

Gert Strobl*

14.1 Introduction

Phase behaviour of polymer blends is controlled by two competing factors. The first one is the entropy of mixing which always supports miscibility. The second one is given by the local Gibbs free energy of the interacting segments. Usually it opposes miscibility and becomes favourable only in case of the presence of specific forces [1].

Driven by the thermal motion concentrations of the blend components fluctuate in space and time about the mean values. These fluctuations can be described as a superposition of statistically independent waves. They become excited by random thermal forces and decay with characteristic relaxation times τ_q depending on the wave vector q. The mean squared amplitudes of the concentration waves can be directly determined. They show up in the structure factor S(q) obtained in scattering experiments.

The relaxation times associated with the molecular dynamics in thermal equilibrium govern also the kinetics of structure changes after temperature jumps. Temperature jumps can be performed within the one-phase region or from the mixed state into the two-phase region. In the first case one observes a change in the amplitude of the concentration fluctuations. The reestablishment of thermal equilibrium occurs with the relaxation times effective at the new temperature. In the second case unmixing processes are induced. The initial mechanism of phase separation depends upon the location in the phase diagram. Near to the bimodal unmixing starts with a nucleation which is then followed by growth. This process needs an activation. No activation is required below the spinodal curve. Here the mixed state becomes intrinsically unstable against small concentration fluctuations, which grow in amplitude. Spinodal decomposition was first theoretically treated by Cahn and Hilliard [2] and Cook [3]. Original theories were devised for metal alloys and other low molec-

* Fakultät für Physik, Albert-Ludwigs-Universität, D-7800 Freiburg

ular weight blends. They were later adapted to the polymer case by de Gennes [4] and Binder [5]. Theories give relaxation times or growth rates as a function of q. They encompass the range of diffusive motions at low q as well as intramolecular relaxations at higher q's.

Experimental studies of the kinetics of relaxation and unmixing can be conducted using X-ray, light-, or neutron-scattering. After performing a temperature-jump relaxation or unmixing is followed by time-resolved scattering experiments. Near to the glass transition or at sufficiently low q kinetics becomes sufficiently slow to enable time-dependent measurements. The paper is organized as follows. It begins with a theoretical section, which deals in short terms with the Flory-Huggins theory of polymer phase behaviour and gives a summary of results concerning the equilibrium concentration fluctuations and the kinetics of relaxation and spinodal decomposition. In the experimental part time-resolved small angle X-ray scattering measurements are presented. They were performed after T-jumps within the one-phase region and also during the initial stages of spinodal decomposition.

14.2 Theory

14.2.1 Phase Behaviour of Polymer Blends

The phase behaviour of polymer mixtures is usually discussed on the basis of the Flory-Huggins theory. The Gibbs free energy of mixing is set up of two parts

$$\Delta G_{mix}/kT = -T\Delta S_t + \Delta G_{loc}. \tag{1}$$

Here ΔS_t describes the "translational" entropy of mixing of the polymer molecules given by

$$\Delta S_t/k = (\Phi/N_A)\ln\Phi + ((1-\Phi)/N_B)\ln(1-\Phi) \tag{2}$$

where Φ, $1\text{-}\Phi$ denote the volume fractions of the components A, B with degrees of polymerization N_A, N_B. ΔG_{loc} gives the change in the local Gibbs free energy which is approximated in the Flory-Huggins theory by a mean-field expression:

$$\Delta G_{loc} = \Phi(1-\Phi)\,\chi. \tag{3}$$

The parameter χ characterizes the local interaction per unit. According to Equ. (1) miscibility can be understood as being dependent on a balance of the

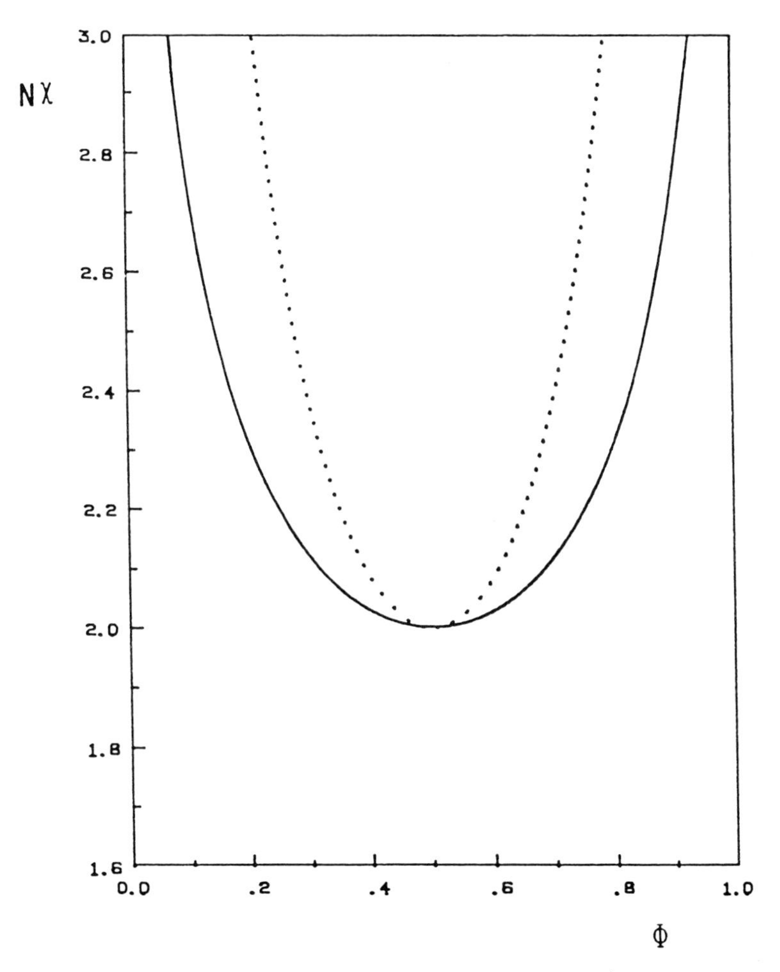

Fig. 14.1: (Φ, χ)-phase diagram of symmetric polymer mixture (N: degree of polymerization).

favourable translational entropy of mixing ΔS_t and the changes ΔG_{loc} in local interactions.

Starting from Equs. (1)–(3) a universal phase diagram in terms of χ and Φ can be derived. In the special case of a "symmetric" blend $N_A = N_B = N$ it shows the form given in Fig. 14.1; A miscibility gap occurs for $\chi > \chi_c = 2/N$;

for $\chi < \chi_c$ there is complete miscibility. The usual (Φ,T)-phase diagram follows from Fig. 14.1 by introducing the T-dependence of the interaction parameter (T). As discussed by Patterson [6], $\chi(T)$ can be set up by two contributions:

$$\chi(T) = \Delta H_{mix}/RT + C_p(T)\,\tau_{12}^2/2R \tag{4}$$

the first one being associated with the contact energy difference between unlike and like segments, i.e. the enthalpy of mixing ($\Delta H_{mix} < 0$ for exothermal mixtures), the second one following from the difference in thermal expansion, i.e. in "free volume", between the two components (τ_{12} is related to the difference in expansion coefficients). Fig. 14.2 shows in a schematic drawing the temperature dependence $\chi(T)$ for endothermal and exothermal mixtures. Miscibility gaps occur for $\chi > 2/N$. Hence, one expects for exothermal systems one miscibility gap at high temperatures (at a lower critical solution temperature, LCST), and for endothermal system ($\Delta H_{mix} > 0$) two miscibility gaps, one at low (UCST, "upper critical solution temperature") and another at high temperatures. Of course, this behaviour describes the ideal case. In reality, freezing of structures at the glass transition or decomposition at high temperatures in many cases prevents actual observation of a theoretically anticipated mixing or unmixing.

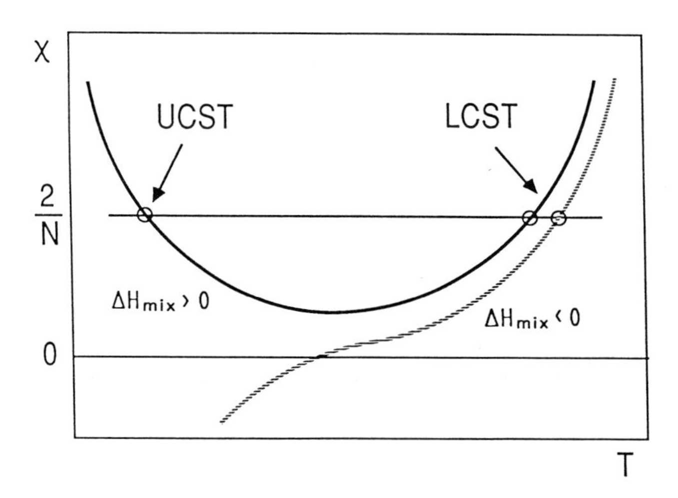

Fig. 14.2: T-dependence of interaction parameter χ for exothermal and endothermal mixtures.

14.2.2 Concentration Fluctuations in Thermal Equilibrium

The mean squared amplitudes $<\delta \Phi_q{}^2>$ of concentration fluctuations with wave vector q which are equal to the structure factor S(q) can be described by de Gennes' RPA (random phase approximation) – expression (Ref. [7]):

$$S^{-1}(q = (4\pi\sin\theta)/\lambda) = \frac{1}{N\Phi} f_D^{-1}(q^2R_A^2) + \frac{1}{N(1-\Phi)} f_D^{-1}(q^2R_B^2) - 2\chi \quad (5)$$

Equ. (5) refers again to a symmetric polymer mixture ($N_A = N_B = N$). f_D denotes the Debye-structure factor of ideal noninteracting chains:

$$f_D = \frac{2}{q^2R^2} (1 - (1 - \exp(-q^2R^2)/q^2R^2). \quad (6)$$

If the chain sizes R_A and R_B are not too different, Equ. (5) may be replaced by the shorter expression

$$S^{-1}(q) = \frac{1}{N\Phi(1-\Phi)} f_D^{-1}(q^2R_\Phi^2) - 2\chi \quad (7)$$

where R_Φ denotes a mean radius of gyration defined as

$$R_\Phi^2 = (1 - \Phi) R_A^2 + \Phi R_B^2. \quad (8)$$

After a thermal excitation concentration waves decay with characteristic q-dependent relaxation times

$$\delta\Phi_q(t) \sim \exp - t/\tau_q. \quad (9)$$

Generally relaxation times decrease with increasing q. Different expressions have been derived for τ_q, the latest result being that obtained by Binder (Ref. [5]):

$$\tau_q^{-1} \simeq \frac{6}{\tau_R} q^2R^2 \left(1 - \frac{\chi}{\chi_S} f_D (q^2R^2)\right). \quad (10)$$

Here τ_R denotes the time required by a chain molecule to diffuse a distance comparable to its own size; χ_S gives the interaction parameter at the spinodal curve (compare Fig. 14.1)

$$\chi_S = \frac{1}{2\,N\Phi(1-\Phi)}\,. \tag{11}$$

Relaxation kinetics can be studied by temperature jump experiments. A change of temperature from T_0 with $\chi = \chi_0$ to T_f with $\chi = \chi_f$ leads to a change of the equilibrium structure factor from

$$S_{\chi_0}(q) = \frac{1}{N\Phi(1-\Phi)}\,f_D^{-1}\,(q^2R^2) - 2\,\chi_0\,. \tag{12}$$

to

$$S_{\chi_f}(q) = \frac{1}{N\Phi(1-\Phi)}\,f_D^{-1}\,(q^2R^2) - 2\,\chi_f\,. \tag{13}$$

The kinetics of transition is governed by a simple first order equation of motion (Ref. [5]):

$$\frac{dS}{dt}\,(q,t) = -\,2\tau_q^{-1}\,(S(q,t) - S_{\chi_f}(q))\,. \tag{14}$$

Analysis of time-resolved scattering experiments based upon Equ. (14) enables a determination of relaxation times.

14.2.3 Kinetics of Unmixing

Transfer of a polymer blend from a temperature in the one-phase region to temperatures within the miscibility gap induces phase separation. Published work so far is mainly concerned with the case of spinodal decomposition. For temperatures within the region enclosed by the spinodal curve wave-like concentration fluctuation with q-values below a critical q_o do not decay as in thermal equilibrium, but become unstable and increase in amplitude exponentially with time.

Proceeding from the theoretical treatment of Binder [5] which includes the effect of random thermal forces, it has been shown [8] that structure evolution during the initial stages of spinodal decomposition is governed by the same relaxation equation as that valid for the T-jump experiment within the one-phase region, namely Equ. (14). Only the meaning of the function S_χ (q) is altered. It is still defined by Equ. (13), but represents now a "virtual" structure factor rather than a true equilibrium structure factor. S_χ (q) shows negative values for q's below a critical q_c. Equ. (10), giving the q-dependence of relaxation times, remains valid as well. For points below the spinodal curve, it is $\chi/\chi_s > 1$. Hence, relaxation rates become negative for small q's, again for q below q_c. Negative relaxation rates indicate exponential growth rather than a decay of concentration waves:

$$\delta\Phi_q \sim \exp Rt \tag{15}$$

with a growth rate $R(q) = -\tau_q^{-1} > 0$. The q-dependence of R, as given by Equ. (10), shows a maximum around $q \simeq q_c/\sqrt{2}$. The corresponding concentration wave dominate structure formation during the early stages of spinodal decomposition.

Equs. (14), (13) and (10) provide a complete description of the kinetics of relaxation and unmixing. Model calculations have been performed [8]. Figs. 14.3 and 14.4 show the evolution of the structure factor during spinodal decomposition calculated for two quenching experiments. Starting in the one-phase region at $N\chi = 1.0$, quenching goes to a point near to spinodal ($N\chi = 2.05$, Fig. 14.3) or to a point more far away ($N\chi = 2.5$, Fig. 14.4). As expected, structure formation is much slower near to the spinodal and dominated by concentration waves with lower q.

Experimental studies of unmixing are restricted by the given limits in time resolution. Structure formation has to be rather slow and should occur on the time scale of minutes. This can be accomplished for two different situations:

(I) Studies can be conducted near to the critical temperature. Here diffusion becomes sufficiently slow due to the small thermodynamic driving force ("critical slowing down"). In order to observe the dominating concentration fluctuations with long wavelengths, light scattering has to be applied (Refs. [9–12] give some examples).

(II) If samples can be quenched to temperatures near T_g kinetics become again slow, now as a result of the low mobility. Here the dominating waves are located at higher q's and can be studied by X-rays. At higher q's relaxation corresponds to intramolecular conformational changes rather than to a diffusion of whole molecules. In the following examples for the second case are presented.

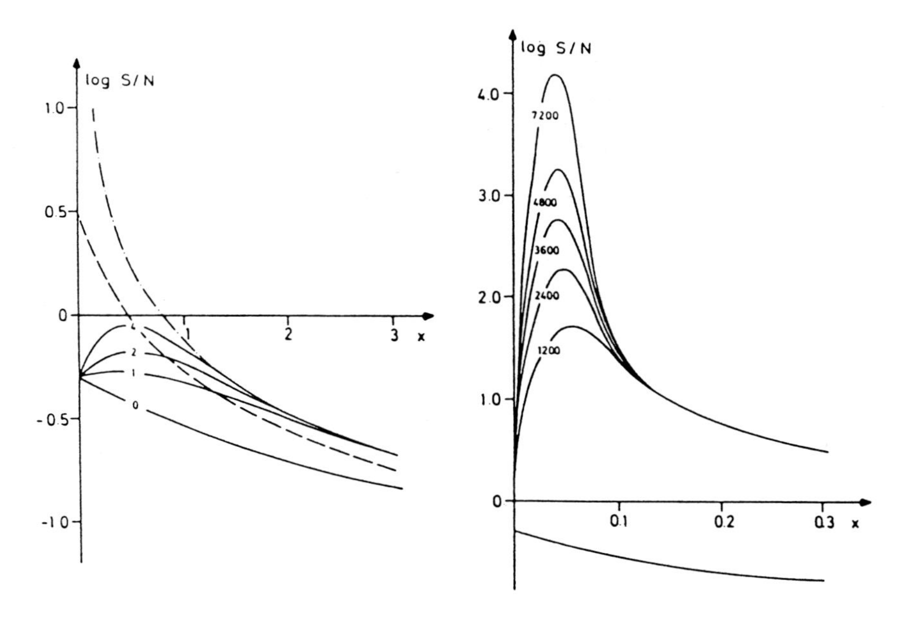

Fig. 14.3: Evolution of structure factor after a quench $N\chi = 1.0 \rightarrow N\chi = 2.05$ (model calculation; times given in units $\tau_R/6$).

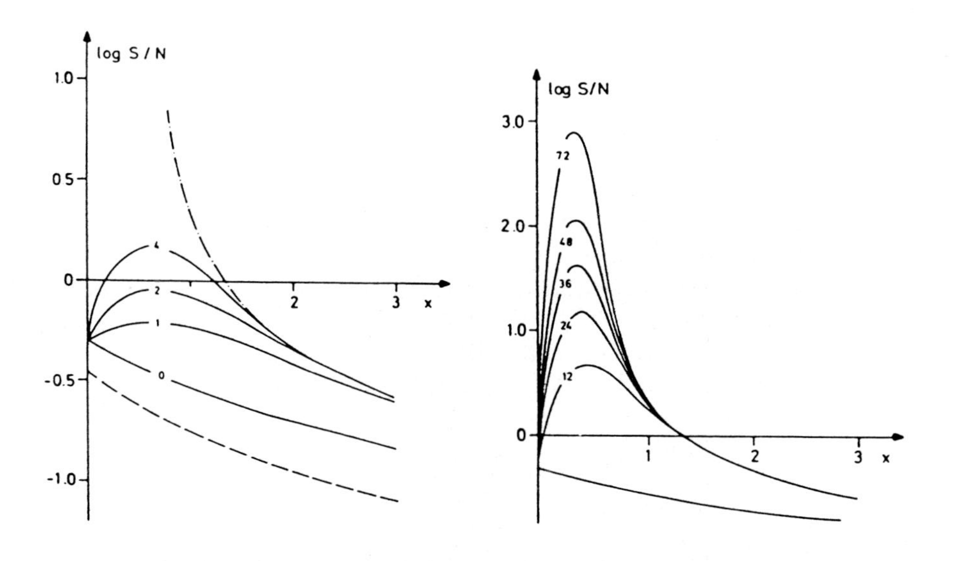

Fig. 14.4: Evolution of structure factor after a quench $N\chi = 1.0 \rightarrow N\chi = 2.5$.

14.3 Small Angle X-Ray Scattering Studies

14.3.1 Polystyrene/Poly(styrene-co-bromostyrene) Blends

X-ray scattering experiments were performed on mixtures of polystyrene and poly(styrene-co-bromostyrene). This is an endothermal mixture ($\Delta H_{mix} > 0$) which can exhibit a lower miscibility gap [13]. The location of the UCST can be continuously shifted and by that conveniently adjusted through the choice of the molecular weight and the degree of bromination x_B. For symmetric blends with an identical degree of polymerization N for both components, as they can be prepared from a polystyrene and its partially brominated derivative, the critical temperature T_c depends upon the product Nx_B^2, increasing with N or x_B^2. For a molecular weight $M \approx 20.000$ ($N \approx 200$) and degrees of bromination $x_B =$ 0.30–0.35 T_c is located in the range 80 °C–230 °C, below those temperatures where chemical decomposition sets in. Quenching an originally homogeneous sample from the one-phase region to temperatures below the binodal induces phase separation. Depending upon the composition, different patterns are observed. Figs. 14.5 and 14.6 present two optical micrographs, one exhibiting spherical precipitates ($\Phi(PS) = 0.38$), the other ($\Phi(PS) = 0.52$) showing interconnected domains. The latter structure suggests phase separation by spinodal decomposition. The phase separation turned out to be thermally reversible.

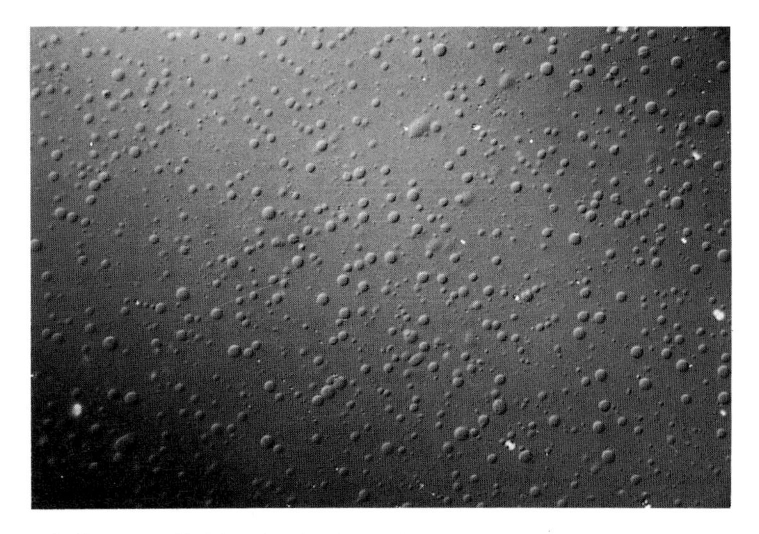

Fig. 14.5: PS/P(S-co-BrS)-blends. Optical micrograph showing spherical precipitates.

Fig. 14.6: PS/P(S-co-BrS)-blends. Optical micrograph showing interconnected domains.

SAXS-experiments were conducted on two mixtures, one using a copolymer with $x_B = 0.31$ ("CPI", $T_c \approx 110\,°C$) the other using a copolymer with $x_B = 0.33$ ("CPII", $T_c \approx 155\,°C$). Measurements were performed with the help of a Kratky-camera. Absolute intensities were obtained after applying a desmearing procedure and determining the primary beam intensity using a standard sample.

14.3.2 Static Structure Factors

Fig. 14.7 presents the static structure factors measured for PS/CPI (0.5/0.5)-mixtures at a series of temperatures between 130 °C and 190 °C. Mixtures were homogeneous throughout this temperature range. Fig. 14.8 gives the corresponding Zimm-plots I^{-1} versus q^2.

As expected, decreasing the temperature, i.e. approaching the spinodal, leads to a general increase of intensities. Fig. 14.8 shows that the changes can be described as a result of the increase in the interaction parameter χ. Curves $I^{-1}(q^2)$ appear to be shifted along the abszissa, the slope remaining essentially constant, just as predicted by Equ. (7). Similar results were also obtained by other authors, at first by Shibayama et al. [14] in a neutron scattering experiment on PS/PVME – blends.

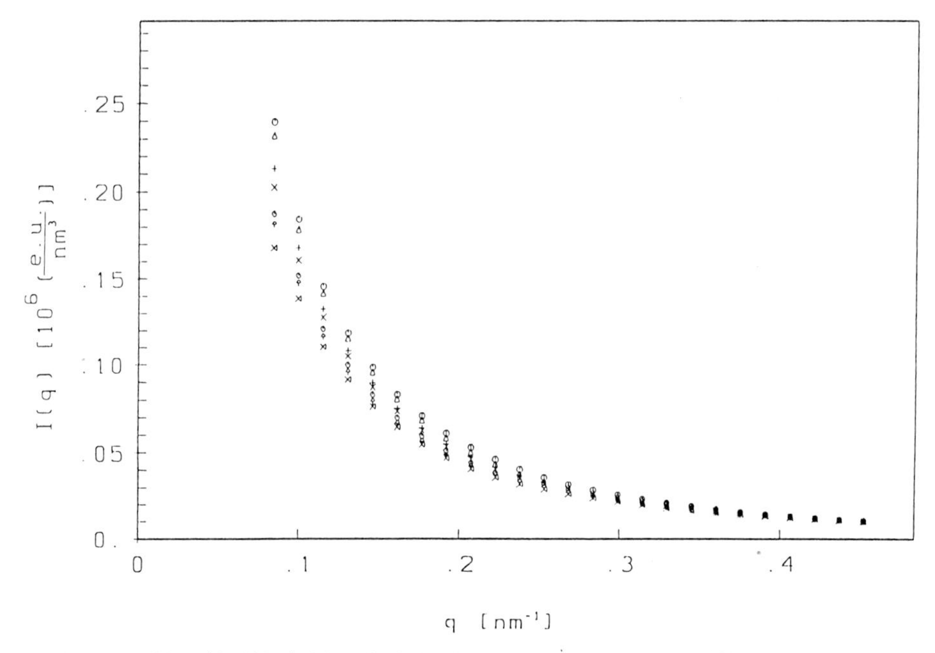

Fig. 14.7: PS/CPI(0.5/0.5)-blend. SAXS-curve measured at different temperatures between 130 °C and 190 °C in 10 °-steps.

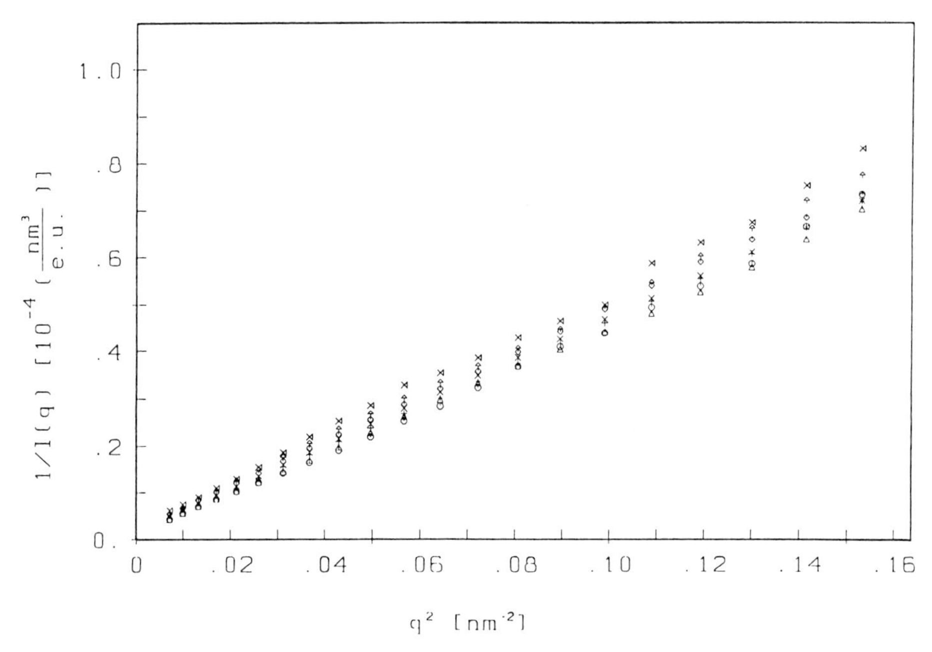

Fig. 14.8: Zimm-plot of curves shown in Fig. 14.7.

Curves can be fitted to the RPA-expression Equ. (7). Fig. 14.9 was obtained for the structure factor at 150 °C and demonstrates, that the representation is satisfactory. By application of the equation

$$I(q) = (\Delta\eta)^2 V_z\, S(q), \tag{16}$$

which relates measured absolute scattering intensities I(q) with the structure factor S(q) of the theory ($\Delta\eta$ denotes the electron density difference between the blend components; V_z specifies the volume of the lattice cell, arbitrarily chosen as being equal to the volume occupied by one styrene unit), the mean radius of gyration, R_Φ, can be derived. One obtains a value $R_\Phi^2 = 16.1\ nm^2$, which indicates a considerable expansion of the brominated polymer compared to polystyrene ($R^2_{PS} = 11\ nm^2$).

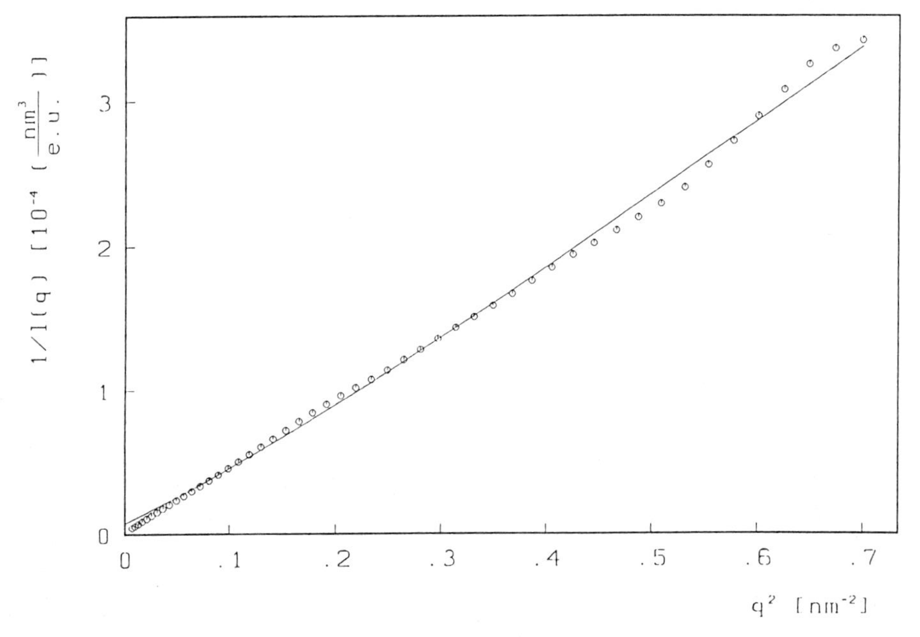

Fig. 14.9: Fit of structure factor by formula given by random phase approximation (data obtained at 150 °C).

14.3.3 Relaxation after T-Jumps within the One-Phase Region

Kinetics of relaxation was investigated by quenching samples, which were first kept at 190 °C, to temperatures near to T_g. The transition from the structure factor S_{χ_o} ($\chi_o = (T = 190\,°C)$) to the new equilibrium values was followed in time-resolved SAXS-experiments [15]. Figs. 14.10 and 14.11 show the results of measurements performed at T = 117 °C and 123 °C. Intensities grow with time for all q's. Structure relaxation occurs on the time scale of minutes, with relaxation times, which increase with decreasing annealing temperature.

According to theory change of structure factors during relaxation should be controlled by the equation of motion Equ. (14). For the initial condition

$$S(q, t = 0) = S_{\chi_o}(q)$$

its solution is given by

$$S(q, t) - S_{\chi_f}(q) = (S_{\chi_o}(q) - S_{\chi_f}(q))\exp - 2t/\tau_q \tag{17}$$

or

$$\ln (S_{\chi_f}(q) - S(q,t)) / (S_{\chi_f}(q) - S_{\chi_o}(q))$$

$$= \ln \Delta S(q,t) / \Delta S(q,0) = - 2\, \tau_q^{-1}\, t. \tag{18}$$

Here S_{χ_f} denotes the equilibrium structure factor at the annealing temperature as obtained through an extrapolation of the static curves. Plots according to Equ. (18) using the data of Figs. 14.10 and 14.11 are shown in Figs. 14.12 and 14.13. Obviously Equ. (17) is invalid. Relaxation does not follow a simple exponential law. Behaviour is clearly nonlinear. Kinetics become slower with increasing time. Equilibrium structure factors are not reached even after long annealing times. Deviations of the limiting values observed after long times from the equilibrium values increase with decreasing q and decreasing temperature. In the given situation further analysis of kinetic data can only be based on the initial slopes of the relaxation curves in Figs. 14.12 and 14.13 which yield "initial relaxation times". The results of the evaluation are presented in Fig. 14.14 which includes also a measurement at 120 °C. The solid lines are fits of the data using Equ. (10). The quality of the fits appears satisfactory. The values derived for the basic relaxation time τ_R are

T = 117 °C : 350 s
T = 120 °C : 150 s
T = 123 °C : 60 s.

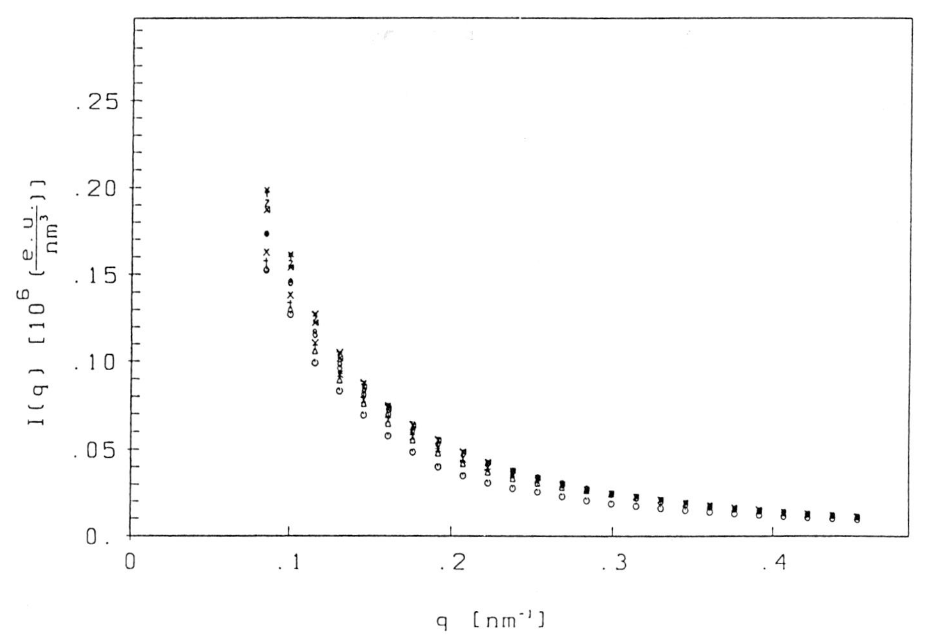

Fig. 14.10: PS/CPI(0.5/0.5)-blend. Time dependence of structure factor after T-jump from 109 °C to 117 °C. Curves measured initially (o), and after 150s (△), 350s, 600s, 900s, 1500s, 2700s, 4500s, 8900s, 16900s (x).

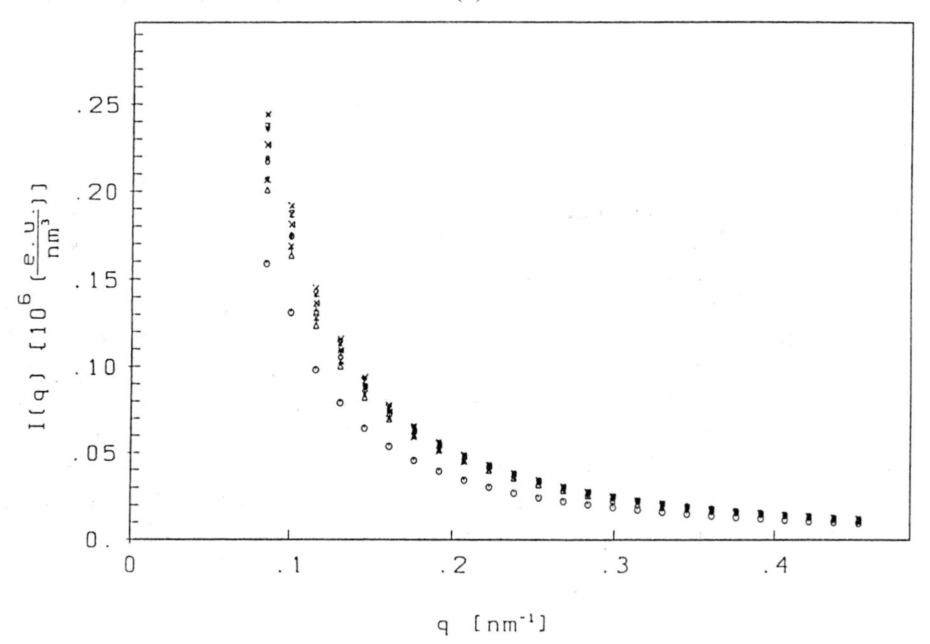

Fig. 14.11: Time dependence of structure factor after T-jump from 190 °C to 123 °C. Times of measurements as in Fig. 14.10.

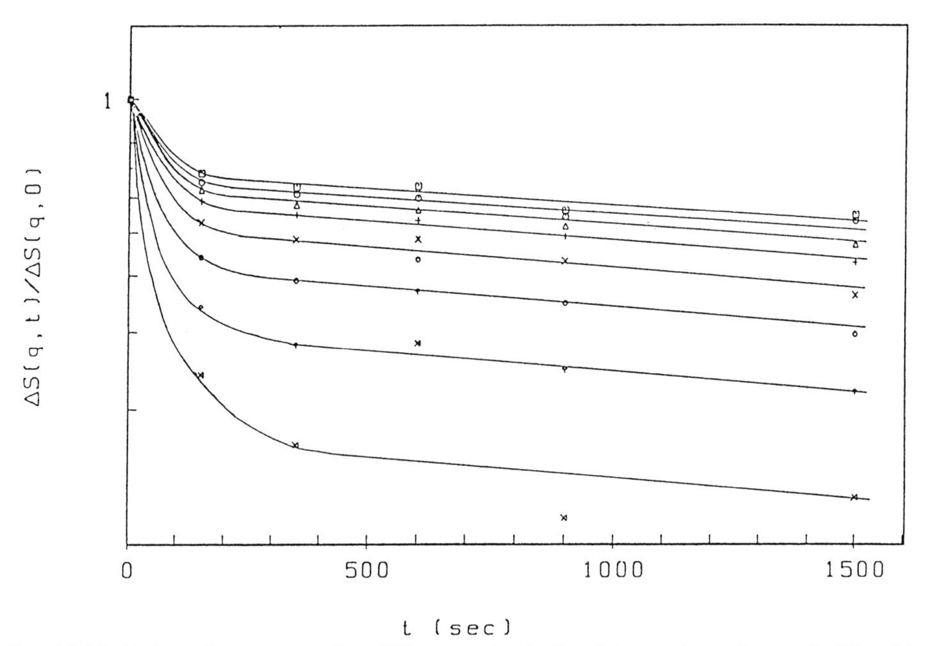

Fig. 14.12: Relaxation curves for different q's derived from data shown in Fig. 14.10 (117 °C). q = 0.84 (□), 0.98, 1.15, 1.30, 1.46, 1.61, 1.76, 1.92 (▷◁) 10^{-1} nm^{-1}.

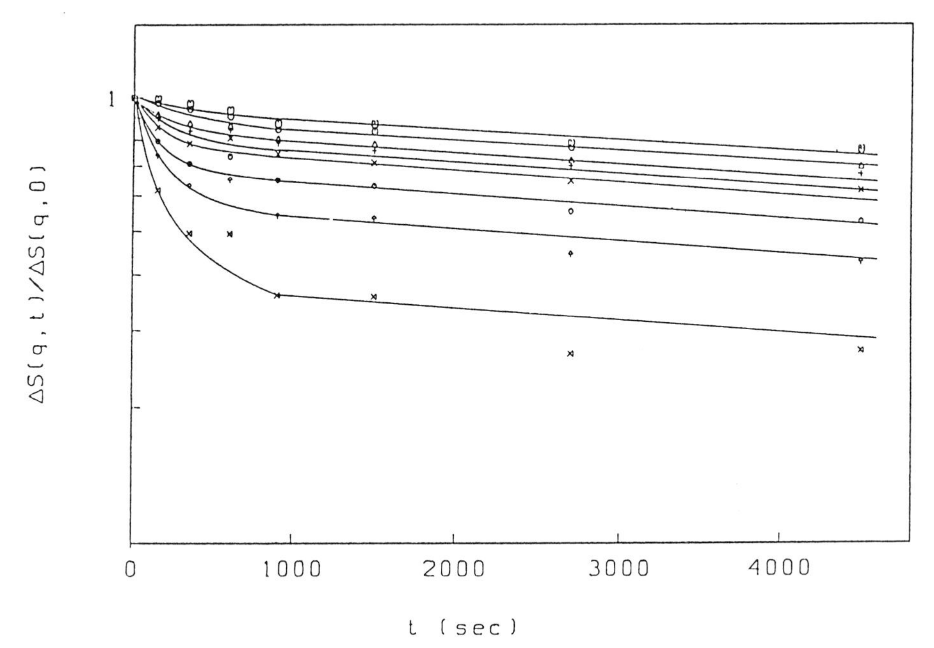

Fig. 14.13: Relaxation curves for different q's derived from data shown in Fig. 14.11 (123 °C); q-values as in Fig. 14.12.

The dependence on temperature is nonlinear, as it is qualitatively expected at temperatures near to T_g.

It is certainly the adjacency of T_g, which leads to the observed deviations from an exponential relaxation behaviour. We shall come back to this point later on, reporting first observations with similar character in a study of spinodal decomposition conducted in the same temperature region.

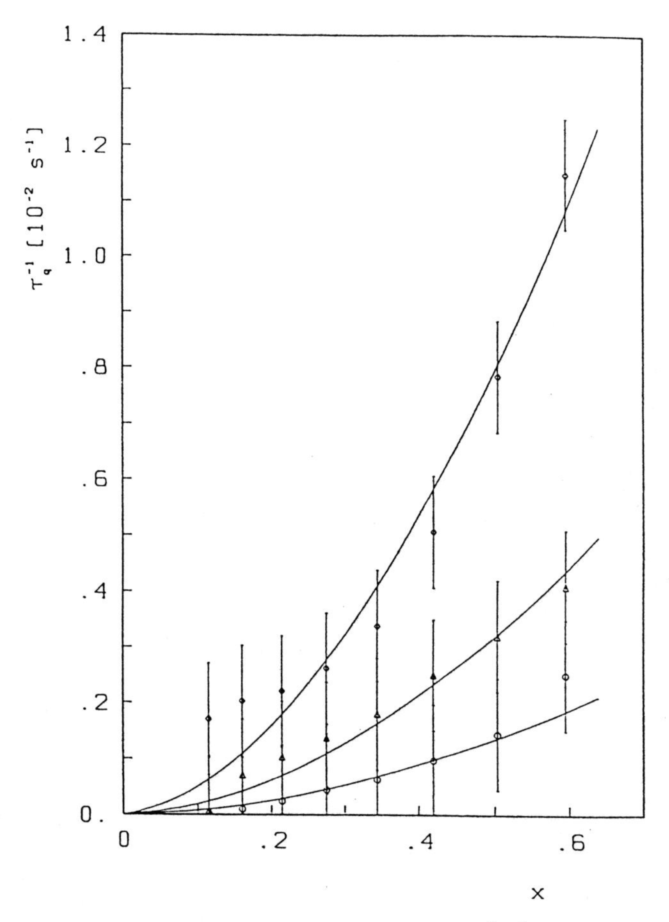

Fig. 14.14: Initial relaxation rates dependent upon $x = q^2 R^2$ for T = 117 °C (0), 120 °C (△) and 123°C (◇). Fit to theoretical result Equ. (10).

14.3.4 Kinetics of Spinodal Decomposition

A PS/CPII(0.53/0.47)-mixture with a critical temperature around 155 °C was used for a study of spinodal decomposition [16]. A sample was quenched from the one-phase region (170 °C) to 124 °C and the structure evolution during phase separation followed by time-resolved SAXS-experiments. Fig. 14.15 presents the observations. During the initial stages of development (Fig. 14.15A) an increase of the intensity is observed for all wave vectors q. For longer times of annealing the tendency changes. After passing over a maximum intensities decrease with time (Fig. 14.15B). Parts C and D of Fig. 14.15 depict this behaviour using Zimm plots $I^{-1}(q^2)$. First, the scattering intensities tend toward the virtual structure factor S_χ. It is given by the broken line in Fig. 14.15C and has been obtained by an appropriate extrapolation on the basis of measured static structure factors. Fig. 14.15D includes in addition to the initial curve (a) and the curve for t = 4 x 10^3s, where scattering intensities

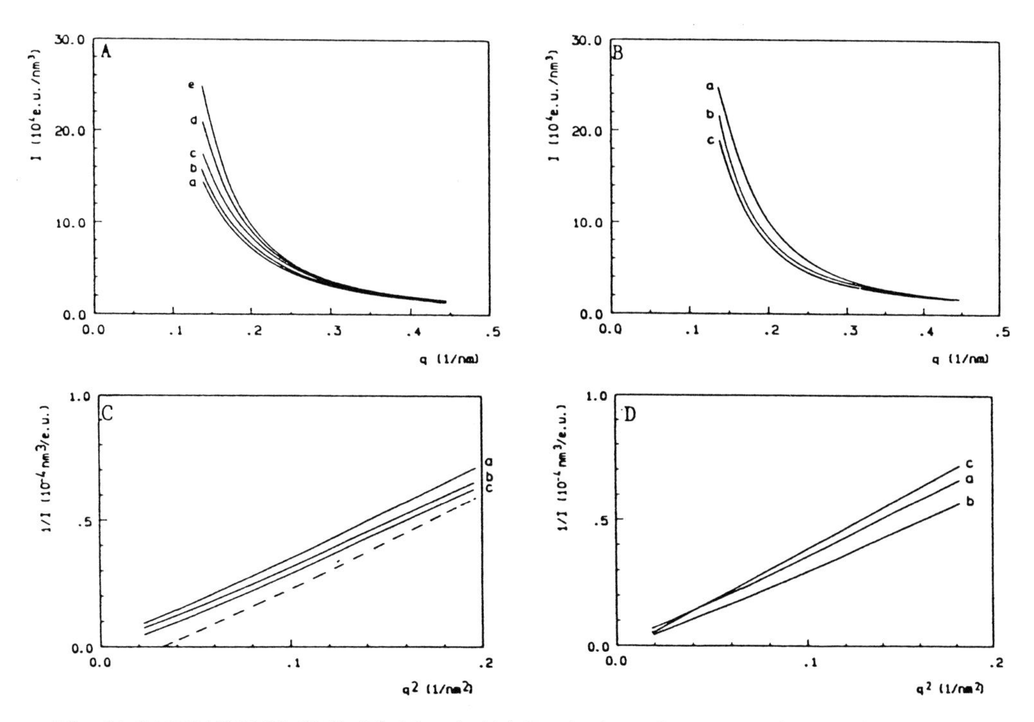

Fig. 14.15: PS/CPII(0.53/0.47)-blend. (A) Evolution of structure factor after a T-jump from 170 °C (one-phase region) to 124 °C (two-phase region). Curve a shows the original scattering curve (170 °C). Annealing times curve b, 10^2; curve e, 4 x 10^3. (B) Structure factor after 4 x 10^3s (curve a) and 5.8 x 10^4s (curve c). (C) Time dependence of reciprocal structure factor s. Broken line indicates virtual structure factor. (D) Time dependence of reciprocal structure factor: curve a, Os; curve b, 4 x 10^3s; curve c, 10^5s.

reached their maxima (b), the Zimm plot of the final scattering curve of the equilibrium two-phase structure (c). It was measured after an annealing time of 27.8 h.

The general behaviour corresponds to that anticipated from theory. During the initial stages of structure development an increase of intensities is observed for all wave vectors q. As expected from the equation of motion, Equ. (14), and its general solution, Equ. (17), the scattering curve tends with time toward the virtual structure factor. This becomes particularly clear in the Zimm plots of Fig. 14.15C. Validity of the equation of motion Equ. (14) is restricted to the early stages of structure formation. The later stages are controlled by ripening processes which finally lead to the new two-phase equilibrium. Therefore the initial increase of the scattering intensities has to be followed by a decrease. The scattering curve measured for the two-phase equilibrium at 124 °C is a superposition of the structure factors of the two individual phases.

For a derivation of relaxation times, Equ. (18) can be used as in the previous chapter. Fig. 14.16 presents a corresponding plot. Again, as in the case of the T-jump experiments performed within the one-phase region, plots are curved rather than straight, indicating nonexponential relaxation behaviour.

In fact, the nonlinearity of relaxation behaviour as it was observed in both experiments does not come unexpected. Experiments were conducted near to the glass transition region. The glass (α-) relaxation itself, when observed in mechanical or dielectric experiments, is always nonlinear, decays being described by the Kohlrausch-Williams-Watts formula rather than by a simple

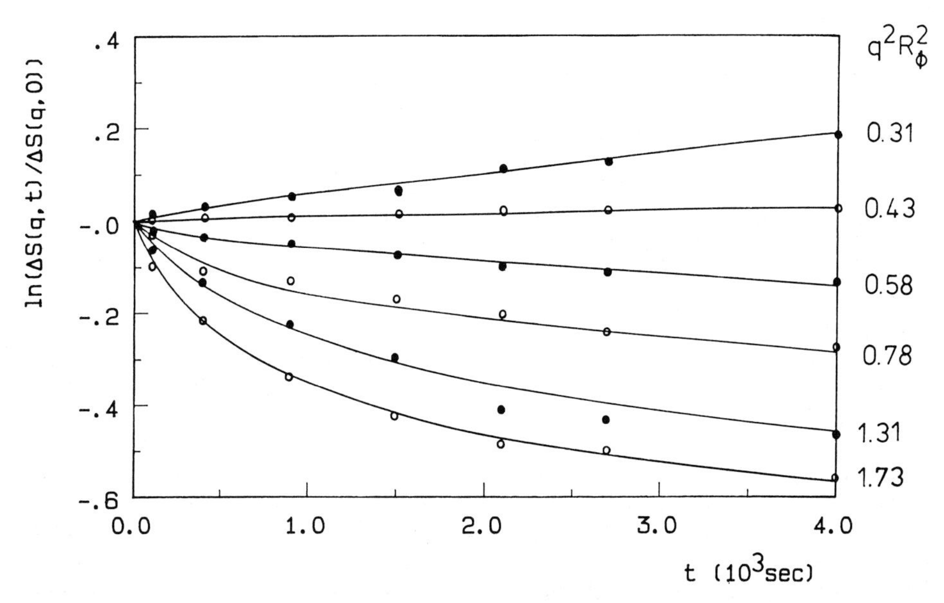

Fig. 14.16: Relaxation curves for different q's derived from data in Fig. 14.15.

exponential law. If the χ-parameter depends upon an internal structure variable like the free volume, fluctuations and relaxation of concentrations and internal structure variables become coupled. Then both would behave nonexponentially. Even if the intrinsic nonexponentiality of the α-process is not directly carried over to the concentration fluctuations, coupling leads to a superposition of two combined relaxation modes with different relaxation times. At short times the quick interdiffusion process dominates, at long times the slow volume relaxation becomes visible. A theoretical model describing this behaviour has been worked out recently by Jäckle and Pieroth [17]. Furthermore, mobilities could also depend on some slowly relaxing internal structure variable, which would also result in a stretching of relaxation curves. At the present state of knowledge, a discrimination between the different possible effects appears unfeasible.

14.4 References

[1] D. R. Paul, S. Newman (eds.): Polymer Blends, Academic Press, New York 1978, Vol. 1, Chs. 1–3.
[2] J.W. Cahn: Metastability, instability, and the dynamics of unmixing in binary critical systems, Trans. Metall. Soc. AIME 242 (1968) 166.
[3] H. Cook: Brownian motion in spinodal decomposition, Acta Metall. 18 (1970) 297.
[4] P. G. de Gennes: Dynamics of fluctuations and spinodal decomposition in polymer blends, J. Chem. Phys. 72 (1980) 4756.
[5] K. Binder: Collective diffusion, nucleation, and spinodal decomposition in polymer mixtures, J. Chem. Phys. 79 (1983) 6387.
[6] D. Patterson, A. Robard: Thermodynamics of polymer compatibility, Macromolecules 11 (1978) 690.
[7] P. G. de Gennes: Scaling Concepts in Polymer physics. Cornell University Press, Ithaca/N.Y. 1979, Ch.4.
[8] G. R. Strobl: Structure evolution during spinodal decomposition of polymer blends, Macromolecules 18 (1985) 558.
[9] J. Gilmer, N. Goldstein, R. S. Stein: Light-scattering studies of phase-demixing of polystyrene/poly(o-chlorostyrene) blends, J. Polym. Sci., Polym. Phys. Ed. 20 (1982) 2219.
[10] H. C. Snyder, P. Meakin, S. Reich: Dynamical aspects of phase separation in polymer blends, Macromolecules 16 (1983) 757.
[11] R. Hill, P. Tomlins, J. S. Higgins: A preliminary study of the dynamics of phase separation in oligomeric polystyrene-polybutadiene blends, Polymer 26 (1985) 1708.
[12] T. Izumitani, T. Hashimoto: Slow spinodal decomposition in binary liquid mixtures of polymers, J. Chem. Phys. 83 (1985) 3694.

[13] G. R. Strobl, J.T. Bendler, R. P. Kambour, A. R. Shultz: Thermally reversible phase separation in poly(styrene-co-bromostyrene) blends, Macromolecules 19 (1986) 2683.
[14] M. Shibayama, H. Yang, R. S. Stein, C. Han: Study of miscibility and critical phenomena of deuterated polystyrene and hydrogenated poly (vinyl methyl ether) by small-angle neutron scattering, Macromolecules 18 (1985) 2179.
[15] G. R. Strobl, G. Urban: Structure relaxation after temperature jumps in homogeneous polystyrene/poly(styrene-co-bromostyrene) blends, Colloid Polymer Sci. 266 (1988) 398.
[16] H. Meier, G. R. Strobl: Small-angle X-ray scattering study of spinodal decomposition in polystyrene/poly(styrene-co-bromostyrene) blends, Macromolecules 20 (1987) 649.
[17] J. Jäckle, M. Pieroth: Theory of structural evolution in viscoelastic binary liquid mixtures after a temperature jump, Z. Physik B (1988).

15 Collective Dynamics in Polymeric Liquids as Measured by Quasi-elastic Light- and Neutron Scattering

Erhard W. Fischer, Bernd Ewen, and Gerhard Meier*

15.1 The Dynamics of Density Fluctuations as Measured by Quasi-elastic Light Scattering

15.1.1 The Theory of Light Scattering from an Isotropic Viscoelastic Medium

The scattering of light in a macroscopic isotropic medium arises from local fluctuations $\delta\underline{\underline{\varepsilon}}(r,t)$ of the tensor of the dielectrical constant around its mean value $\underline{\underline{\varepsilon}}$. By means of the photon correlation spectroscopy we measure the time correlation function of this quantity. The scattered electric field $E_f\,(R, \vec{q}\,,t)$ at a large distance R from the scattering volume V depends on the scattering vector $\vec{q} = \vec{k}_i - \vec{k}_f$, where $\vec{k}_i$, $\vec{k}_f$ denotes initial and final wave vectors respectively. For light scattering $|\vec{k}_i| \approx |\vec{k}_f|$ and therefore $|\vec{q}| = \frac{4\pi n}{\lambda} \sin\frac{\theta}{2}$. As usual, the spatial Fourier transformation of the fluctuations of ε is defined by [1]

$$\delta\underline{\underline{\varepsilon}}(\vec{q},t) = \frac{1}{V}\int d^3r \exp\,(i\vec{q}\vec{r})\delta\underline{\underline{\varepsilon}}(\vec{r},t) \tag{1}$$

* Max-Planck-Institut für Polymerforschung, Postfach 3148, D-6500 Mainz

Taking into account the different tensor components we can write the time autocorrelation function of the scattered electrical field $E_f(R,t)$

$$<E_f(R,0,\vec{q})E_f^*(R,t,\vec{q})>$$

$$= \frac{k_f^4 |E_0|^2}{(4\pi R\varepsilon_0)^2} <\delta\varepsilon_{if}(-\vec{q},0)\delta\varepsilon_{if}(\vec{q},t)> \exp i\omega t \qquad (2)$$

Where i,f denotes initial and final respectively. Now several cases can be discussed whether the polarization is parallel or perpendicular to the z-axis. For convenience only the case $\delta\varepsilon_{zz}(\vec{q},t)$ which gives the so-called VV scattering (polarized scattering) is considered.

We would now like to connect the correlation function of scattered light – related to the correlation function of fluctuations of the dielectric constant of the material – with dynamic mechanical properties. In order to do so [2], we consider a system being in thermodynamic equilibrium which is optically isotropic. We start with the definition of the autocorrelation function of the components of the dielectric tensor.

$$G_{\alpha\beta}(\vec{q},t) = < \delta\varepsilon_{\alpha\beta}(\vec{q},t)\delta\varepsilon_{\alpha\beta}(-\vec{q},0)>$$

$$= \frac{1}{V^2}\int dV \int dV' \exp(-i\vec{q}(\vec{r}-\vec{r}')<\delta\varepsilon_{\alpha\beta}(\vec{r},t)\delta\varepsilon_{\alpha\beta}(\vec{r}',0)> \qquad (3)$$

r,r' are different positions in the scattering volume and α,β can be the axis of a cartesian coordinate system. In the solid medium we can describe $\delta\underline{\underline{\varepsilon}}(\vec{q},t)$ by a symmetric local deformation tensor:

$$\gamma_{\alpha\beta}(\vec{r},t) = \frac{1}{2}\left[\frac{\partial u_\alpha(\vec{r},t)}{\partial r_\beta(\vec{r},t)} + \frac{\partial u_\beta(\vec{r},t)}{\partial r_\alpha(\vec{r},t)}\right] \qquad (4)$$

with $u(\vec{r},t)$ being the displacement vector. For isotropic systems we have

$$\delta\underline{\underline{\varepsilon}}(\vec{r},t) = a_1\gamma_{iso}(\vec{r},t)\delta_{\alpha\beta} + a_2\widetilde{\gamma}_{\alpha\beta}(\vec{r},t) \qquad (5)$$

with

$$\gamma_{iso} = \frac{1}{3}\,\mathrm{Tr}\gamma(\vec{r},t) \quad \text{and} \quad \widetilde{\gamma}_{\alpha\beta} = \gamma_{\alpha\beta} - \frac{1}{3}\,\mathrm{Tr}\gamma\delta_{\alpha\beta}$$

with $\delta_{\alpha\beta}$ being the Kronecker δ.

The proportionality factor a_1 describes the change of ε with respect to a longitudinal compression whereas a_2 accounts for the change due to shear. By this manipulation one sees that we have split the deformation tensor into components along the diagonal which give rise to the isotropic part and off-diagonal elements which give rise to anisotropy.
It is now more convenient to fouriertransform Equ. (5) into $\vec{q}$ space to get

$$\delta\varepsilon_{\alpha\beta}(\vec{q},t) = a_1\gamma_{iso}(\vec{q},t)\delta_{\alpha\beta} + a_2\widetilde{\gamma}_{\alpha\beta}(\vec{q},t) \tag{6}$$

With this result we go back into Equ. (3) and reach after some tensor algebra the result

$$G_{\alpha\beta}(\vec{q},t) = a_1^2 < \gamma_{iso}(\vec{q},t)\gamma_{iso}(-\vec{q},0) > \delta_{\alpha\beta}{}^2$$

$$+ a_2^2 < \widetilde{\gamma}_{\alpha\beta}(\vec{q},t)\widetilde{\gamma}_{\alpha\beta}(-\vec{q},0) > + \text{ terms being zero} \tag{7}$$

Now we define to be

$$G_{iso}(\vec{q},t) = a_1^2 < \gamma_{iso}(\vec{q},t)\gamma_{iso}(-\vec{q},0) > \tag{8a}$$

and

$$G_{VH}(\vec{q},t) = a_2^2 < \widetilde{\gamma}_{\alpha\beta}(\vec{q},t)\widetilde{\gamma}_{\alpha\beta}(-\vec{q},0) > \tag{8b}$$

Then one can show after a lengthy mathematical manipulation that

$$G_{if}(\vec{q},t) = G_{iso}(\vec{q},t)\delta_{if} + \left(1 + \frac{1}{3}\delta_{if}\right)G_{VH}(\vec{q},t) \tag{9}$$

with $i = V$ and $f \in (V,H)$.

What is calculated now first is the quantity $\gamma_{iso}(\vec{q},t)$. We recall from continuum mechanics that

$$\mathrm{Tr}\gamma_{\alpha\beta}(\vec{q},t) = \frac{\Delta\varrho(\vec{q},t)}{\varrho_0} = \frac{-\Delta V}{V_0} \tag{10}$$

which gives for $G_{iso}(\vec{q},t)$:

$$G_{iso}(\vec{q},t) = \frac{1}{9\,\varrho_0{}^2}\left[\left(\frac{\partial\varepsilon}{\partial\varrho}\right)_T\right]^2 <\delta\varrho(\vec{q},t)\delta\varrho(-\vec{q},0)> \tag{11}$$

Thus the isotropic part of the correlation function for the dielectric tensor is related to the autocorrelation function of density fluctuations, as well known. Now we are interested in the question, which thermodynamical and mechanical quantities are related to the density fluctuation correlation function defined for our purpose to be

$$C(\vec{q},t) = \langle\delta\varrho(\vec{q},0)\delta\varrho(-\vec{q},t)\rangle \tag{12}$$

For that reason we introduce the spectral power density of this correlation

$$\Lambda(\vec{q},\omega) = \int_0^\infty e^{i\omega t}C(\vec{q},t)dt \tag{13}$$

and a density susceptibility $\chi(\vec{q},\omega)$ which describes the response of the density $\delta\varrho$ with regard to an external potential $\delta\varphi_e(\vec{q})$ defined by

$$\delta\varphi_e(\vec{q}) = \left[\frac{1}{2\pi}\right]^3 \frac{1}{V}\int_V \varphi_e(\vec{r})e^{-i\vec{q}\vec{r}}\,d\vec{r} \tag{14}$$

The relationship between $\delta\varrho$ and $\delta\varphi$ for the static case is defined by

$$\delta\varrho(\vec{q}) = V\chi(\vec{q},\omega=0)\cdot\delta\varphi_e(\vec{q})/N \tag{15}$$

where N is the number of particles in V. Generalizing this definition for $\omega \neq 0$ a response function $\Phi(t)$ can be obtained from

$$\chi(\vec{q},\omega) = \varrho_0\int_0^\infty e^{i\omega t}\Phi(\vec{q},t)dt \tag{16}$$

which is related to the relaxation function $\Psi(\vec{q},t)$ in the usual way [4]

$$\Phi(t) = -\frac{d\Psi(t)}{dt} \tag{17}$$

$\Psi(\vec{q},t)$ describes the relaxation ore more exactly retardation of the density if the potential $\varphi_e(\vec{q})$ is removed at time $t=0$. From Equs. (16) and (17) one obtains

$$\chi(\vec{q},\omega) = \varrho_0\left\{\varphi(\vec{q},t=0) + i\omega\int_0^\infty e^{i\omega t}\Psi(\vec{q},t)dt\right\} \tag{18}$$

Now we use the fluctuation dissipation theorem [4] in the classical limit

$$\Lambda(\vec{q},\omega) = \frac{1}{\beta}\frac{\chi''(\vec{q},\omega)}{\omega} \tag{19}$$

with $\beta = 1/k_B T$ and $\chi'' = \mathrm{Im}\{\chi\}$

$$\chi'' = \varrho_0 \omega \int_0^\infty \Psi(\vec{q},t)\cos \omega t \, dt \tag{20}$$

Together with the symmetric part of the fluctuations of Equ. (13) it follows from Equ. (19) because of the uniqueness of Fourier transforms

$$\frac{\beta}{V^2} <\delta\varrho(\vec{q},t)\delta\varrho(-\vec{q},0)> = \varrho_0 \Psi(\vec{q},t) \tag{21}$$

The important point of this derivation [5] is that the scattered intensity correlation $G_{iso}(\vec{q},t)$ described by Equ. (11) is related to a mechanical compliance function $\Psi(\vec{q},t)$ rather than to the mechanical modulus function $M(\vec{q},t)$ as it is stated usually [6].

Usually the relaxation function $D(t)$ is introduced in description of visco-elastic liquids [7]. This describes the relaxation of volume, instead of density, if the potential Ψ is applied and not removed at time $t = 0$. Then one obtains

$$\lim_{q\to 0} \Psi(\vec{q},t) = \frac{\varrho_0}{V}(D_e - D(t)) \tag{22}$$

where D_e is the equilibrium bulk compliance $\lim_{t\to\infty} D(t) = D_e = -\frac{1}{V}\left[\frac{\partial V}{\partial p}\right]_e = K_T$

where K_T is the isothermal compressibility.
As final result one obtains from Equ. (21)

$$D_e - D(t) = \frac{\beta}{V\varrho_0{}^2} \lim_{q\to 0} <\delta\varrho(\vec{q},t)\delta\varrho(-\vec{q},0)> \tag{23a}$$

or in terms of the correlation function $G_{iso}(\vec{q},t)$

$$G_{iso}(\vec{q},t) = \frac{a_1^2 V}{9\beta}(D_e - D(t)) \tag{23b}$$

For the anisotropic part in Equ. (9) we get using a similar formalism as before with the use of $J_e = \lim_{t \to \infty} J(t)$:

$$J(t) - J_e = -\frac{\beta}{V} < \gamma_{zy}(\vec{q},t)\gamma_{zy}(-\vec{q},0) > \tag{24}$$

and thus for the correlation function of depolarized scattering

$$G_{VH}(\vec{q},t) = a_2^2 \frac{V}{\beta} (J_e - J(t)) \tag{25}$$

without writing a_2 explicitly.

Taking the foregoing calculation into consideration we can write Equ. (2) now for our purpose in the following way:

$$< E(\vec{q},t)E^*(\vec{q},0) >_{\alpha\beta} = b_1 G_{iso}(\vec{q},t)\delta_{\alpha\beta} + \left(1 + \frac{1}{3}\delta_{\alpha\beta}\right) b_2 G_{VH}(\vec{q},t) \tag{26}$$

with $\alpha = V$, and $\beta \in (V,H)$, and b_1, b_2 being properly chosen constants.

We take now further into account that via the Siegert relation [1]

$$< I(0)I(t) > = < I >^2 + f| < E(0)E^*(t) > |^2 \tag{27}$$

– valid for the case of homodyne detection and the fact that the scattered light has a gaussian amplitude distribution – we can connect the autocorrelation function of the scattered field with the actual measured autocorrelation function of the intensity $I(\vec{q},t)$. We introduce now the second order correlation function to be

$$g_{\alpha\beta}^{(2)}(\vec{q},t) = \frac{< I(\vec{q},0)I(\vec{q},t) >_{\alpha\beta}}{< I >_{\alpha\beta}^2} \tag{28}$$

Recalling Equ. (27) the normalized time correlation function of the scattered field:

$$g_{\alpha\beta}^{(1)}(\vec{q},t) = \frac{< E(\vec{q},0)E^*(\vec{q},t) >_{\alpha\beta}}{< I >_{\alpha\beta}} \tag{29a}$$

can be calculated from the measured intensity correlation function $g_{\alpha\beta}^{(2)}(\vec{q},t)$ by

$$g_{\alpha\beta}^{(1)}(\vec{q},t) = \frac{1}{\sqrt{f}} \sqrt{g_{\alpha\beta}^{(2)}(\vec{q},t) - 1} \tag{29b}$$

The function is normalized to its value at $t = 0$. The factor f is a subject of either calibration or lengthy calculation [8].

15.1.2 The Comparison between Theory and Experiment

The foregoing theoretical result in mind (Equ. 23) we have been very interested in comparing data from quasielastic light scattering with those from dynamic mechanical measurements. For that we need data of the frequency or time dependent longitudinal compliance D(t) which is unfortunately not available. But data of the frequency dependent compressional compliance $B^*(\omega)$ is reported in literature [9]. To discuss the difference between D(t) and B(t), it is more convenient first to consider the relationship among the various dynamic compliances $D^*(\omega)$, $B^*(\omega)$ and $J^*(\omega)$, which are the one-sided Fourier transforms of D(t), B(t) and J(t) respectively:

$$D^* = \frac{B^* J^*}{J^* + \frac{4}{3} B^*} \tag{30}$$

For simplicity we have not written explicitly out the frequency dependence. Above T_g in the melt the dynamic shear compliance J^* is greater than the compressional compliance by at least one order of magnitude [7]. It is thus appropriate to write Equ. (30) in a series expansion of $\frac{B^*}{J^*}$:

$$D^* \cong B^* \left[1 - \frac{4}{3} \left(\frac{B^*}{J^*} \right) + \ldots \right] \tag{31}$$

For the reported data $B^*(\omega)$ by McKinney and Belcher [9] of poly(vinylacetate) (PVAc) the factor B^*/J^* is in the order of 10^{-3} [10], so the replacement of D^* by B^* is justified. Since in the experimental results on PVAc by McKinney and Belcher [9] the relaxation strength in the real part is more marked than in $B''(\omega)$ we choose $B'(\omega)$ [11, 12] to evaluate B(t) by means of

$$B(t) = \frac{2}{\pi} \int_0^\infty \frac{B'(\omega)}{\omega} \sin \omega t \, d\omega \tag{32}$$

This is essential because the light scattering result is measured in the time domain. At this stage now we have to discuss briefly in what way we have to treat the light scattering result in order to compare properly with the mechanical data.

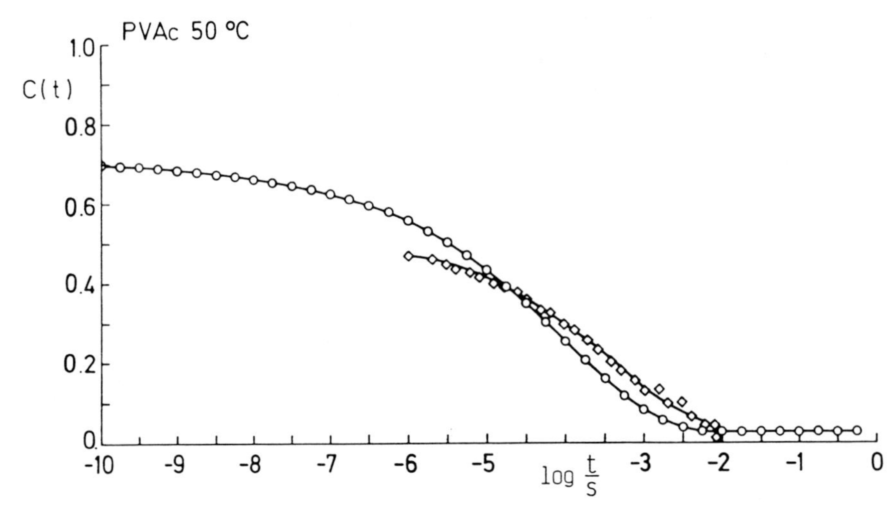

Fig. 15.1: Correlation functions C(t) of density fluctuations of poly(vinylacetate) plotted against log t. Mechanical data calculated according to Equ. (32) (○) and quasi-elastic light scattering result (◇).

Since PVAc is a polymer with very low depolarized scattered intensity ($I_{VH} \approx 0$) a measure of the VV component gives us directly the isotropic part of the correlation function as given by Equ. (23). Thus the second term on the right hand side of Equ. (26) drops and we can use Equ. (29) to analyze the measured density autocorrelation function. Doing that, while taking the relaxation strength $\alpha = (B_0 - B(t))/B_\infty$ (B_0 from mechanics, B_∞ for light scattering from an independent Brillouin experiment) into account, we have plotted in Fig. 15.1 the two correlation functions. The agreement is good especially if one takes into consideration that no adjustable parameters have been used. An experimental verification of Equ. (25) is still missing. It is been found experimentally that there is a strong dynamic coupling between the measured correlation function measured in VV and VH geometry with the result that the correlation functions have very similar time dependence [3, 13, 14]. But a proper determination of $G_{iso}(t)$ and $G_{VH}(t)$ independently is possible by performing a VH scattering experiment.

15.1.3 Data Evaluation of PCS Results

Experimentally it was found, that the measured field correlation function $g^{(1)}(t)$ caused by the density fluctuation cannot be described by a single exponential decay. Such a case holds for concentration fluctuations which will be discussed for the case of binary polymer mixtures in section 15.2 later on.

Density fluctuations have the property of decaying slower than a single exponential and the best known three – parameter fit to describe this feature is a stretched exponential, a so-called Kohlrausch-Williams-Watts (KWW) function [15].

$$g^{(1)}(t) = a \cdot \exp\left[-\left[\frac{t}{\tau_{kww}}\right]^{\beta}\right] \tag{33}$$

with $0 < \beta \leq 1$ as a measure of the width of the distribution. This KWW function is widely used in physics and there are several approaches to interpret its parameters within a meaningful physical picture [16]. The discussion is lively and has not yet ended. One possible approach is the assumption, that the deviation from the single exponential decay is caused by the superposition of relaxation processes with various retardation times. Therefore we used an inverse Laplace transformation technique to get the distribution of retardation times directly from the $g^{(1)}(t)$ curve. The first attempt to obtain $L(\ln\tau)$ was historically by Mehler [17], who used the algorithm by Pike and Ostrowsky [18] to solve for the distribution $L(\ln\tau)$ via

$$g^{(1)}(t) = \int_{-\infty}^{+\infty} \exp\left(-\frac{t}{\tau}\right) L(\ln\tau)\, d\ln\tau \tag{34}$$

The disadvantage of this algorithm is the very difficult determination of the eigenfunctions and values of the kernel of Equ. (34) in such a way that the solutions oscillate or do not match. This situation was succesfully overcome by a program called "Contin" developed by Provencher [19] which is based on the same mathematical formalism but avoids the possible mismatch by using a regularization parameter. The reader is referred to contact the original literature for the detailed discussion of this parameter. In an excellent review by Stock and Ray [20] the use of Contin for the related problem of analyzing the correlation functions from solutions of bimodal polymers was shown to be the most up-to-date way to face the problem.

We have modified the Contin program so that it can easily be used for very broad retardation time distributions [3]. Whether the inverse Laplace transformation gives the same results as the representation of $g^{(1)}(t)$ via Equ. (33) was checked with formulas giving analytical expression for $L(\ln\tau)$ if $g^{(1)}(t)$ is expressed by Equ. (33) [21]. We have compared computer simulated data and real data from various samples with varying KWW-β parameters [22] and found that only in cases where β is very small ($\beta < 0.3$) the mathematical description KWW versus Contin differs. But as already stated the use of Contin does not involve any functional form of the time dependence of $g^{(1)}(t)$. Furthermore in the case mentioned above for very small β parameters we have found that the analysis of a correlation function with the inverse Laplace transformation does bring more physical information about the system under study as will be discussed later in the section 15.1.5.

For any time correlation function C(t) which is continuously decreasing an average correlation time $\bar{\tau}$ can be defined by [1]

$$\bar{\tau} = \int_0^\infty C(t)dt \tag{35a}$$

which in the special case of KWW yields

$$\bar{\tau} = \frac{\tau_{kww}}{\beta} \Gamma(\beta^{-1}) \tag{35b}$$

with Γ being the Gamma function. If $L(\ln\tau)$ is determined by Eq. (34) an average $\overline{\ln\tau}$ can be calculated by

$$\overline{\ln\tau} = \int_{-\infty}^{+\infty} \ln\tau L(\ln\tau) d\ln\tau \tag{36}$$

The difference between $\bar{\tau}$ and $\overline{\log\tau}$ is less that 0.5 decades for $\beta \geq 0.5$. For a further discussion see [22].

15.1.4 Connection between Molecular and Collective Dynamics

Since the PCS method is applicable in a long time range (10^{-6} sec $< \tau <$ 10^3 sec) and therefore very suitable for dynamic phenomena related to the glass transition, the comparison with other dynamic methods will give us valuable information about the relationship between the molecular and the collective dynamics and therefore contribute towards the understanding of the nature of the glass transition. Here of course the dielectric relaxation spectroscopy plays a major role. In the literature for a series of materials the dynamics of slowly relaxing density fluctuation have been studied by both methods mentioned above [23]. The motion under study here is classified as the α-motion or primary relaxation process because it is related to the glass transition of the polymer and has a Williams-Landel-Ferry (WLF) temperature dependence (see Equ. 41). Usually good agreement has been found between the activation parameters (WLF-constants), characterizing the temperature dependence of the dynamics. On the other hand, the absolute values of the relaxation times at a given temperature were found to be different [23]. This difference has not been investigated up to now. We have proposed a new method of data analysis to compare the results of the two methods adequately [24]. We have chosen the polymer poly(methyl-phenyl-siloxane) PMPS for our comparison because it may serve as a model substance for a chain molecule which has been used by a great variety of different other methods to study dynamical processes. The references known to the authors concerning this point are listed in Ref. [24].

Dealing with dielectric relaxation the frequency dependent complex permittivity $\varepsilon^*(\omega) = \varepsilon'(\omega) - i\varepsilon''(\omega)$ is related to the dipole moment time correlation function $\Phi(t)$ by a one-sided Fourier or pure imaginary Laplace transformation via

$$\frac{\varepsilon^*(\omega)-\varepsilon_\infty}{\varepsilon_0-\varepsilon_\infty} = \int_0^\infty \exp(-i\omega t)\left[\frac{-d\Phi(t)}{dt}\right]dt \tag{37}$$

where ε_0 and ε_∞ are the limiting low- and high frequency permitivities respectively. As shown by Williams [25] the total dipole moment time correlation function $\Phi(t)$ is given by

$$\Phi(t) = \frac{\sum_{i,j}^{N} <\mu_i(0)\mu_j(t)>}{\sum_{i,j}^{N} <\mu_i(0)\mu_j(0)>} \tag{38}$$

where $\mu(t)$ denotes the elementary dipole moment in a polymeric chain at time t. Since $\varepsilon'(\omega)$ and $\varepsilon''(\omega)$ are related via a Kramer-Kronig relation we have used ε'' to extract $\Phi(t)$. This is done by a half-sided cosine transformation

$$\Phi(t) = \frac{2}{\pi}\int_0^\infty \frac{\varepsilon''(\omega)}{\varepsilon_0-\varepsilon_\infty}\,\frac{\cos\omega t}{\omega}\,d\omega \tag{39}$$

To describe the frequency dependence of $\varepsilon''(\omega)$ analytically the simplest case of dielectric relaxation involves the Debey function, which does not give a sufficient description of the relaxation behaviour in polymeric melts. We have found that only the model by Havriliak-Negami [26] describes the complex permittivity of PMPS within the experimental accuracy adequately. The equation reads

$$\varepsilon^*(\omega) = \varepsilon_\infty \frac{\varepsilon_0-\varepsilon_\infty}{(1+(i\omega\tau)^\alpha)^\gamma} \tag{40}$$

with $0 < \alpha, \gamma \leq 1$. This ansatz fits the asymmetry of the loss curve independently of its width. In Fig. 15.2 we show some exemplaric results for measured $\varepsilon''(\omega)$ curves together with the fits of Equ. (40) to the data. From the description of the experimental $\varepsilon''(\omega)$ data by Havriliak-Negami parameters we have

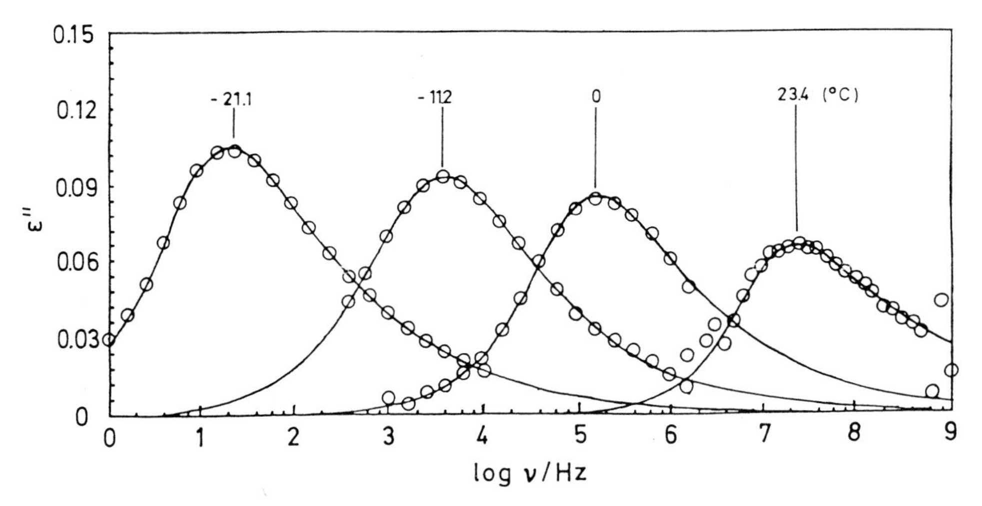

Fig. 15.2: Dielectric loss ε" of poly(phenylmethylsiloxane) plotted against log ν. The full lines are results of a fit of the Havriliak-Negami model (Equ. 40) to the data (○).

solved Equ. (39) numerically and have fitted a KWW stretched exponential according to Equ. (33) to the result. By this procedure we avoid of course the simple $\omega\tau = 1$ condition, valid for the maximum of the $\varepsilon''(\omega)$ curve of a Debey process, to get a meaningful mean-relaxation time. Even the formula for a $\bar{\tau}$ in the formalism of a Hăvriliak-Negami model is not easy to obtain. Thus we have circumvented the problem by switching to the time domain. Here a $\bar{\tau}$ is easily obtained by Equ. (35).

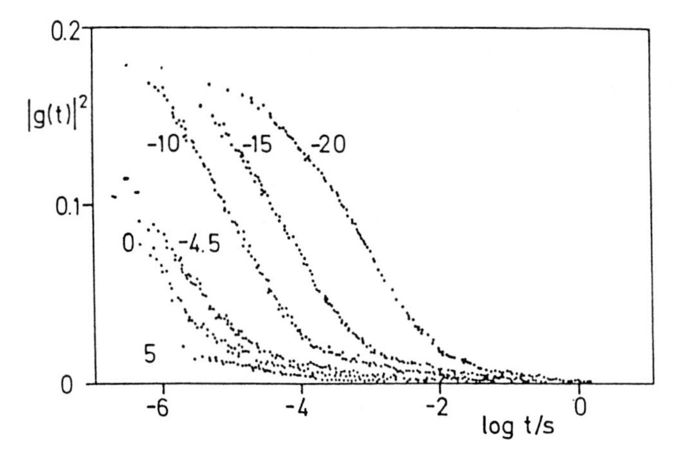

Fig. 15.3: Density autocorrelation functions $|g^{(1)}(t)|^2$ of poly(phenylmethylsiloxane) plotted against log t.

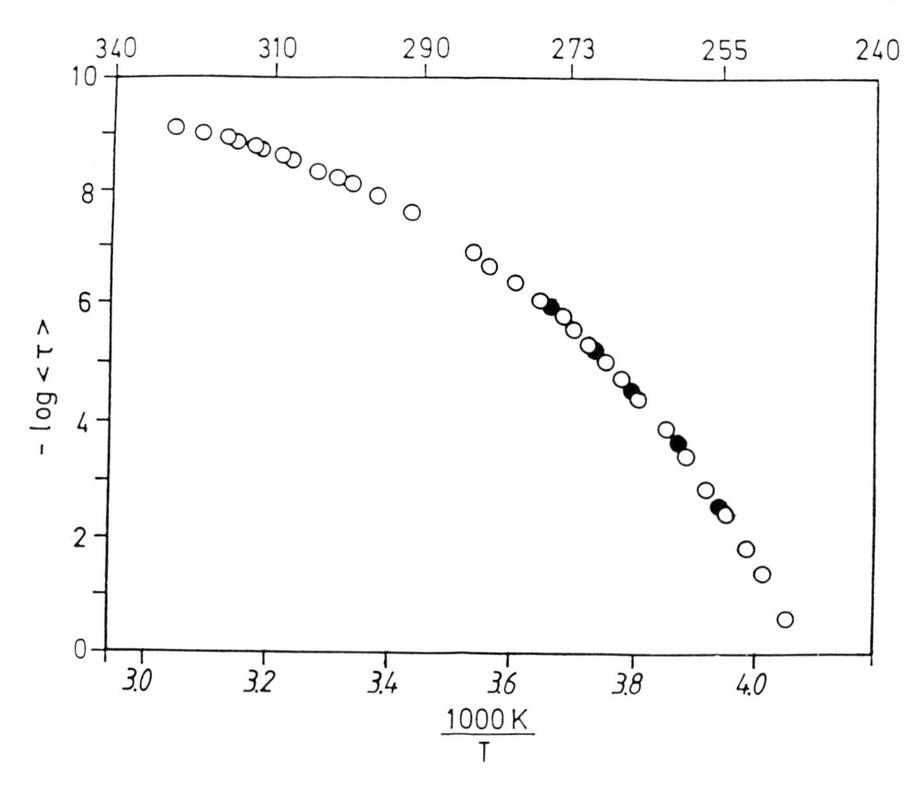

Fig. 15.4: Activation plot of mean relaxation times $\bar{\tau}$ for the α-process of PPMS comparing the dielectric relaxation results (○) with the PCS results (●).

Now we are in a condition to compare with the light-scattering result since data here is usually taken in the time domain. In Fig. 15.3 we show the results of the PCS measurements at different temperatures. (The T_g of the sample is $-29\,°C$). A fit of Equ. (33) to this data leads to β parameters which are very close to those we obtained from KWW fit to the dielectric $\Phi(t)$ curve via Equ. (39). Also the mean times extracted are found to be identical. Fig. 15.4 shows this result in an activation plot.

Thus we conclude that the molecular dynamics of reorienting molecular dipoles is highly controlled by its environment as the collective dynamics completely controls the motion on a molecular level.

The result of Fig. 15.4 immediately leads to another problem which is connected with the description of the given curve by the Williams-Landel-Ferry equation. Usually in a region $T_g + 10K < T < T_g + 100K$ the equation holds to describe the temperature dependence of the dynamics response of a system [7]. The equation reads

$$\log a_T = -\frac{C_1(T-T_0)}{C_2+T-T_0} \tag{41}$$

Here a_T is a shift factor ($a_T = \omega/\omega_0$). C_1 and C_2 are constants for a given system. T_0 is a reference temperature. The WLF parameters C_1 and C_2 are usually given with respect to $T_0 = T_g$ as reference temperature. The WLF equation is a mathematical formulation of the temperature – frequency superposition principle.

Consequently we have tried to construct a master curve for all the data from Fig. 15.3 with T = 0 °C as reference temperature, which we show in Fig. 15.5. The evaluated C_1 and C_2 parameters from the masterplot representation agree with those from a fit of Equ. (41) to the data points given in Fig. 15.4 thus indicating that no particular mathematical model is needed to put the results in the framework of a dynamic viscoelastic picture. This is in complete agreement with the findings outlined in the beginning of this chapter that the light scattering spectrum is connected with a dynamic mechanical viscoelastic property for which, as commonly believed, the time-frequency superposition principle is valid.

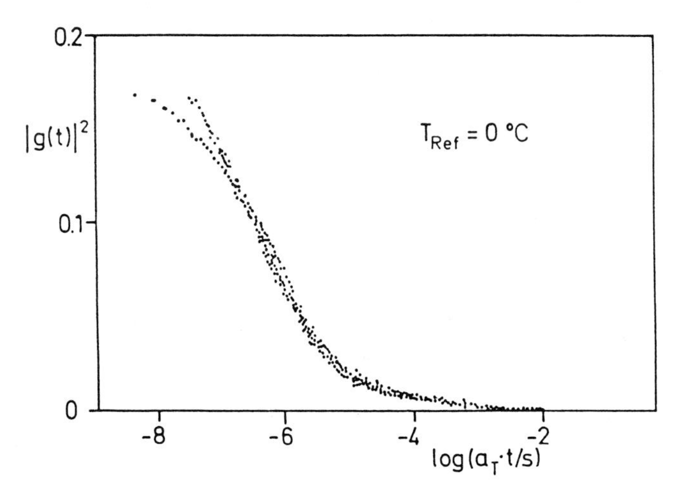

Fig. 15.5: Masterplot of the correlation functions from Fig. 15.3 showing the applicability of the time-temperature superposition principle.

15.1.5 Separation of Primary and Secondary Relaxation Processes in Polymethacrylates

We would now like to report results of the series of alkylmethacrylate polymers (PnRMA) with R being methyl(M) and hexyl(H). This is a very interesting class of polymers which are useful to study the influence of the secondary β-relaxation on the α-process where the β-relaxation is attributed to the motion of the side group. These findings are derived from studies of the dielectric relaxation [10] and dynamic mechanical measurements [27] indicating that it is becoming more and more difficult, especially by mechanical methods, to distinguish between the two modes with increasing side group length. In case, one uses dielectric relaxation, the situation is better since the dipole moment is in the side group. On the other hand the influence of the α-relaxation usually understood as main chain relaxation, involving many monomeric segments, is becoming less pronounced. Thus, light scattering may be useful to clarify these points [28]. Historically PMMA was the first polyalkylmethacrylate which had been studied with PCS [29–31]. However, the early papers failed to report consistent results, probably due to the poor sample quality and due to the much restricted time range of the correlators used. A more recent attempt [32, 33] to encompass the broad distribution of retardation times is to use a data handling by means of the inverse Laplace transformations via Equ. (34). Results of this procedure for a PMMA of molecular weight of 10^7 g/mol are shown in Fig. 15.6.

One clearly sees two maxima in the distribution curve at a temperature of 128° C, which is well above T_g = 107° C, the occurrence of the α- and β-relaxation peak. For PHMA(T_g = –5° C) the separation was also possible [34] as

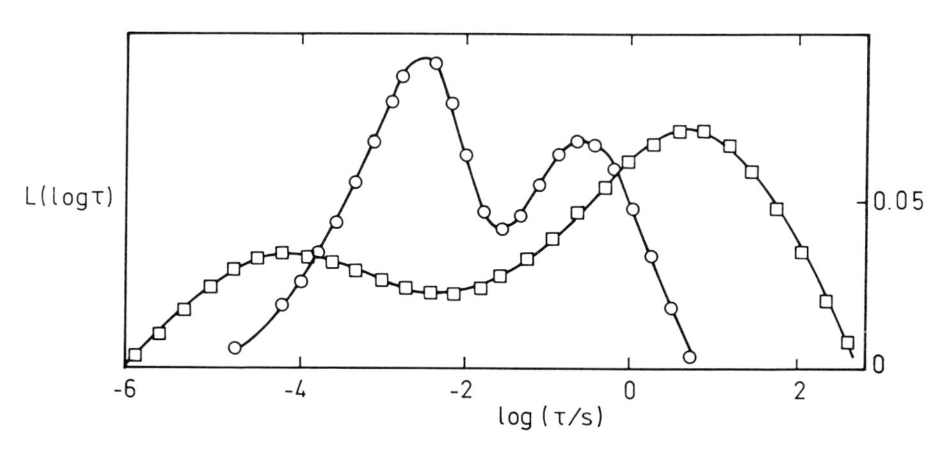

Fig. 15.6: Comparison of the retardation time spectra L (log τ) against log τ for PMMA (□) and PHMA (○) at a temperature equidistant to T_g (ΔT = 20K).

shown in Fig. 15.6. But the relative difference in logτ between the maxima of the two modes is much smaller, for the same temperature difference $T-T_g$ = 20K. This indicates that for the longer side chain polymers the two relaxations are more coupled to each other. For comparison of various polymethacrylates in Fig. 15.7, the mean relaxation times of both the α- and β-relaxation are plotted versus 1/T [35, 36]. Here, we notice the less pronounced differences in τ for temperatures of measurement equidistant to T_g for increasing side chain length, which therefore put the results of Fig. 15.6 in the framework of interpretation of the relaxation behaviour of this homologous series of polymers. A dynamic mechanical experiment, however, just measures [34] for PHMA the value of $\log(\tau/s) \simeq 2$ for T = 15° C without differentiating between α and β modes. Usually, WLF parameters are extracted from the activation plot and compared with those derived from other methods especially dynamic mechanical methods. For those cases where the separation of modes is clear, good agreement has been found [23, 33], whereas for longer side group polymers the full set of activation parameters can only be obtained from PCS measurements. Since for PMMA a lot of relaxation data even in the high frequency regime has been published, we further would like to compare photon correlation data with existing mechanical data, remembering the connection between density autocorrelation function and longitudinal compliance. D(t) is usually very difficult to extract from a mechanical experiment. In practice, it is usually the shear modulus (or compliance) or Young's modulus that is measured

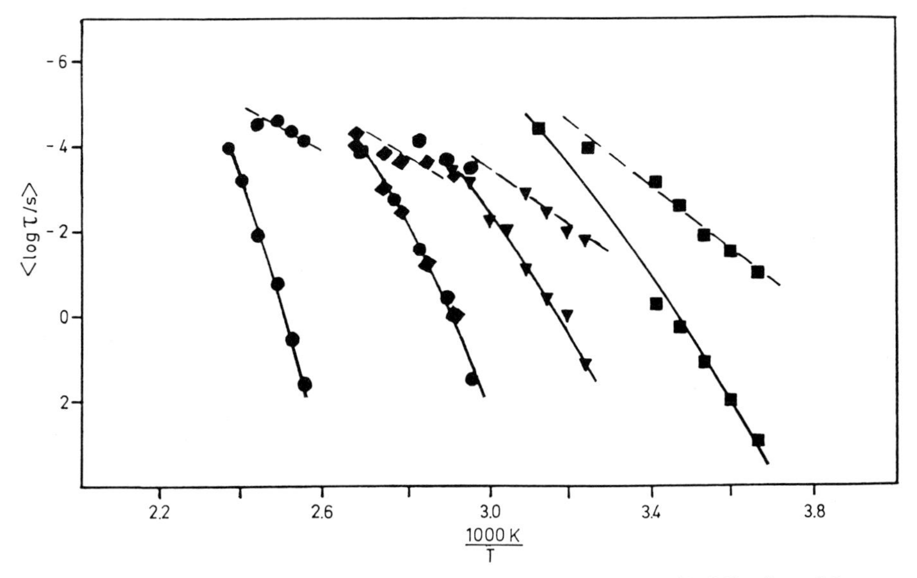

Fig. 15.7: Activation plot of relaxation times for the α process (full line) and β process (broken line) of PMMA (●); PEMA (◇) [35]; PBMA (▽) [36] and PHMA (■).

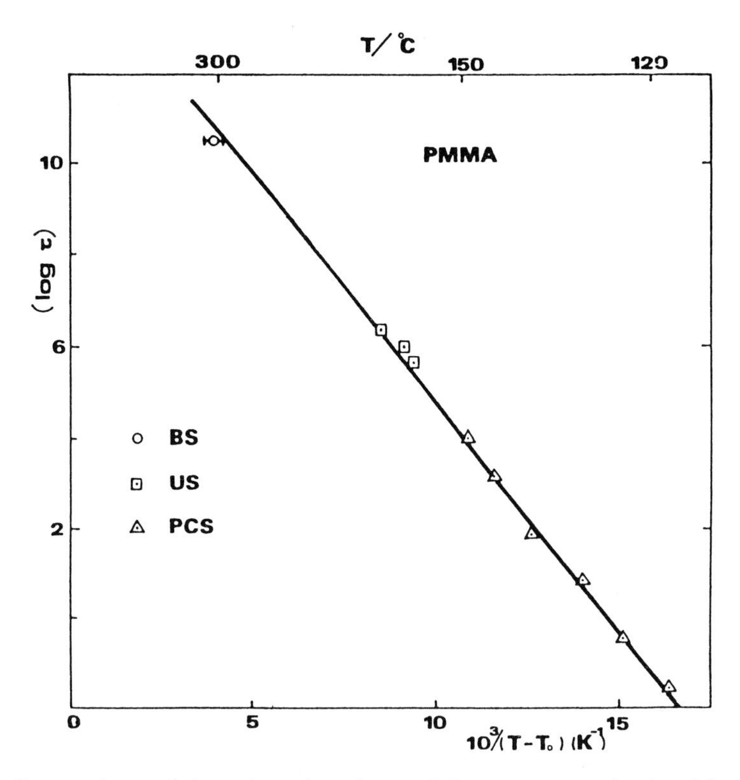

Fig. 15.8: Comparison of the relaxation times of the α process obtained from photon correlation (△), ultrasonic absorption (□) [37] and Brillouin scattering (○) [38].

mechanically [7]. For PMMA, however, the relaxation of the longitudinal modulus M* (= 1/D*) has been investigated by using ultrasonic [37] and Brillouin scattering techniques [38]. The time 1/ω (ω being the sound frequency) for the temperature T_{max} at which the sound attenuation has its maximum, together with the PCS results from Ref. [33], are plotted in Fig. 15.8. For the replacement of D(t) by M(t) see Ref. [39].

The value of T_{max} for the hypersonic absorption obtained from the Brillouin spectra is a crude estimate [38]. Nevertheless the three sets of data covering 12 decades in time can be well represented by using a constant C_2 value from the WLF equation indicating that the description of the relaxation behaviour in terms of the free volume concept is an adequate picture.

15.2 The Dynamics of Concentration Fluctuations in a Compatible Binary Polymer Blend

In the foregoing chapter we have dealt with the dynamics of density fluctuations. The dynamics of concentration fluctuation in compatible binary polymere mixtures, however, can also be measured with the PCS method. In light scattering we are concerned with large scale fluctuations and microscopic relaxation times such that we can write a phenomenological diffusion equation (Φ: volume fraction of one component)

$$\frac{\partial}{\partial t}\,\delta\Phi(\vec{r},t) = -\,D(\Phi)\nabla^2\delta\Phi(\vec{r},t) \tag{42}$$

Where $\delta\Phi = \delta\Phi_A = -\delta\Phi_B$ and $D(\Phi)$ is the mutual diffusion coefficient which may depend on Φ. The light scattering is determined by the Fourier components $\varepsilon_q(t)$ of the fluctuations of the dielectric constant

$$I \sim < \varepsilon_q(t)\ \varepsilon_q(0) > = \left(\frac{\delta\varepsilon}{\delta\Phi_A}\right)^2 < \Phi_q(t)\Phi_{-q}(0)> \tag{43}$$

Where $\Phi_q(t)$ is the Fourier transform of $\delta\Phi(\vec{r},t)$.
From the diffusion Equ. (42) follows

$$\Phi_q(t) = \Phi_q(0)\exp(-Dq^2t) \tag{44}$$

and

$$S(q,t) = <\Phi_q(t)\Phi_{-q}(0)> \tag{45a}$$

$$= S(q)\cdot\exp(-Dq^2t) \tag{45b}$$

$S(q)$ and $S(q,t)$ are the static and dynamic structure factor respectively and q is given by $|\vec{q}|$.

The quantities D and $S(q)$ are directly accessible by a light scattering experiment. What is needed is a connection between these quantities and those given by the thermodynamics of irreversible processes where the particle current $J_A(\vec{r},t)$ is considered to be driven by a chemical potential gradient $\nabla(\mu_A - \mu_B)$ rather than by a density gradient like in Equ. (42). Working this out [40] one ends up in a simple relation between the Onsager oefficient Ω and measurable quantities

$$\Omega = S(0) \cdot D \tag{46}$$

with S(0) being the extrapolation of S(q) for $q \rightarrow 0$.

There are various theoretical approaches regarding the calculation of Ω from microscopic models which disagree one from the other. The relaxation times τ of concentration fluctuations, given by $\tau = (Dq^2)^{-1}$, cannot be obtained from a phenomenological treatment, there is need for a microscopic model [41–44].

We have performed our measurements on relatively low molecular weight polymers and refer this case to the mean field treatment of Brochard and de Gennes [43] where the symmetric case of unentangled polymer chains has been considered. The result which was in a similar form obtained by Binder [44] can be written as:

$$D = 2\,\Phi\,(1-\Phi)\left\{\chi_s(N,\Phi)-\chi(T)\right\}W^o \tag{47}$$

Where

$$2\chi_s = (\Phi_A N_A)^{-1} + (\Phi_B N_B)^{-1}$$

with $N_{A,B}$ being the degrees of polymerization and

$$S(0)^{-1} = 2(\chi_s - \chi(T)) \tag{48}$$

There is no explicit temperature dependence in Equ. (47) but it is normally assumed to occur in the interaction parameter $\chi = \chi(T)$. The microscopic transition rate W^o is determined by the single component Rouse diffusion coefficient

$$D_R = \frac{W^o}{N} \tag{49}$$

We now recall Equ. (48) together with Equ. (46) to end with

$$\Omega = \Phi(1-\Phi)W^o \tag{50}$$

which means that the degree of polymerization cancels out. As a consequence the mutual diffusion coefficient should not depend on molecular weight if the thermodynamic driving force for mixing is large. That means if $2N|\chi|\Phi(1-\Phi) \gg 1$. One then obtains

$$D = 2\Phi(1-\Phi)|\chi|W^{o} \tag{51}$$

Finally the question arises what microscopic rate constant W^* has to be used in the non-symmetric case instead of W^o in Equ. (50). This case has been considered by Binder [44] and the result can still be written in the general form (cf. Equ. (47)), but now

$$\frac{1}{S(q=0)} = \frac{1}{N_A\Phi_A} + \frac{1}{N_B\Phi_B} - 2\chi \tag{52}$$

and $\Omega = \Phi_A\Phi_B W^*(\Phi)$
where

$$\frac{1}{W^*(\Phi)} = \frac{\Phi_A}{W_B{}^o} + \frac{\Phi_B}{W_A{}^o} \tag{53}$$

This result is in contradiction to the formula derived by Sillescu [41] and Kramer [42] who give for $W^*(\Phi)$

$$W^*(\Phi) = \Phi_A W_B{}^o + \Phi_B W_A{}^o \tag{54}$$

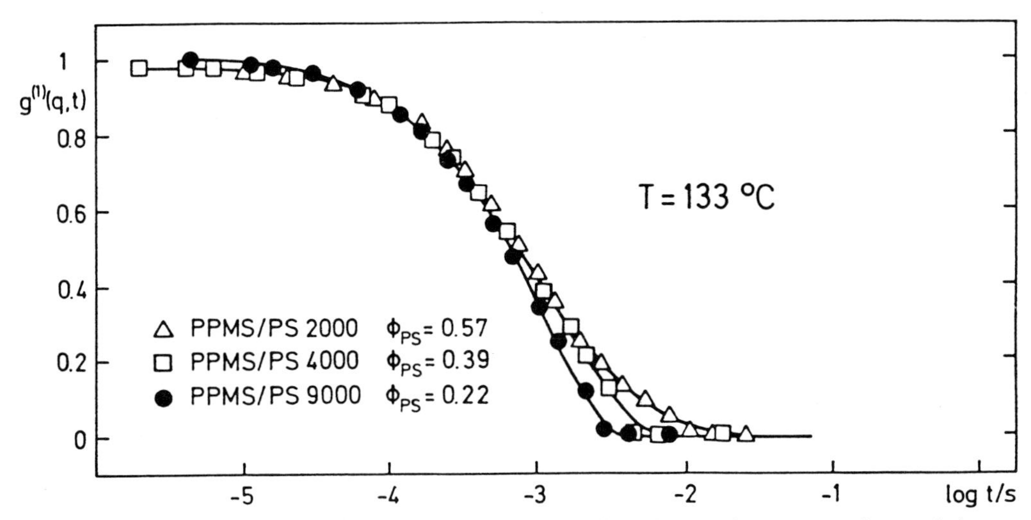

Fig. 15.9: Net correlation function of concentration fluctuations versus log t of three different blends of PPMS/PS with varying Φ_{PS} and M_{PS}.

It was our intention to explore the possibilities of experimentally distinguishing between these two approaches of a "series" or "parallel" addition of the W* by means of quasielastic light scattering on a compatible binary polymer blend.

The experiments were carried out with mixtures of poly(phenyl-methyl-siloxane)(PPMS) (M_w = 2600 g/mol; $M_w/M_n \sim$ 1.6) and polystyrene (M_w = 2000, 4000, 9000 g/mol; $M_w/M_n \sim$ 1.03) [40, 45]. The experimentally determined correlation functions (examples are shown in Fig. 15.9) were fitted by an almost single exponential decay and the measured time τ^{-1} was found to scale with q^2, as shown in Fig. 15.10, therefore the collective diffusion coefficient D is well defined by the experiment. The sake of Fig. 15.9 is furthermore to prove the prediction of Equ. (50), that is the independence of τ with varying molecular weight.

In Fig. 15.9 the molecular weight of PS has been changed. The different glass transition temperatures of the mixtures are compensated by varying Φ to a constant T_g of the mixture accordingly. Obviously the prediction of Equ. (50) is

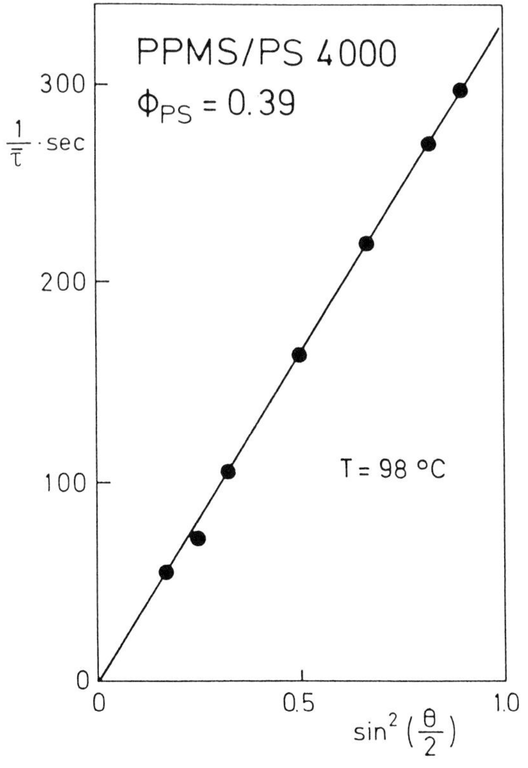

Fig. 15.10: Relaxation rate τ^{-1} versus the square of the scattering angle for a PPMS/PS blend.

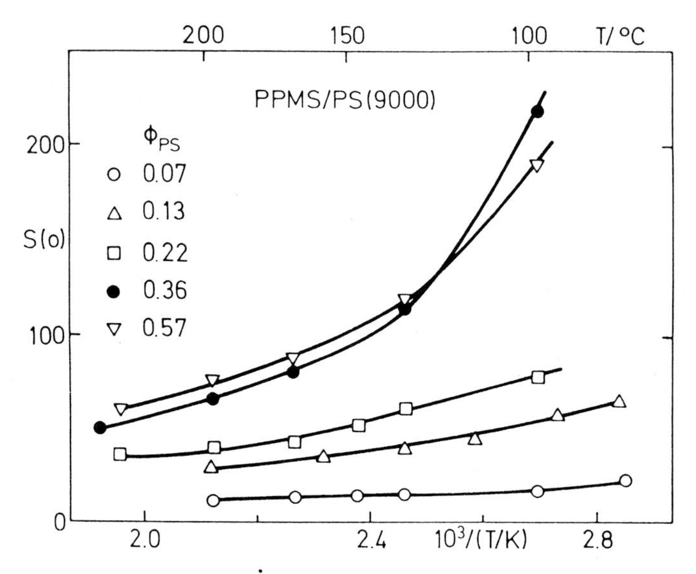

Fig. 15.11: The variation of the static structure factor S(q = 0) with temperature for PPMS/PS blend at five different volume fractions for PS with M_w = 9000 D. The lines are to guide the eye.

nicely fulfilled. The static structure factor in the thermodynamic limit S(0) can be obtained by extrapolation $q \rightarrow 0$ and some results are shown in Fig. 15.11 in dependence on Φ_{PS}.

The system has an upper critical solution temperature and therefore a behaviour as shown is qualitatively expected. The experimental data can be used to determine the Rouse rates W* by means of Equs. (46) and (52). This situation is complicated because we have an explicit Φ-dependence of W* described by Equs. (53) or (54) but also an implicit Φ-dependence because of the additional Φ-dependence of the $W_A{}^o$ and $W_B{}^o$ caused by the great differences in the glass transition temperatures of the two polymers used. In pure systems the equation

$$W_A{}^o = \nu_A \exp \frac{B_A}{T - T_{0A}} \tag{55}$$

usually holds (the same for index B). The same ansatz we assume to hold for the mixtures:

$$\frac{1}{W^*(\Phi)} \exp \left[- \frac{B(\Phi)}{T - T_0(\Phi)} \right] = \frac{\Phi}{\nu_B{}^o} + \frac{1-\Phi}{\nu_A{}^o} \tag{56}$$

for the Binder result and

$$W^*(\Phi) \exp\left[-\frac{B(\Phi)}{T-T_0(\Phi)}\right] = \Phi v_B{}^o + (1-\Phi) v_A{}^o \qquad (57)$$

for the Sillescu result.

Now it was experimentally found that B and C_2 (C_1, C_2 WLF constants; $B = C_1C_2$) were independent on Φ, so we can use Equs. (56) and (57) to distinguish between the two models because at a constant $T-T_g$ the exponential term is constant. Thus in Fig. 15.12 a plot of W^* and W^{*-1} versus Φ_{PS} is shown which favours the mean field Binder, Brochard, de Gennes result [43, 44].

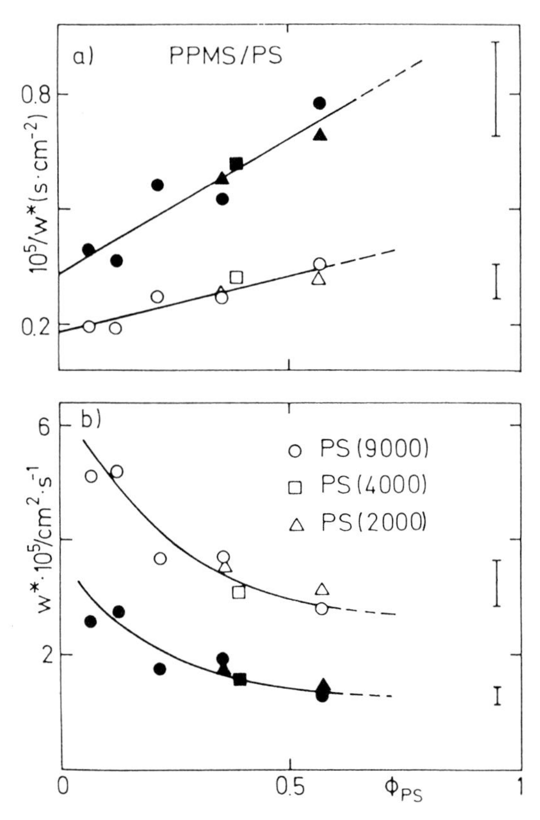

Fig. 15.12: The variation of the microscopic Rouse rates W^* with volume fraction of PS. In Fig. 15.12 a the quantity W^* according to Equ. (53), in Fig. 15.12 b W^* according to Equ. (54). The symbols refer to different molecular weights as indicated. Values are plotted for constant temperature differences $T-T_O(\Phi)$ of 330 K (hollow symbols) and 280 K (solid symbols).

15.3 Rayleigh-Brillouin Spectroscopy of Structurally Relaxing Liquids

15.3.1 Inelastic Scattering from a Liquid Consisting of Linear Chain Molecules

The study of structural relaxation processes in the GHz region is usually performed by measuring the Rayleigh-Brillouin spectrum. This is via the Wiener-Khintchin theorem connected with the Fourier transform of the autocorrelation function of the density fluctuations as given by Equ. (11).

We have seen that a rigorous calculation of the density autocorrelation function in the long time limit leads to quantities which can directly be compared with dynamic mechanical properties of the material. For the spectrum (high frequency case) an analytical expression was derived by Lin and Wang [46]. They used a linear response theory of a Mori-Zwanzig type to calculate $I_{VV}(\vec{q},\omega)$ for a visoelastic medium assuming a coupling of density fluctuation to the fluctuation of the longitudinal momentum and energy densities. The result for the Rayleigh-Brillouin spectrum is rather complex. The theory predicts a four peak structure, consisting of an unshifted Rayleigh peak, a symmetrically shifted Brillouin doublet and an unshifted peak known as Mountain peak, whose width is determined by the structural relaxation time τ_s (time to relax the longitudinal modules), the unrelaxed adiabatic sound frequency ω_s and the unrelaxed longitudinal (or bulk) modules $M = \varrho_0(R-1)\omega_s^2$ with $R = \omega_\infty^2/\omega_s^2$ being the relaxation strength. The expressions for the Brillouin shift f_B and the Brillouin linewidth Γ_B in the framework of this theory are

$$f_B = \pm\,\omega_s \left\{ 1 - \frac{1}{2}\left(\frac{\alpha_2}{\omega_s^2}\right)\frac{\omega_s^2\tau_s^2}{1+\omega_s^2\tau_s^2} \right\} \tag{58}$$

$$\Gamma_B = \frac{\alpha_2\tau_s^2}{1+\omega_s^2\tau_s^2} + (1-\gamma^{-1})D_T q^2 \tag{59}$$

with $\gamma = C_p/C_v$ and D_T the thermal diffusivity.

From the experiment f_B and Γ_B can be obtained and a fit of Equs. (58) and (59) can be used to describe the measured f_B and Γ_B at all temperatures. By that procedure the determining parameters of the Mountain line are accessible.

We have used this theory to check the particular observation that in the melt of a linear chain molecule (n-tetracosane) the sound velocity depends linearly on temperature but the slope displays a gradual increase above the melting point [47]. It was suggested by Krüger [48–50] that this non-linear behaviour is associated with a mesomorphic phase transition. In order to clarify the nature of this finding we have used DOS (di-2ethylhexl-sebacate) [47], a chain molecule similar to the one mentioned above but somewhat sterically hindered in order to considerably lower the melting point (T_m = −48 °C) and by that to extend the accessible temperature range for the Rayleigh-Brillouin spectra. Fig. 15.13 gives the results of the measurements in terms of the temperature dependencies of f_B and Γ_B.

We particularly note a maximum in Γ_B at about T = 270 K. This behaviour for our DOS system is similar to the findings in viscoelastic polymeric and non-polymeric liquids [51–53]. There these effects have been shown to be due to structural relaxation. Consequently we have fit the theory of Lin and Wang to our data. The full lines in Fig. 15.13 are the results of such a fit. To give further a consistency check regarding the five fit parameters of the theory we have computed the apparent Landau-Placzek ratio (LPR): $= I_{central} / 2I_{Brillouin}$ of DOS from these parameters and compared that with our measurements. In Fig. 15.14 we show our experimental results for the LPR together with the prediction of the theory (full line).

Thus we conclude that a relaxation theory describes the experimental data of f_B,Γ_B and LPR in a broad temperature interval and that a mesomorphic phase transition does not play any role.

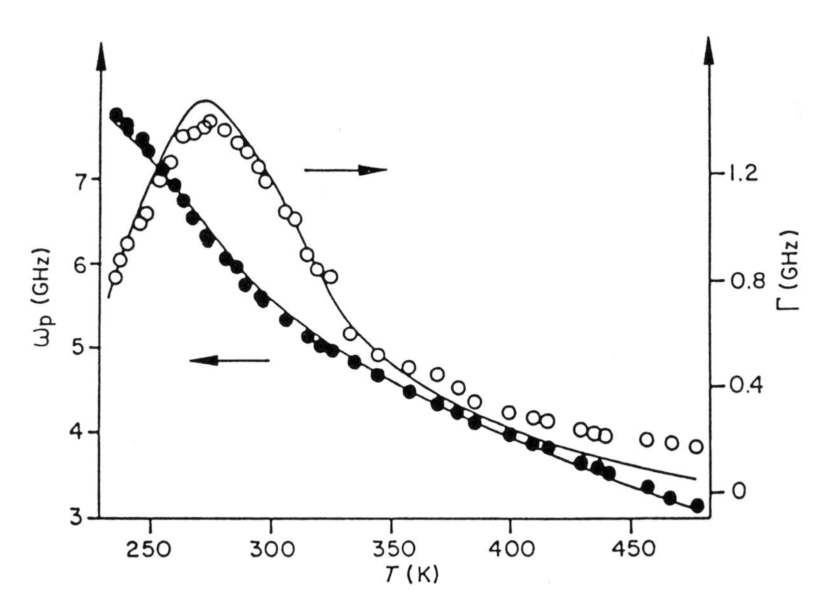

Fig. 15.13: Temperature dependence of f_B (●) and Γ_B (○) for DOS. The solid lines were calculated from theory [46].

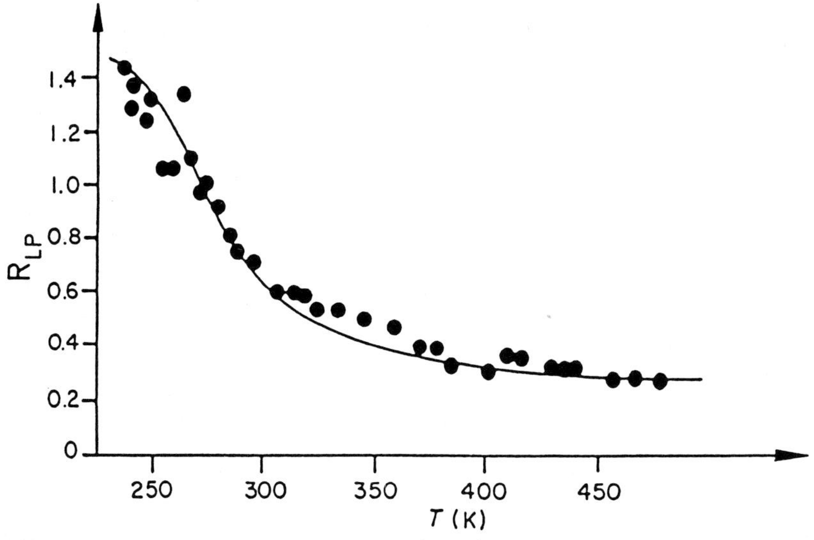

Fig. 15.14: Apparent Landau-Placzek ratio (LPR) for DOS as a function of temperature. Points were determined from experimental spectra, the solid line was obtained from the theoretical spectra.

15.3.2 Hypersonic Relaxation in a Viscoelastic Non-polymeric Liquid

Since the theory by Lin and Wang has been worked out for a viscoelastic liquid we have further investigated the temperature dependence of f_B and Γ_B for a nonpolymeric glass forming liquid: o-terphenyl [54]. In Fig. 15.15 f_B and Γ_B are plotted versus T together with the result of the Lin-Wang fit as a full line. Again the theory describes adequately the dynamic properties of the material in the high frequency range. From o-terphenyl a lot is known about the structural relaxation processes from other methods [55], especially very thoroughly investigated by quasielastic light scattering [3]. The temperature dependence of this structural relaxation (the α-relaxation mode) in the vicinity of T_g should show up at elevated temperatures in the time range of the Rayleigh-Brillouin spectrum. This simply reveals the fact that the time scale of the dynamic response of a system depends on the time scale of the experiment to measure it. Thus the maximum in Γ_B in Fig. 15.15 reflects the condition $\omega_B^{max} \cdot \tau = 1$ with ω_B being the phonon velocity at maximum loss. The time τ should thus lie on the extrapolation of the data points measured by PCS in the long-time regime [39]. In Fig. 15.16 mean times are plotted versus the reciprocal temperature. The full line is a fit of the WLF equation to the PCS data points. The extrapolation meets the Brillouin point nicely, futhermore the structural relaxation time τ from the Lin-Wang fit follow also the fitted WLF line which is in complete agreement with the other dynamical findings.

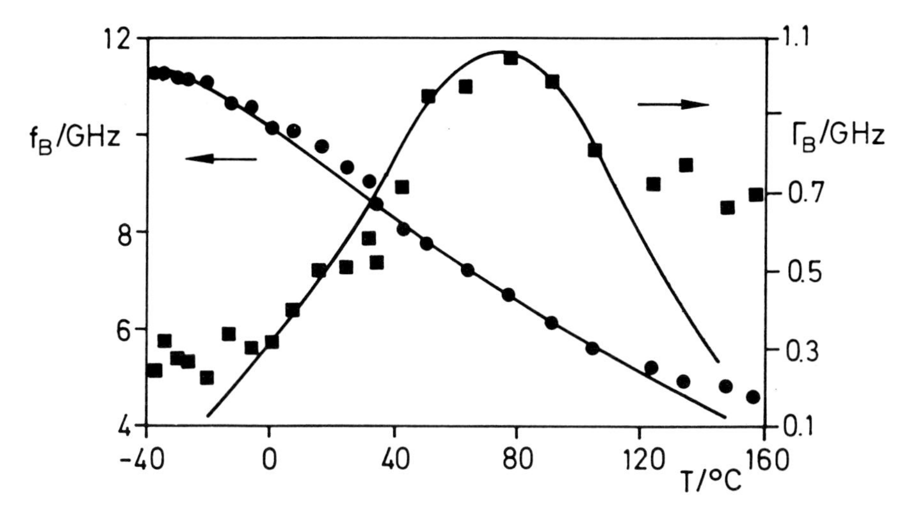

Fig. 15.15: Temperature dependence of f_B (●) and Γ_B (■) for o-terphenyl. The solid lines were calculated from theory [46].

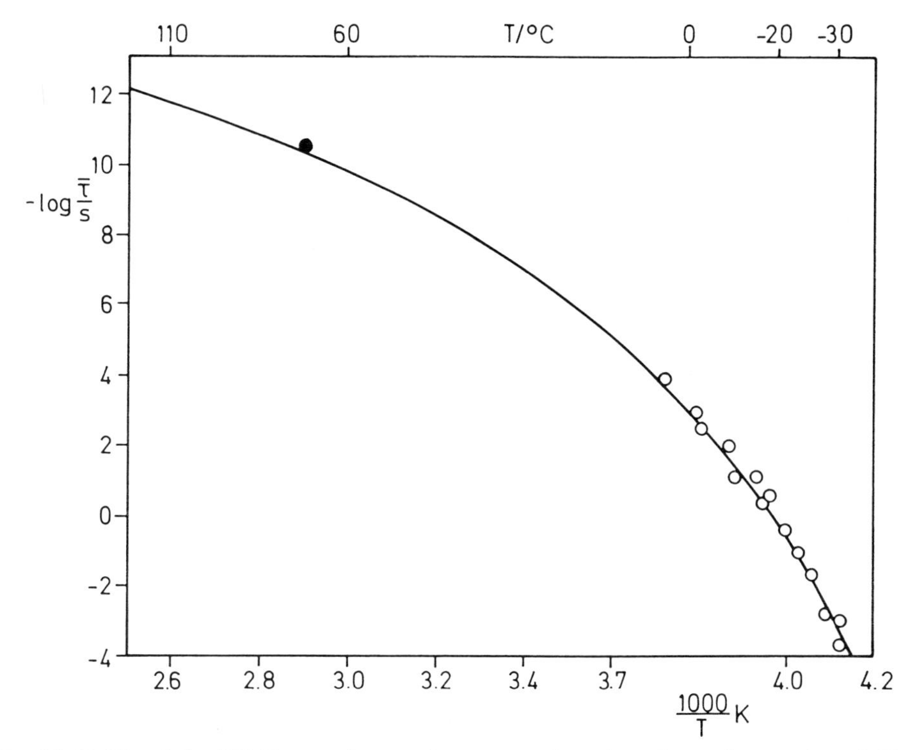

Fig. 15.16: Fit of the WLF equation to the mean relaxation time of o-terphenyl. Data from PCS (○) and Brillouin spectroscopy (●).

E.W. Fischer, B. Ewen, and G. Meier

15.4 Collective Fluctuations of Chain Segments in Dense Liquid Polymer Systems as Studied by Neutron Spin Echo Spectroscopy

Similar to the PCS the quasi-elastic neutron scattering provides the unique chance to study the dynamic properties of polymeric systems simultaneously in space and time [56, 57]. In addition this technique has the advantage that information on the collective single chain relaxation is not restriced to dilute solutions but may also be obtained from dense systems if the method of labelling by hydrogen-deutron exchange is used. Especially the recently developed neutron spin echo (NSE) spectroscopy gives access to motional processes which occur on time scales between 0.5 and 40 ns and length scales reaching up to 50 Å. In the following the principles of this method are described and some applications are presented dealing with segmental fluctuations in concentrated solutions and melts of linear chains and polymer networks.

15.4.1 Neutron Spin Echo

The unique feature of NSE is its ability to determine energy changes of neutrons occurring during a scattering in a direct way [58–60]. NSE measures the neutron velocities of the incoming and scattered neutrons utilizing the larmor precessions of the neutron spin in an external guide field. Since the neutron spin vector acts like the hand of an "internal clock" attached to each neutron which stores the result of the velocity measurement on the neutron itself, this measurement is performed for each neutron individually. Therefore, the incoming and outgoing velocities of one and the same neutron can be compared directly and a velocity difference measurement becomes possible. Thus, energy resolution and monochromatization of the primary beam or the proportional neutron intensity are decoupled and an energy resolution in the order of 10^{-5} can be achieved with an incident neutron spectrum of 20 % bandwidth.

The basic experimental set up of a neutron spin echo spectrometer is shown in Fig. 15.17. A velocity selector in the primary neutron beam selects a wavelength of about 20 % full width half maximum. The spectrometer offers primary and secondary neutron flight paths, where guide fields $\vec{H}$ and $\vec{H}'$ can be applied. At the beginning of the first path a supermirror polarizer polarizes the neutrons in direction of propagation. A first $\pi/2$ coil turns the neutron spins

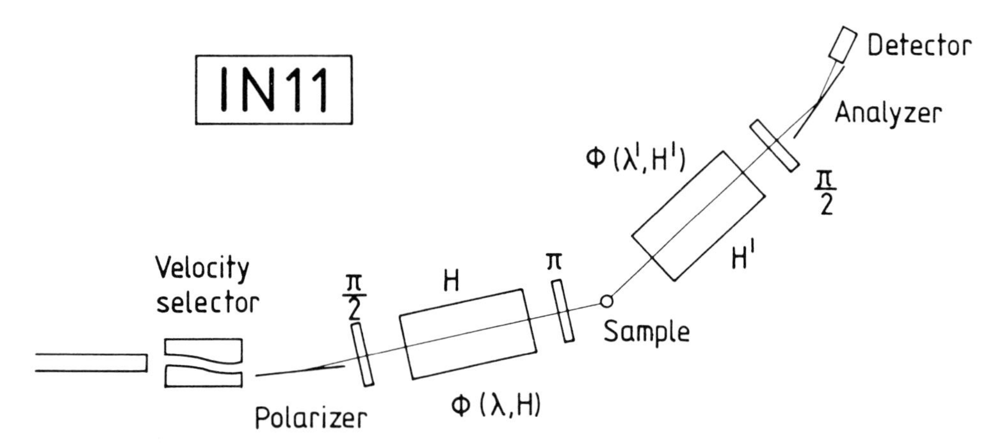

Fig. 15.17: Schematic drawing of the neutron spin echo spectrometer IN|1|1 at the ILL.

into the x-direction perpendicular to the neutron momentum h$\underline{k}$. Starting with this well-defined initial condition the neutrons commence to precess in the applied guide field Without the action of the π-coil each neutron performs a phase angle $\Phi \sim \lambda \int$ Hds. Since the wavelengths are distributed over a wide range, in front of the second π/2-coil the phase angle will be different for each neutron and the beam will be completely depolarized. A π-coil positioned at the half value of the total field integral avoids this effect: on its way to the π-coil the neutron may pass an angle $\Phi_1 = 2\pi n + \Delta\Phi_1$. The action of the π-coil transforms the angle $\Delta\Phi_1$ to $-\Delta\Phi_1$. In a symmetric set up (both field integrals before and after the π-coil are identical) the neutron spin turns by another phase angle $\Phi_2 = \Phi_1 = 2n\pi + \Delta\Phi_1$. The spin transformation at the π-coil just compensates the residual angles $\Delta\Phi_1$ and in front of the second π/2-coil the neutron spin points again into x-direction independent of its velocity. Finally, the second π/2-coil projects the x-component of the polarization in the z-direction and then at the supermirror analyzer the total polarization is recovered. The experimental set up is spin focussing: similar to NMR spin echo methods in front of the second π/2-coil for each spin separately the phase is focussed to its initial value.

In the spin echo spectrometer IN11 realized at the Institut Laue-Langevin, Grenoble, the sample is positioned near the π-coil (Fig. 15.17). With exception of losses due to field inhomogeneties in the case of elastic scattering the polarization remains preserved. If the neutron energy is changed due to inelastic scattering at the sample, the neutron wavelength is modified from λ to $\lambda' = \lambda + \delta\lambda$. Then the phase angle Φ_1 and Φ_2 do not compensate each other and the second π/2-coil projects only the x-component of the polarization into the z-direction which passes afterwards through the analyzer. Apart from resolution corrections the final polarization P_f is then related to initial polarization P_i by:

$$P_f = P_i \int_{-\infty}^{+\infty} S(\vec{q},\omega) \cos\omega t \, d\omega \tag{60}$$

where the scattering function $S(\vec{q},\omega)$ is the probability that during scattering at a certain momentum transfer hq an energy change hω occurs. Furthermore, we have introduced the time variable $t \sim \lambda^3 H$.

From Equ. (60) it is realized that the NSE is a Fourier method and essentially measures the real part of the intermediate scattering function $S(\vec{q},t)$. The Fourier time thereby is proportional to λ^3 and the applied guide field H. A spin echo scan is performed by varying the guide field and thereby studying the intermediate scattering function $S(\vec{q}, t)$ at different Fourier times. Finally, we notice that the use of a broad wavelength band introduces a further averaging process containing an integration over the incident wavelength distribution $I(\lambda)$

$$P_f = P_i \int_{0}^{\infty} I(\lambda) S(\vec{q}(\lambda), t(\lambda)) d\lambda \tag{61}$$

This averaging process obscures somewhat the relationship between P_f and $S(\vec{q},t)$. For many relaxation processes, however, where $S(\vec{q},t)$ is a function of $q^2 t^n$ with $n \leq 1$, the smearing of Equ. (61) becomes weak in λ and is of no practical importance.

15.4.2 Segmental Diffusion in Polymer Melts and Concentrated Solutions

In contrast to dilute solutions the properties of chain molecules in concentrated solutions and in the melt are governed by intermolecular interactions. Although, from a static point of view, these interactions are present on all length scales, screening the excluded volume interactions, from a dynamic point of view they behave differently on smaller and larger spatial dimensions. This can be seen from the molecular mass (M) dependence of the zero shear viscosity η and the center of mass diffusion coefficient D, which exhibit a crossover from $\eta \sim M$ to $\eta \sim M^{3.4}$ and from $D \sim M^{-1}$ to $D \sim M^{-2}$ respectively, when a certain value of M_c is exceeded [7].

A model which can account for these effects is the tube or reptation model [61–64]. There, each arbitrarily chosen chain molecule is assumed to be surrounded by a virtual tube of diameter d_t which imposes topological constraints (entanglements) on the lateral displacement of the chain and force it to reptate like a snake along its own contour. From such a mechanism, $\eta \sim M^{3.0}$ and $D \sim M^{-2}$ is derived. However, on small length scales (or short time scales) the motion of the polymers is not affected by the entanglements and the relaxation may be treated in the framework of the Rouse model [65]. In the Rouse model, the many chain system is represented by an ensemble of gaussian chains. Each chain experiences entropic restoring as well as stochastic forces and interacts with the other only via local frictional forces. When the inertial term is neglected, the related equations of motion [65] lead to a spectrum of relaxation rates $1/\tau_j^R$

$$\frac{1}{\tau_j^R} = \frac{6\pi^2 k_B T\, l^2}{<R_0^2>^2 f} j^2 = \frac{1}{\tau_1} j^2 \quad j = 1,2,\ldots N \tag{62}$$

(k_B Boltzmann constant, T temperature, $<R_0^2>$ mean square end to end distance, f/l^2 friction coefficient per segment length squared, N total numbers of segments per chain).

As outlined above, NSE gives direct access to the intermediate scattering function $S(\vec{q},t)$ which for coherent scattering is given by the correlation function

$$S(\vec{q},t) = \frac{1}{N} \sum_j \sum_k < \exp\left\{ -i\, \vec{q}\, (\vec{r}_k(t) - \vec{r}_j(o)) \right\} > \tag{63}$$

$\vec{r}_k(t)$ and $\vec{r}_j(o)$ are the positions of segments k and j at time t and time zero. The brackets $<\ldots>$ denote the thermal average. In the Gaussian approximation Equ. (63) becomes

$$S(\vec{q},t) = \frac{1}{N} \sum_j \sum_k \exp\left\{ -\frac{q^2}{2} <(\vec{r}_k(t) - \vec{r}_j(0))^2> \right\} \tag{64}$$

Equ. (64) shows that the incoherent scattering function, which is related to the self correlation function (k = j), is directly determined by the mean square displacement of the polymer segments. Sometimes the coherent as well as the incoherent scattering of functions can be approximated by

$$\frac{S(\vec{q},t)}{S(\vec{q},0)} = \exp\left\{ -\text{const.}\, (\Omega(\vec{q})t)^n \right\} \tag{65}$$

The parameters in Equ. (65) are the characteristic frequency in Fourier space $\Omega(q)$ and the lineshape parameter n.

In the case of the Rouse model which is assumed to be valid for spatial dimensions smaller than the tube diameter d_t or $qd_t > 1$ and short times $t \ll \tau_1^R$ the coherent scattering law is given by [66]

$$S(\vec{q},t) = \int_0^\infty du \exp\left\{ -u-(\Omega_R t)^{1/2} g\left(u(\Omega_R t)^{1/2}\right)\right\} \tag{66}$$

with

$$g(y) = \frac{2}{\pi} \int_0^\infty dx \frac{1-e^{x^2}}{x^2} \cos xy \tag{67}$$

and the characteristic frequency

$$\Omega_R = \frac{1}{12} k_B T \frac{l^2}{f} q^4 = \frac{1}{36} W l^4 q^4 \tag{68}$$

$W = 3\, k_B T/fl^2$ being the elementary relaxation rate.

In the asymptotic limit $\Omega_R t \gg 1$ Equ. (66) has the form of Equ. (65) with $n = \frac{1}{2}$, $\Omega(q) \equiv \Omega_R$ and const. $= 2/\sqrt{\pi}$. In contrast, the incoherent scattering function of the Rouse model is determined by the line shape parameter $n = \frac{1}{2}$ in the whole $\Omega_R t$-range and given by

$$S(\vec{q},t) = \exp\left\{ -\frac{12}{\sqrt{\pi}} (\Omega_R t)^{1/2} \right\} \tag{69}$$

The q^4 power law of the characteristic frequencies and the dependence of the scattering laws (66) and (69) on only one scaling variable $(\Omega_R t)^{1/2}$ are typical for the Rouse behaviour.

For spatial dimensions r larger than d_t or $qd_t < 1$, where the reptation model is assumed to provide an adequate microscopic description of the dynamic behaviour, an explicit expression of the coherent scattering law is also available for the long time behaviour $(t > d_t^4/(l^4 W)$ [67].

$$S(\vec{q},t) = 1 - \frac{q^2 d^2}{36} t \left[1 - \exp(\Omega_R t) \operatorname{erf} (\Omega_R t)^{1/2} \right] \tag{70}$$

(erf is the error function)

For $\Omega_R t<1$ Equ. (70) is approximated by Equ. (65), with $n = ½$ and $\Omega(q) \sim q^8$. The corresponding incoherent scattering law leads directly to the form of Equ. (65) with the line shape parameter $n = ¼$ and $\Omega(q) \sim q^8$.

An alternative approach to account for topological constraints was recently proposed by Ronca [68]. His treatment has the advantage that it allows the calculation of the coherent scattering function in the cross-over regime, where the Rouse behaviour changes into the entanglement controlled reptation motion. If this scattering law is plotted vs. $(\Omega_R t)^{1/2}$ and compared with the predictions of the Rouse model (Equ. 66), a q-dependent splitting occurs, whereas a universal curve is obtained for the Rouse dynamics (see Fig. 15.18).
NSE-experiments were performed on three different systems of polydimethylsiloxane (PDMS) melts at T = 373 K or T = 473 K. Samples I and II were blends of deuterated and protonated PDMS molecules (M_w^H = 16 500 g/mol, M_w^D = 15 500 g/mol, M_w/M_n = 1.20; M_w^H = 95 000 g/mol, M_w^D = 142 000 g/mol, M_w/M_n = 1.06, where the concentrations of the protonated species were 10 and 13 %, respectively.) In contrast, sample II was a blockcopolymer composed of deuterated PDMS sequences (M_w^D = 5 700 g/mol and M_w^D/M_w^D = 1.8) and sequences of 8 protonated dimethylsiloxane monomers, the total molecular mass being 100 000 g/mol. In such a sample the scattering from each label arises from its contrast to deuterated surroundings and thus is coherent in nature. However, due to their random distribution along the chains, the scattering from the different labels does not exhibit constructive interferences. Consequently, as long as the internal motions within one label are not seen, the self correlation function is observed [69].

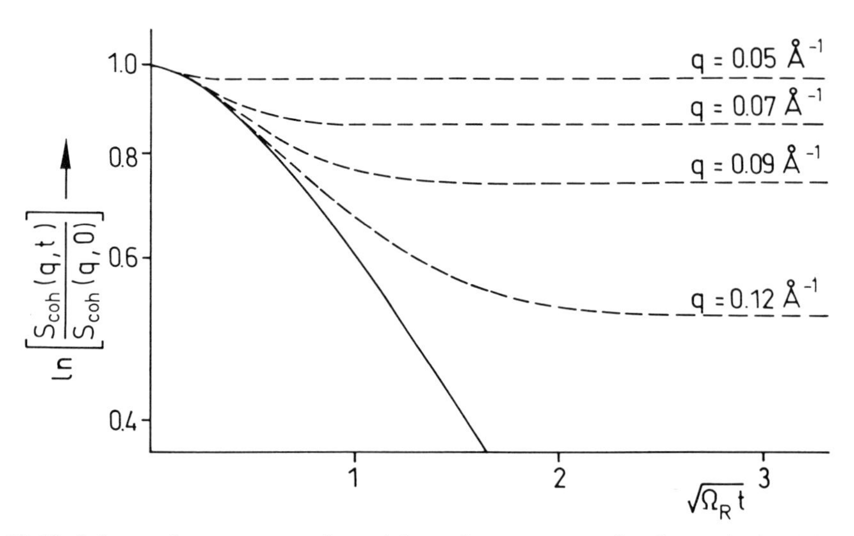

Fig. 15.18: Schematic representation of the coherent scattering laws, derived from the Rouse (———) and the reptation model (----), respectively, (Rouse rate: $Wl^4 = 2 \cdot 10^{13}$ Å^4 s^{-1}; tube diameter: d_t = 47 Å).

In addition neutron scattering experiments were performed at T = 373 K on a mixture of protonated and deuterated PDMS ($M_w^{H,D}$ = 60 000 g/mol c_D = 10 %) diluted by d-chlorobenzene (sample IV). The overall polymer concentrations $0.5 \leq c \leq 1$ were well above the so-called entanglement concentration $c_e \equiv \varrho\, M_c/M$. Since there is no scattering contrast between the deuterated solvent and polymer species the scattering data of these concentrated solutions again gives direct access to the single chain behaviour.

In Fig. 15.19 the scattering value of the low molecular weight sample (I), obtained in the q-range between 0.03 and 0.08 $Å^{-1}$, are shown in a scaling plot S(q,t) vs. $(\Omega_R t)^{1/2}$, using a semilogarithmic scale. As expected, all data are located on a universal curve and in excellent agreement with the predictions of the Rouse model.

The line shape analysis of the different spectra, obtained from the high molecular weight samples (II) yields line shape parameters n, which are all very close to ½, characteristic for the reptation model. However, if the data are plotted in a similar graph as Fig. 15.19, they again lie on a master curve and no q-dependent splitting can be observed (see Fig. 15.20).

Whereas at smaller $(\Omega_R t)^{1/2}$ -values the experimental results are in complete agreement with the predictions of the Rouse model, at larger values of the scaling variable pronounced deviations can be detected. Obviously these deviations, which can be interpreted as a slowing down of the segmental diffusion, have to be attributed to the existence of topological constraints, although they

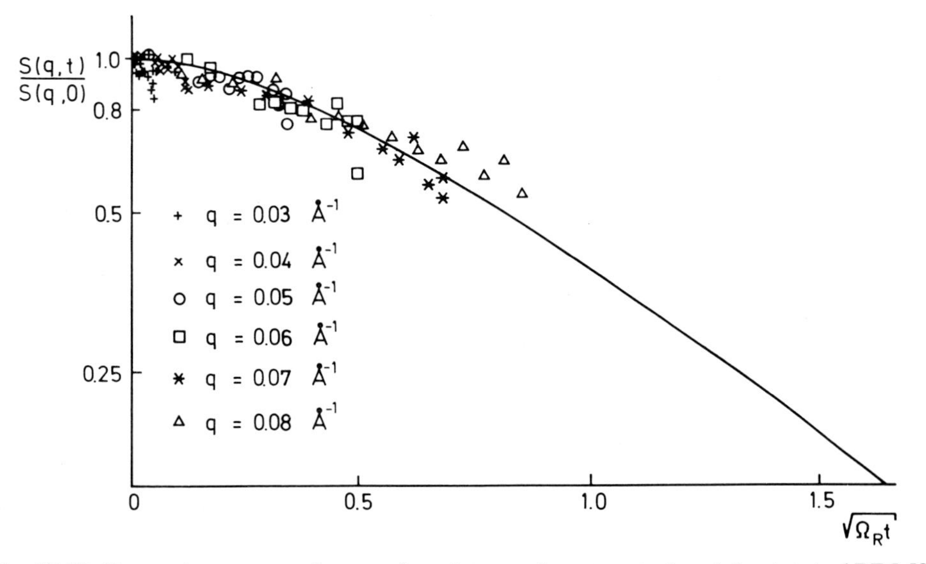

Fig. 15.19: Dynamic structure factor of a mixture of protonated and deuterated PDMS chains ($M<M_c$) at T = 373 K as dependent on the scaling variable $(\Omega_R t)^{1/2}$. Solid line: coherent scattering function of the Rouse model (Equ. 66) with $Wl^4 = 1.59 \cdot 10^{13}\, s^{-1}\, Å^4$.

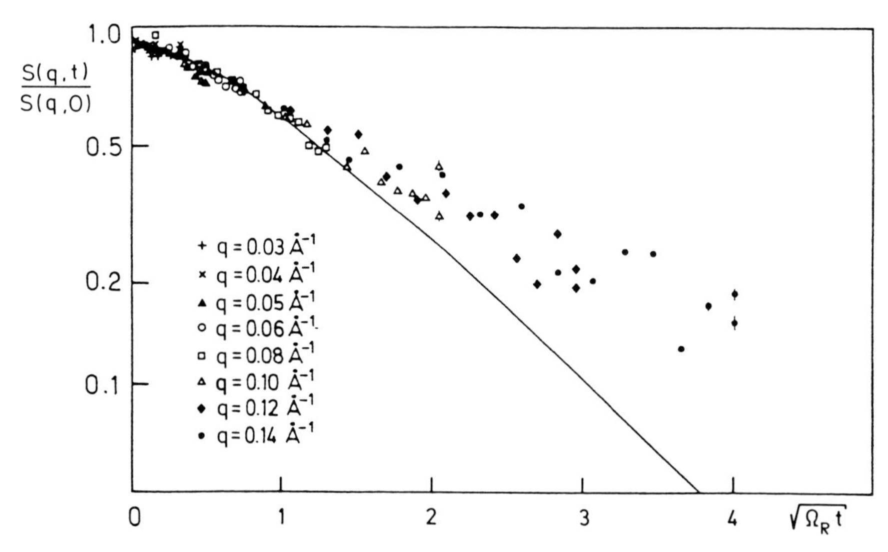

Fig. 15.20: Dynamic structure factor of a mixture of protonated and deuterated PDMS chains ($M>M_c$) at 473 K, as dependent on the scaling variable $(\Omega_R t)^{1/2}$. Solid line: coherent scattering function of the Rouse model (Equ. 66) with $Wl^4 = 3.59 \cdot 10^{13}\,s^{-1}\,Å^4$.

do not show up in the typical features of the reptational motion (see Fig. 15.18) and do not allow extracting an additional length scale, explicitly taken into account in the tube model.

In contrast, the data obtained from sample III do not show any deviations from the Rouse behaviour, indicating that the self correlations are probably less sensitive to topological constraints than the pair correlations (see Fig. 15.21).

Within the regime of validity the Rouse model also fits very accurately to the fine details of the master curves (see. Figs. 15.19–15.21), which on a semilogarithmic scale are bent in the case of pair correlations or coherent scattering (see Equ. 66) and straight in the case of self correlations or incoherent scattering (see Equ. 69).

The only adjustable parameter of the Rouse model is Wl^4 or finally the friction coefficient per mean segment length f/l^2 (see Equ. 68). From simultaneous fits Wl^4 is determined to $1.59 \cdot 10^{13}\,s^{-1}\,Å^4$ (samples I and II, T = 373 K), $1.73 \cdot 10^{13}\,s^{-1}\,Å^4$ (sample III, T = 373 K) and $3.9 \cdot 10^{13}\,s^{-1}\,Å^4$ (sample II, T = 473 K). With respect to the friction coefficient a direct comparison between microscopic scattering data and macroscopic data, obtained for example from dynamic mechanical relaxation, is impossible, since with the latter method the mean square end to end distance $< R_0^2 >$ has to be measured independently from another experiment (see Equ. 62). However, both f/l^2 values are in satisfying agreement.

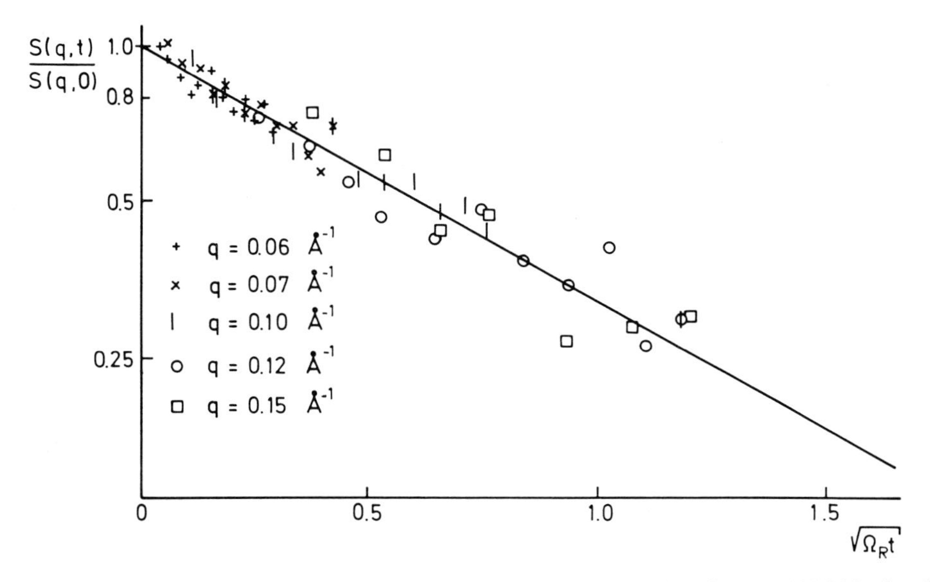

Fig. 15.21: Dynamic structure factor of deuterated PDMS chains (M_w = 100 000 g/mol) with random labels along the chains at T = 373 K, as dependent on the scattering variable $(\Omega_R t)^{1/2}$. Solid line: incoherent scattering function of the Rouse model (Equ. 69) with $Wl^4 = 1.75 \cdot 10^{13}\ s^{-1}\ Å^4$.

Since convincing indications for local reptation were not found on length scales reaching up to 40 Å, even in the melt, it is not surprising that such a mechanism is not observed in concentrated solutions (sample IV). Although the line shape parameters n again are very close to ½, characteristic for the reptation mechanism, the characteristic frequencies $\Omega(q)$ follow the q^4 power law, as predicted by the Rouse model (see Equ. 68).

In Fig. 15.22 the related segmental friction coefficients per segment length squared $f(c)/l^2$, as obtained by fitting all relaxation curves at fixed concentrations with the scattering law of the Rouse model (Equ. 66) are plotted vs. concentration on a double logarithmic scale.

In the case of non-entangled solutions the Rouse model relates $f(c)/l^2$ with the viscosity of the solution by

$$\eta(c) = \eta_0 + \frac{L}{36\ M} < R_0^2 >^2 \varrho \frac{f(c)}{l^2} \qquad (71)$$

In order to compare microscopic and macroscopic parameters, zero shear viscosity measurements on PDMS (M ≅ 7 400 g/mol) in chlorobenzene were performed at T = 373 K. The results are also shown in Fig. 15.22.

Both $f(c)/l^2$ values, the microscopic ones, deduced from the scattering experiments which covered time scales of 10^{-8} to 10^{-9} s and length scales of 10 to 40 Å, and those derived from macroscopic viscosity measurements, exhibit the same concentration dependence, which is close to c^2. The magnitudes differ by a factor less than 2. This agreement between macroscopic measurements on non-entangled solutions and melts, respectively, and microscopic measurements on macroscopically entangled systems may be taken as further evidence that entanglement constraints do not affect significantly the segmental diffusion on the time and length scales accessible to NSE.

Summarizing the scattering results from melts of linear PDMS chains and concentrated solutions, one can conclude that no unambiguous indications for the reptational mechanism can be identified, even if the critical molecular mass M_c is exceeded by factor of more than three. The simple Rouse model provides a general good description of the experimental data with respect to the coherent scattering of short chains as well as to the incoherent scattering of long chains. With respect to the coherent scattering of long chains this agreement is restricted to short time scales, whereas q-independent deviations become pronounced at larger time scales.

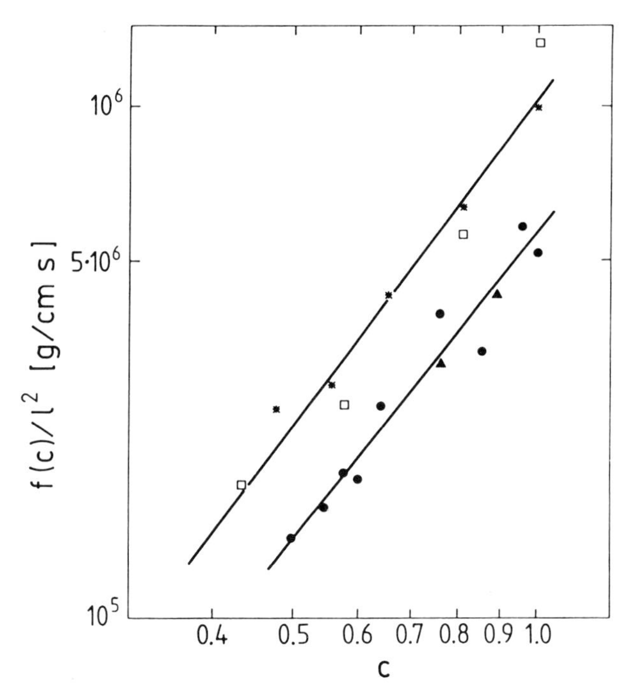

Fig. 15.22: Segmental friction coefficient f(c) per mean square segment length l^2 of PDMS as a function of concentration: ● derived from NSE measurements on macroscopically entangled systems, using the Rouse model; □ derived from viscosity measurements on macroscopically non entangled systems; *data from the literature.

15.4.3 Unattached Polymer Chains in Permanent Networks

The aim to make topological constraints perhaps more effective in the q-t range, experimentally accessible by NSE, was followed up by the idea to trap labelled chains of high molecular weight in permanent networks of small mesh size and to investigate their segmental diffusion behaviour.

For these kind of investigations PDMS model networks were used. They were prepared by end-linking deuterated bifunctional precursor polymers ($M_w^D = 5500$ g/mol, $M_w^D / M_n^D = 1.8$) with four-functional protonated cross-links, each of them containing 22 protons. By this method of synthesis the requirements of uniform cross-linking density and uniform effective functionality are best fulfilled [70]. The molecular mass of the protonated trapped PDMS molecules was 100 000 g/mol.

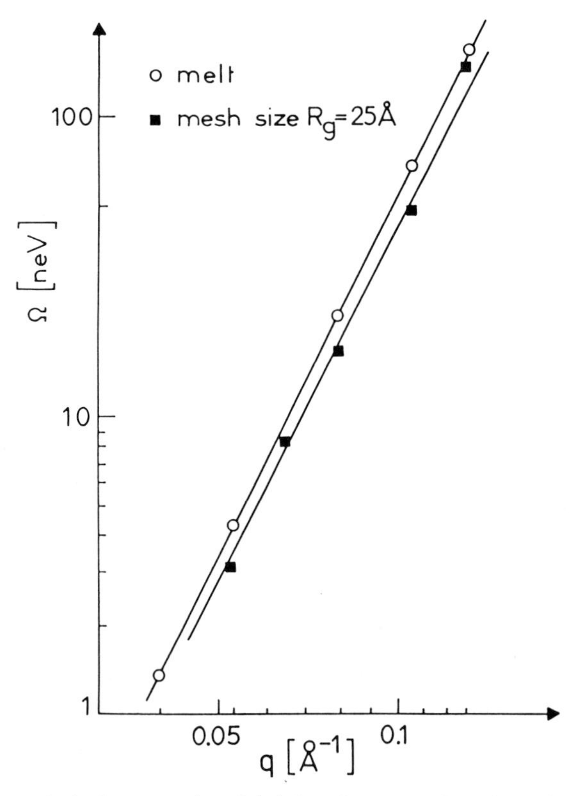

Fig. 15.23: Characteristic frequencies $\Omega(q)$ for the PDMS melt and the chains trapped in a PDMS model network. The solid lines display the $\Omega \sim Q^4$ behaviour, characteristic for the Rouse model.

The meshsize was determined from an equivalent network, prepared from a mixture of protonated and deuterated precursors. The radius of gyration of the mesh strands was found to be 25 Å, which is identical with the radius of gyration of the uncross-linked precursors.

The NSE experiments were performed at T = 373 K under the same conditions as the measurements on the melts. Fig. 15.23 compares the characteristic frequencies Ω(q) obtained from fitting the spectra of the high molecular mass and the network sample individually or simultaneously with the coherent scattering law of the Rouse model (Equ. 66). As in the case of melt the q^4 power law is found and the absolute values of Ω(q) are in nice agreement. These results indicate that the permanent cross-links do not introduce additional topological constraints on the segmental diffusion of the long labelled chains, unattached to the network surroundings. This is quite surprising, since the distance of topological neighbouring cross-links is by a factor of 2 smaller than the distance between macroscopically effective entanglements of the melt, which can be estimated from literature data of the height of plateau modulus. However, for a final judgement of the situation, the plateau modulus of both samples should be measured under identical conditions, which is still not completed.

15.4.4 Diffusive Motions of Cross-links

From the results reported in the previous section, it looks less probable that the cross-links of a polymer network can be treated as fixed obstacles with respect to segmental diffusion of unattached, trapped long chains. However, up to now no direct microscopic information on the kind and extent of the cross-links dynamics is available, although all theories on the network behaviour have to make more or less far reaching assumptions on it.

For this reason, NSE experiments were performed on PDMS model networks at T = 373 K. The network characteristics were the same as those described before. The samples differed only by the absence of unattached labelled chains. For comparison a melt of the corresponding end labelled precursor polymers was also investigated. In order to look for correlations between the cross-links, SANS experiments were performed. In the q-range $0.01<q<0.3$ Å^{-1}, the intensity profile exhibits nearly no q-dependence. A slight decay towards larger q is probably due to the form factor of the individual cross-links which have a radius of gyration of about 10 Å. Since we do not observe any correlations between different cross-links, their scattering can be treated as originating from a single junction and it reveals the self correlation function of the cross-link.

Fig. 15.24 compares the obtained relaxation curves for the cross-links and the chain ends for two different q-values. Without any analysis a strong reduction of the cross-link mobility compared to that of the chain ends is obvious. A

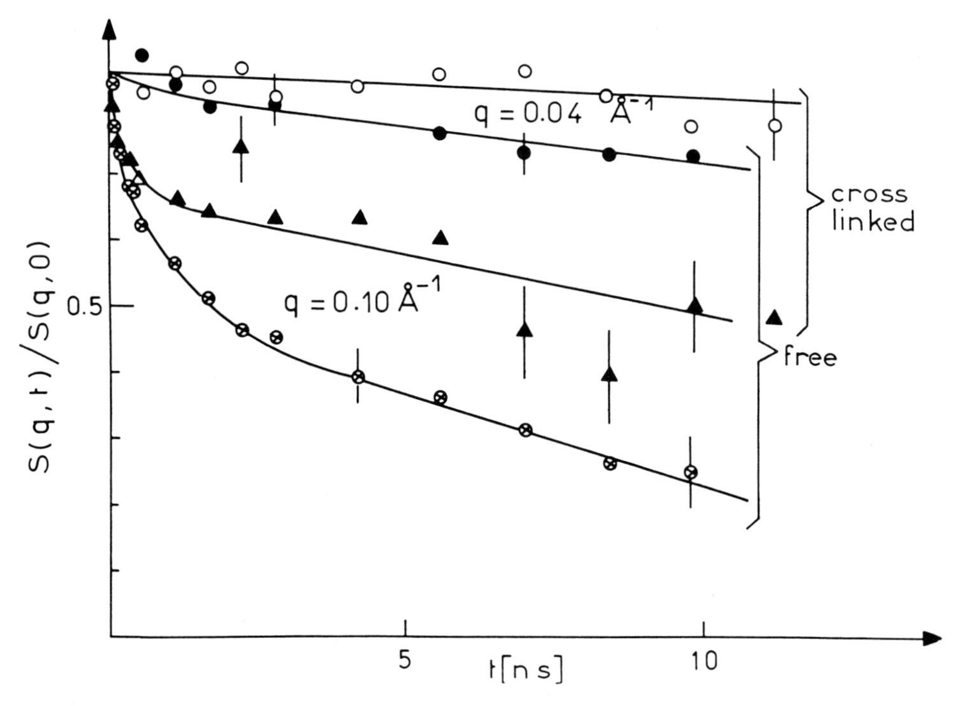

Fig. 15.24: NSE spectra from linear end labelled PDMS chains (M_w = 5 500 g/mol) at T = 373 K. The solid lines are guides for the eyes.

closer inspection also shows that the lineshape of both curves differs. While $S(\vec{q},t)$ from the chain ends decays continuously, $S(\vec{q},t)$ from the cross-links appears to decay faster for shorter than for longer times. This difference in lineshape is quantified in Fig. 15.25 where the lineshape parameter of both samples is plotted vs. q.

The measured line shape parameter for the end labelled chain is in close agreement with the $n = \frac{1}{2}$ prediction of the self-correlation function of a Rouse chain (see Equ. 69). The observation of an initially faster decaying relaxation curve for the cross-links translates into a significantly smaller time exponent n. A fit with Equ. (69) yields the Rouse parameters $Wl^4 = (3.8 \pm 0.4) \cdot 10^{13}$ Å^4 s^{-1} and $(0.62 \pm .06) \cdot 10^{13}$ Å^4 s^{-1} for the chain ends and cross-links, respectively.

These findings have to be compared with the results of a theoretical treatment provided by Warner [71]. This author showed that within the frame work of the Rouse model the characteristic frequency for the ends of linear chains doubles and is reduced by a factor $2/\Phi$ for Φ-fold connected segments (e. g. centers or star-shaped molecules or cross-links.) While the Rouse parameters of the chain ends and the labels along the chains are in good agreement with these predictors, the Rouse parameter of the cross-links is 30 % more reduced than expected for 4-functional junctions.

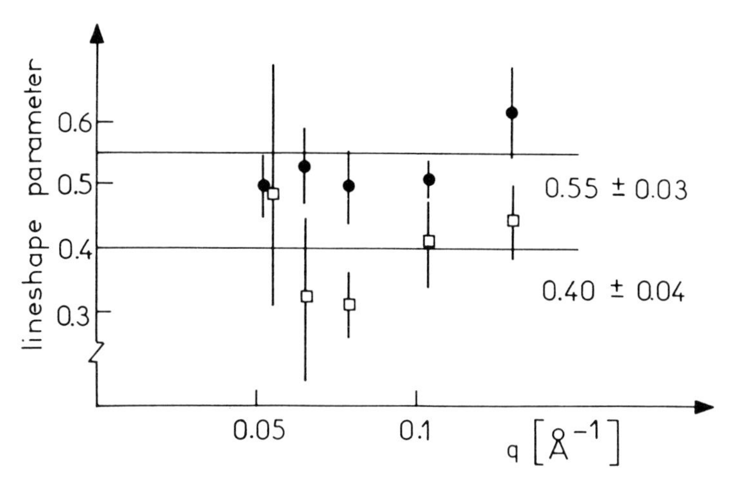

Fig. 15.25: Lineshape parameters obtained from the NSE spectra of the end-labelled PDMS chains (●) and the labelled cross-links (□) according to Equ. (65).

Both discrepancies in time scale and in lineshape suggest a more refined analysis of the cross-link data. Thus in a second step we consider spatial limitations for the cross-link motion. Other than the motion of a labelled chain ends and of individually labelled chain segments, which may essentially probe the whole sample volume, the spatial fluctuation of a cross-link being part of a macroscopic network is strongly confined in space. We assume that for asymptotic times the probability to find a junction at a distance r from its equilibrium position is given by a Gaussian profile $\exp(-r^2/\sigma^2)$. In this case, the self-correlation function of the cross-link motion does not decay to zero for infinite times, since even then there is a finite probability to find the cross-link at position r close to its initial position. Thus, besides a component decaying in time it contains a time independent part. The Fourier transform of this spatial limitations appears as a constant part in the intermediate scattering function. In the language of quasielastic neutron scattering this contribution is called elastic incoherent structure factor (EISF) [72]. Assuming that the dynamics of a cross-link still remains Rouse-like, Equ. (69) transforms to:

$$S(\vec{q},t) = \exp\left\{-\frac{q^2\,\sigma^2}{4}\right\} +$$

$$\left[1-\exp\left\{-\frac{q^2\,\sigma^2}{4}\right\}\right]\exp\left\{-\frac{12}{\sqrt{\pi}}\,(\Omega t)^{1/2}\right\} \tag{72}$$

Fig. 15.26 presents the fit of Equ. (72) to the cross-link data.

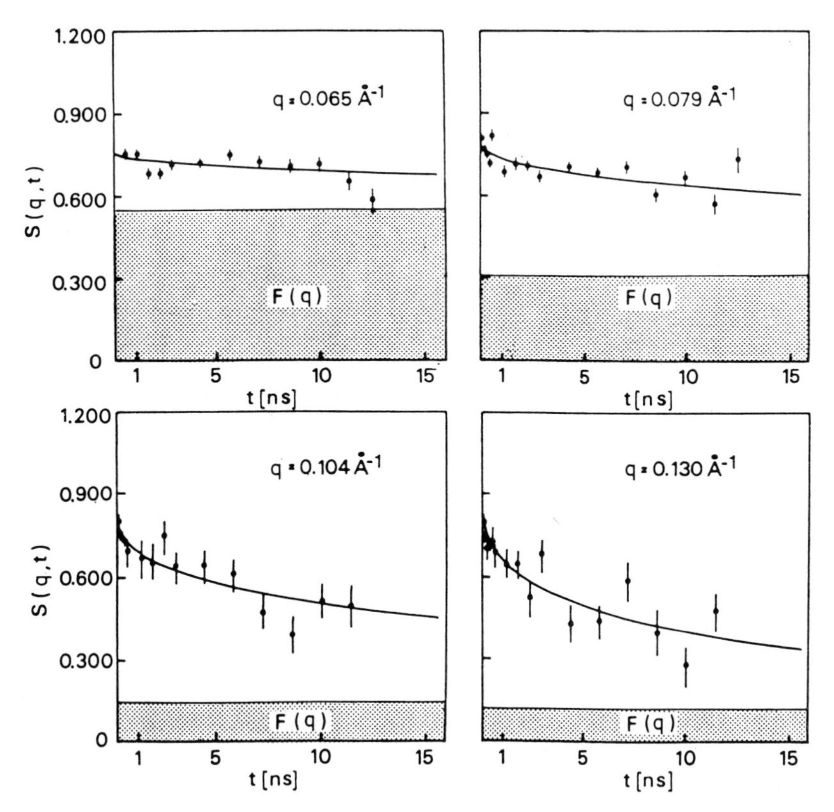

Fig. 15.26: NSE spectra for labelled cross-links at 4 different Q-values. Included is a result of a fit with a sum of a constant EISF contribution F(Q) and a Rouse relaxation spectrum (Equ. 72). Note that the NSE spectra do not approach 1 for $t \to 0$. This effect is related to fast relaxation processes of the deuterated mesh which have not been subtracted.

The shaded area displays the constant EISF contribution to $S(\vec{q},t)$. A combined fit to all spectra yields $Wl^4 = 0.84 \cdot 10^{13}$ Å^4 s^{-1} and $\sigma = 24.5 \pm 1.5$ Å. Displayed are fits to single spectra with the rate fixed to the value obtained from the joint fit. The obtained σ-values fluctuate between 22.6 and 28.1 Å and demonstrate the consistency of the description. The obtained Rouse parameter amounts to nearly exactly half the value of the randomly labelled melt in very good agreement with the prediction of a $2/\Phi$ reduction.

These measurements for the first time allowed an experimental access to the microscopic extend of cross-link fluctuations. We now compare the outcome with the prediction of the so-called phantom network model [73, 74]. This model by James and Guth is the simplest and its formulation by Flory [75] is also the most elegant network model allowing fluctuations of cross-links. It assumes freely intersecting chains with forces acting only on pairs of junctions.

In this model under stress the average junction positions are affinely deformed and each cross-link performs Gaussian spatial fluctuations around its average position. The width of this distribution is predicted to:

$$<\sigma^2> = 4/3\Phi <R_E^2> \tag{73}$$

where $<R_E^2>$ is the mean square end to distance between cross-links. Taking $<R_E^2> = 6<R_g^2>$ and inserting the measured value of $R_g = 25$ Å, the phantom network predicts a fluctuation range of $\sigma = 35$ Å, while the experimental values range between $\sigma = 23$ Å and $\sigma = 28$ Å. Qualitatively theoretical prediction and experimental results agree, quantitatively the smaller experimental fluctuation ranges suggest that in a real network additional constraints are active which reduce the fluctuation range below the phantom network prediction.

15.5 Conclusions

In the first section we have reviewed the use of time resolved quasi-elastic light scattering (photon correlation spectroscopy) for analyzing dynamic processes in bulk viscoelastic systems. Since the q-vector for light scattering is very small usually a continuum mechanical treatment is appropriate to interpret the measured density autocorrelation function. Thus we present a theory which connects the latter quantity with the longitudinal bulk compliance and perform a quantitative comparison between experiment and theory supporting it.

The time dependence of density autocorrelation functions is often described by fractional exponential decay functions. The use of an inverse Laplace transformation technique to directly analyze the correlation functions gives more useful information about the system than a stretched exponential by avoiding any functional form of the correlation function. The capabilities of the transformation technique is demonstrated on a series of polyalkylmethacrylates where the separation of primary and secondary relaxation processes was possible.

The theoretical result in mind which relates the density correlation function with mechanical compliance data we show that the time-frequency superposition principle, widely used in the treatment of mechanical data, can also be applied for photon correlation data.

The comparison between light scattering and dielectric relaxation in bulk viscoelastic systems shows that the collective dynamics governs the molecular one since both the distribution of retardation times and the characteristic times are identical for a model system when studied with either method.

The further extension of the photon correlation regime to shorter times is possible by performing Rayleigh-Brillouin spectroscopy. Using linear response theory the proper analysis of Brillouin linewidth and shift as a function of temperature of a melt of linear chain molecules led to the conclusion that in such liquids mesomorphic phase transition does not play any role.

By connecting the long time (quasi-elastic scattering) with the hypersonic regime a relaxation time range of more than 12 decades is covered. We have demonstrated that the temperature dependence of the relaxation times of a viscoelastic fluid in that range is consistently described by using the free volume concept and show further the applicability of recent theories by analyzing the structural relaxation in the GHz region.

Apart from the analysis of the dynamics of density fluctuations it is a well known fact that quasi-elastic light scattering can be used to measure the time decay function of concentration fluctuations. The usual application of dilute solutions has been extended for the first time to solution of compatible binary polymer blends in order to decide between two different theoretical models describing the microscopic dynamics. The data analysis shows that a parallel and not a series connection between the individual Rouse mobilities of the two polymer species in the mixtures is consistent with experimental results.

In the last section, a comprehensive review is given on investigations of segmental diffusion in uncross-linked and cross-linked dense polymer systems by using neutron spin echo spectroscopy. In particular, we address the role of topological constraints which are believed to cause a reptational behaviour of motion on larger length scales. However, segmental diffusion of labelled long linear PDMS chains ($M > M_c$), and $c > c_e$ surrounded either by equivalent linear chains or trapped in a four functional PDMS model network of small mesh size, does not exhibit the indications of a reptational mechanism of motion although the probed length and time scales reached up to 50 Å and 40 ns, respectively. The incoherent scattering data are in quite good agreement with the predictions of the Rouse model, which does not take account for an explicit topological hindrance, but replaces all intermolecular interaction by local friction forces. Especially the coherent scattering data from the systems of linear chains certainly show deviations from the Rouse behaviour at larger time scales, which are comparable with a cross over to a reduced Rouse relaxation rate, but we have no indications for the occurrence of an additional internal length scale (tube diameter) characteristic for the reptation model.

Investigations on the dynamics of labelled cross-links in four functional networks show that the diffusive motions are Rouse-like and limited in space. For the first time the spatial extension of the junction fluctuations was determined experimentally and found to be rather extended. These results support the ideas underlying the phantom network model.

Acknowledgements

The authors gratefully acknowledge the contributions of Ch. Becker, J.-U. Hagenah, R. Oeser and D. Richter. We are further especially indebted to W. Hess for his work on the relation between density fluctuations and mechanical compliance. We also thank Th. Vilgis for helpful discussions.

15.6 References

[1] B. J. Berne, R. Pecora: Dynamic Light Scattering, Wiley, New York 1976.
[2] W. Hess unpublished results.
[3] J.-U. Hagenah, Ph. D. Thesis, Mainz 1988.
[4] R. Kubo, M. Toda, N. Hashitsume: Statistical Physics, Springer 1985.
[5] C. H. Wang, E.W. Fischer, J. Chem. Phys. 82 (1985) 632.
[6] G. D. Patterson, in R. Pecora (ed.): Dynamic Light Scattering, Plenum Press (1985).
[7] J. D. Ferry Viscoelastic Properties of Polymers, Wiley, New York 1980.
[8] L. Giebel, Diplom Thesis, Mainz 1986.
[9] J. E. McKinney, H.V. Belcher, J. Res. Nat. Bur. Stand 67 (1963) 43.
[10] N. G. McCrum, B. E. Read, G. Williams: Anelastic and Dielectric Effects in Polymer Solids, Wiley, New York 1967.
[11] G. Meier, J.-U. Hagenah, C. H. Wang, G. Fytas, E.W. Fischer, Polymer 28 (1987) 1640.
[12] C. H. Wang, G. Fytas, E.W. Fischer, J. Chem. Phys. 82 (1985) 4332.
[13] i. e. G. Fytas, C. H. Wang, D. Lilge, Th. Dorfmüller, J. Chem. Phys. 75 (1981) 4747.
[14] B. Momper, Diplom Thesis, Mainz 1987.
[15] i. e. G. D. Patterson, Adv. Poly. Sci. 48 (1983) 175.
[16] i. e. A. K. Rajagopal, K. L. Ngai, in K. L. Ngai, G. Wright (eds.): Relaxation in Complex Systems, Naval Research Lab., Wash., DC 1984.
[17] K. Mehler, Ph. D. Thesis, Mainz 1983.
[18] N. Ostrowsky, D. Sornette, P. Parker, E. R. Pike, Optica Acta 28 (1981) 1059.
[19] S.W. Provencher, Comput. Physics Commun. 27 (1982) 213; ibid. 27 (1982) 229.
[20] R. S. Stock, W. H. Ray, J. Polym. Sci. Polym. Phys. Ed. 23 (1985) 1393.
[21] H. Pollard, Bull, Am. Math. Soc. 52 (1946) 908.
[22] J.-U. Hagenah, G. Meier, G. Fytas, E.W. Fischer, Polym. J. 19 (1987) 441.
[23] For a comparison of experimental data see eg: G. Fytas, A. Patkowski, G. Meier, Th. Dorfmüller, J. Chem. Phys. 80 (1984) 2214.
[24] D. Boese, B. Momper, G. Meier, F. Kremer, J.-U. Hagenah, E.W. Fischer, Macromolecules 22 (1989) 4416.
[25] G. Williams, Chem. Soc. Rev. 7 (1977) 89.
[26] S. Havriliak, S. Negami, Polymer 8 (1967) 161.
[27] J. Heijboer, in J. A. Prins (ed.): Proc. Inter. Conf. on Physics of Non-Crystalline Solids, North Holland Publ. 1985.

[28] G. Meier, in Th. Dorfmüller, G. Williams (eds.): Molecular Dynamics and Relaxation Phenomena in Glasses, Lecture Notes in Physics, Springer 1987.
[29] D. A. Jackson, E. R. Pike, J. G. Powles, J. M. Vaughan, J. Phys. C6 (1973) L55.
[30] C. Cohen, V. Sankur, C. J. Pings, J. Chem. Phys. 67 (1977) 1436.
[31] T. A. King, M. F. Treadaway, Chem. Phys. Lett. 50 (1977) 636.
[32] G. Fytas, C. H. Wang, E.W. Fischer, K. Mehler, J. Polym. Sci. Polym. Phys. Ed. 24 (1986) 1859.
[33] G. Fytas, C. H. Wang, E.W. Fischer, Macromolecules 21 (1988) 2253.
[34] L. Giebel, G. Meier, G. Fytas, E.W. Fischer, submitted to J. Polym. Sci. Polym. Phys. Ed.
[35] G. Fytas, in Physical Optics of Dynamic Phenomenon and Processes, in B. Sedlacek (ed.): Macromolecular Systems, W. De Gruyter 1985.
[36] G. Meier, G. Fytas, Th. Dorfmüller, Macromolecules 17 (1984) 957.
[37] R. Kono, J. Phys. Soc. Jpn. 15 (1970) 718.
[38] B.Y. Li, D. Z. Jiang, G. Fytas, C. H. Wang, Macromolecules 19 (1986) 778.
[39] G. Meier, E.W. Fischer, in D. Richter, T. Springer (eds.): Polymer Motion in Dense Systems, Springer 1988.
[40] M. G. Brereton, E.W. Fischer, G. Fytas, U. Murschall, J. Chem. Phys. 86 (1987) 5174.
[41] H. Sillescu, Makromol. Chem., Rapid Comm. 5 (1984) 519.
[42] E. Kramer, P. Green, C. Palmstrøm, Polymer 25 (1986) 473.
[43] F. Brochard, P. G. de Gennes, Physica 118A (1983) 289.
[44] K. Binder, J. Chem. Phys. 79 (1983) 6387.
[45] U. Murschall, E.W. Fischer, Ch. Herkt-Maetzky, G. Fytas, J. Polym. Sci. Polym. Lett. 24 (1986) 191.
[46] Y. H. Lin, C. H. Wang, J. Chem. Phys. 70 (1979) 681.
[47] T. G. Oh, E.W. Fischer, G. P. Hellmann, T. P. Russel, C. H. Wang, Polymer 27 (1980) 261.
[48] J. K. Krüger, Solid State Comm. 30 (1979) 43.
[49] J. K. Krüger, L. Peetz, W. Wildner, M. Pietralla, Polymer 21 (1980) 620.
[50] J. K. Krüger, L. Peetz, M. Pietralla, H. G. Unruh, Colloid Polym. Sci. 259 (1981) 715.
[51] C. H. Wang, Y. H. Lin, D. R. Jones, Mol. Phys. 37 (1979) 287.
[52] Y. Higashigaki, C. H. Wang, J. Chem. Phys. 74 (1981) 3175.
[53] G. D. Patterson, J. Polym. Sci. 15 (1977) 455.
[54] Ch. Becker, Diplom Thesis, Mainz 1988.
[55] Th. Dries, F. Fujara, M. Kiebel, E. Rössler, H. Sillescu, J. Chem. Phys. 88 (1988) 2139.
[56] D. Richter, B. Ewen, in R. Pynn and A. Skjeltorp (eds.): Scaling Phenomena in Disordered Systems, NATO ASI Series B. Physics Vol. 133, Plenum Press 1985.
[57] B. Ewen, D. Richter, Festkörperprobleme 27 (1987) 1.
[58] F. Mezei, in F. Mezei (ed.): Neutron Spin Echo, Springer 1980.
[59] F. Mezei, Z. Phys. 255 (1972) 146.
[60] J. B. Hayter, J. Penfold, Z. Phys. B35 (1979) 199.
[61] P. G. de Gennes, J. Chem. Phys. 55 (1971) 572.
[62] S. F. Edwards, J. M. V. Grant, J. Phys. A6 (1973) 1169.
[63] M. Doi, S. F. Edwards, J. Chem. Soc. Farad. Trans. 2 274 (1978) 1789, 1802, 1818.
[64] M. Doi, J. Polym. Sci., Polym. Phys. Ed. 18 (1980) 1005; 21 (1983) 667.
[65] P. E. Rouse, J. Chem. Phys. 21 (1953) 1272.

[66] P. G. de Gennes, Physics (USA) 3 (1967) 37.
[67] P. G. de Gennes, J. Chem. Phys. 72 (1980) 4756; J. Phys. 42 (1981) 735.
[68] G. Ronca, J. Chem. Phys. 79 (1983) 1039.
[69] R. Oeser, B. Ewen, D. Richter, B. Farago, Phys. Rev. Lett. 60 (1988) 1041.
[70] E. Herz, P. Rempp, W. Borchard, J. Polymer Sci. 2 (1978) 105.
[71] M. Warner, J. Phys. C., Solid State Physics 14 (1981) 4985.
[72] T. Springer, in Springer Tracts in Modern Physics, Vol. 64, Springer 1972.
[73] H. M. James, J. Chem., Phys. 15 (1947) 669; 21 (1953) 1039.
[74] H. M. James, E. Guth, J. Chem., Phys. 15 (1947) 669; 21 (1953) 1039.
[75] P. J. Flory, Proc. Roy. Soc. A351 (1976) 51.

16 Structure and Properties of Semicrystalline Polymers

Erhard W. Fischer*, Manfred Stamm*,
and Ingrid G. Voigt-Martin**

16.1 Introduction

Semicrystalline polymers resemble a large class of materials which are of significant industrial importance. They can show quite different appearance and physical behaviour which depends besides chemical differences largely on the detailed morphology or molecular arrangement of the chains. "Perfect" polymer crystals are virtually non-existent but can only be prepared by e. g. solid state polymerization or under extreme conditions. Most crystalline polymers are semicrystalline and consist of amorphous and crystalline regions. Crystallization conditions and physical treatment can largely influence e. g. the degree of crystallinity, the shape of crystallites and the trajectory of chains within this morphology. In the following we will describe methods which enable us to learn something about the structural and conformational details of these materials which might help to understand some of their physical properties.

* Max-Planck-Institut für Polymerforschung, Postfach 3148, D-6500 Mainz
** Institut für Physikalische Chemie der Universität Mainz, Jakob-Welder-Weg, D-6500 Mainz

16.2 Morphology of Semicrystalline Polymers

16.2.1 Comparison of Different Techniques

Semicrystalline polymers are composed of alternating amorphous and crystalline regions. The detailed architecture and arrangement of those regions depends on various parameters like molecular weight and its distribution, crystallization temperature, branching of the chains and many more [1, 2]. An analysis and understanding of this morphology, however, is strongly needed, since many physical properties of the semicrystalline material are related to morphological features. There are well established techniques in order to obtain information on morphology in different ways. "Direct" pictures of amorphous and crystalline regions can be obtained by transmission electron microscopy, but special staining or etching techniques have to be elaborated for visualizing details of the lamellar structure. Small angle X-ray scattering, on the other hand, can be used on bulk samples without special contrasting. The density difference between crystalline and amorphous regions is mostly large enough to produce a peak in the scattering pattern. The position of the peak is related to the mean repeating distance between neighbouring lamellae and to the electron density distribution within the sample. The obtained long period is an average over the whole sample, and techniques for obtaining information on the amorphous and crystalline phase behaviour will be described in the following chapters. Other techniques like Raman spectroscopy, Brillouin scattering, density measurements, differential calorimetry etc. are sensitive to different aspects of the crystalline – amorphous morphology and generally average over the sample in a different way. Some results and relations will be discussed in more detail later on.

16.2.2 Small Angle X-Ray Scattering (SAXS)

The method of small angle X-ray scattering (SAXS) has been widely used to investigate the morphology of semicrystalline polymers. The use of the technique and its application to polymers is described in many references (for a general review see e.g. [1, 3–6]). It is nowadays a standard procedure to obtain the so-called long spacing of a semicrystalline polymer from SAXS. One mostly uses a Kratky slit collimation system to obtain the slit-smeared scattering curve which has to be desmeared by a mathematical procedure [7, 8]. The position of the peak is a measure of the one-dimensional periodicity of the crystalline and amorphous regions. Besides this long period, other parameters can be extracted from the scattering curve by suitable analysis. Thus by an appropriate integration of the scattered intensity the "scattering invariant" Q can be obtained, which is directly related to the mean square fluctuation of electron density, irrespective of special features of the structure. From the tail of the scattering pattern information about the specific inner surface O_S of the interphase between the phases and the size of the interfacial region can be obtained.

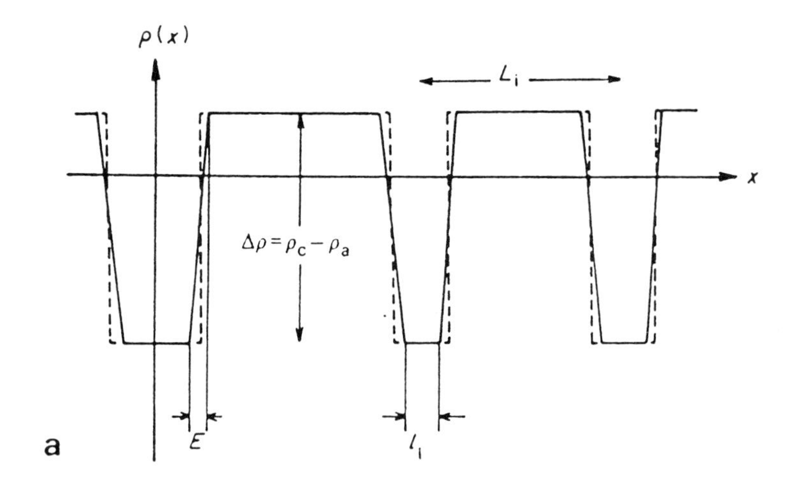

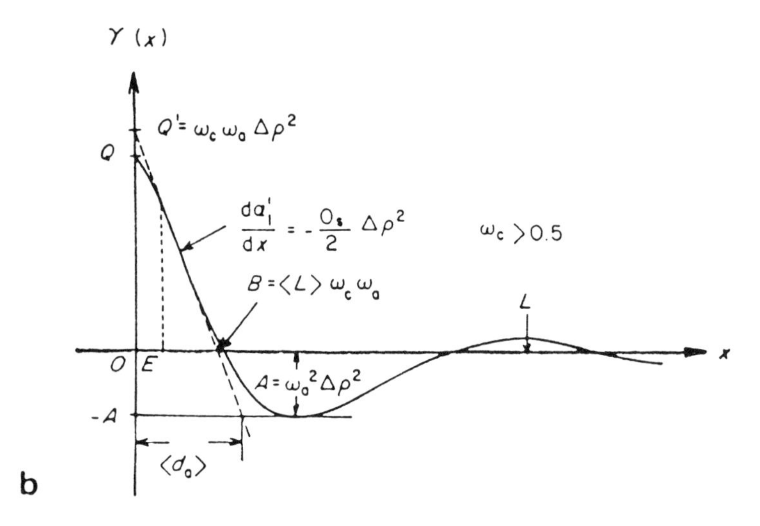

Fig. 16.1: (a) Model of the electron density distribution, $\varrho(x)$, normal to the lamellar surfaces for a lamellar structure of a semicrystalline polymer (crystallinity $w_c > 0.5$); (b) corresponding electron density correlation function $\gamma(x)$ (Q invariant, w_c = crystallinity = $(1 - w_a)$, $\Delta\varrho$ electron density difference between crystalline and amorphous regions, O_s specific surface, $<d_a>$, mean thickness of amorphous regions, L long spacing) (Ref. [6, 9]).

An analytical technique was developed by Strobl and coworkers [9] at Mainz, which permits a direct analysis of the scattering pattern compared to the more conventional techniques. By Fourier analysis an electron density correlation function γ is obtained which can be interpreted in terms of the above mentioned parameters (Fig. 16.1).

$$\gamma(x) = \int_0^\infty 4\,\pi s^2 I(s) \cos 2\pi s x \, ds \tag{1}$$

where s is the magnitude of the scattering vector $s = (2/\lambda)\sin\theta$, 2θ denoting the scattering angle and λ the wavelength. $\gamma(x)$ describes spatial correlations of electron density fluctuation

$$\gamma(x) = \langle [\varrho(x_0) - \langle\varrho\rangle]\,[\varrho(x_0 + x) - \langle\varrho\rangle] \rangle \tag{2}$$

$\varrho(x)$ and $\langle\varrho\rangle$ denote electron density and mean electron density normal to the lamellae (see Fig. 16.1a) and it is spatially averaged over the sample.

The evaluation of the electron density correlation function is especially helpful when changes in certain parameters with sample treatment are investigated. This will be demonstrated in the following chapters for the investigation of crystallization and melting of semicrystalline polymers as well as for the development of a model for stress induced crystallization in elastomers.

16.2.3 Electron Microscopy (EM)

Electron microscopy is also a technique which is commonly used for the investigation of semicrystalline polymers. For a review see e. g. [2,10] and references therein. New experimental techniques which include electron energy analysis, improved spatial resolution, scanning techniques and X-ray microanalysis promise further insight into many areas. In the following we will discuss mostly transmission electron microscopy which has been developed especially in the case of polyethylene to a high degree of perfection. The large variety of morphological structures which are developed depending on molecular weight and crystallization conditions is indicated in Fig. 16.2a–16.2d. For linear polyethylene it was demonstrated that there is a dramatic irreversible increase in both long spacing and crystal thickness at a specific crystallization temperature (Fig. 16.3a). It was further shown that at high crystallization temperatures thick, roof-shaped crystals having large lateral dimensions are formed. Analysis of the apex angle indicated the surface planes to be {101}, {302}, {201} and {301} corresponding to molecular tilts with respect to the surface normal of 18°, 27°, 34° and 46°. Quenching, on the other hand, was shown to give rise to curved crystals of shorter lateral dimensions. In a subsequent study of the molecular weight dependence, it was shown that there is a significant difference between quenched and isothermally crystallized material. For quenched linear polyethylene, the crystal thickness d_c is virtually independent of molecular weight while the long spacing L increases slightly. Furthermore, the thickness distributions are narrow. The situation is entirely different for isothermally crystallized material, where there is an extremely broad distribution in both d_c and L (Fig. 16.3b,c). In this case, it is necessary to distinguish between d_{cmin} and d_{cmax} as well as L_{min} and L_{max}. While both d_{cmin} and L_{min} behave as in the quenched case, d_{cmax} and L_{max} increase rapidly up to $M_w \sim 10^5$ and then decrease again for larger molecular weights. These features indicate that isothermal thickening [12] leads to a broad thickness distribution, which is however, increasingly restricted at higher molecular weights due to entanglements. Following these investigations it became clear that the answer to the long standing problem relating to different numerical values as obtained by small angle X-ray scattering, Raman longitudinal acoustic mode spectroscopy and electron microscopy was to be sought in the significant differences occur-

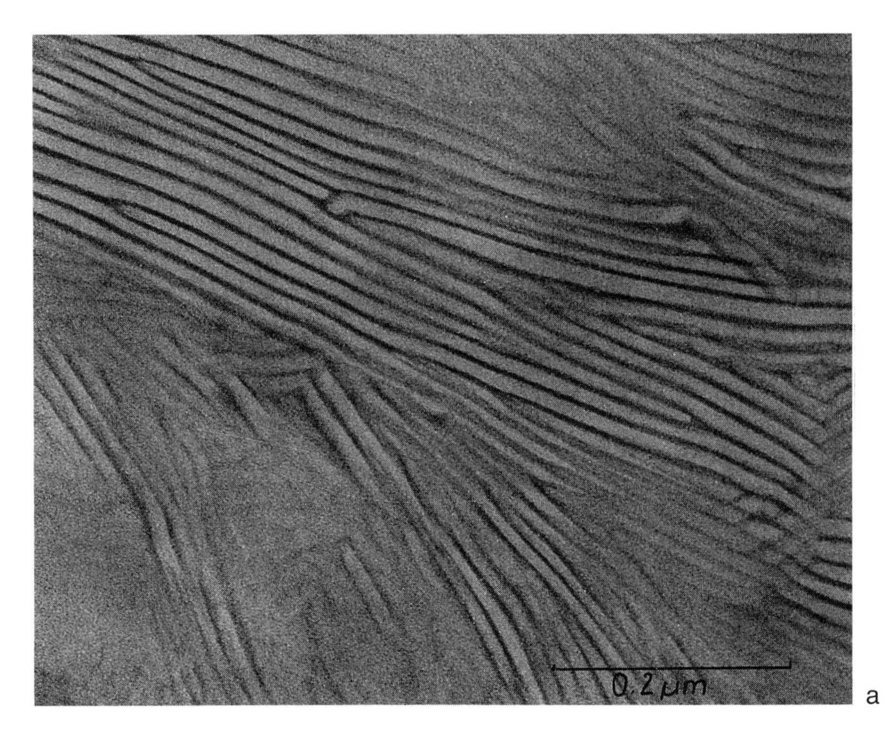

Fig. 16.2: Transmission electron micrographs of polyethylene samples of different molecular weight crystallized from the melt at different temperatures: (a) M = 5.6 x 10^3, quenched, (b) M = 1.89 x 10^5, quenched, (c) M = 5.6 x 10^3, T_c = 127 °C, and (d) M = 6 x 10^6, T_c = 130 °C.

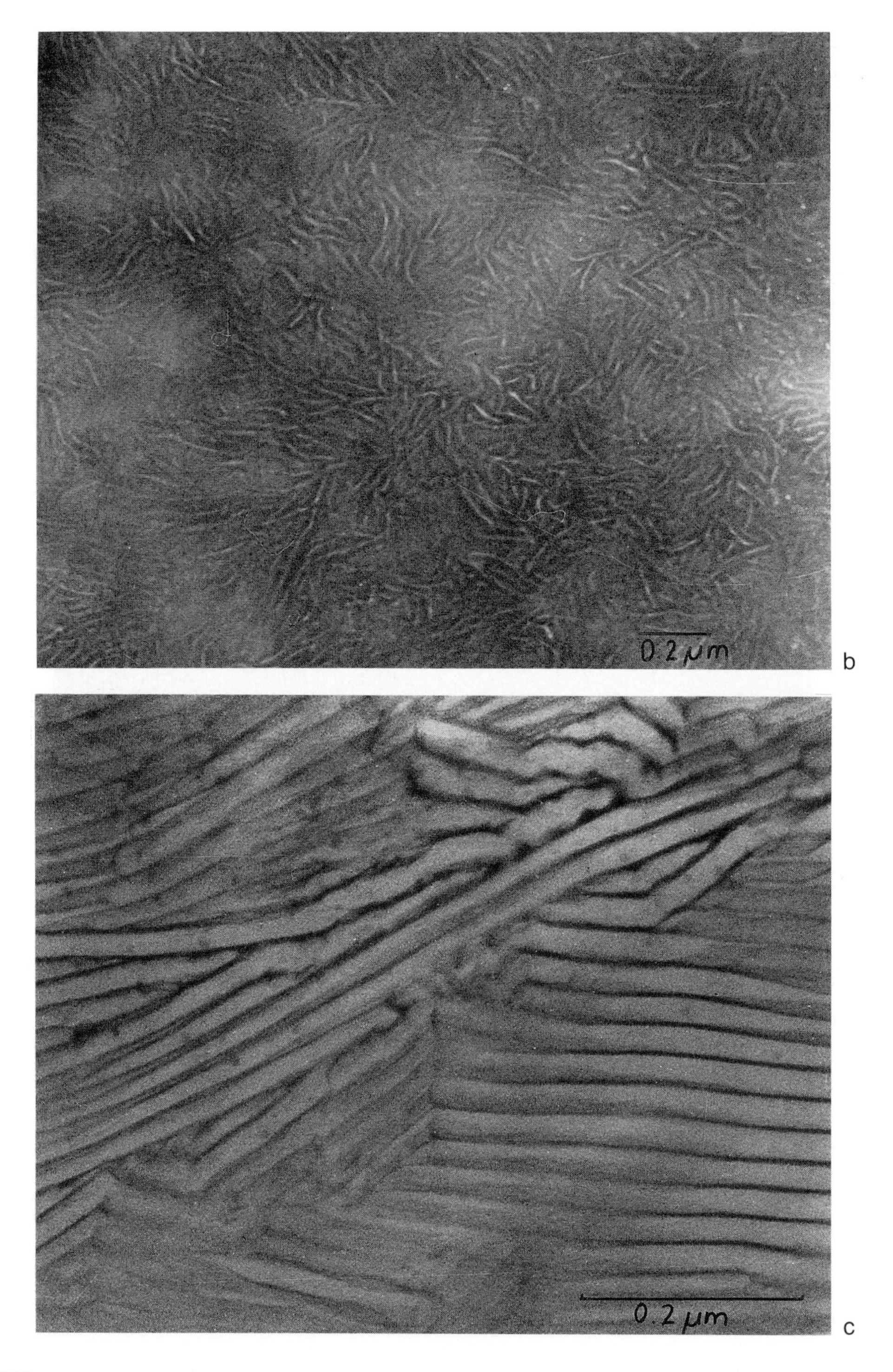
0.2 µm
b
0.2 µm
c

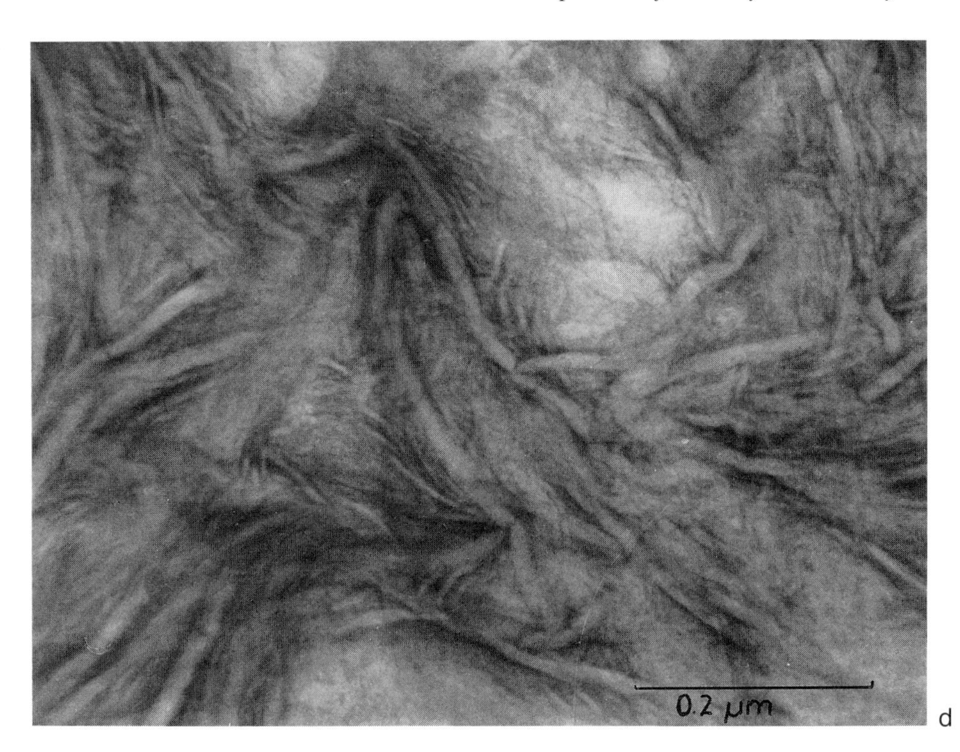

d

ing in the nature of these distributions [10]. Therefore investigations were undertaken to clarify this problem.

In order to obtain a narrow crystal thickness distribution in linear polyethylene, it is necessary to ensure that (a) crystallization at T_c is fast and virtually complete before cooling the samples in order to prevent isothermal thickening and – b) cooling is fast enough to prevent the formation of thinner crystals at a lower temperature. This can be achieved with linear polyethylene in the intermediate molecular weight range ($M_w \sim 10^4$–10^5) by crystallizing at a temperature where the crystallization rate has a maximum and by fast quenching. Such samples form extended lamellar crystals with thin amorphous zones. From the histograms obtained by numerical evaluation of the electron micrographs, a mean value of 215 Å with a relatively small standard deviation of 30 Å is obtained for the crystal thickness (Fig. 16.3a, vertical bars). The Raman LAM spectrum which measures the extended chain length, is identical in shape to the histogram but shifted to slightly larger values. Its mean value is 245 Å. Corrections for an average molecular tilt of 30 Å (Fig. 16.4a, continuous curve) leads to perfect correspondence between the two distributions.

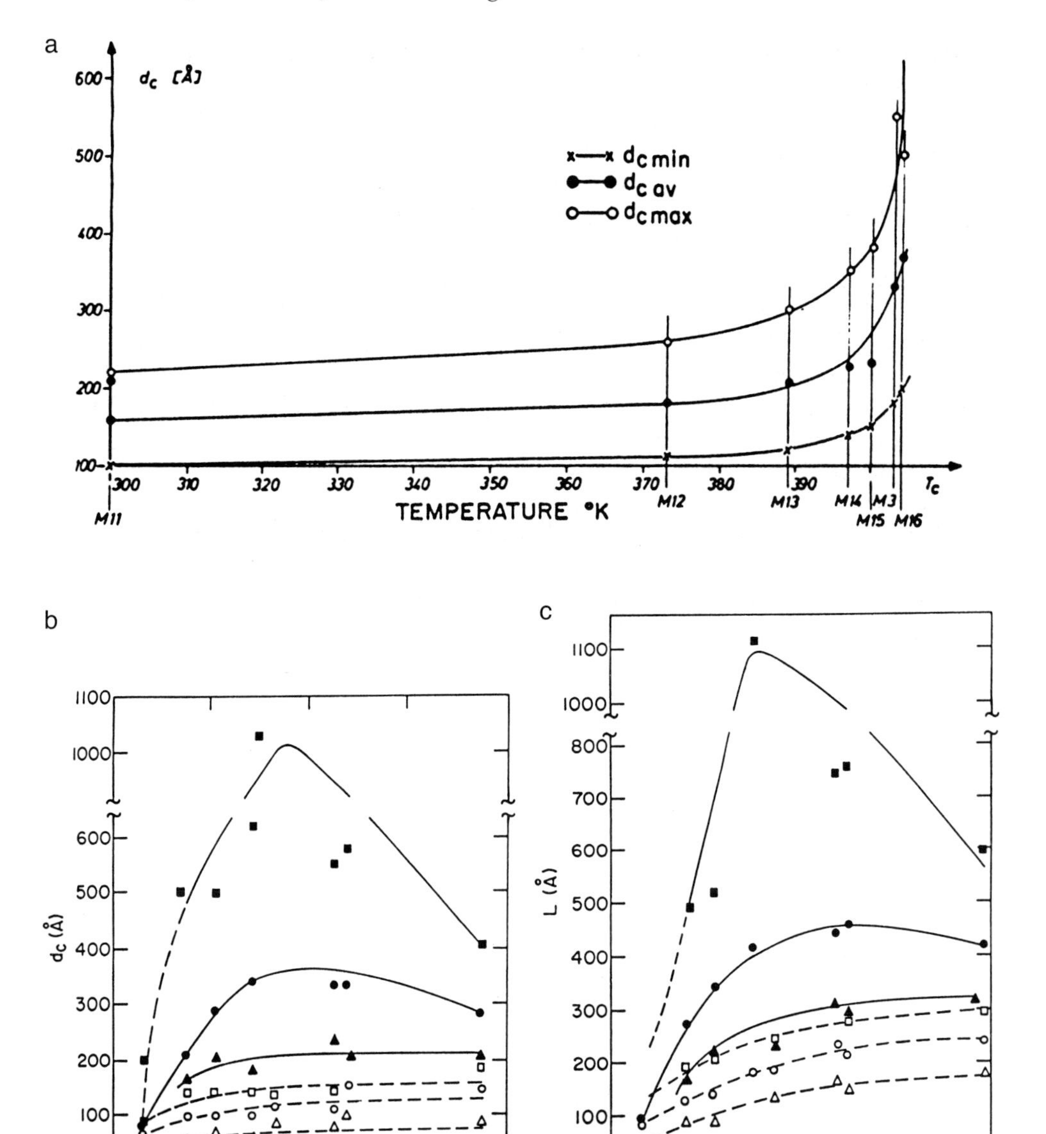

Fig. 16.3: (a) Dependence of long spacing on crystallization temperature for different molecular weights (Ref. [11]); (b) crystallite thickness d_c as a function of molecular weight; isothermally crystallized: (■) d_{cmax}, (●) d_{cav}, (▲) d_{cmin}; quenched: (□) d_{cmax}, (○) d_{cav}, (△) d_{cmin}, (Ref. [12]); (c) plot of long spacing L against molecular weight: isothermally crystallized: (■) L_{max}, (●) L_{av}, (▲) L_{min}; quenched: (□) d_{cmax}, (○) d_{cav}, (△) d_{cmin}, (Ref. [12]).

Bimodal thickness distributions are generally obtained when linear PE samples are not fully crystallized at T_c and then quenched. Alternatively, this situation arises with linear samples of a sufficiently low molecular weight to form extended chain crystals at high crystallization temperatures but where crystallization in the extended chain configuration is not complete. In a very low molecular weight sample ($M_w = 1925$, $M_n = 1673$) crystallized isothermally at 117 °C, the bimodal distribution is only weakly (Fig. 16.4b) indicated in the histogram. The desmeared, Lorentz corrected SAXS curve clearly shows two peaks and inspection of the micrograph immediately shows that 2 different populations exist (Fig. 16.4c). The thick lamellae are clearly recognizable but their number is too small to show up clearly in the histogram. The bimodal nature of the distribution is further substantiated by DSC experiment, where two distinct endothermic peaks appear.

Very broad distributions are generally obtained when high molecular weight material is crystallized at high temperatures. Indeed, the small angle X-ray scattering curve shows a continuously decreasing scattering curve which could lead to the erroneous assumption that there are no lamellae. However, these are clearly recognized in the micrographs. After choice of an appropriate base line, the Raman spectrum (not corrected for tilt) reflects the histogram rather well (Fig. 16.4c).

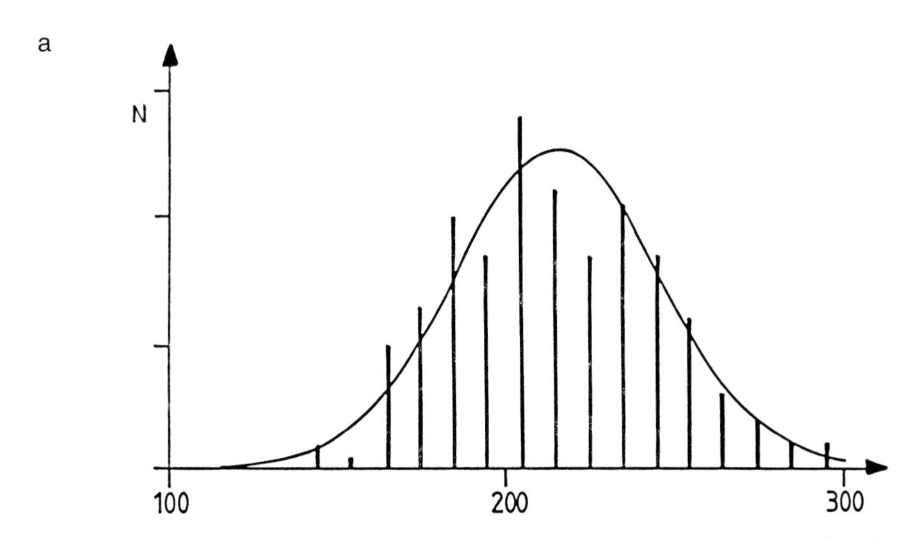

Fig. 16.4: (a) Comparison of crystal thickness values for LPE, $M_w \sim 10^4–10^5$ obtained by numerical evaluation of electron micrographs (vertical bars) and Raman LAM after correction for tilt (continuous curve) (Ref. [10]); (b) comparison between Raman, EM and SAXS for binodal distribution. Mol.wt = 2×10^3 (T_c = 117.3 °C); (c) comparison between Raman, EM and SAXS for extremely broad distribution.

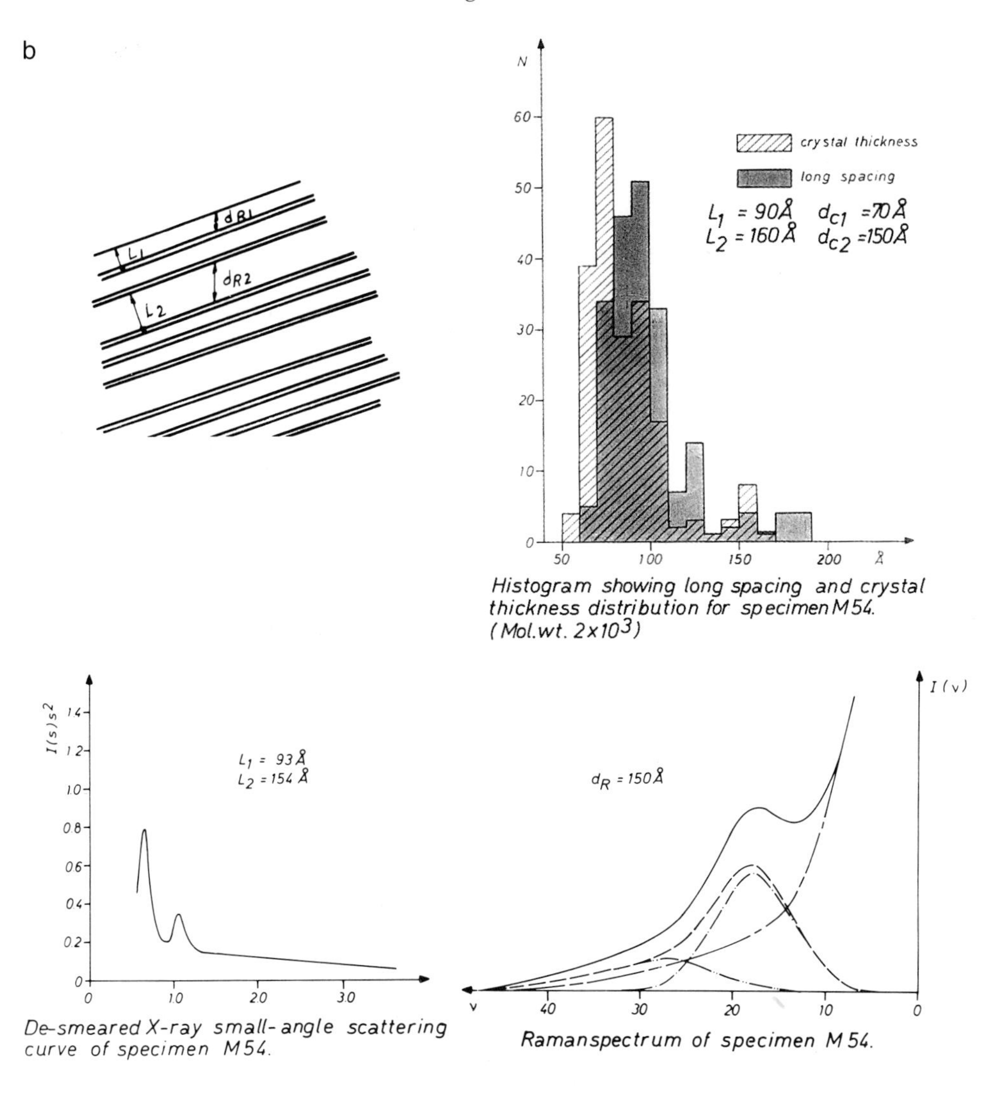

With the aid of a large number of examples, it was shown that the numerical estimation of crystal thicknesses and long spacings is a complex problem. The values which are obtained by different measuring techniques depend not only on details such as crystal morphology (curvature, lateral extension) but in particular on the nature of their thickness distributions.

For ethylene co-polymers [13], the crystal thicknesses and long spacings depend mainly on the branching content and branching distribution (Table 16.1). The thickness distributions are relatively narrow. Significant differences are observed in the lateral dimensions and curvature of the crystals.

c

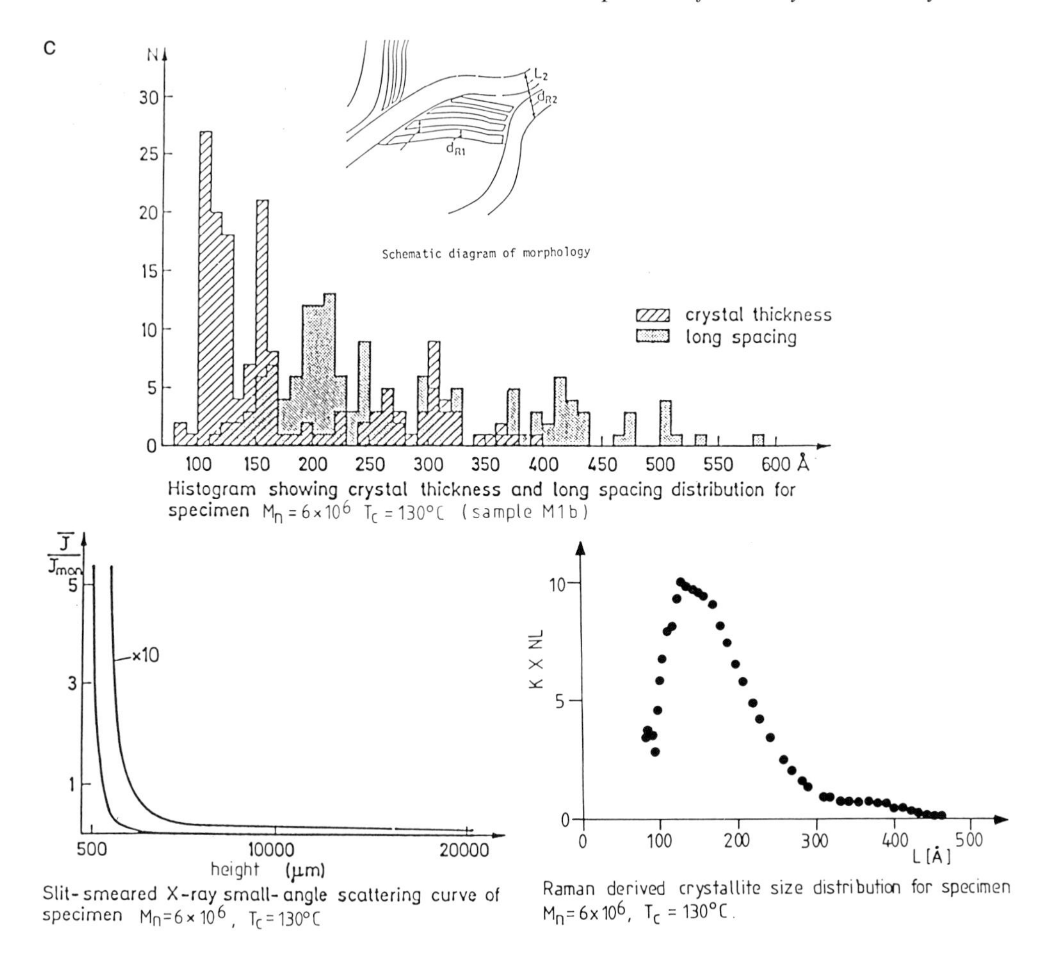

Histogram showing crystal thickness and long spacing distribution for specimen $M_n = 6 \times 10^6$ $T_c = 130°C$ (sample M1b)

Slit-smeared X-ray small-angle scattering curve of specimen $M_n = 6 \times 10^6$, $T_c = 130°C$

Raman derived crystallite size distribution for specimen $M_n = 6 \times 10^6$, $T_c = 130°C$.

16.3 Chain Conformation in the Crystalline State

16.3.1 General Considerations

When we know the morphology of semicrystalline polymers, we can measure the mean separation between crystalline lamellae and possibly also the mean electron density distribution of crystalline and amorphous regions. This, however, still does not tell us anything about the conformation of a single polymer molecule and how it traverses the crystalline and amorphous regions.

Table 16.1: Characteristics of ethylene copolymers.

					Electron micrographs (Å)			Raman LAM (Å)		
Sample type	Sample	Mol % branches	Crystallization conditions	100 $(1\text{-}\lambda)_{\Delta H}$	d_c ± SD	L ± SD	100 (1-λ)	$L^{\cdot}_R$	$L^{\cdot}_R$ cos30°	Crystallite characteristics[a]
Hydrogenated polybutadiene	P108	2.2	FQ	28						Small crystal
			SC	37	70 ± 12			75	65	Medium length, curved lamellae
	HPB 4	3.2	FQ	22.5						Small crystals
			SC	27	40 ± 10					Short, curved, segmented lamellae
	HPB 5	4.5	FQ	18						Very small crystals
			SC	23						Very small crystals
	HPB 6	5.7	SC	18						Very small crystals
Ethylene-vinyl acetate	EVA F9	1.12	FQ	36	70 ± 13	156 ± 25	45	70	61	Short, curved, segmented lamellae
			SC	38	64 ± 13	149 ± 25	47	80	69	Straight, medium length lamellae
	EVA F17	2.73	SC	27						Highly segmented lamellae

Table 16.1: (Continued)

Sample type	Sample	Mol % branches	Crystallization conditions	100 $(1\text{-}\lambda)_{\Delta H}$	Electron micrographs (Å)			Raman LAM (Å)		Crystallite characteristics[a]
					d_c ± SD	L ± SD	100 (1-λ)	$L^{\cdot}_{R}$	$L^{\cdot}_{R}$ cos30°	
Ethylene-vinyl acetate	EVA F35	4.14	SC	20				46		Very small crystals
	EVA A F30		6.06	SC	11					Very small crystals
Ethylene-butene	MA 3 F2	0.42	SC	50	125 ± 22	229 ± 52	55			Well-formed, long lamellae
	MA 4 F2	1.15	SC	45	97 ± 20	217 ± 42	45			Well-formed long lamellae
	MP4X B2	2.64	SC	35	94 ± 13	233 ± 50	40			Well-formed, long lamellae
			FQ	27	56 ± 10	173 ± 60	32	70	61	Short, segmented, curved lamellae
	MP4X A2	4.22	SC	20	35 ± 10	273 ± 125	13	45	39	Long, segmented lamellae
Ethylene-octene	MA 6 F2	0.69	SC	54	130 ± 17	203 ± 30	64			Well-formed, long lamellae
	MA 5 F2	1.49	SC	42	121 ± 16	213 ± 50	56			Well-formed, long lamellae

[a] Very small crystal means lamellae not discerned.

It is well established that a polymer molecule in the melt forms a coil with unperturbed dimensions (see Chapter 17, this Volume). During crystallization this conformation obviously has to change to incorporate the extended chain sequences in the lamellae. There have been various models proposed (see e.g. [14] – and other papers in this volume of the Faraday Discussion) which differ mostly in the amount of chain re-entries into the same lamella. Two examples are shown in Fig. 16.5. The nature of the so-called fold surface and the structure of the amorphous regions is crucial for the understanding of some properties of semicrystalline polymers. Thus, for example, the mechanical strength and toughness largely depends on the number of tie molecules in the amorphous regions. The deformation behaviour, modulus and glass transition change with the conformation and chain orientation within those regions, and generally crystallization theory has quite an impact on the development and improvement of copolymers, high-molecular weight fibres or blended materials of semicrystalline polymers.

16.3.2 Experimental Techniques

There are several techniques which can be used to investigate the chain conformation in the semicrystalline state. Generally, a single chain has to be made "visible" against the background of all the other chains. This is done by full or partial deuteration for NMR, IR and especially neutron scattering experiments. Deuteration offers the possibility of marking a single chain against the others without introducing large thermodynamic changes into the system. This statement is not always true since deuteration sometimes causes a significant change of the melting point, and a tendency of phase separation between deuterated and non-deuterated material is observed. This effect is large in polyethylene, where PE(H) and PE(D) have a melting point difference of approx. 6 degrees [15], but is fortunately smaller in other common polymers.

In NMR investigations [16, 17] a detailed analysis of the second moment in "deuteration dilution experiments" yields evidence against a very regular chain folding in polyethylene, but the method is not very sensitive to the mode of chain folding.

For IR investigations [18–20] the frequency shift of certain rocking and bending modes is shown to depend on the nearest neighbour interactions. In mixtures of deuterated and protonated molecules very regular chain folding along certain crystallographic directions is predicted [18, 19] which, however, is shown in more recent experiments and in refined calculations to be not in contradiction to modified models from neutron scattering results [20].

In another technique [21, 22] the chain loops in the amorphous region are chemically etched with ozone or reacted with halogen. The molecular weight distribution of the fragments is analyzed in terms of tight or loose folding and differences are seen in melt and solution crystallized material.

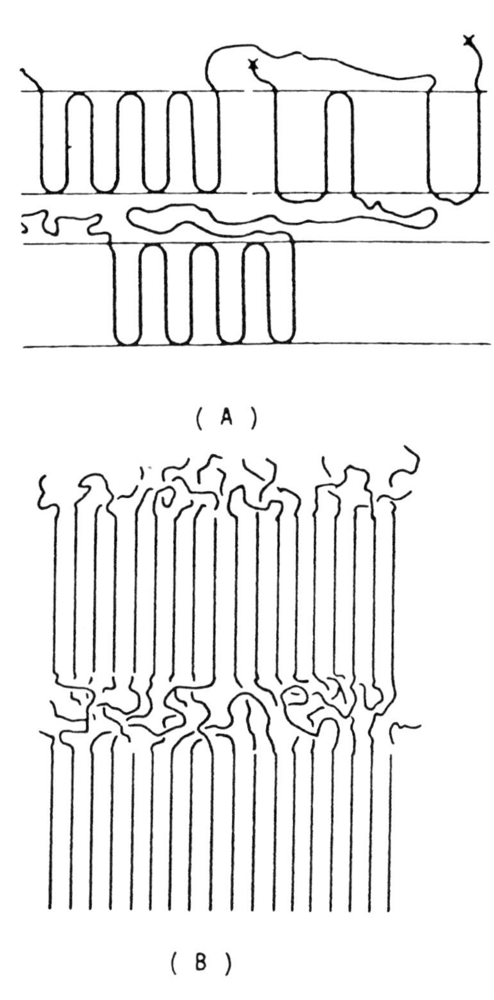

Fig. 16.5: Two examples of proposed models for chain folding in semicrystalline polymers. (A) central core model with 8 adjacent reentry folds in the cores (Ref. [34]), (B) "Switchboard" model (Ref. [14]).

Neutron scattering, on the other hand, is the commonly used method for the investigation of the conformation of single chains in the semicrystalline state since it yields most information about details of the conformation. Like all scattering methods it does not supply the chain conformation itself but only correlation functions. The type of information which can be obtained will be discussed in the following.

16.3.3 Small Angle Neutron Scattering (SANS)

The method of neutron scattering is described in several references (see e.g. [5, 23, 24]) and its application to polymers has also been recently reviewed (see e.g. [25–27]). Early SANS experiments concentrated on the small angle region (see Table 16.2) where information on the radius of gyration of the molecules, i.e. a measure of the overall size of the chains, can be obtained. This region extends to $q \lesssim 0.03$ Å^{-1}, q being the magnitude of the scattering vector ($q = (4\pi/\lambda)\sin\theta$). The problem of chain folding in the semicrystalline polymers, on the other hand, is best resolved in the intermediate angle scattering range (IANS) where typical dimensions ($D \sim 2\pi/q$) smaller than 100 Å will be resolved.

In first SANS experiments on polyethylene the surprising result was obtained [28, 29] that the radius of gyration R_g changes only slightly by going from the melt to the semicrystalline state. To avoid phase separation between H- and D-molecules, the samples were quenched from the melt. The absence of phase separation can be checked by the extrapolation to zero scattering angle where the intercept in a Zimm-plot should yield the correct molecular weight [30]. Similar results are observed for polyethyleneoxide quenched from the melt (Fig. 16.6) [31, 40]. Also shown in Fig. 16.6 is the increase of the long period L with R_g and M_w, an observation which is in contrast to conventional crystallization theory.

A simple explanation for the constancy of R_g during crystallization was attempted with the "Erstarrungsmodell" [32] schematically shown in Fig. 16.7.

Table 16.2: Information obtainable from neutron scattering experiments in various angular ranges: small (SANS), intermediate (IANS) and wide (WANS) angle neutron scattering.

Range	Informations
SANS $0.001 < q/\text{Å}^{-1} < 0.03$	Molecular weight Radius of gyration R_g Segregation effects
IANS $0.03 < q/\text{Å}^{-1} < 0.3$	Average number of stems per "cluster" in one lamella Average distance between crystalline stems of the same molecule in a lamella
WANS $0.3 < q/\text{Å}^{-1} < 5$	Direct correlation function of stems

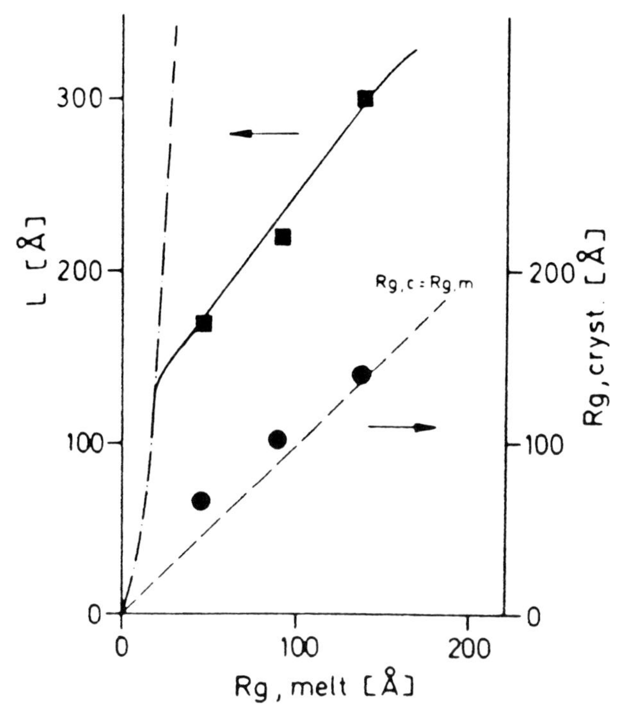

Fig. 16.6: The long spacing L and the radius of gyration $R_{g,cryst}$ of polyethylene oxide crystallized by quenching to $T_c = 40\,°C$ depending on the radius of gyration $R_{g,melt}$ in the melt. R_g values are weight-averages. The broken line gives approximate values L for extended chain crystals [40].

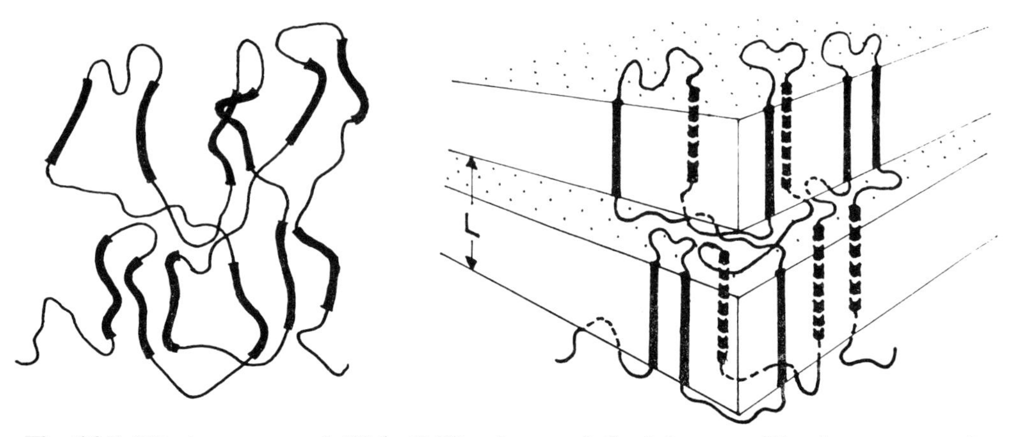

Fig. 16.7: "Erstarrungsmodell" (solidification model) of the crystallization process of chain molecules. The fully drawn sequences of the coil are incorporated into the growing lamellae without long range diffusion or major reorganization of the chain conformation [32]. Note that the dimensions are not drawn in right scale. The length of crystalline stems is much larger than their mutual distances.

In this model it is assumed that crystallization occurs by straightening of suitable chain segments of the coil in the melt and incorporation of those straightened stems into the lamellae. Rearrangement and long range diffusion of the chain is assumed to be largely restricted and only local movements are supposed to occur. Thus the general overall configuration and spatial arrangement of segments is not significantly changed during crystallization and R_g stays constant. Entanglements and defects move into the amorphous regions and the chain trajectory within the fold surface resembles statistical folding.

It has already been mentioned that the comparison of R_g alone gives only very limited information on chain folding. A more critical test is obtained in the intermediate angle range. Various models have been developed with differing amount of adjacency of the crystalline stems and compared with experimental data.

Experiments in that range became possible since it could be shown [32, 33] that mixtures of high concentrations of deuterated and non-deuterated molecules could be used in the measurements, thus producing a significantly higher scattering intensity. If there is no specific thermodynamic interaction between H/H and D/D molecules the differential scattering cross section per unit volume is given by

$$\frac{d\sigma}{d\Omega}(q) = c_D(1-c_D)Kn_wP(q). \tag{3}$$

c_D is the concentration of deuterated molecules, K the contrast factor, n_w the degree of polymerization and P the form factor of the polymer molecule:

$$P(q) = \frac{1}{n_w^2}\left\langle \sum_{ij} \exp(iq\, R_{ij}) \right\rangle, \tag{4}$$

where R_{ij} describes the distance between the H-or D-atoms, respectively, of the molecule and the bracket denotes spatial and configurational averaging.

It is easily seen from Equ. (3) that highest scattering intensity can be obtained for a 50% mixture. A comparison between IANS experiments of that type and e.g. Monte Carlo calculations based on the “Erstarrungsmodell” [32] showed reasonable agreement. Other investigations with alternative model calculations ([34, 14] and other papers in that volume) can reach a similar agreement with their experimental data, so that a rigorous decision between the models is not possible. We therefore used two approaches, (a) going to still higher q-values right into the beginning of the wide angle region and (b) to develop a theoretical approach which is largely independent of model considerations.

Based on model calculations [35] for specific adjacent reentry folding models including e.g. the central core model (see Fig. 16.5a), it could be concluded from experimental data of solution and melt crystallized polyethylene in the WANS region [32, 36] that a significant amount of regular folding of more than 4 adjacent crystalline stems in certain crystallographic directions can be excluded. For folding models in one crystallographic plane a peak should be observable in the WANS pattern which otherwise is forbidden because of crystal symmetry reasons [35]. It is on the other hand quite clear that also for statistical folding models an appreciable amount of adjacent reentry folding is present [32, 27] which is necessary for instance because of density reasons. The density in the amorphous regions is typically 5–15 % smaller than the density in the crystalline regions. Also from a detailed analysis of the plateau level in a Kratky plot [37] a certain amount of adjacent reentry folding can be estimated. Regular adjacent reentry folding can be excluded, however [38].

For further insight into folding statistics a formalism has been developed [39, 40, 27] which permits the extraction of conformational parameters from experimental data without specific model assumptions. It is only assumed (see Fig. 16.8) that the crystalline stems are located within crystalline lamellae and

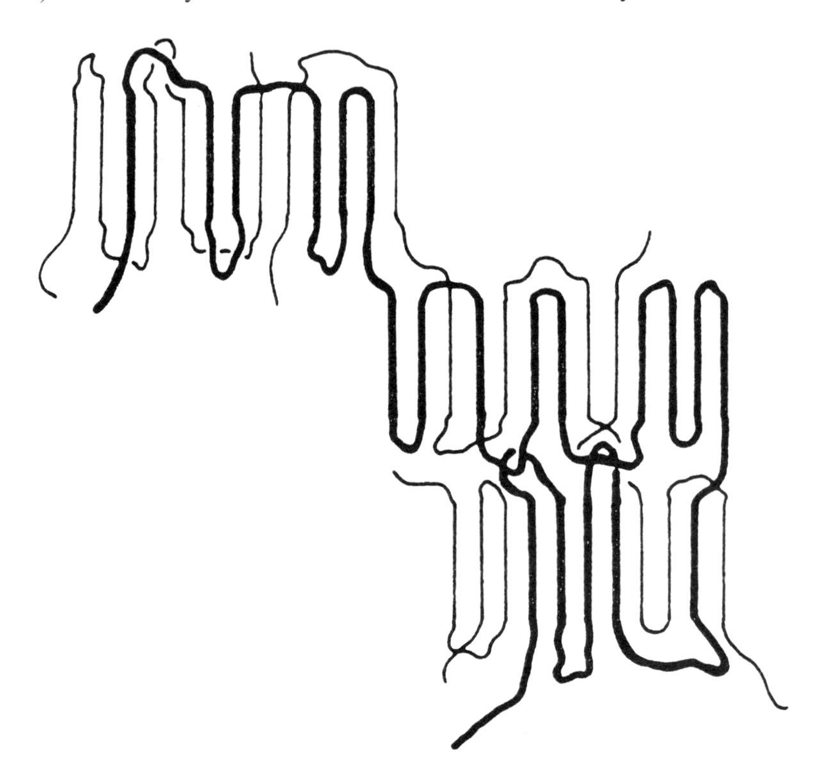

Fig. 16.8: Schematic representation of a single macromolecule traversing different lamellar crystals. Clusters of crystalline stems situated in different lamellae are connected by tie molecules. Note that the drawing is not in correct scale: The length of the crystalline stems is about 50 times larger than their lateral distances.

that distances between stems of clusters in one lamella are much smaller than distances between the clusters in different lamellae. As a consequence scattering from stems in one lamella can be separated from scattering of the whole molecule. At small q-values one thus observes R_g and M_w of the whole molecule, whereas in the IANS range (see Table 16.2) information on clusters is obtained. It can be shown [39] that the reduced intensity J(q) can be written

$$J(q) = \frac{d\sigma/d\Omega}{c_D(1-c_D)K} = n_{St}P_{St}(q)(1+H(q)) \tag{5}$$

n_{St} is the number of monomer units per stem, P_{St} the single stem form factor and H(q) a stem correlation function which contains in the IANS range information on details of the cluster:

$$H(q) \approx (N_c - 1) - \frac{N_c}{2} R_{cc}^2 q^2 . \tag{6}$$

The cluster parameters N_c, the number of stems per cluster, and R_{cc}, the radius of gyration of stem centers in a cluster, can be obtained in a Zimm type plot of $H(q)^{-1}$ vs. q^2. In the case of melt quenched polyethylene [39] one obtains from this analysis (Fig. 16.9) $N_c = 5.2$ and $R_{cc} = 21.5$ Å. There are 2.7 clusters per molecule which are mostly situated in neighbouring lamellae. Also for other polymers like isostatic polypropylene [39], polyethylenoxide [40] and poly-

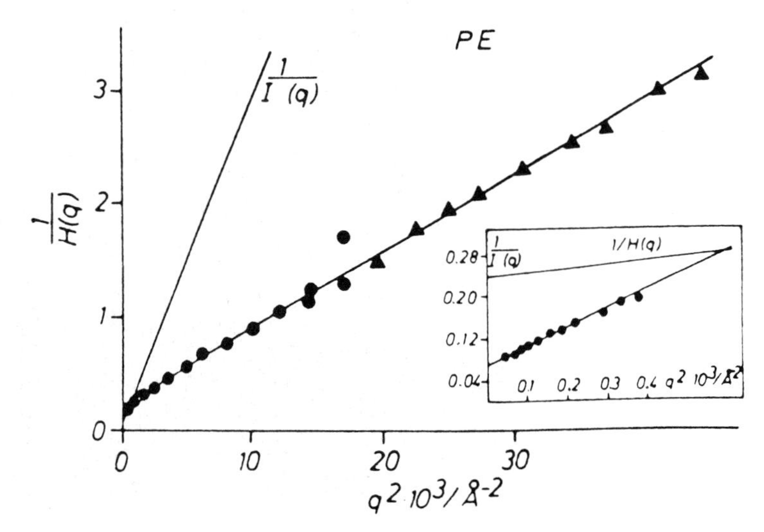

Fig. 16.9: Evaluation of N_c and R_{cc} from the stem correlation function $H(q)^{-1}$ vs. q^2 for polyethylene crystallized by quenching from the melt. Experimental points and triangles are taken from Ref. [29] and Ref. [32], respectively. The insert shows the experimental data in a $J(q)^{-1}$ vs. q-plot.

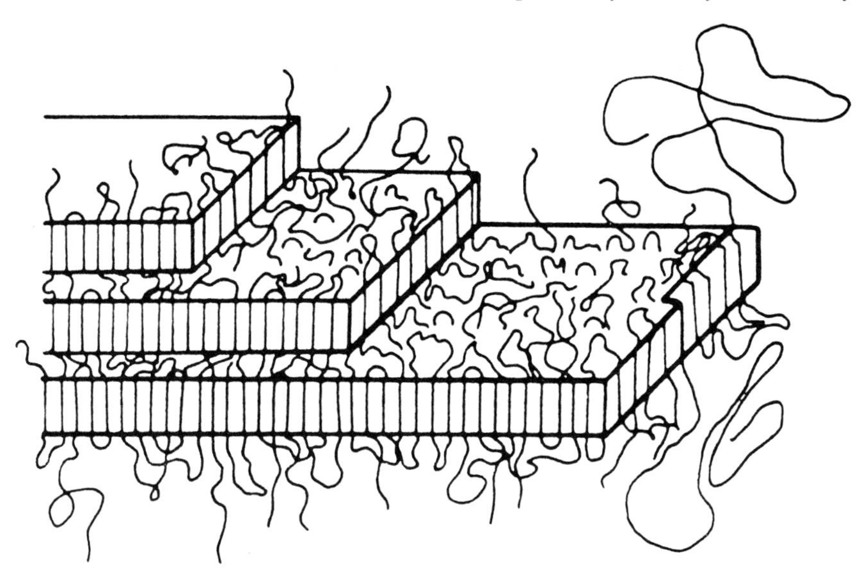

Fig. 16.10: Schematic diagram of the growth of a stack of lamellae in the melt. The growth fronts do not arrive simultaneously at the location of a single molecule.

ethylene terephthalate [41] reasonable results are obtained. The parameters observed are consistent with "near" reentry or statistical folding.

Thus the general cluster model (Fig. 16.8) seems to be reasonable. It can be justified from crystallization considerations (Fig. 16.10). The growth front during crystallization does not generally arrive simultaneously at the position of a molecule, but different parts of the molecule are incorporated at different times in different lamellae. The growth of a cluster within one lamella might be stopped by kinetic hinderances caused by entanglements and other molecules.

So far we have been discussing melt crystallized material mostly. Similar neutron experiments have been performed on solution crystallized semicrystalline polymers [32, 37, 42] and various models have been developed. On the basis of IANS data on polyethylene we have proposed a "stacked-sheets" model [32] where crystallization of the stems in adjacent crystalline planes along the (110)-crystallization direction is assumed. Based on this model the decrease in R_g and the scattering in the IANS range can be explained which is significantly different from the melt. Very similar "superfolding" models [14, 37] have been proposed. Again folding to "near" sites and not necessarily to the adjacent site along a crystallographic folding direction is assumed. Thus the chain conformation in semicrystalline polymers depends strongly on crystallization conditions and is significantly different in solution and melt crystallized material. Details of the models are still under discussion and especially the chain conformation within the amorphous regions is still not completely understood.

16.4 Crystallization and Melting

16.4.1 Investigation of the Crystallization and Melting Process

In chapter 16.2 we have already mentioned the dependence of the properties and morphological appearance of melt crystallized polymers on crystallization temperature, molecular weight and molecular weight distribution. It is on the other hand also well known that for a semicrystalline material of given morphology the SAXS-intensity increases drastically with temperature while its crystallinity decreases upon heating to the melting point (Fig. 16.11) [43]. This phenomenon is called "premelting" and various explanations have been given. In Fig. 16.11 we observe for instance that the position of the small angle peak shifts to smaller values and thus the long spacing increases with annealing temperature. This behaviour is explained in a model [44, 45] where partial

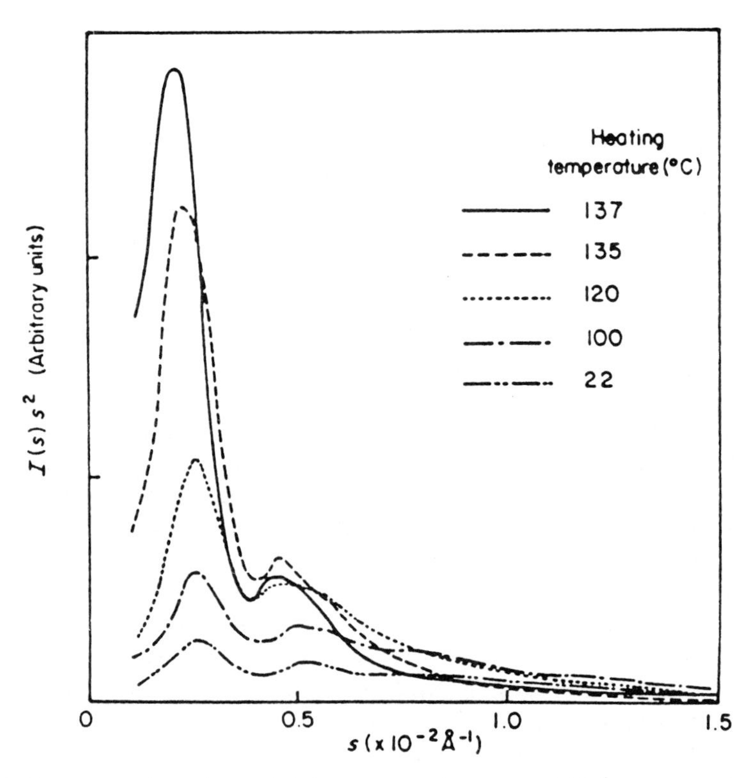

Fig. 16.11: Desmeared small angle X-ray scattering curve $I(s)s^2$ vs. s obtained during heating of low density polyethylene crystallized at 125 °C.

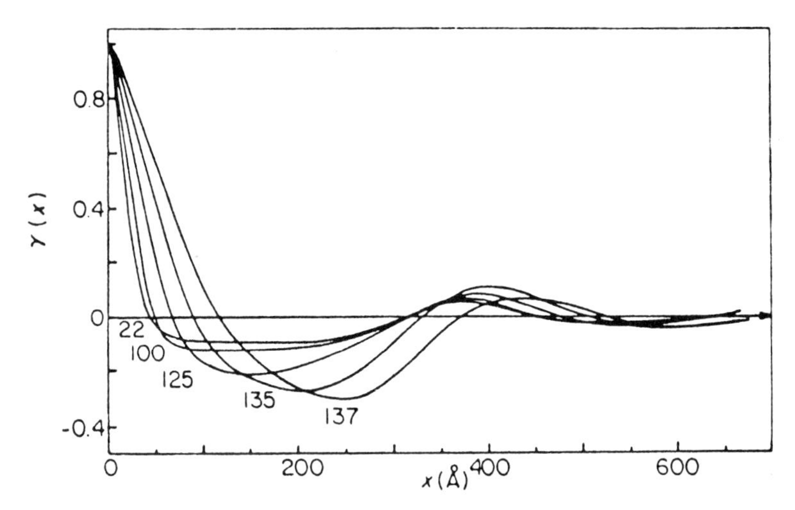

Fig. 16.12: Electron density correlation function γ(x) obtained during heating of low density polyethylene crystallized at 125 °C. Heating temperatures are marked on the curves in degrees Celsius.

melting of small lamellae within the lamellar stacks is assumed. Thicker lamellae are believed to be more stable against melting and melt at higher temperatures only. Besides this process another effect plays an important role. It can be shown [44, 45] that for thermodynamic reasons a continuous boundary melting takes place which affects all lamellae in the same way. The boundary between crystalline and amorphous regions melts first and the thickness of the amorphous layers increases with temperature [46, 47].

The application of the correlation function analysis described in chapter 16.1 gives further insight into this problem since the change of all the relevant parameters during temperature treatment can be followed. The electron density correlation function of the SAXS intensity of linear polyethylene crystallized at 125 °C is given in Fig. 16.12 [43]. A detailed analysis of the heating and cooling process reveals the following features: (a) the average thickness of the amorphous regions increases during heating, while the average thickness of the crystalline regions is constant or slightly decreases simultaneously, (b) this change in the amorphous regions is reversible at temperatures below the initial crystallization temperature T_c of the sample, and (c) is irreversible above T_c. Thus it can be concluded that the surface melting model describes the premelting behaviour of polyethylene at least below T_c and that in that region the amorphous phase seems to be in an equilibrium state. Experiments at different crystallization temperatures are in accordance with that mechanism.

Which one of the two mentioned processes governs the melting behaviour of semicrystalline polymers depends on the chemical structure, on the overall crystallinity and on the crystallization temperature. A study on low density

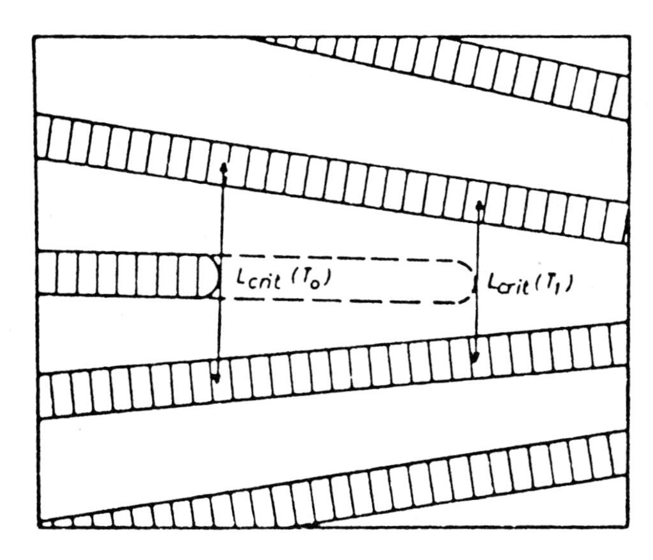

Fig. 16.13: Lateral growth model for lamella between tilted neighbours (Ref. [48]).

polyethylene by both SAXS and electron microscopy show a different behaviour [48]. Here (a) the average thickness of amorphous layers increases with annealing temperature while the crystal thickness remains essentially constant because branches conglomerate at the crystal surfaces, and (b) this change is reversible. Both observations are in contrast to linear polyethylene. A model is shown in Fig. 16.13 where lateral melting/growth of lamellae during heating and cooling is postulated. The model is based on the experimental result that the thickness of the amorphous regions depends critically on annealing temperature in branched materials.

16.4.2 Stress Induced Crystallization

Up to now we have not considered the actual crystallization process and its dependence on time. Crystallization from the isotropic melt is mostly described by the nucleation and growth model (see e.g. [1] and references therein). It is assumed that in a first step a crystalline nucleus is formed. As shown in Fig. 16.14a, this nucleus increases in size during crystallization by incorporating chains at the crystal surface. The density of the crystalline phase does not change during progressing crystal growth. Deviations from this behaviour have been observed during stress induced crystallization [49, 50] which can be understood in terms of a spinodal demixing of chain defects and stretched sequences. In contrast to the nucleation and growth model the phase boundary is not sharp and the density shows a continuous periodic modulation as depicted in Fig. 16.14b. The amplitude increases with time to reach a maximal and minimal value for the crystalline and amorphous regions, respectively. Starting from that point deviations from the ideal spinodal behaviour are expected. This process is formally very similar to the spinodal decomposition of polymer mixtures.

A detailed analysis of the time dependence of the SAXS behaviour offers a possibility of resolving the kinetics and mechanism of stress induced crystallization. The use of synchrotron scattering also allows an investigation of fast crystallization processes in real time. We have investigated the stress induced crystallization of polyethyleneterephthalate (PET), peroxide-cross-linked natural rubber (NR) and cis-1,4-polybutadiene (PB) [51]. PB is in the melt a rubbery material which can be elastically stretched to an extension $\lambda = L/L_0 \sim 6$. L and L_0 are the length of the sample with and without applied stress, respectively. Samples are first melted in the extended state and then quickly cooled to the crystallization temperature. SAXS pattern are taken with a time resolution of 5 s. The electron density correlation function is shown in Fig. 16.15 for various crystallization times. At short times the peak position and corresponding long period stays constant and only the intensity of the scattering curve increases with time. One can define a q-dependent amplification factor which shows typical spinodal behaviour and the electron density difference increases with time. At longer times a different type of behaviour is observed (Fig. 16.15b). New lamellae are formed but the lamellar thickness stays constant. The small angle peak shifts with time and the long period decreases. This behaviour can be explained by lateral growth of the lamellae (Fig. 16.13). Similarly, during melting, first whole lamellae melt without a change in lamellar thickness, whereas at later times a spinodal process is observed. Also oriented PET and NR samples show at certain crystallization temperatures this spinodal crystallization behaviour [51–53].

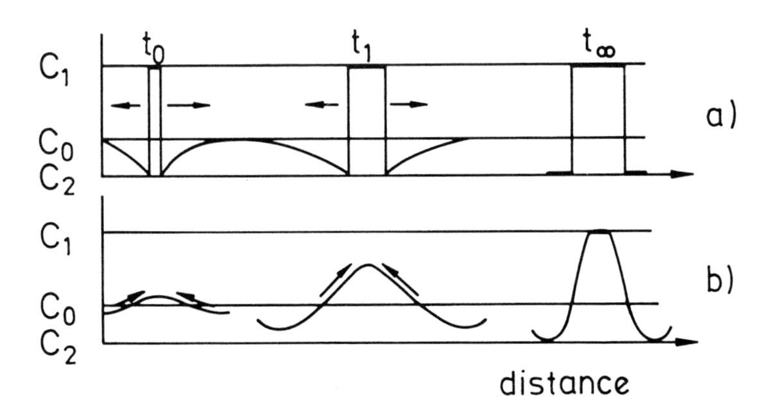

Fig. 16.14: Electron density distribution at different stages of stress induced crystallization, a) nucleation and growth model, b) spinodal crystallization model.

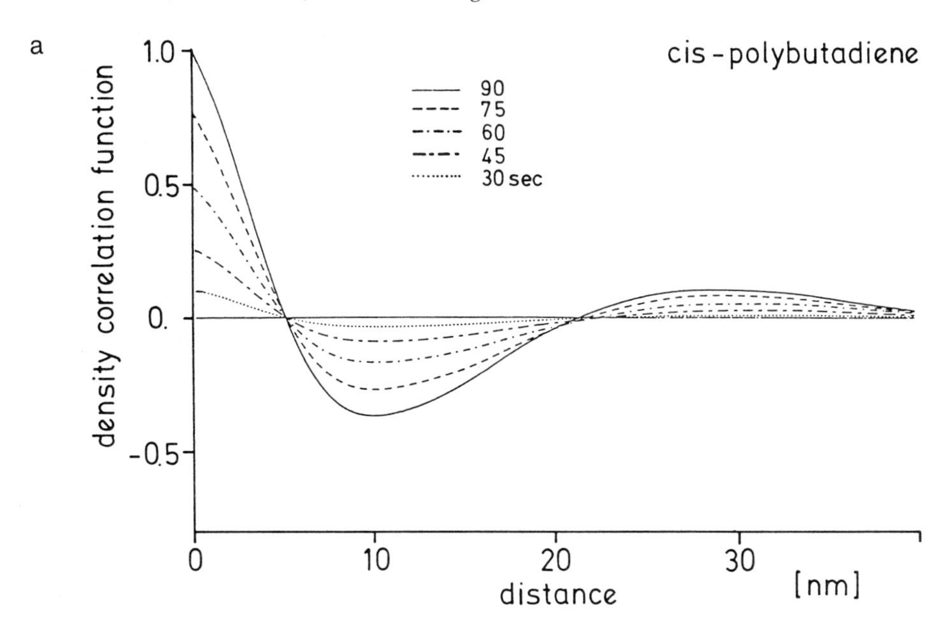

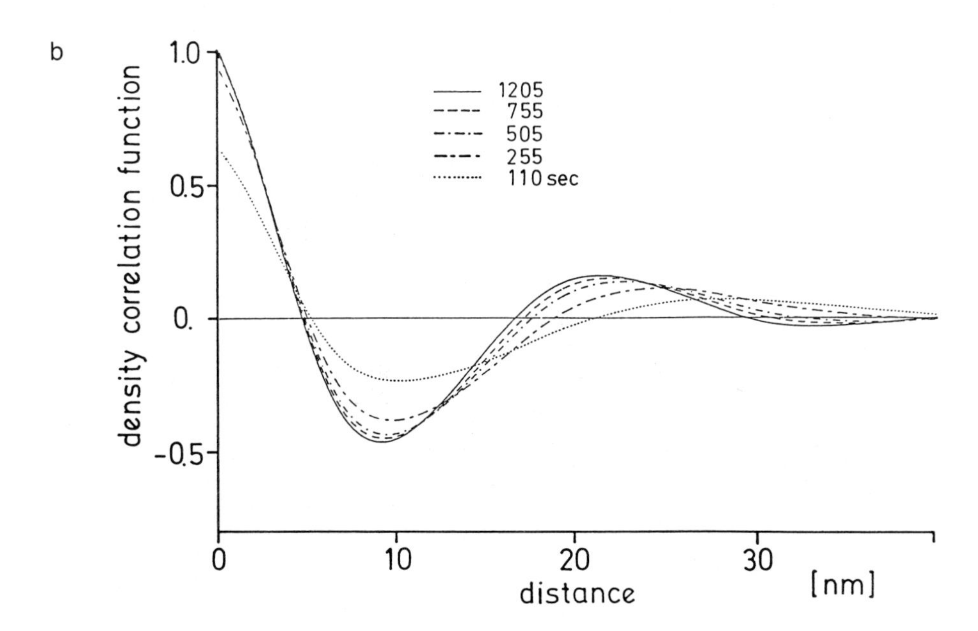

Fig. 16.15: Development of the electron density distribution function γ during stress induced crystallization of cis-polybutadiene (extension ratio $\lambda = 5.6$, crystallization temperature $T_c = -4.5\,°C$). Shown are early (a) and later stages (b) of crystallization. The time is given in seconds.

16.5 Blockcopolymers with Crystalline Blocks

16.5.1 Morphology and Models

Usually blockcopolymers consist of two or more amorphous blocks. Cases are known, however, where one block forms a crystalline or paracrystalline phase. Examples of such systems are the polyurethane-elastomers, which are used extensively in industrial applications. These materials are described to a first approximation by a two phase model shown in Fig. 16.16 [54]. Depending on relative composition and compatibility, various morphologies and mechanical

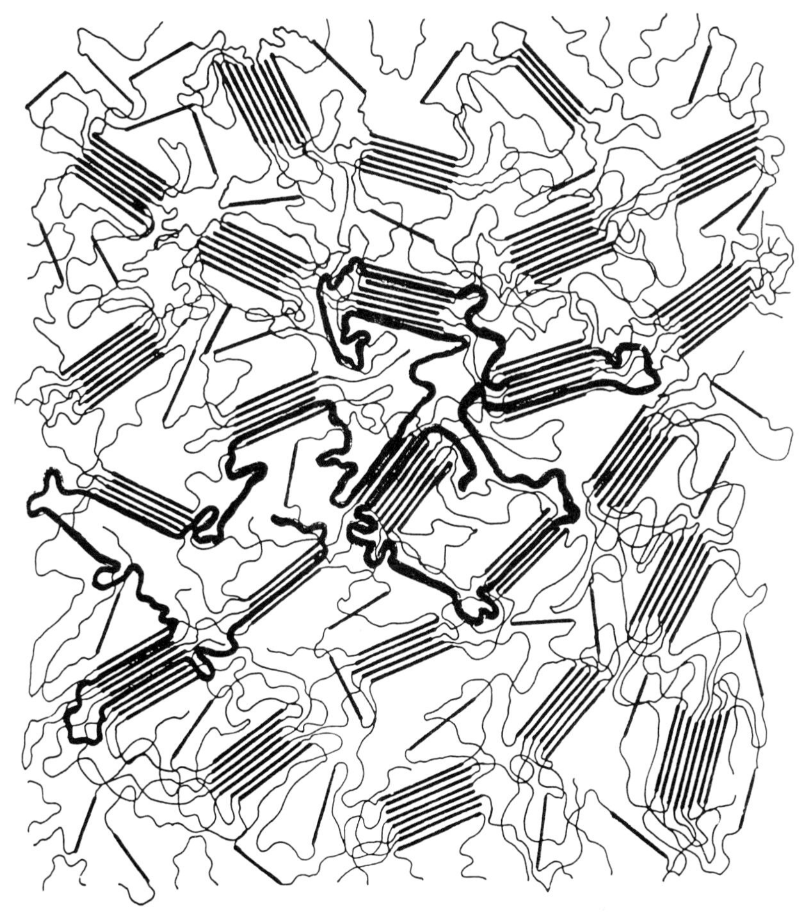

Fig. 16.16: Schematic model of the domain structure and conformation of a block copolymer with crystallizable blocks (Ref. [54]).

properties can be achieved. If the amorphous component is above its glass transition temperature, elastic behaviour can be observed. In the model of Fig. 16.16 the crystalline hard segments form small and highly disturbed crystals and act as physical cross-linking points between the amorphous soft segments. The components are clearly phase separated in that case which is consistent with small and wide angle X-ray scattering experiments. Nevertheless, detailed models on structure, conformation and the relation to mechanical behaviour are still missing. For this purpose a detailed investigation on model polymers was performed [54].

16.5.2 SAXS and SANS Investigations

For small angle X-ray and neutron investigations segmented polyurethane elastomers with monodisperse hard segments were prepared [54]. Hard or soft segments were selectively deuterated to achieve different contrasts. SAXS and SANS investigations show a strong morphology peak (Fig. 16.17) which can be attributed to the regular morphology of hard and soft segments (see Fig. 16.16). It can be estimated that approximately 60 % of hard segments are situated in small domains which have cylindrical shape (diameter 3 nm, height 6 nm). The phase boundary between hard and soft segments seems to be relatively sharp. By contrast matching the morphological scattering can be made

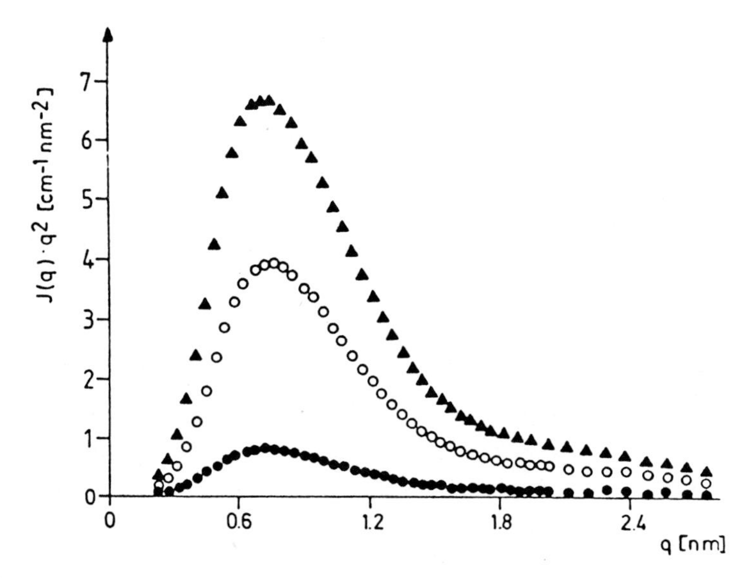

Fig. 16.17: Neutron small angle scattering curves for polyurethane blockcopolymers. ● fully H-chains, ○ deuterated soft segments and ▲ deuterated hard segments (Ref. [54]).

negligibly small. The conformation of soft segments shows nearly unperturbed coil dimensions and also the conformation of the whole chain can be explained on the basis of a random-walk model with isotropic orientation of the stiff hard segments.

During deformation the morphology changes in two steps: at small deformations a four-point type SAXS diagram is observed which changes to a layer-line type diagram at larger deformations. The change of the chain conformation during deformation can be explained by an affine model where the microscopic deformation follows the macroscopic sample at small elongations. A gliding mechanism is assumed at larger deformations.

Using this specific model polymer some structural information on block copolymers could be obtained. A detailed understanding especially of the deformation behaviour in this two phase system is still not possible since various parameters change during the process. Thus in particular the phase boundaries and compositions might change significantly. Further investigations and a combination of different techniques are still necessary.

16.6 Conclusions

The morphology and the structure related properties have been studied extensively in the Sonderforschungsbereich 41 by a broad variety of methods: Electron and optical microscopy, small angle X-ray scattering, neutron scattering in the small and intermediate angular range, Raman spectroscopy and differential calorimetry. There were three main fields of interest:

- The lamellar morphology in semicrystalline polymers
- The chain conformation and the nature of the amorphous regions
- The mode of crystallization and the process of partial melting during heating of the sample.

With regard to the first topic dependence of morphology on molecular weight and weight distribution has been demonstrated. It has been shown that by combination of various methods a number of relevant structure parameters can be evaluated.

The chain conformation can be studied by neutron scattering. It turned out that in many cases the radius of gyration does not change significantly during crystallization, therefore the "solidification model" was proposed. The arrangement of crystalline stems belonging to the same molecule within one lamella can be evaluated from the scattering patterns in the intermediate angular range. The results show that crystallization from the melt generally does not lead to a regular chain folding with adjacent reentry over a large range.

Regarding the decrease of crystallinity with rising temperature two different mechanisms play a role: The melting of small crystals and a reversible increase in the thickness of the amorphous layers ("boundary melting").

Crystallization takes place in the isotropic melt by a "nucleation and growth" process. On the other hand in an oriented melt the crystallization mode can be described by a spinodal defect clustering.

16.7 References

[1] J. H. Magill: Morphologenesis of Solid Polymer Microstructures, in J. M. Schulz (ed.): Treatise on Mat. Sci. Techn., Vol. 10A, Academic Press, New York 1977.
[2] D. C. Bassett: Principles of Polymer Morphology, Cambridge University Press, 1981.
[3] L. E. Alexander: X-ray Diffraction Methods in Polymer Science, Krieger Publ. Comp., Malakar 1985.
[4] A. Guinier, G. Fournet: Small-Angle Scattering of X-rays, John Wiley Publ., New York 1955.
[5] M. Hoffmann, H. Krömer, R. Kuhn, Polymeranalytik II, Thieme, Stuttgart 1977.
[6] G. C. Vonk, in O. Glatter, O. Kratky: Small Angle X-ray Scattering, Academic Press, London 1982.
[7] O. Glatter, in O. Glatter, O. Kratky: Small Angle X-ray Scattering, Academic Press, London 1982.
[8] G. R. Strobl, Kolloid Z. u. Z. Polym. 250 (1972) 1039.
[9] G. R. Strobl, M. Schneider, J. Pol. Sci., Phys. 18 (1980) 1343.
[10] I. G. Voigt-Martin, Adv. Pol. Sci. 67 (1985) 194.
[11a] I. G. Voigt-Martin, L. Mandelkern, J. Pol. Sci., Phys. 19 (1981) 1769.
[11b] I. G. Voigt-Martin, L. Mandelkern, J. Pol. Sci., Phys. 22 (1984) 1901.
[12] G. Stack, L. Mandelkern, I. Voigt-Martin, Macromolecules 17 (1984) 321.
[13] I. G. Voigt-Martin, L. Mandelkern, J. Pol. Sci., Phys. 24 (1986) 1283.
[14] D.Y. Yoon, P. J. Flory, Faraday Discuss. Chem. Soc. 68 (1979) 288.
[15] F. C. Stehling, E. Ergos, L. Mandelkern, Macromolecules 4 (1971) 672.
[16] R. Voelkel, H. Sillescu, Macromolecules 12 (1979) 162.
[17] K. M. Naiaraiar, E.T. Samulski, R. I. Cukier, Nature 275 (1978) 527.
[18] J.H. C. Ching, S. Krimm, Macromolecules 6 (1975) 894.
[19] X. Jing, S. Krimm, J. Pol. Sci., Phys. 20 (1982) 1155.
[20] S. J. Spells, A. Keller, D. M. Sadler, Polymer 25 (1984) 749.
[21] G. N. Patel, A. Keller, J. Pol. Sci., Phys. 13 (1975) 2259; 2275.
[22] T. Oyama, K. Shiokawa, Y. Kawamura, Pol. J. 9 (1977) 1.
[23] J. Schelten, R.W. Hendricks, J. Appl. Cryst. 11 (1978) 297.
[24] J. S. Higgins, R. S. Stein, J. Appl. Cryst. 11 (1978) 346.
[25] L. H. Sperling, Pol. Eng. Sci. 24 (1984) 1.
[26] G. D. Wignall, in Mark/Bikales/Overberger/Menges (eds.): Encyclopedia of Pol. Sci. Eng., Vol. 10, John Wiley Publ., 1987.
[27] E.W. Fischer, Makromol. Chem., Makromol. Symp. 20/21 (1988) 277.
[28] G. Lieser, E.W. Fischer, K. Ibel, J. Pol. Sci. 13 (1975) 29.

[29] J. Schelten, D.G.H. Ballard, G.D. Wignall, G.W. Longman, W. Schmatz, Polymer 17 (1976) 751.
[30] J. Schelten, G.D. Wignall, D.G.H. Ballard, G.W. Longman, Polymer 18 (1977) 1111.
[31] J. Kugler, U. Struth, R. Born, E.W. Fischer, K. Hahn, in preparation.
[32] M. Stamm, E.W. Fischer, M. Dettenmaier, P. Convert, Faraday Discuss. Chem. Soc. 68 (1979) 263.
[33] W. Gawrisch, M.G. Brereton, E.W. Fischer, Polymer Bull. 4 (1981) 687.
[34] J.D. Hoffmann, C.M. Guttmann, E.A. Di Marzio, Faraday Discuss. Chem. Soc. 68 (1979) 177.
[35] M. Stamm, J. Pol. Sci., Phys. 20 (1982) 235.
[36] G.D. Wignall, L. Mandelkern, C. Edwards, M. Glotin, J. Pol. Sci., Phys. 22 (1982) 245.
[37] D.M. Sadler, in I. Hall (ed.): The Structure of Crystalline Polymers, Elsevier Publ., Barking 125.
[38] J.D. Hoffmann, C.M. Guttmann, E.A. Di Marzio, Polymer 22 (1981) 597.
[39] E.W. Fischer, K. Hahn, J. Kugler, U. Struth, R. Born, M. Stamm, J. Pol. Sci., Phys. 22 (1984) 1491.
[40] E.W. Fischer, Polym. J. 17 (1985) 307.
[41] K.P. McAlea, J.M. Schultz, K.H. Gardner, G.D. Wignall, Macromolecules 18 (1985) 447.
[42] D.M. Sadler, A. Keller, Science 19 (1979) 265.
[43] Y. Tanabe, G.R. Strobl, E.W. Fischer, Polymer 27 (1986) 1147.
[44] E.W. Fischer, Koll. Z., Z. Polym. 231 (1969) 458.
[45] H.G. Zachmann, Koll. Z., Z. Polym. 216 (1967) 180.
[46] P.J. Flory, Trans. Faraday Soc. 51 (1955) 848.
[47] H.G. Kilian, Koll. Z., Z. Polym. 231 (1969) 534.
[48] G.R. Strobl, M.J. Schneider, I.G. Voigt-Martin, J. Pol. Sci., Phys. 18 (1980) 1361.
[49] J. Petermann, J.M. Schultz, J. Mat. Sci. 13 (1978) 2188.
[50] J. Petermann, R.M. Gohil, J.M. Schultz, R.W. Hendricks, J.S. Lin, J. Pol. Sci., Phys. 20 (1982) 523.
[51] Q. Fan, M. Stamm, E.W. Fischer, in preparation; E.W. Fischer, M. Stamm, Q. Fan, R. Zietz, ACS Polymer Preprints 30 (2) (1989) 291.
[52] R. Günther, Thesis, Mainz 1981.
[53] O. Fischer, Diploma thesis, Mainz 1984.
[54] U. Struth, Thesis, Mainz 1986.

17 Chain Conformation and Local Order in Amorphous Polymers

Erhard W. Fischer* and Ingrid G. Voigt-Martin**

17.1 Introduction

The structure of an amorphous polymer is characterized by the absence of a regular three-dimensional arrangement of molecules or subunits of molecules extending over distances which are large compared to atomic dimensions, i. e., there is no long-range order. However, owing to the close packing of the particles in the condensed state, a certain regularity of the structure exists on a local scale, denoted as short-range order.

Dealing with such amorphous systems there is no general theory available which permits evaluation of the partition function and description of the properties of the amorphous bulk material to a good approximation. Therefore models have to be used, and it is well known that in the theories of the liquid state [1, 2] two different kinds of models have been developed approaching the problem either from the crystal side or from the gas side. That means, attempts have been made to describe the kind of disorder which distinguishes a liquid from an ideal crystal and, on the other hand, theories on the change in properties of gas with increasing density (i. e. virial expansions) have been developed.

The situation is very similar to the case of high polymers. Two kinds of models for the structure of amorphous polymers in the bulk have been proposed: In the "coil model" it is assumed that the material is homogeneous in structure and that the configuration statistics of a single molecule in the melt or glassy state are the same as those of an unperturbed molecule in solution [3, 4]. Secondly the various "bundle models" are based on the assumption of domains with nematic liquidcrystal like arrangements of the macromolecules [5–7]. Fig. 17.1 demonstrates some of the proposed models.

*Max-Planck-Institut für Polymerforschung, Postfach 3148, D-6500 Mainz
**Institut für Physikalische Chemie der Universität Mainz, Jakob-Welder-Weg, D-6500 Mainz

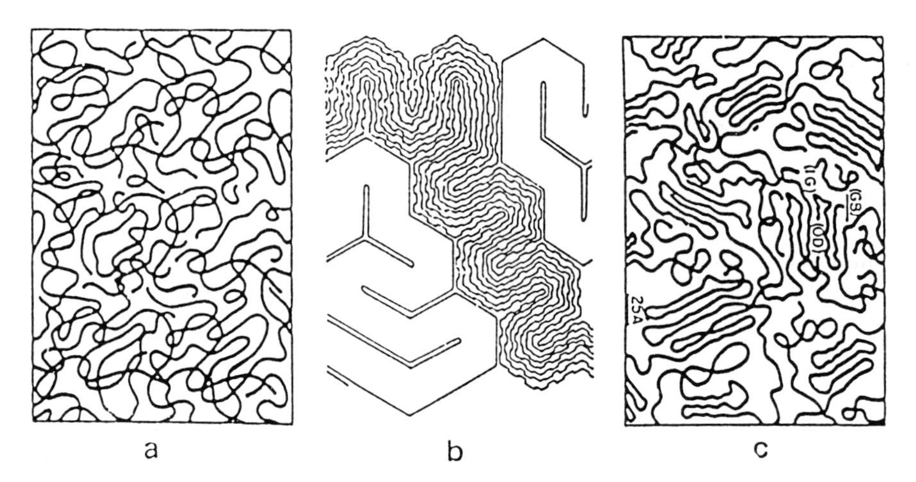

Fig. 17.1: Schematic models of the structure of amorphous polymer: (a) interpenetrating coils [3, 4]; (b) meander model [7]; (c) folded-chain fringed miscellar grain model [6].

The coil model follows, according to P. J. Flory [8], "from considerations of a theoretical nature which are at once simple and virtually incontrovertible". He concluded [4]: "A polymer molecule situated in a medium consisting of other polymer molecules of the same kind, or in a medium of unlike polymer molecules with which it mixes athermally, will occupy spatial configurations coinciding with those calculated in the random flight approximation". Newer theoretical considerations [9, 10] dealing with a whole range of concentrations are consistent with the assumption that the molecular conformations in the bulk and in a Θ-solvent are identical.

The concept of a "granular" structure of amorphous polymers (see Fig. 17.1c) is mainly based on electron microscopical observations, which will be discussed later. The meander model [7] (Fig. 17.1b) starts from the assumption of molecular bundles [5] as a thermodynamically stable form. This point of view is mainly justified by qualitative packing density considerations [11, 12], from which it was concluded, that a coiled chain conformation is incompatible with the macroscopic density of an amorphous polymer, which is only 5–15 % lower than the density in the crystalline state.

Recent Monte-Carlo simulations have shown, however, that in a lattice completely filled with chains, ideal chain conformations are also observed. Detailed microscopic models of cooperative motions in dense polymer systems developed in recent years [13] enable dynamic systems to be simulated in which space is completely filled by polymer chains. Results of such simulations [14] show that when ideal flexible chains are considered the equilibrium structure is very close to that postulated in the Flory model. Even the initially perfectly

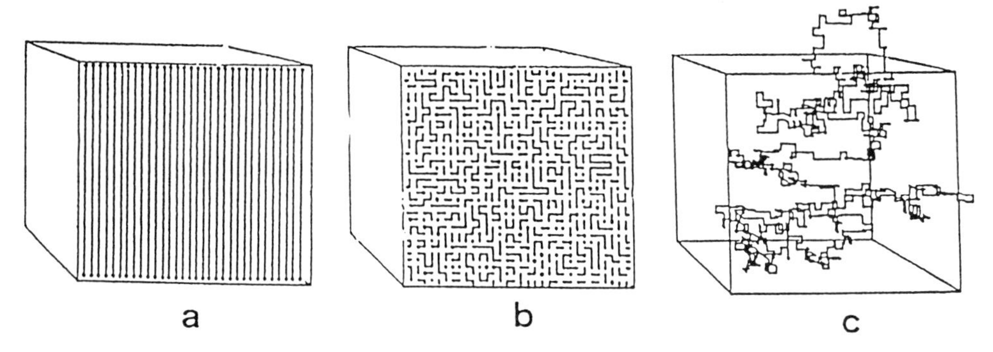

Fig. 17.2: Illustrations for the results of Monte Carlo simulations in densely packed systems [13, 14]; (a) starting state of order, (b) two-dimensional cut of the completely disordered state, (c) one exemplary chain out of the dense system (chain lengths 512 units).

ordered dense system of chains shown in Fig. 17.2a can reach the amorphous equilibrium (Fig. 17.2b), in which chains are coiled like the exemplary chain shown in Fig. 17.2a. There is no indication of any kind of deviations from the ideal random chain conformation in such systems unless some additional interactions are introduced as for example chain stiffness or chain closure to a ring [15].

Considerable effort has been spent in order to solve the problem of order and chain conformation in the amorphous state. Accordingly several review articles [16–18], textbooks [19, 20] and conference reports [21, 22] deal with this topic. In the following those experiments which have been performed within the framework of the Sonderforschungsbereich 41 will be described without going into the experimental details.

17.2 Chain Conformation

The conformation of macromolecular chains in bulk can be determined by neutron scattering on mixtures of deuterated and undeuterated chains.

This technique was introduced over a decade ago [23, 24] and represents a very significant experimental development in polymer science. Several review articles [25–27] outline the principles and the results obtained so far. Therefore we restrict ourselves to a description of the basic quantities in neutron scattering which will be used in the following.

If $\underline{k}_o$ and $\underline{k}$ are the wavevectors of the incident and scattered wave, then the scattered intensity $I(\underline{q})$ can be expressed as a function of the scattering vector.

$$\underline{q} = \underline{k} - \underline{k}_{\geqslant} \tag{1}$$

(or momentum transfer $\Delta\underline{p} = \overset{+}{h}\,\underline{q}$). For elastic scattering $\underline{k}$ and $\underline{k}_o$ are equal in magnitude, hence

$$|\underline{q}| = (4\pi/\lambda)\sin\Theta \tag{2}$$

where 2Θ is the scattering angle. The scattered intensity normalized by the intensity of the incident beam is given by

$$I(q) = \sum_{i,j}^{N} b_i b_j \exp\left(i\underline{q}(\underline{r}_i - \underline{r}_j)\right) \tag{3}$$

where b_i,b_j are the scattering lengths of the nuclei i and j situated at $\underline{r}_i$ and $\underline{r}_j$. Since the spin states of the nuclei are not correlated to the $\underline{r}_i$ for all cases which we shall consider here, one obtains an incoherent background scattering

$$I_{incoh}(q) = N(<b^2> - <b>^2) \tag{4}$$

and a coherent scattering contribution, which can again be described by Equ. (3), but now the b's are the coherent scattering lengths defined by the coherent scattering cross section

$$\sigma_{coh} = 4\pi<b>^2 = 4\pi b^2_{coh} \tag{5}$$

(Averaging is performed over the spin states of the same type of nuclei.) Most of the applications of the elastic neutron scattering to polymers are based on the large difference of b_{coh} between ^1H and deuterium: $b(H) = 0.374 \cdot 10^{-12}$cm, $b(D) = 0.667 \cdot 10^{-12}$cm.

The first motivation for using small angle neutron scattering (SANS) was to check Flory's hypothesis [4] experimentally which stipulates that polymer molecules in their bulk amorphous state should behave like ideal unperturbed coils in solution under Θ-conditions. SANS is the only technique available for investigating the conformation of a polymer molecule in the bulk by means of a straightforward structure method because a scattering contrast $K_N \alpha (m|b_H - b_D|)^2$ between molecules can be introduced by substituting m deuterium for hydrogen per monomer. In analogy to light scattering the results can be evaluated according to the Debye-Zimm scattering equation

$$K_N c_D / I(q) = (\overline{M}_w \cdot P(q))^{-1} + 2A_2 c_D \tag{6}$$

where $\overline{M}_w$ is the weight-average molecular weight, A_2 is the second virial coefficient, c_D is volume fraction of deuterated polymer and P(q) is the form factor of the single molecule

$$P(q) = N_w^{-2} < \sum_{i,j}^{n_W} \exp\left[i\underline{q}(\bar{r}_i - \bar{r}_j) \right] \tag{7}$$

with the approximation for small q's:

$$P(q)^{-1} \approx 1 + (1/3)q^2 R_g^{\ 2} \tag{8}$$

where R_g is the radius of gyration (n_w is weight-average degree of polymerization).

Later on it was recognized [28–31] that under certain assumptions the measurements can be performed with concentrated mixtures of deuterated and normal polymers. Then the form factor P(q) can be evaluated from

$$I(q) = c_D(1-c_D)K_N \cdot n_w \cdot P(q) \tag{9}$$

SANS experiments have been carried out on numerous polymeric systems, for references see [27]. In general the experimental data supported Flory's hypothesis and satisfactory agreement was found between the radii of gyration in the bulk amorphous state and those of dilute solutions under Θ-conditions. As an example for the application of Equ. (9) Fig. 17.3 shows the results for the case of polycarbonate [32]. From those measurements a radius of gyration R_g could be derived in agreement with the theoretical value [33].

The results from the small q-range do not give information about the chain conformation at small distances. As pointed out in Chapter 16 of this book it was observed that the radius of gyration R_g for many crystalline polymers is the same as in the molten (amorphous) state. Therefore the finding that R_g is equal to the unperturbed dimensions does not rule out the possibility that appreciable fractions of the molecules exhibit a stretched conformation with significant parallelism, as proposed for "bundle" models. In order to test the local conformation of the chains, measurements have to be extended to a q-range $1/R_g < q < 1/D$, where D is some measure of the dimensions of a single monomer unit.

The form factor defined by Equ. (7) is given for the case of a Gaussian coil by the well-known Debye equation

$$P(q) = \frac{2}{v^2}(v - 1 + 1^{-v}) \tag{10}$$

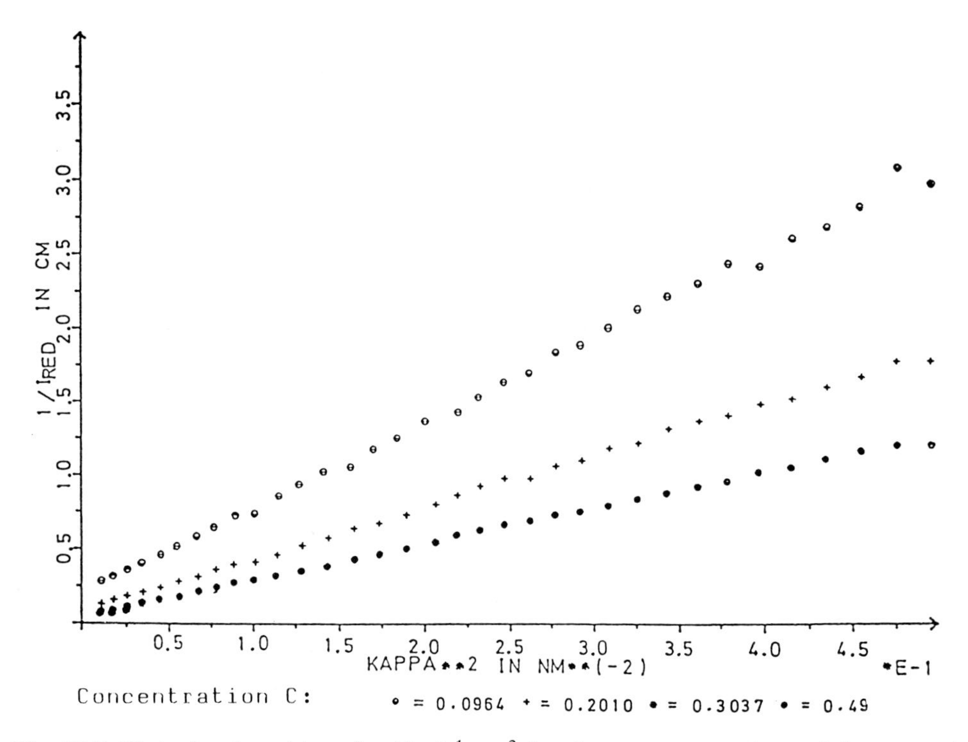

Fig. 17.3: Plot of reduced itensity $(I_{red})^{-1}$vs q^2 for three concentrations of deuterated polycarbonate (M_w = 49.000) in an H-matrix [32].

where $v=R_g^2q^2$. In the submolecular range the plot $P(q)q^2$ yields a plateau, the height of which is proportional to $2/R_g^2$ ("Kratky plot"). At larger q values deviations are expected for the real polymer chain, since the Gaussian statistics are no longer valid for short chain sequences. In Figs. 17.4 and 17.5 two examples are demonstrated. For polystyrene the Debye approximation holds up to about $q=4\ nm^{-1}$. On the other hand in the case of polycarbonate (Fig. 17.5), experiments which extended to a value of about $10\ nm^{-1}$ (corresponding to distances of about 8 Å) shows the Debye equation is verified only for a rather small q-range [31]. Subsequently at larger q values, appreciable deviations from the random coil conformation are both observed and calculated. The apparent discrepancy at large q values is due to the fact that in this region the density fluctuations are no longer negligible, or in other words the incompressibility assumption used in deriving Equ. (9) breaks down [34]. Nevertheless one can conclude that Flory's hypothesis is proved over a large q-range by neutron scattering.

More recent work has been concerned with chain conformation in binary mixtures of compatible polymers [35, 36]. As an example [37] we refer to our

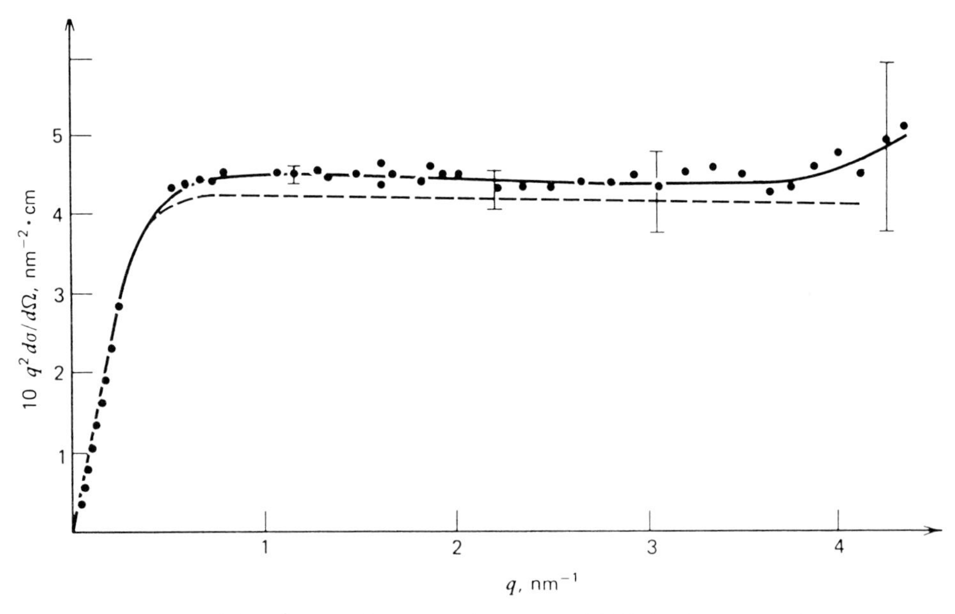

Fig. 17.4: Plot of P(q) · q^2 vs q of protopolystyrene (5 %) in deuteropolystyrene. The dashed line represents equilibrium [24].

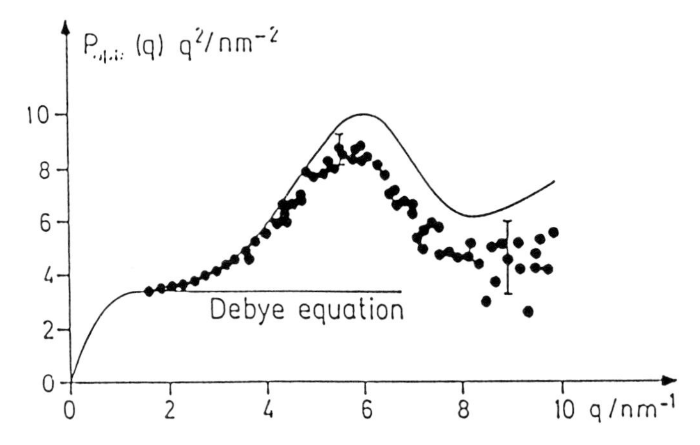

Fig. 17.5: Neutron scattering by amorphous polycarbonate in the intermediate q-range; ● scattering data P_{app} q^2 vs q [34]; — theoretical curve [33].

small angle neutron scattering experiments with homopolymer blends of polystyrene/polyparamethylstyrene and blockcopolymers of the same type. With measurements of this kind the temperature dependence of the Flory-Huggins interaction parameter can be determined. The SANS method is a powerful tool for determining phase diagrams (spinodal temperatures). The temperature dependence of the χ parameter could be expressed by $\chi(T) = A + B/T$. Furthermore, a good agreement with theoretically predicted spinodal value for $(\chi N)_s$ was achieved. A comparison of the temperature coefficient of the interaction parameter χ from these measurements indicated a composition and molecular weight dependence.

17.3 Short-range Positional Order

Diffraction patterns in the wide angle range for the purpose of obtaining radial distribution functions have been obtained by X-rays, neutrons, and electrons using coherent elastic scattering methods. Information about atomic distributions is obtained through interferences arising from phase differences that are due to path differences.

For atoms of one kind only the intensity (or the scattering cross section) is given by an expression of the type

$$I(q) = \sum_{m} \sum_{n} f_m(q) f_n(q) \exp(i\underline{q} \cdot \underline{r}_{mn}) \tag{11}$$

where $f_m(q)$ and $f_n(q)$ are the form factors for atoms m and n, and $\underline{r}_m$ and $\underline{r}_n$ are their positions. The intensity will depend on the frequency with which a given atomic distance r_{mn} occurs. For an isotropic liquid with a local density $\varrho(r)$ and an average density ϱ_o the coherent scattered intensity for only one kind of atom becomes

$$I(q) = N f(q)^2 \left\{ 1 + \int_0^\infty \frac{\sin qr}{qr} [\varrho(r) - \varrho_o] 4\pi r^2 dr \right\} \tag{12}$$

This equation shows that for every distance r in real space, there is a sin qr/qr damped wave in reciprocal space, so that the intensity distribution is a superposition of sine waves corresponding to each distance in real space. Details in the expression for the intensity distribution depend on the scattering radiation used.

If the atoms are not all of one kind the following relationship is obtained

$$4\pi r^2 \sum_m \sum_n P_m \frac{b_m b_n}{(b)^2} \varrho_{nm}(r) = 4\pi r^2 \varrho_o \frac{(b)^2}{(b^2)} + \frac{2r}{\pi} \int_0^\infty q\left[\frac{I}{Nb^2} - 1\right] \sin rq \, dq \quad (13)$$

where N is the total number of atoms irradiated; P_m equals N_n/N, the fractional number of atoms of type m; b_m and b_n are the scattering amplitudes of atoms m and n, respectively; and ϱ_{nm} is the density of atoms n around an atom m.

An example of pair distribution functions obtained by electron diffraction from molten polyethylene [38] is shown in Fig. 17.6. In principle, the technique is straightforward and has been used to obtain radial distribution functions for a number of amorphous polymers using X-rays, neutrons, and electrons. However, in practice each scattering method presents considerable experimental difficulties. The main difficulties are a correct determination of the background scattering and the fact that intensities are experimentally attainable only in a limited q-range, so that the integration, instead of extending from 0 to ∞ extends only from q_{min} to q_{max}.

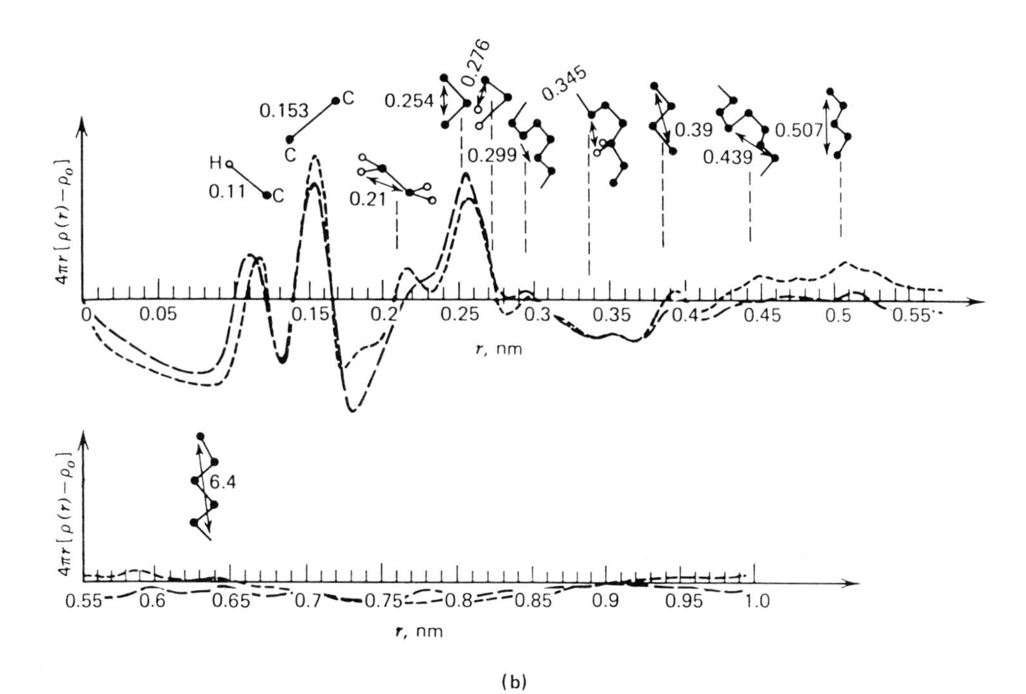

Fig. 17.6: Experimental correlation functions for polyethylene films at 140° C (. . .) and 200° C (–) [38].

The experimentally obtainable information from monochromatic X-rays and electrons lies in the wide angle range of reciprocal space (small distances in real space). The methods differ in that X-ray experiments suffer from a cutoff at rather small values of q_{max}, whereas electron diffraction measurements suffer from a cutoff at rather large values of q_{min}. Neutron scattering experiments can be organized so as to cover a very wide range in reciprocal space but suffer from the disadvantage that they require partial deuteration.

The radial distribution curves which have been obtained for many polymers [38–41] were interpreted as indicating the presence of ordered structures or a lack of them, depending on the polymer and the extent to which the investigators had been successful in eliminating experimental artifacts. The scattering patterns of nonoriented and oriented samples are found to be very similar both for atactic and syndiotactic poly-(methyl methacrylate) [42] and can be explained on the basis of all trans-sequences. Regular conformation persisted over 16 bonds for the syndiotactic polymer.

Very distinct maxima at 0.5 nm intervals in pair distribution functions of molten polyethylene observed in electron diffraction data were attributed by some authors to intermolecular distances [43, 44]. A model was proposed according to which chain segments are parallel within regions as large as 5 nm. Such oscillations in the r.d.f. are almost certainly an artifact, since electron diffraction, due to the q_{min} limitation, does not in principle give information

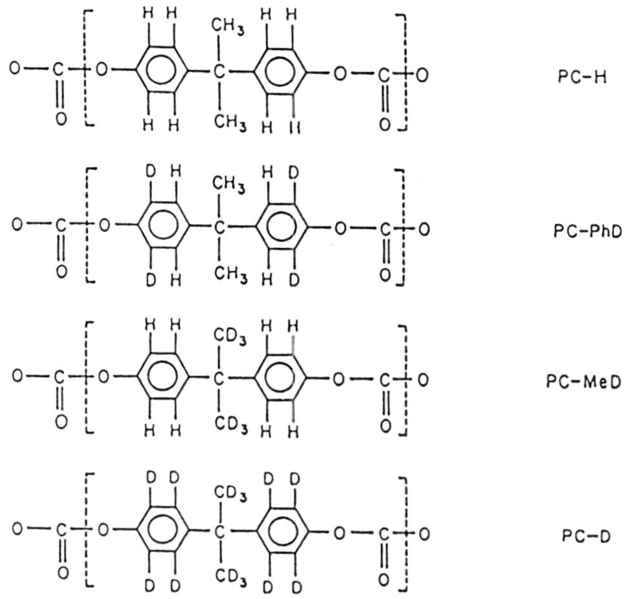

PC-H

PC-PhD

PC-MeD

PC-D

Fig. 17.7: Structure of the isotopic isomers of PC [48].

beyond 1 nm in real space. Order in amorphous polymers was again proposed after experiments on crystalline polyethylene which had been "amorphized" by γ irradiation [44]. Other authors did not find correlations of these dimensions in polyethylene [38, 42, 45]. A broad maximum in the radial distribution function between 0.4–0.7 nm was attributed to short-range intermolecular correlations while all the superimposed sharp maxima could be attributed to intramolecular distances [38]. Whereas this broad intermolecular distribution made order to the extent suggested in Ref. [43, 44] unlikely, it was considered that the pair distribution function is not a suitable tool for enabling a decision to be made between details concerning conformation [38]. However, in a semi-quantitative manner, comparative work on flexible and rigid chains [42] indicated that the former tend to adopt a random conformation, while the latter lie parallel over distances well in excess of their diameter.

Instead of using radial distribution functions one can evaluate the scattering patterns directly in reciprocal space. Rather often there exists a convenient partitioning of the structural information into that which arises from interchain or intrachain correlations [46]. One may also calculate the scattering curves for certain models by means of Equ. (12) modified for atoms with different scattering factor f(q). Comparison with experimental results is difficult, since the information contained in the X-ray patterns is rather restricted. A very promising way to broaden extensively the range of available information is the isotopic substitution technique well-known in low molecular weight liquids [47]. This method has been applied to polycarbonate of bisphenol A [48].

The isotopic isomers of Fig. 17.7 show rather different neutron scattering curves (see Fig. 17.8). Model calculations were performed using the Debye equation

$$I(q) = \sum_{m,n} b_m b_n \frac{\sin q r_{mn}}{q r_{mn}} \tag{14}$$

where m and n denote different atoms, b_m and b_n are their neutron scattering lengths, r_{mn} is their mutual distance and q is the absolute value of the scattering vector. Scattering curves were calculated for all four isotopic isomers of PC of all model units considered. Very good agreement was obtained for models consisting of four parallel segments each consisting of two monomer units with an intersegmental distance of 5.3 Å. There is no evidence for a more far reaching "order" in polycarbonate [49].

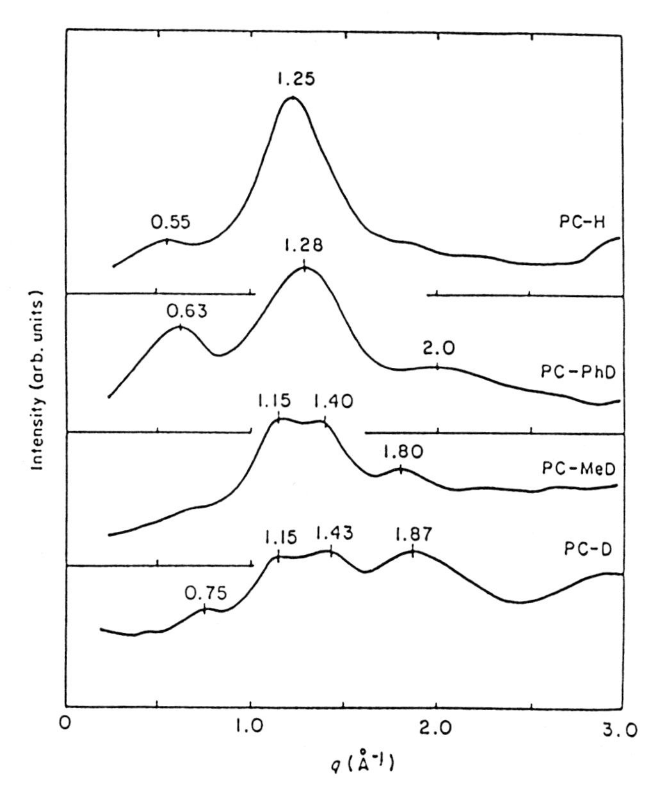

Fig. 17.8: Neutron scattering curves of amorphous PC-H, PC-PhD, PC-MeD and PC-D [48].

17.4 Direct Structural Observation

Unlike the experiments described previously [38] where the pair distribution function was obtained by analyzing the electron intensity distribution in reciprocal space, structural information can also be obtained from direct electron microscopical evidence in real space. The electron microscope image is produced in a two stage process, involving the production of a diffraction pattern (Fourier transform of the object function) and subsequently the production of an image (Fourier transform of the diffraction function). The Fourier transform of this image function consists of two parts, a phase and amplitude contribution. The transform of the phase contrast image T^{i}_{phase} is related to the transform of the object T^{o} by

$$T^{i}_{phase}\ (q,\Theta) = -T^{o}\ (q,\Theta)\ A(\alpha)\ f(\alpha)\ \sin\ \chi(\alpha) \tag{15}$$

and the transform of the amplitude contrast image T^{i}_{ampl} is related to the transform of the object T^{o} by

$$T^{i}_{ampl}\ (q,\ \Theta) = -T^{o}\ (q,\Theta)\ A(\alpha)\ F(\alpha)\ Q(\alpha)\ \mathrm{Cos}\ \chi(\alpha) \tag{16}$$

where q is the scattering angle, λ the electron wavelength, Θ the azimuthal coordinate, A(α) is the aperture function and

$$\chi(q) = \frac{2\pi}{\lambda}\left[-\frac{1}{4}\ C_s\ q^4 + \frac{1}{2}\ \Delta f\ q^2\right] \tag{17}$$

where C_s is the spherical abberation of the lens, Δf is the defocus value which leads to the desired phase shift χ(q).

Phase contrast has been used to obtain information about amorphous materials [50] but the results, which were interpreted as indicating order in amorphous polymers, were severely criticized [51]. Contrast variations in the electron microscopic image depend in a complicated way on various aspects of scattering theory and the microscope transfer function, so that the observed effects can easily be misinterpreted. Conclusions about molecular conformation should not be drawn unless detailed calculations have been performed and various parameters affecting the image are carefully checked. Experiments, in which the phase contrast function is used in order to obtain information about partially disordered polymers are now in progress and are described in Chapter 9 this book.

Alternatively amplitude contrast can be used to obtain information about amorphous polymers. In this case one component in a mixture of polymers should contain atoms of a higher atomic number. This may be inherent in the polymer or it can be induced by chemical staining. Inherent amplitude scattering was used in order to study the kinetics of phase separation in a polystyrene/ polyvinylmethylether blend [52]. For deep quenches it was found that the size of the phase separated domains in the spinodal region increased linearly with time, implying that hydrodynamic effects control the rate of growth of the domains in the time scale and temperature range under consideration. The growth velocities and approximate diffusion coefficients were calculated. These studies enabled the phase diagrams to be determined and the three thermodynamic regions (a) stable, (b) metastable and (c) unstable to be distinguished. Consideration of the thermodynamics involved showed that the experimental spinodal curves could be reproduced theoretically only if the difference in contacting surface area of the two molecules was accounted for [53].

17.5 Orientational Order

A distinguishing feature among the proposed models for the amorphous state is the amount of orientational order. The random coil models only require local intrachain correlations of segments, but all the bundle models or meander models are based on the assumption of extended intersegmental correlations of various sizes. Light scattering responds to such structural features in the case of the depolarized scattering if the chain segments themselves are sufficiently optically anisotropic. The optical anisotropy δ_o of the repeat units of the chain is given as

$$\delta_o = \alpha_{\parallel} - \alpha_{\perp} \tag{18}$$

where $\alpha_{\parallel}$ and $\alpha_{\perp}$ are the optical polarizabilities with regard to some optical main axis. For macroscopically isotropic samples, the orientation correlations are usually expressed in terms of the orientation correlation function [1, 2, 54, 55]

$$f_o(r) = \left< \frac{1}{2} (3 \cos^2\Theta_{ij}(r) - 1) \right> \tag{19}$$

$\Theta_{ij}(r)$ is the angle between the vectors i and j, characterizing the scattering units located at a distance r. The depolarized light-scattering intensity, due only to orientation fluctuations, is

$$H_v(q) = \frac{16\pi^4}{15\lambda^4} \left(\frac{<n>^2 + 2}{3} \right) \delta_0^2 N_{cm}^3 \, 4\pi \int_0^\infty f_\Theta(r) \frac{\sin qr}{qr} r^2 dr \tag{20}$$

where N_{cm}^3 is the number of repeat units per cm^3, and $<n>$ is the average refractive index. Equ. (20) can be derived as a special case of a random orientation correlation model [56]. The same results also follow from another treatment [57]. In this case the internal field was taken into account by thěLorentz approximation. In the case of light scattering, $(qr) \ll 1$ for $f_\Theta(r) \neq 0$, and it follows

$$H_v = \frac{16\pi^4}{15\lambda_o^4} \left(\frac{<n>^2 + 2}{3} \right)^2 N_{cm}^3 \, \delta_o^2 4\pi \int_0^\infty f_\Theta(r) r^2 dr \tag{21}$$

The correlation range is thus expressed in terms of a correlation volume V_c

$$V_c = 4\pi \int_0^\infty f_\Theta(r) r^2 dr \tag{22}$$

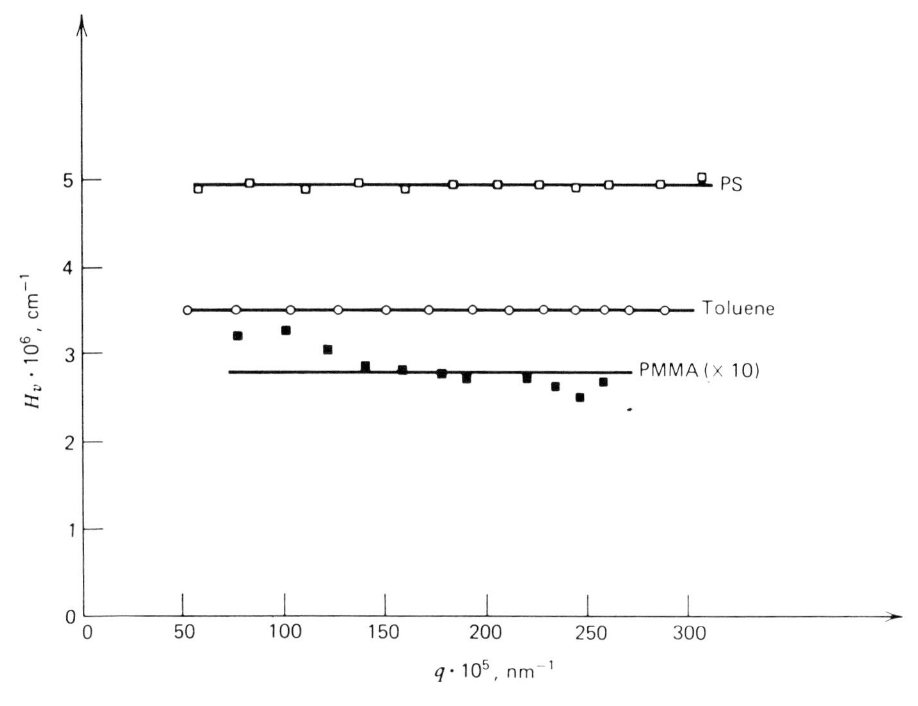

Fig. 17.9: H_v component of light scattered by polystyrene, toluene, and poly(methyl methacrylate) [18].

If $f_\Theta(r)$ is defined by a step function within the correlation range

$$f_\Phi(r) = \begin{cases} 1 \text{ within } V_c \\ 0 \text{ otherwise} \end{cases} \tag{23}$$

we can define a quantity p which gives the number of parallel repeating units in the volume V_c where $p = N_{cm}^3 \cdot V_c$. The quantity $\delta = \delta_o \sqrt{p}$ can then be defined as the effective optical anisotropy of the monomer unit. It is a measure of the average orientation correlation both along the chains and with respect to units of other chains.

All the results obtained so far for amorphous polymers, such as polycarbonate [58–63], polystyrene [59–63] and polydimethylsiloxane [58] show that the H_v component is independent of scattering angle, confirming Flory's results for polypropylene and polystyrene. As an example, in Fig. 17.9 the measured data for polystyrene and poly(methyl methacrylate) are plotted [59]. For comparison, the scattering by toluene is also shown. In each case only a very small depolarized scattered intensity is oberved. The value for δ^2 in the glassy state agrees surprisingly well with the value measured in solution [65], and in addition, the calculated value of δ^2 is in reasonable agreement [66]. δ^2 was evaluated on the basis of rotational isomeric state theory using a value of $\delta^2 = 2800 \times 10^{-50}$ cm^6. Therefore, the results of depolarized light-scattering measurements of these polymers show no indication of parallelization of neighbouring molecules in the bulk material. In order to demonstrate how parallelization would affect the values of H_v, the light scattering of 4-methoxybenzylidene-4'-n-butylaniline (MBBA) in the isotropic phase is shown in Fig. 17.10. A comparison with the PS data shows two important differences: First, even at high temperature the scattered intensity of MBBA is two orders of magnitude larger, and second, there is a strong temperature dependence of H_v and accordingly of p.

A weak tendency for parallelization in the condensed state of chain molecules is observed in the case of n-paraffin melts [67, 68, 70] as well as in poly(-phenylmethylsiloxane) [71]. In the latter case, the whole spectrum arising from inherent anisotropy was measured as a function of temperature and concentration. In Fig. 17.11 the effective anisotropy is plotted vs temperature after correction for the inelastic depolarized scattering.

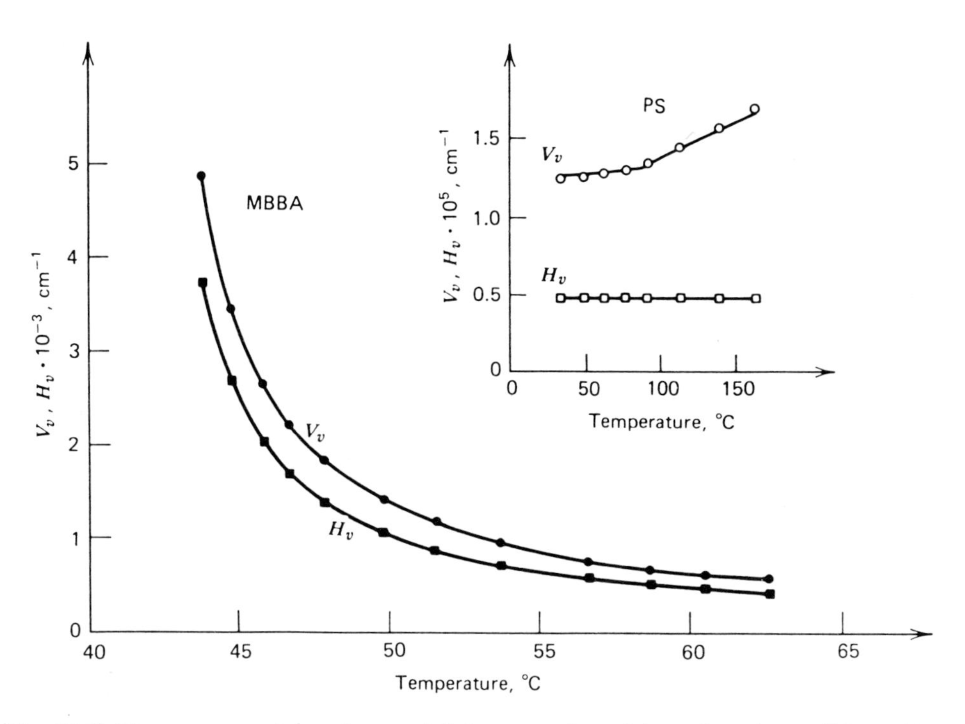

Fig. 17.10: Temperature dependence of light scattering of 4-methoxybenzylidene-4'-n-butylaniline (MBBA) in the isotropic phase compared with behaviour of polystyrene [18].

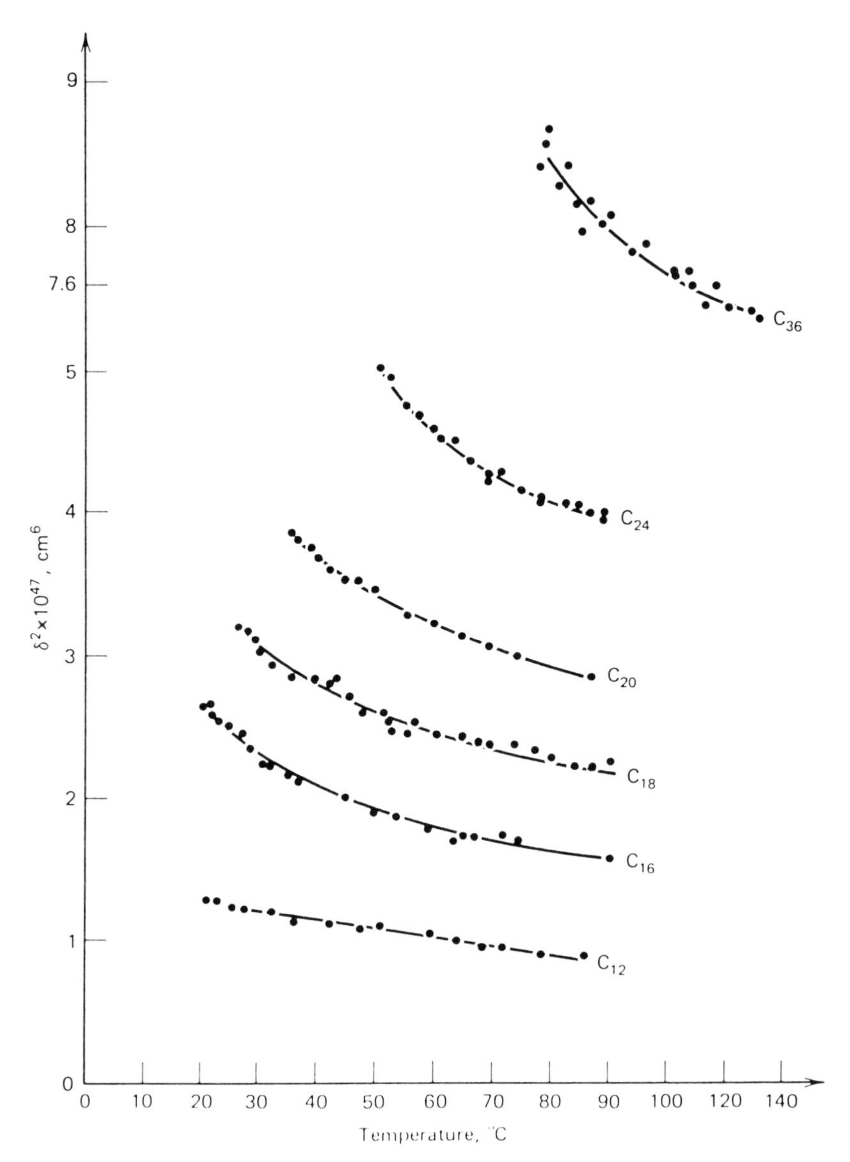

Fig. 17.11: Temperature dependence of the effective optical anisotropy δ^2 of n-paraffins [67].

A phenomenological description of the orientation correlation in paraffin melts has been made [71]. The ratio of the anisotropy from the bulk to the dilute solution increases with chain length and levels off when n = 16. This orientational order displays a temperature dependence as predicted by the de Gennes theory of pretransitional short-range orientational correlations.

17.6 Density Fluctuations

Thermal density fluctuations are related to the integral over the pair correlation function or to the scattering law in the limit of $q \to 0$. Thus the theoretical prediction for the equilibrium state is that the small angle X-ray or neutron scattering is directly related to the isothermal compressibility. The scattering curve should be approximately constant for small values of q, allowing an extrapolation of the scattering curve $q \to 0$. Deviations are expected to occur near a critical point, where the compressibility diverges and where the correlation length of the thermal density fluctuations also diverges. In practice it has been observed that in the case of small angle X-ray scattering a strong increase of the scattered intensity occurs with decreasing values of q [72, 73]. The correlation length obtained from this scattering curve is about 5–25 nm. Earlier investigators frequently interpreted such data on the basis of a two-phase structural model, incorporating different orientational order and therfore also different densities. However, a series of experiments carried out on polymers such as poly(ethylene terephthalate), poly(methyl methacrylate), and polycarbonate indicated that the total scattering obtained was caused by the superposition of scattering components from additives such as pigments, stabilizers, etc., and of the scattering due to pure thermal density fluctuations. It has become possible to break down the total small angle scattering, so that the thermal density fluctuation component could be determined. It was found that the experimental results agree closely with the theoretical predictions, provided that the temperature is kept above the glass-transition temperature. In this case, the fluctuations decrease with decreasing temperature, since both the absolute temperature and the isothermal compressibility decrease.

At the glass transition, the isothermal compressibility changes in a stepwise manner, and based on the expression for the equilibrium state, the thermal density fluctuations are expected to behave similarly. Fig. 17.12 shows that this is not the case. The temperature coefficient of the density fluctuations changes discontinuously at the glass transition temperature; the magnitude of the fluctuations, however, behave continuously. It has been observed that in a limited temperature range the thermal density fluctuations can be expressed approximately by

$$\frac{\langle \delta N \rangle^2}{\langle N \rangle} = \varrho_N k T \chi_T(T_g) \qquad (24)$$

where $\chi_T(T_g)$ is the isothermal compressibility in the equilibrium state at the glass-transition temperature. Various interpretations have been given for the deviations of the thermal density fluctuations from the original compressibility equation. They are all based on the fact that the glassy state is a nonequilibrium state. In terms of the thermodynamic description this means that order para-

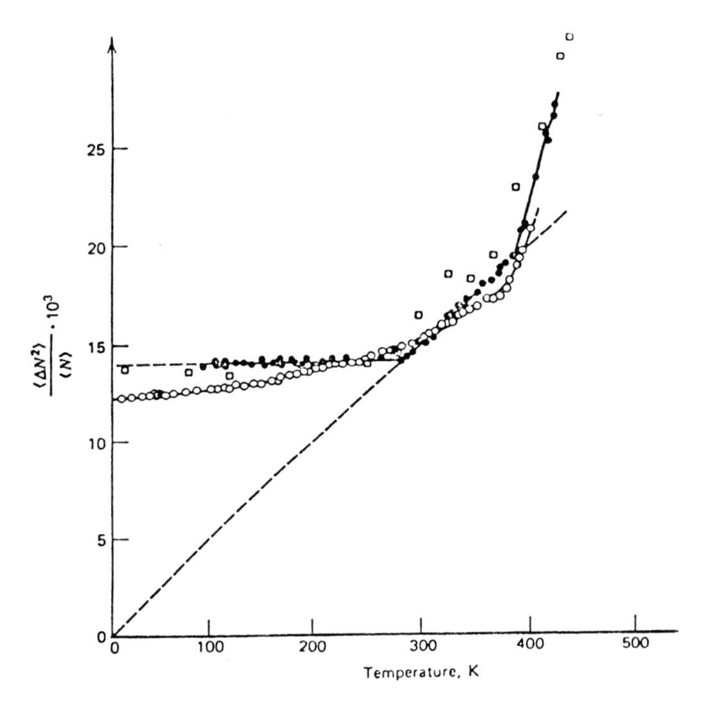

Fig. 17.12: Density fluctuations of poly(methyl methacrylate) (PMMA), measured in different experiments: O, X-ray scattering (Ruland); ●, X-ray scattering (Wendorff); □, elastic neutron scattering, entropy contribution [73].

meter z_i and, as the conjugate quantities, affinities A_i, must be introduced to characterize the glassy state

Equilibrium state — Nonequilibrium state

$$A_i = -\left(\frac{\partial G}{\partial z_i}\right)_{T,\,p,\,N,\,zj} = 0 \qquad A_i = -\left(\frac{\partial G}{\partial z_i}\right)_{T,\,p,\,N,\,zj} \neq 0 \tag{25}$$

where G is the Gibbs free energy.

One way of accounting for the excess density fluctuations is in terms of order parameter fluctuations which, however, does not readily give any insight into the mechanism

$$<\partial z_j^2> = kT/(\partial A_j/\partial z_j)_{T,V,N} \tag{26}$$

Some have argued that such nonpropagating contributions of thermal density fluctuations are mainly responsible for their variation with temperature

[72–74] whereas others have stressed the strong influence of propagating fluctuations, that is, of phonons [76, 77]. This problem cannot be solved by elastic scattering experiments alone, since they do not allow a separation of the total fluctuations into propagating and nonpropagating components. In principle, this is possible from inelastic neutron scattering data [73] but the determination of the mean square value of density fluctuations by neutron scattering poses a big problem.

In order to determine thermal density fluctuations, the coherent scattering law without any incoherent background scattering must be determined. This is accomplished by measuring the thermal density fluctuations both in a system containing protons and in a system containing deuterons. The experimental results obtained in this way agreed closely with those obtained from small angle X-ray scattering [73].

17.7 References

[1] P. A. Egelstaff: An Introduction to the Liquid State, Academic Press, London 1967.
[2] F. Kohler: The Liquid State, Verlag Chemie, Weinheim 1972.
[3] P. J. Flory: Principles of Polymer Chemistry, Cornell University, 1953.
[4] P. J. Flory, J. Chem. Phys. 17 (1949) 303.
[5] V. A. Kargin, J. Polymer Sci. 30 (1958) 247.
[6] G. S.Y. Yeh, J. Macromol. Sci. B6, 451 (1972) 465
[7] W. Pechhold, S. Blasenbrey, Koll. Z. u. Z. Polymere 241 (1970) 955.
[8] P. J. Flory: Statistical Mechanics of Chain Molecules, J. Wiley, New York 1969, p. 35.
[9] P. G. de Gennes, J. Phys. (Paris) Lett. L55 (1975) 36
[10] J. P. Cotton, M. Nierlich, F. Boué, M. Daoud, B. Farnoux, G. Jannink, R. Duplessix, C. Picot, J. Chem. Phys. 65 (1976) 1101.
[11] R. E. Robertson, J. Phys. Chem. 69 (1965) 1575.
[12] R. E. Robertson, Annual Review of Mat. Sci. 5 (1975) 173.
[13] T. Pakula, Macromolecules 20 (1987) 679.
[14] T. Pakula, S. Geyler, Macromolecules 20 (1987) 2909.
[15] T. Pakula, Polymer 28 (1987) 1293.
T. Pakula, S. Geyler, Macromolecules 21 (1988) 1665.
S. Geyler, T. Pakula, Makromol. Chem., Rapid Comm. 9 (1988) 617.
[16] I. G. Voigt-Martin, J. H. Wendorff: Encyclopedia of Polymer Science and Engineering, Vol. 1, 2nd ed., J. Wiley & Sons, 1986.
[17] G. S.Y. Yeh, Critical Rev. in Macromol. Sci. 1 (1972) 173.
[18] E.W. Fischer, M. Dettenmaier, J. Non-Cryst. Sol. 31 (1978) 181.
[19] P. G. de Gennes: Scaling Concepts in Polymere Physics, Cornell University Press, Ithaca N.Y. 1979.
[20] L. H. Sperling: Introduction to Physical Polymer Science, J. Wiley, New York 1986.

[21] Organization of Macromolecules in the Condensed Phase, Faraday Discussions Royal Soc. Chem. (1979) 68.
[22] S. E. Kleinath et al. (eds.): Order in the Amorphous State of Polymers, Plenum Press, New York 1987.
[23] R. G. Kirste, W. Kruse, J. Schelten, Makromol. Chem. 162 (1972) 299; H. Benoit, J. P. Cotton, D. Decker, B. Farnoux, J. S. Higgins, G. Jannink, R. Ober, C. Picot, Nature (London) 245 (1973) 13.
[24] G. D. Wignall, D. G. H. Ballard, J. Schelten, Eur. Polym. J. 9 (1973) 965; G. D. Wignall, D. G. H. Ballard, J. Schelten, Eur. Polym. J. 10 (1974) 861.
[25] R.W. Richards, in Developments in Polymer Characterisation, Ed. J.V. Dawkins, Applied Science Publ., 1978.
[26] C. Picot, in R. A. Pethrick, R.W. Richards (eds.): Static and Dynamic Properties of the Polymeric Solid State, D. Reidel Publ. Co., 1982.
[27] G. D. Wignall, in H. Mark et al. (eds.): Encyclopedia of Polymer Science and Engineering, Vol. 10, J. Wiley, New York 1987, 112.
[28] M. Stamm, E.W. Fischer, M. Dettenmaier, P. Convert, Faraday Discuss. Chem. Soc. 68 (1979) 263.
[29] P. G. de Gennes: Scaling Concepts in Polymer Physics, Cornell University Press, Ithaca and London 1979, p. 67.
[30] A. Z. Akcasu, G. C. Summerfield, S. N. Jahshan, C. C. Han, C.Y. Kim, H. Yu, J. Polym. Sci., Polym. Phys. Ed. 18 (1980) 863.
[31] W. Gawrisch, M. G. Brereton, E.W. Fischer, Polymer Bull. 4 (1981) 687.
[32] W. Gawrisch, Ph. D. Thesis, University of Mainz 1979.
[33] D.Y. Yoon, P. J. Flory, Polym. Bull. 4 (1981) 693.
[34] B. Z. Jiang, E.W. Fischer, K. Hahn, K. J. Kuhn, unpubl. results.
[35] J. Jelenic, R. Kirste, B. J. Schmitt, S. Schmitt-Strecker, Macromol. Chem. 180 (1979) 2057.
[36] G. Wignall, R.W. Hendrichs, W. C. Koehler, J. S. Lin, M. P. Wai, E. L. Thomas, R. S. Stein, Polymer 22 (1981) 886.
[37] W. G. Jung, E.W. Fischer, Makromol. Chem., Makromol. Symp. 16 (1988) 281.
[38] I. G. Voigt-Martin, F. C. Mijlhoff, J. Appl. Phys. 47 (1976) 3942.
[39] H. G. Kilian, K. Bouke, J. Polym. Sci. 58 (1962) 311.
[40] R. Lovell, A. Windle, Polymer 17 (1976) 488.
[41] G. Natta, P. Corradini, Makromol. Chem. 16 (1955) 77.
[42] R. Lovell, G. Mitchell, H. Windle, Disc. Faraday Soc. 68 (1979) 46.
[43] Yu. K. Ovchinnikov, G. S. Markova, V. A. Kargin, Polym. Sci. USSR 11 (1969) 369.
[44] A. Odajima, S. Yamane, O. Yoda, I. Kuriyama, Rep. Prog. Polym. Phys. Japan 29 (1978) 213.
[45] G.W. Longmann, G. D. Wignall, R. P. Sheldon, Polymer 20 (1979) 1063.
[46] G. R. Mitchell, in ref. [22].
[47] D. Zeidler, Angew. Chem. 92 (1980) 700; Z. Phys. Chem. (NF) 133 (1982) 1.
[48] L. Cervinka, E.W. Fischer, K. Hahn, B.-Z. Jiang, G. P. Hellmann, K.-J. Kuhn, Polymer 28 (1987) 1287.
[49] G. Longmann, R. Sheldon, G. Wignall, J. Mat. Sci. 11 (1976) 1339.
[50] D. Uhlman, Disc. Faraday Soc. Chem. Soc. 68 (1979) 87.
[51] E. Thomas, E. J. Roche, Polymer 20 (1979) 1413; 22 (1981) 341.
[52] I. G. Voigt-Martin, K. H. Leister, R. Rosenau, R. Koningsveld, J. Polym. Sci., Polym. Phys. 24 (1986) 723.

[53] R. Koningsveld, L. A. Kleintjens, J. Polym. Sci., Polym. Symp. 61 (1977) 221.
[54] E.W. Fischer, J. H. Wendorff, M. Dettenmaier, G. Lieser, I. Voigt-Martin, Polym. Prepr. 15 (1975) 8.
[55] W. B. Street, K. E. Gubbins, Ann. Dev. Phys. Chem. 28 (1977) 373.
[56] R. S. Stein, P. R. Wilson, J. Appl. Phys. 33 (1962) 1914.
[57] J. J. v. Aartsen, in A. J. Chompff and S. Newman (eds.): Polymer Networks, Plenum Press, New York 1971.
[58] M. Dettenmaier, H. H. Kausch, Coll. Polym. Sci. 259 (1981) 209.
[59] M. Dettenmaier, E.W. Fischer, Koll. u. Z. Polym. 251 (1973) 922.
[60] M. Dettenmaier, E.W. Fischer, Makromol. Chem. 177 (1975) 1185.
[61] M. Dettenmaier, Ph. D. Thesis, Mainz 1975.
[62] G. D. Patterson, ACS Polym. Prepr. 18 (1977) 713.
[63] M. Dettenmaier, Progr. Coll. Polym. Sci. 66 (1979) 169.
[64] H. J. Hölle, R. G. Kirste, B. R. Lehnen, M. Steinbach, Progr. Coll. Polym. Sci. 58 (1975) 30.
[65] E. G. Ehrenburg, E. P. Piskareva, I.Y. A. Poddubnyi, J. Polym. Sci. C42 (1973) 1021.
[66] A. E. Tonelli, Y. Abe, P. J. Flory, Macromolecules 3 (1970) 303.
[67] E.W. Fischer, M. Dettenmaier, J. Non. Crystall. Sol. 31 (1978) 181.
[68] G. D. Patterson, P. J. Flory, J. Chem. Soc. Faraday Trans. II, 68 (1972) 1098.
[69] P. Tancrede, B. Bothorel, P. de St. Romain, D. Patterson, J. Chem. Soc. Faraday Trans. II, 73 (1977) 15.
[70] E.W. Fischer, G. Strobl, M. Dettenmaier, M. Stamm, N. Steidle, Faraday Disc. Chem. Soc., 68 (1979) 26.
[71] J.T. Bendler, Macromolecules 10 (1977) 162.
[72] J. H. Wendorff, E.W. Fischer, Koll. Z. u. Z. Polym. 251 (1973) 876.
[73] R. Hoffmann, Doctoral Theses, University of Mainz 1981.
[74] A. Renninger, G. Wicks, D. Uhlmann, J. Polym. Sci. Phys. 13 (1975) 1247.
[75] W. Ruland, Progr. Coll. Polym. Sci. 57 (1975) 192.
[76] J. Rathje, W. Ruland, Coll. Polym. Sci. 254 (1976) 358.
[77] W. Wiegand, W. Ruland, Progr. Coll. Polym. Sci. 66 (1979) 355.

18 Theory of Dense Polymer Systems

Kurt Binder, Kurt Kremer, Ingeborg Carmesin, and Alla Sariban*

We review several aspects of the statics and dynamics of dense polymer systems. Within a mean-field approximation the spinodal decomposition of polymeric alloys is investigated. Here we also include the effects of polydispersity and shear. This investigation then is extended to the problem of surface wetting of polymeric mixtures. The phase diagram is calculated and it is predicted that the thickness of the wetting layer grows with the logarithm of the time.

These general investigations are complemented by scaling and Monte Carlo investigations of polymer systems. In particular the spinodal decomposition of ternary systems on a lattice, containing A-chains, B-chains and vacancies is investigated by Monte Carlo. There we use a grand canonical algorithm, where the chains can exchange their identity. Apart from these global aspects of statics and dynamics of macromolecular systems, we also consider dynamic properties on time and length scales smaller than the chain diameter. To do this we test several aspects of the reptation theory by a Monte Carlo and scaling analysis of a chain in a straight tube, and a molecular dynamics study of polymer melts.

* Institut für Physik der Universität Mainz, Postfach 3980, D-6500 Mainz
(Present address of K. Kremer: IFF, KFA Jülich, Postfach 1913, D-5170 Jülich)
(Present address of A. Sariban: Institut für Physikalische Chemie, Technische Hochschule Darmstadt, D-6100 Darmstadt)

K. Binder, K. Kremer, I. Carmesin, and Alla Sariban

18.1 Introduction

Since the discovery of De Gennes [1] that the statistical properties of long flexible macromolecules in dilute solution can be mapped to the statistical mechanics of magnetic phase transitions, which implied that techniques such as scaling concepts [1], renormalization group methods [2, 3] etc. can readily be applied to polymers, the theoretical understanding of polymers in dilute and semidilute solutions has greatly advanced [1–3].

On the other hand, dense polymer systems are much less well understood, apart from the fact that the chain configurations are essentially Gaussian [1], and hence not so interesting from the theoretical point of view. Interesting problems, however, still occur with respect to the dynamics of such dense systems – although the reptation model (or tube model, respectively) provides a framework to discuss many experiments and many of its implications have been worked out in detail [4], it still involves many unsolved questions and lacks a rigorous microscopic foundation. And nontrivial problems relating to static properties of dense polymer systems are encountered when we consider polymer mixtures, when the local concentration enters as additional microscopic variable. To a large extent, the understanding of phase diagrams of such mixtures is still based on the primitive Flory-Huggins mean field theory [5], which by now is more than 40 years old.

In this chapter, we review work on the theory of dense polymer systems which is related to all these questions: in the following section we first discuss polymer mixtures with nonuniform concentration, as they form via spinodal decomposition [6–10] or due to preferential adsorption near walls [11, 12]. Also rather specialized extensions, such as the effect of polydispersity [13] and the effect of shear [14, 15] on phase separation kinetics, and the growth in thickness of wetting layers on initially non-wet walls will be briefly mentioned in the second and third part of this section [16, 17].

The third section of this chapter then is devoted to computer simulation studies which shed light on some of the problems mentioned above. A Monte Carlo study of symmetrical polymer mixtures including vacancies [18–25] clarifies the accuracy of some approximations made in the Flory-Huggins theory. A Monte Carlo study of a chain in a fixed straight tube [26, 27] tests the validity of the reptation model description, and already indicates the high mobility of chain ends. This insight proves useful in the analysis of polymer melts, where the predictions of the reptation model are verified for inner monomers of the chain [28], and thus a longstanding controversy about the evidence for or against reptation from computer simulation is clarified. Chapter 18.4 finally contains a brief summary of our conclusions.

18.2 Mean Field Theories of Polymer Mixtures

We consider a mixture of two polymer species, denoted as A and B, with "chain lengths" N_A, N_B: i. e., the numbers of subunits of size σ_A, σ_B, the latter being defined such that the end-to-end distances satisfy the standard relations $R_A^2 = \sigma_A^2 N_A$, $R_B^2 = \sigma_B^2 N_B$. We are interested in a coarse-grained description only, considering time scales large enough that all fluctuations concerning internal degrees of freedom of the chains have died out, the only variable of interest (which is so slow that it is not necessarily at full equilibrium) is the volume fraction $\Phi(\vec{r},t)$ taken by species A at time t in a volume region centered at $\vec{r}$. In this description, the thermodynamics of the mixture is controlled by a free energy functional ΔF describing the free energy excess of the mixture [6, 29, 30]

$$\frac{\Delta F}{k_B T} = \int d\vec{r} \left\{ f[\Phi(\vec{r})] + \frac{1}{36} \frac{a^2}{\Phi(1-\Phi)} [\nabla\Phi(\vec{r})]^2 \right\}, \quad (1)$$

where the free energy density (divided by temperature T and Boltzmann's constant k_B) $f(\Phi)$ is taken in the standard Flory-Huggins form involving the Flory-Huggins χ parameter [5]

$$f(\Phi) = \frac{\Phi}{N_A} \ln\Phi + \frac{1-\Phi}{N_B} \ln(1-\Phi) + \chi\Phi(1-\Phi). \quad (2)$$

Finally, the characteristic length a (lattice spacing of the Flory-Huggins lattice) is expressed in terms of σ_A, σ_B as

$$a^2/[\Phi(1-\Phi)] = \sigma_A^2/\Phi + \sigma_B^2/(1-\Phi). \quad (3)$$

Equ. (3) is justified by the random phase approximation [1], which implies for the collective structure factor describing the scattering from concentration fluctuations under wave vector $\vec{q}$

$$S_T^{coll}(\vec{q}) \equiv \langle |\Phi_{\vec{q}}|^2 \rangle_T = \left[\frac{1}{\Phi_o S_A(\vec{q})} + \frac{1}{(1-\Phi_o) S_B(\vec{q})} - 2\tilde{\chi}(\vec{q}) \right]^{-1}, \quad (4)$$

with $\tilde{\chi}(\vec{q})$ being the fourier transform of the effective interaction, and $S_A(\vec{q})$, $S_B(\vec{q})$ being single-chain static coherent structure factors. For chains obeying Gaussian statistics, one gets in terms of the well-known Debye function $f_D(x) = (2/x)\{1-[1-\exp(-x)]/x\}$ that

$$S_A(q) = N_A f_D\left(\frac{1}{6}N_A\sigma_A^2 q^2\right),\ S_B(q) = N_B f_D\left(\frac{1}{6}N_B\sigma_B^2 q^2\right). \tag{5}$$

Expanding for small q yields (note $f_D(x) \approx 1-x/3$ for small x) with $\tilde{\chi}(q) \approx \tilde{\chi}(0)(1-q^2r_o^2/6)$, r_o being the "effective" range of interaction:

$$\begin{aligned}[S_T^{coll}(q)]^{-1} \approx (\Phi_o N_A)^{-1} + [(1-\Phi_o)N_B]^{-1} + \sigma_A^2 q^2/[18\Phi_o] + \\ + \sigma_B^2 q^2/[18(1-\Phi_o)] - 2\tilde{\chi}(0) + \tilde{\chi}(0)q^2r_o^2/3.\end{aligned} \tag{6}$$

On the other hand, the fluctuation relation $[S_T^{coll}(q)]^{-1} = \partial^2(\Delta F(k_BT)/\partial\Phi_q\,\partial\Phi_{-q})$ where $\Phi_q = \int \exp(i\vec{q}\cdot\vec{r})[\Phi(\vec{r},t) - \Phi_o]d\vec{r}$ is the fourier transform of the concentration fluctuation around the average concentration Φ_o allows to calculate $S_T^{coll}(q)$ from Equs. (1) and (2). Consistency with Equ. (6) is found if $2\tilde{\chi}(0) \equiv \partial^2[\chi\Phi(1-\Phi)]/\partial\Phi^2$ and if Equ. (3) is postulated {neglecting the last term involving r_o in Equ. (6), which is small in most – but not all [31] – cases}. This consideration also shows that the free energy functional is only useful for small concentration gradients, $a^2(\nabla\Phi)^2 \ll N_A^{-1}$, $a^2(\nabla\Phi)^2 \ll N_B^{-1}$. For more rapid concentration variation this long wavelength approximation is inaccurate, and a more complicated theory [32] is required. In addition, also the validity of Equ. (2) is questionable; but this problem is deferred to the next section. At this point, we shall assume Equs. (1–3) as a working hypothesis and consider various applications. Note that $\tilde{\chi}(0) = \chi$ if χ is independent of concentration [8].

18.2.1 Spinodal Decomposition of Polymer Alloys

Spinodal decomposition occurs if a polymer blend is suddenly quenched from a state in the one-phase regime (where a macroscopically homogeneous blend is thermodynamically stable) to a state within the spinodal curve $\tilde{\chi}(0) = \chi_s(\Phi)$, defined from $[S_T^{coll}(q=0)]^{-1} = 0$, i. e.

$$2\chi_s(\Phi_o) = (N_A\Phi_o)^{-1} + [N_B(1-\Phi_o)]^{-1}. \tag{7}$$

For such a homogeneous state with $\tilde{\chi}(0) > \chi_s(\Phi_o)$, to which this quench leads, $S_T^{coll}(q) < 0$, for a whole range of wavevectors. For a symmetrical mixture ($N_A = N_B = N$, $\sigma_A = \sigma_B = \sigma$) one finds

$$0 < q < q_c, \; q_c = 2\pi/\lambda_c \cong \{(18/N\sigma^2)[1-\chi_s(\Phi_o)/\tilde{\chi}(0)]\}^{1/2},$$
$$\tilde{\chi}(0)/\chi_s(\Phi_o) - 1 \ll 1, \tag{8}$$

i. e. the critical wavelength λ_c is of the order of the coil size off the spinodal but diverges when the spinodal is approached. In the opposite limit, however,

$$q_c \approx (2/\sigma)[6\Phi(1-\Phi)\tilde{\chi}(0)]^{1/2} \text{ for } \tilde{\chi}(0) \gg \chi_s(\Phi_o),$$

i. e. λ_c would be smaller than the coil size for a very deep quench and the long wavelength approximation breaks down [8].

This instability of the homogeneous state inside the spinodal curve implies that long wavelength concentration fluctuations (with wavelengths $\lambda > \lambda_c$) will spontaneously grow and thus phase separation starts with an exponential growth of such fluctuations. The dynamics of this process is called spinodal decomposition [6–10]. It is theoretically described by a continuity equation which links $\Phi(\vec{r},t)$ to a concentration current $\vec{j}_\Phi(\vec{r},t)$ and a stochastic force $\eta(\vec{r},t)$,

$$\partial\Phi(\vec{r},t)/\partial t = -\vec{\nabla} \cdot \vec{j}_\Phi(\vec{r},t) + \eta(\vec{r},t); \tag{9}$$

the current $\vec{j}_\Phi(\vec{r},t)$ is related to the local chemical potential difference $\mu(\vec{r},t)$ between segments of species A and B, V being the total volume [6]

$$j_{\vec{\Phi}}(\vec{r},t) = -\frac{1}{V}\int \frac{\Lambda(\vec{r}-\vec{r}')}{k_B T} \nabla' \mu(\vec{r}',t) d\vec{r}'. \tag{10}$$

The stochastic force $\eta(\vec{r},t)$ is related to the fourier transform $\Lambda_{\vec{q}}$ of $\Lambda(\vec{r}-\vec{r}')$ by a fluctuation-dissipation formula, $\eta_{\vec{q}}(t)$ being the fourier transform of $\eta(\vec{r},t)$,

$$\langle \eta_{\vec{q}}(t)\eta_{-\vec{q}}(t') \rangle = 2\Lambda_{\vec{q}} q^2 \delta(t-t'). \tag{11}$$

Using $\mu(\vec{r},t) = \delta(\Delta F)/\delta\Phi(\vec{r},t)$ one finds from Equs. (1),(9),(10) in *linearized* approximation

$$\frac{\partial}{\partial t}\Phi_{\vec{q}}(t) = -r(\vec{q})\Phi_{\vec{q}}(t) + \eta_q(t), \tag{12}$$

with the rate factor $r(\vec{q}) = \Lambda_q q^2 \{ f''(\Phi_o)/k_BT + a^2q^2/[18\Phi_o(1-\Phi_o)] \}$. Equs. (11) and (12) then yield [8]

$$S(\vec{q},t) = \langle |\delta\Phi_{\vec{q}}(0)|^2 \rangle \exp[-2r(\vec{q})t] + \\ + \{ f''(\Phi_o)k_BT + a^2q^2/[18\Phi_o(1-\Phi_o)] \}^{-1} \{ 1 - \exp[-2r(\vec{q})t] \}. \tag{13}$$

Since inside the spinodal $f''(\Phi_o) < 0$, we see that $r(\vec{q}) < 0$ for $q < q_c$ and thus Equ. (13) explicitly shows the exponential growth of the scattering intensity with time.

Of course, the question must be raised under which conditions the linearization involved in deriving Equ. (12) is valid. This can be answered by a generalized Ginzburg criterion: one must require

$$\langle \{ \delta\Phi(\vec{r},t) \}^2 \rangle_{T,\lambda_\psi} \ll [\Phi_o - \Phi_{sp}(\tilde{\chi}(0))]^2, \tag{14}$$

i. e. the mean square concentration fluctuation averaged over a (d – dimensional) volume λ_c^d must be less than the squared concentration distance from the spinodal. Equ. (14) yields [8, 33], with $q_{max} = q_c/\sqrt{2}$ the wavevector of maximum growth

$$\exp[-2r(q_{max})t] \ll N^{(d-2)/2} \left[1 - \frac{\chi_{crit}}{\tilde{\chi}(0)} \right]^{(4-d)/2} \cdot \\ \cdot \left[\frac{\Phi_o}{\Phi_{sp}(\tilde{\chi}(0))} - 1 \right]^{(6-d)/2} . \tag{15}$$

Equ. (15) is rather restrictive – it never could be satisfied for spinodal decomposition of two-dimensional films, consistent with computer simulations [34]; also one must stay off the critical point χ_{crit}, i. e. in the mean field critical regime which satisfies the condition

$$N^{(d-2)/2} (1 - \chi_{crit}/\tilde{\chi}(0))^{(4-d)/2} \gg 1.$$

Also, one must stay off the spinodal curve – in fact, the singular behaviour at the spinodal curve is rounded over a concentration region of width proportional to $N^{-1/3}$ (in $d = 3$ dimensions). In this latter region, a gradual transition from spinodal decomposition to nucleation occurs, which is not yet fully understood.

Deeply in the unstable regime, where Equ. (15) holds, one thus has a regime of times $\{ t < (\ln N)/|r(q_{max})| \}$ where the intensity of small-angle scattering increases exponentially in time, as described by Equ. (13). For small q the normalized growth rate $r(q)/q^2$ can be written in form of a linear function of q^2

("Cahn plot") $r(q)/q^2 = [\Lambda_o|f''(\Phi_o)|/k_BT](1-q^2/q_c^2)$. However, even in this regime pronounced deviations from linearity of this Cahn plot can occur, due to two reasons: (i) finite quench rate at which one moves from the initial temperature to the final temperature, rather than the idealized instantaneous quench considered so far [35]. (ii) Coupling of the concentration $\Phi(\vec{q},t)$ to another slow variable (e. g. structural relaxation near a glass transition) [36]. These mechanisms have been analyzed by detailed model calculations [35, 36].

18.2.2 Effects of Polydispersity and Shear

Comparison between experiment (for reviews see [37, 38]) and the above theory is somewhat hampered by the fact that real polymers always are polydisperse. A generalization [13] of the above theory is possible only when rather restrictive assumptions are made. However, the problem is still simple with respect to the static aspects: Equs. (1) and (2) are replaced by

$$\Delta F/k_BT = \int d\vec{r}\left\{f[\vec{\Phi}(\vec{r})] + \sum_{i=1}^{m_A} a_i^2(\vec{\Phi})[\nabla\Phi_i^A(\vec{r})]^2 + \right.$$
$$\left. + \sum_{i=1}^{m_B} b_i^2(\vec{\Phi})[\nabla\Phi_i^B(\vec{r})]^2\right\}, \tag{16}$$

where $\Phi_i^{A(B)}$ is the volume fraction of the i'th fraction of polymer A(B) of chain length N_i^A (N_i^B) and we assure there are m_A, m_B such fractions. Assuming also a volume fraction Φ_v of vacancies and allowing for three Flory-Huggins parameters χ_{AB}, χ_{AV} and χ_{BV}, the free energy density f which depends on $\vec{\Phi} \equiv (\Phi_1^A, \ldots, \Phi_{m_A}^A, \Phi_1^B, \ldots, \Phi_{mB}^B, \Phi_V)$ is

$$f[\vec{\Phi}] = \sum_{i=1}^{m_A} (\Phi_i^A/N_i^A)\ln\Phi_i^A$$
$$+ \sum_{i=1}^{m_B} (\Phi_i^B/N_i^B)\ln\Phi_i^B + \Phi_V\ln\Phi_v + \chi_{AB}\Phi_A\Phi_B + \chi_{AV}\Phi_A\Phi_V + \tag{17}$$
$$\chi_{BV}\Phi_B\Phi_V,$$

where $\Phi_A \equiv \sum_{i=1}^{m_A} \Phi_i^A$, $\Phi_B \equiv \sum_{i=1}^{m_B} \Phi_i^B$, $\Phi_A + \Phi_B + \Phi_v = 1$. Again the coefficients

$a_i^2(\vec{\Phi}), b_i^2(\vec{\Phi})$ of the gradient terms in Equ. (16) are constructed such that Equs. (15)–(17) yield $S_T^{coll}(\vec{q})$ in agreement with the random phase approximation,

$$a_i^2(\Phi) = \sigma_A^2/(36\Phi_i^A),\ b_i^2(\Phi) = \sigma_B^2/(36\Phi_i^B), \tag{18}$$

for $\Phi_V \to 0$ which is the limit of interest here. For a polydisperse mixture, $S_T^{coll}(\vec{q})$ can be expressed in terms of the standard averages $\overline{N}_A^W$, $\overline{N}_B^W$, $\overline{N}_A^Z$, $\overline{N}_B^Z$ over the molecular weight distribution as [39]

$$\begin{aligned}\left[S_T^{coll}(\vec{q})\right]^{-1} = \left[\Phi_A \overline{N}_A^W\right]^{-1} + \left[\Phi_B \overline{N}_B^W\right]^{-1} - 2\chi_{AB} + \\ + \frac{q^2}{18}\left[\frac{\overline{N}_A^Z}{\overline{N}_A^W}\frac{\sigma_A^2}{\Phi_A} + \frac{\overline{N}_B^Z}{\overline{N}_B^W}\frac{\sigma_B^2}{\Phi_B}\right].\end{aligned} \tag{19}$$

Assuming then that all polymer fractions $\Phi_i^{A(B)}$ and the number of lattice sites and hence also Φ_V are strictly conserved one has continuity equations [13]

$$\begin{aligned}\frac{\partial \Phi_i^A(\vec{r},t)}{\partial t} + \vec{\nabla}\cdot\vec{j}_i^A(\vec{r},t) = 0,\ i=1,\ldots m_A\ ;\ \frac{\partial \Phi_i^B(\vec{r},t)}{\partial t} \\ + \vec{\nabla}\cdot\vec{j}_i^B(\vec{r},t) = 0,\ i=1,\ldots m_B\ ;\ \frac{\partial \Phi_V(\vec{r},t)}{\partial t} + \nabla\cdot\vec{j}_V(\vec{r},t) = 0\ ,\end{aligned} \tag{20}$$

where the current densities of the i'th fraction of polymer A(B) and vacancies V are denoted as $j_i^{A(B)}(\vec{r},t)$ and $j_V(\vec{r},t)$ respectively. In the long wavelength limit, where the nonlocal nature of Onsager coefficients can be neglected, the generalization of the constitutive Equ. (10) is assumed as [13]

$$\vec{j}_i^A(\vec{r},t) = -\frac{\lambda_i^{AA}\left[\vec{\Phi}(\vec{r},t)\right]}{k_B T}\nabla\left[\mu_i^A(\vec{r},t) - \mu_V(\vec{r},t)\right];\ i=1,\ldots m_A, \tag{21a}$$

$$\vec{j}_i^B(\vec{r},t) = -\frac{\lambda_i^{BB}\left[\vec{\Phi}(\vec{r},t)\right]}{k_B T}\nabla\left[\mu_i^B(\vec{r},t) - \mu_V(\vec{r},t)\right];\ i=1,\ldots m_B, \tag{21b}$$

where λ_i^{AA}, λ_i^{BB} are suitable Onsager coefficients. Note that Equs. (21) explicitly neglect offdiagonal terms $\lambda_{i\,j}^{AB}$, $\lambda_{i\,j}^{BA}$ relating $j_i^{A(B)}$ to $\nabla(\mu_j^A - \mu_V)$ or $\nabla(\mu_j^B - \mu_V)$, respectively: even in the framework of a lattice model, this is a poor approximation, as recent Monte Carlo work [40] shows. In addition, the lattice model cannot properly include "bulk flow" which is believed to make important contributions to polymer-polymer interdiffusion [41, 42]. If one nevertheless accepts Equ. (21) as a model case, and uses Equ. (16) to obtain the chemical potential gradients needed in Equ. (21), one finds that the interdiffusion mode near the spinodal curve can still be written as in the monodisperse case as a product of an effective Onsager coefficient $\Lambda^{eff}_{q=0}$ times q^2 and a thermodynamic factor, the latter being again $[S_T^{coll}(\vec{q})]^{-1}$ {Equ. (19) in our case}. For the case where single-chain dynamics is described by the Rouse model, one finds ($\lambda_i^{AA} = \gamma_{AA}\Phi_V\Phi_i^A$, $\lambda_i^{BB} = \gamma_{BB}\Phi_V\Phi_i^B$, with γ_{AA}, γ_{BB} independent of chain length, for $\Phi_V \to 0$, $\Phi_i^A \to 0$, $\Phi_i^B \to 0$)

$$\left[\Lambda^{eff}_{q=0}\right]^{-1} = \Phi_V^{-1}\left\{\gamma_{AA}^{-1}\frac{1}{\Phi_A}\frac{\overline{N}_A^Z}{\overline{N}_A^W} + \gamma_{BB}^{-1}\frac{1}{\Phi_B}\frac{\overline{N}_B^Z}{\overline{N}_B^W}\right\}, \tag{22}$$

while for the case of the reptation model higher moments of the molecular weight distribution would enter [13]. But in both cases the interdiffusion Onsager coefficient $\Lambda^{eff}_{q=0}$ is controlled by the slowly diffusing species, while effects due to bulk flow imply the opposite case, the faster diffusing species would dominate [41]. This controversy between "slow mode theory" and "fast mode theory" has found a lot of interest [42], but is difficult to clarify experimentally. Note also that neither of these theories is expected to be exact, not even for a rigid lattice model and a monodisperse case, where Equ. (22) can be written as $\Lambda^{eff}_{q=0} = \Lambda_{AA}\Lambda_{BB}/(\Lambda_{AA} + \Lambda_{BB})$, whereas the correct result is [40] $\Lambda^{eff}_{q=0} = (\Lambda_{AA}\Lambda_{BB} - \Lambda_{AB}^2)/(\Lambda_{AA} + 2\Lambda_{AB} + \Lambda_{BB})$, involving the off-diagonal term $\Lambda_{AB} = \Lambda_{BA}$. These problems about interdiffusion also hamper the quantitative understanding of the growth of fluctuations during spinodal decomposition [13].

As a last point of this subsection we consider polymer mixtures under shear flow [14, 15], where the velocity field $\vec{v}$ can be written in terms of the shear rate $\dot{\gamma}$ as $\vec{v} = \dot{\gamma} y \vec{e}_x$ (x,y,z being cartesian coordinates and $\vec{e}_x$ a unit vector in x-direction). The flow orients and stretches the polymer coils, and thus already the single chain structure factors $S_A(\vec{q})$, $S_B(\vec{q})$ get anisotropic [13, 14]:

$$S_A(\vec{q}) = N_A\left\{1 - \frac{1}{18}N_A\sigma_A^2\left[q^2 + \frac{2\pi^2}{15}\tau_1^A\dot{\gamma}q_xq_y + \frac{4\pi^4}{315}(\tau_1^A\dot{\gamma})^2q_x^2\right] + 0(q^4)\right\} \tag{23}$$

where τ_1^A is the largest relaxation time of the Rouse spectrum for A-chains (an analogous expression holds for B-chains, of course). Via the random phase approximation, it is implied that the effective free energy functional for mixtures in a steady-state shear flow must have an anisotropic gradient energy term [14, 15]:

$$\frac{\Delta F}{k_B T} = \int \Big\{ f(\Phi) + \frac{q^2}{36\Phi(1-\Phi)} \cdot \Big[(\vec{\nabla}\Phi)^2 + \frac{2\pi^2}{15}\dot{\gamma}\bar{\tau}_1 \frac{\partial \Phi}{\partial x}\frac{\partial \Phi}{\partial y} + \frac{4\pi^4}{315}(\dot{\gamma}\bar{\bar{\tau}}_1)^2 (\frac{\partial \Phi}{\partial x})^2 \Big] \Big\} d\vec{r} \,, \tag{24}$$

where $f(\Phi)$ still is given by Equ. (2) and a^2 by Equ. (3) but there are anisotropic corrections involving "average" Rouse times $\bar{\tau}_1$, $\bar{\bar{\tau}}_1$ defined by $[\sigma_A^2/\Phi + \sigma_B^2/(1-\Phi)]\,\bar{\tau}_1 \equiv \sigma_A^2 \tau_1^A/\Phi + \sigma_B^2 \tau_1^B/(1-\Phi)$ and $[\sigma_A^2/\Phi + \sigma_B^2/(1-\Phi)]\,(\bar{\bar{\tau}}_1)^2 = \sigma_A^2(\tau_1^A)^2/\Phi + \sigma_B^2(\tau_1^B)^2/(1-\Phi)$. The dynamics of concentration fluctuations then is described by the generalized continuity equation

$$\frac{\partial}{\partial t}\Phi(\vec{r},t) + \vec{\nabla}\cdot[\Phi(\vec{r},t)\,\vec{v}(\vec{r},t)] = -\vec{\nabla}\cdot\vec{j}_\Phi(\vec{r},t) + \eta(\vec{r},t) \tag{25}$$

instead of Equ. (9), and Equ. (10) is still assumed but the effective chemical potential difference $\mu(\vec{r},t) = \delta(\Delta F)/\delta\Phi(\vec{r},t)$ using Equ. (24) also becomes anisotropic. Thus the scattering intensity in the q_x–q_y-plane no longer is isotropic. Some details of this complicated behaviour are discussed in [14, 15].

18.2.3 Wetting of Polymer Mixtures

If a polymer mixture is in contact with a wall, the forces between the wall and the two species of monomers will differ in general: this tendency of preferential adsorption of one species at the wall due to energetic considerations competes, of course, with the "local" entropy of mixing (near the wall). This competition may give rise to the wetting transition: In the nonwet state of the surface, the enhancement of the volume fraction $\Phi(z)$ of the preferred species near the wall decays to its bulk value $\Phi_\infty \equiv \Phi(z\to\infty)$ at a distance z from the wall, which distance is of the order of the gyration radius (or smaller). In the wet state of a surface (which can occur only if the bulk mixture is in a two-phase coexistence condition), at $\Phi_\infty = \Phi^{(1)}_{coex} < \Phi^{(2)}_{coex}$, $\Phi(z)$ first decays from its value Φ_1 right at the wall to the value $\Phi^{(2)}_{coex}$ corresponding to the other side of the coexistence curve, and at a large distance from the wall there is another interface separating the two coexisting phases, i. e. the high concentration phase $\Phi^{(2)}_{coex}$ attached to the wall, and the bulk phase with $\Phi_\infty = \Phi^{(1)}_{coex}$.

Such phenomena (experimental evidence for this "surface enrichment" in polymer mixtures now is starting to become available [43]) can again be described by a free energy functional of the type as constructed in Equs. (1),(16),(24): but now we have an anisotropy due to the wall, directions parallel and perpendicular to it are no longer equivalent, and in addition one needs a localized perturbation at $z = 0$ describing the forces due to the wall. Disregarding concentration variations parallel to the wall, one then postulates [11, 12], A being the area of the wall

$$\frac{\Delta F}{Ak_BT} = \int_0^\infty dz \left[f(\Phi) + \frac{a^2}{36\Phi(1-\Phi)}\left(\frac{d\Phi}{dz}\right)^2\right] - \mu_1\Phi_1 - \frac{1}{2}g\Phi_1^2, \quad (26)$$

where the local surface perturbation was assumed a simple quadratic function of the surface concentration Φ_1, with μ_1,g being suitable coefficients. Minimizing then the surface excess free energy with respect to Φ_1 yields a boundary condition at the surface,

$$\frac{a^2}{18\Phi_1(1-\Phi_1)}\left.\frac{d\Phi}{dz}\right|_{z=0} = -\mu_1 - g\Phi_1, \quad (27)$$

whereas the concentration profile is described by [12]

$$\frac{a^2}{36\Phi(1-\Phi)}\left(\frac{d\Phi}{dz'}\right)^2\Bigg|_{z'=0}^{z'=z} = f(\Phi(z)) - f(\Phi_1) \ . \quad (28)$$

Equ. (28) is solved by

$$\frac{6z}{a} = \int_{\Phi_1}^{\Phi_{(z)}} d\Phi / \{\Phi(1-\Phi)[f(\Phi) - f(\Phi_\infty)]\}^{1/2}, \quad (29)$$

whereas Equ. (27) with the help of Equ. (28) can be rewritten as

$$\pm\left[\frac{a}{3}\,\frac{f(\Phi_1)-f(\Phi_\infty)}{\Phi_1(1-\Phi_1)}\right]^{1/2} = -\mu_1 - g\Phi_1 \ , \quad (30)$$

and the solution yielding the minimum of the surface excess free energy has to be taken. A second-order wetting transition occurs when the solution of Equ. (30) reaches the value $\Phi_1^{crit} = \Phi_{coex}^{(2)}$. The "surface susceptibility" $\chi_{11} \equiv (\partial\Phi_1/\partial\mu_1)_T$ then exhibits a jump singularity, whereas at the tricritical transition χ_{11} diverges [12]. One finds that such a tricritical transition occurs for $\Phi_\infty^t = -(a/g)/\sqrt{18N}$, $\mu_1^t = -g + a/\sqrt{18N}$. Thus for large N the model mostly predicts second-order rather than first-order wetting, if the coefficients g, μ_1 are of order unity (while χ must be of order 1/N to allow for partial compatibility). Fig. 18.1 shows the surface phase diagram predicted by the model.

Again this model can be extended to a study of the dynamics of fluctuations. So far this has been worked out for the case of non-conserved order parameter only [16], which is not relevant for polymer mixtures, but related to studies of wetting phenomena in the Ising model [17]. A particularly interesting question concerns again quenching experiments, where one starts the system in a non-wet state of the surface and quenches it to a state where the surface should be wet. Thus one finds [16] that the thickness of the film D(t) grows with the time after the quench according to a law $D(t) \propto \text{const} + \ln t$. Simulations [17] are consistent with this law but one finds that mean field theory [16] underestimates the prefactor in this logarithmic growth law. Since the conservation of the order parameter (the volume fraction in the case of a polymer mixture) has the effect to produce further slowing down of long wavelength fluctuations, a similarly slow growth is expected for wetting layers in polymer mixtures, too.

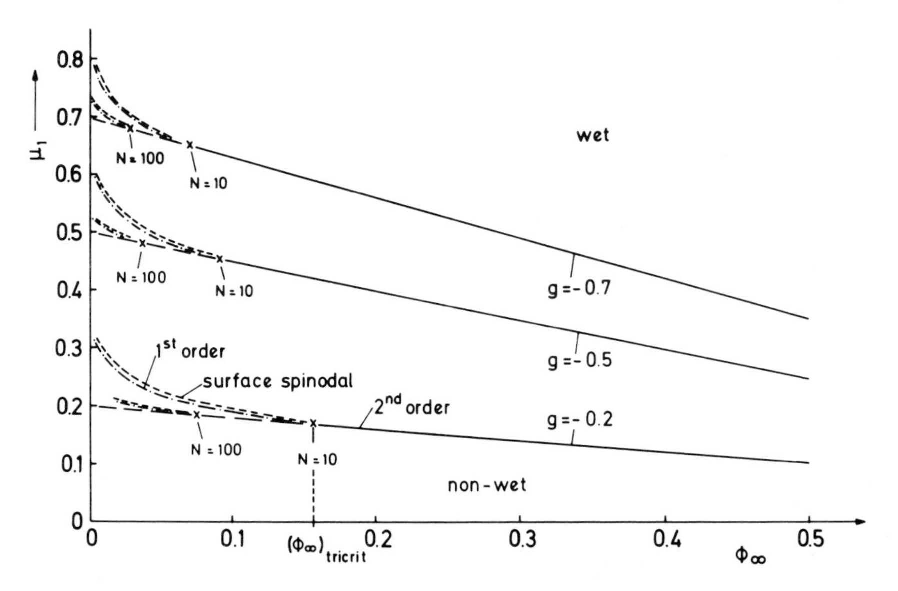

Fig. 18.1: Surface phase diagram in the (μ_1, Φ_∞) plane for three values of g. The region where the surface is non-wet (at small μ_1) is separated from the wet region by a phase boundary which describes the wetting transition. For $\Phi_\infty > \Phi_\infty^t$ (second-order wetting) this is just the straight line $\mu_1^{crit} = -g(1-\Phi_\infty)$. The region of first order wetting is shown for symmetrical mixtures with $N_A = N_B = N = 10$ and $N = 100$, respectively, and denoted by dash-dotted curves. In this regime metastable wet and nonwet phases are possible up to the stability limits ("surface spinodals") denoted by broken curves. Assuming that μ_1 and g are essentially independent of temperature T, variation of T essentially means variation of Φ_∞. From Schmidt and Binder [12].

Finally we emphasize that our treatment always implies the validity of the long wavelength approximation, which requires interfacial widths of the order of the coil size (or larger). This holds for $\Phi_\infty \gtrsim 0.1$ but not for smaller values of Φ_∞: then the interfacial profile is characterized by two lengths [44], namely a length $L \propto \sigma/\sqrt{x}$ in the center of the profile and the correlation length $\zeta \approx \sigma \sqrt{2N}/6$ in the wings of the profile. A quantitatively reliable description of wetting phenomena in this regime on the basis of the Hong-Noolandi integral equation formulation [32] has recently been derived [45].

18.3 Scaling and Computer Simulation Investigations

So far we discussed static and dynamic properties on the level of a general mean field picture. This approach turns out to be very useful for very long chains and small concentration differences. However, deep in the spinodal regime the concentration of the minority phase compared to the majority phase might be small. For finite chain lengths with and without solvent this can cause consequences which go beyond the mean field description. On the other hand, for small concentration-differences (small quenches) one approaches (for small chains) the critical point region where mean field again is expected to fail. In order to analyze such problems we performed a Monte Carlo simulation of a polymer-polymer-solvent system [18–25], as discussed in Chapter 18.3.1. There we only treat dynamic properties on time scales of many diffusion times of the individual chains. However, there are still many open questions concerning the dynamics of long polymer chains in a melt on shorter time scales. The most favoured but also most questioned model is the reptation model of de Gennes and Edwards [4, 46]. This model describes the motion of the chain as a motion along a coarse-grained primitive path, so that the chains perform a snakelike motion forward and backward along their contour. We approached this problem from two sides. First we performed a Monte Carlo and scaling investigation of the chain in a straight tube [26, 27]. There we especially discuss the effects of the mobile chain ends and the various time scales for the coherent structure factor S(q,t) depending on the q-value under consideration. The second approach was made by performing a large scale molecular dynamics simulation of a polymer melt [28, 47]. This simulation turned out to be the first case, where one was able to directly show that long chains in a melt are moving as the reptation concept suggests.

18.3.1 Phase Separation of A-chain – B-chain – Solvent Systems

As discussed in the previous chapter, the Flory-Huggins theory makes use of several serious approximations. These approximations come from the fact that the simple mean field approach, as discussed in Chapter 18.2, does not take into account any interaction coming from the configurational properties of the polymers. However, it is not clear at all that the number of AA, BB and AB contacts is only determined by the overall average concentration in the system. Within the Flory-Huggins approximation the Gibbs free energy of mixing is written [5, 6] as in Equ. (2), yielding:

$$\chi = z\,\frac{(2\varepsilon_{AB} - \varepsilon_{AA} - \varepsilon_{BB})}{2k_BT}\;. \tag{31}$$

z is the coordination number of the underlying lattice and the ε are the interaction parameters of AA, BB and AB neighbours. Strictly speaking, Equ. (31) only makes sense for $z\to\infty$, since except for end monomers, always 2 neighbours are occupied by the subsequent monomers of the same chain. In the presence of solvent Equs. (2) and (31) have to be modified in order to take these interactions into account (see below).

Let us, however, first consider a system with a small number of vacancies in order to check the general validity of the Flory-Huggins theory. To do this we performed a Monte Carlo simulation of polymers on the simple cubic lattice with a vacancy density of 20% [18–20]. It should be noted that for lattice polymers this is considered a very high polymer density [48]. For simplicity we confine the study to symmetrical mixtures $N_A = N_B = N$. We used periodic boundary conditions and lattices with a side length of $L = 8$ to 20 lattice units. The chain lengths ranged from $N = 4$ to $N = 32$ monomers. The simulation was performed by the standard kink-jump algorithm. Although this algorithm is somewhat slower than other methods, see e. g. [48], we used this approach since it allows for an analysis of the dynamics as well. The energies were taken as $\varepsilon_{AA} = \varepsilon_{BB} = \varepsilon$, $\varepsilon_{AB} = e_{AV} = \varepsilon_{BV} = \varepsilon_{VV} = 0$ for nearest neighbour contacts on the lattice. The samples were prepared at $\varepsilon = 0$ ($T = \infty$) and then quenched to a finite temperature. This alone, however, would be a very expensive and inefficient way to determine the coexistence curve, since one simulates a canonical ensemble. This is not appropriate for use in the miscibility gap at temperatures of $T \lesssim T_c$. Thus we used a grand canonical algorithm to determine the coexistence curve. To do this we treat the chemical potential difference $\Delta\mu$ as an independent variable. With the symmetry $N_A = N_B = N$ we then allow "moves" $A\to B$, $B\to A$. With the additional energy $\pm\Delta\mu N$ in the transition probability W we can generate systems at an arbitrary volume fraction $\Phi_B(\Phi_A)$. In practice such a "move" is attempted for an arbitrarily chosen chain every one to fifty time steps. Note that one time step is given by one attempted move per monomer of the total system. The transition probability is then determined by the standard Metropolis sampling [49]. The initial system then always contained only B-chains. Here a time step is given by one attempted move per monomer of the whole system. The relative excess of one species defines the order parameter m (n being the number of chains)

$$m = \frac{\Delta n}{n} = \frac{n_B - n_A}{n_B + n_A} . \tag{32}$$

The volume fractions for the considered symmetrical systems are given by $\Phi_{B,1}^{coex} = \Phi_{crit}(1-m)$, $\Phi_{A,1}^{coex} = \Phi_{crit}(1+m)$ with $\Phi_{crit} = (1-\Phi_V)/2$.

We are especially concerned with the validity of the mean field theory. Following the Flory-Huggins theory for the ternary system of A,B-chains and vacancies V we can write instead of Equs. (2) and (3)

$$\frac{\Delta G^{tern}}{k_B T} = \frac{\Phi_A \ln\Phi_A}{N_A} + \frac{\Phi_B \ln\Phi_B}{N_B} + \chi_{AB}\Phi_A\Phi_B + \\ + \Phi_V \ln\Phi_V + \chi_{BV}\Phi_B\Phi_V + \chi_{AV}\Phi_A\Phi_V . \tag{33}$$

For the symmetrical case $N_A = N_B = N$, and $\chi_{AV} = \chi_{BV} = 0$ we can rewrite Equ. (33) with the normalized densities $\tilde{\Phi}_A = \Phi_A/(1-\Phi_V)$, $\tilde{\Phi}_B = \Phi_B/(1-\Phi_V)$ and $\tilde{\Phi}_A + \tilde{\Phi}_B = 1$ as

$$\frac{\Delta G^{tern}}{k_B T} = (1-\Phi_V)\left\{\frac{\tilde{\Phi}_A \ln\tilde{\Phi}_A + \tilde{\Phi}_B \ln\tilde{\Phi}_B + \ln(1-\Phi_v)}{N} + \right. \\ \left. + \Phi_V \ln\Phi_V + \chi_{AB}(1-\Phi_V)\tilde{\Phi}_A\tilde{\Phi}_B \right\} . \tag{34}$$

Since Φ_V is a constant quantity it is obvious that with the identification $\chi \equiv \chi_{AB}(1-\Phi_V)$ and $\Delta G \equiv \Delta G^{tern}/(1-\Phi_V)$, Equ. (34) describes a system equivalent to the standard system of Equ. (2), where $\Phi_V = 0$. Using (34) and (32) we find for the chemical potential differences $\Delta\mu = \mu_A - \mu_B$ and the order parameter m

$$\frac{\Delta\mu}{k_B T} = \frac{1}{N} \ln \frac{1+m}{1-m} - \chi_{AB}(1-\Phi_V)m \\ m = \tanh\left[\frac{\chi_{AB}}{\chi_C} m + \frac{N\Delta\mu}{2\,k_B T}\right] , \tag{35}$$

with the critical $\chi_C = 2/[N(1-\Phi_V)]$. Near the critical point ($\chi_{AB}/\chi_C - 1 \ll 1$, $\Delta\mu \ll 1$) $\Delta\mu$ can be expanded in powers of the order parameter up to m^3 (note that $\Delta\mu$ is an odd function of m!) leading to the standard mean field expressions and exponents, e. g.

$$\Phi_{B,1,2}^{coex} = \frac{1-\Phi_V}{2}\left[1 - \sqrt{3}\left(\frac{\chi_{AB}}{\chi_C} - 1\right)^{1/2}\right] ,$$

giving for the majority phase:

$$\left.\begin{array}{l}\Phi_{B,2}^{coex} - \Phi_B^{crit} = \hat{B}\,(1-T/T_C)^{\beta} \\ \text{with} \\ \beta = 1/2,\ \hat{B} = \sqrt{3}(1-\Phi_V)/2.\end{array}\right\} \tag{36}$$

For the other exponents one also gets the mean field result $\delta = 3$, $\gamma = 1$. Similarly the critical amplitudes can be given exactly. The limitations of this mean field picture are expected to be given by the Ginzburg criterion discussed in Equ. (14). Within the critical region one expects the Ising universality class exponents to describe the system. These are [50]

$$\beta \cong 0.32,\ \gamma \cong 1.24,\ \delta \cong 4.8. \tag{37}$$

In addition the correlation length ξ of fluctuations diverges according to

$$\xi \propto |(T-T_c/T_c|^{-\nu} \tag{38}$$

with an exponent $\nu \cong 0.63$ which also differs significantly from the mean field value $\nu = 1/2$.

The first aim of the theoretical investigation is to estimate the coexistence curve, the critical temperature and the critical amplitudes. To achieve this, one has to make use of finite size scaling. Both the cumulants of the order parameter distribution and the variation of the peak in the specific heat [20] give the same critical temperature. Combining all analysis yields e. g. $\frac{k_B T_c(N=32)}{\varepsilon} = 31.5 \pm 0.3$. Already the extrapolation to $L \to \infty$ shows that $\nu = 1/2$ does not fit the data. This becomes much more pronounced in a full finite size scaling analysis. There we get that the order parameter m multiplied by $L^{\beta/\nu}$ is only a function of the scaling variable

$$\left(\frac{T-T_c}{T_c}\right) L^{1/\nu} .$$

Fig 18.2 shows a scaling plot including a test of the mean field case. Clearly the mean field exponents give a much worse common curve than the plot using Ising exponents. Similarly one expects $m \propto (\Delta\mu)^{1/\delta}$ with $\delta = 3$ (mean field) and $\delta \cong 4.8$ (Ising). Again the data can only be understood if one takes Ising exponents. The same result was found for the fluctuations in the order parameter which is given by the collective structure factor $S(k)$ at $k = 0$. The maximum of $S(k = 0)$ should scale as $L^{\gamma/\nu}$. Again the data nicely confirm the Ising universality class. Thus there is no region where the mean field theory describes the system satisfactorily. In this short review we confined the discussion to the $N = 32$ case since this is the system which was expected to exhibit mean field behaviour most clearly. The systems of shorter chains then also clearly show Ising-like behaviour.

With Equ. (36) we can construct the coexistence curve. Since $\beta \cong 0.32$ instead of $\beta = 1/2$ the coexistence curve is much flatter than the usually plotted parabolic curve. The unexpected strength of the deviations to mean field suggested that also real polymers should show this. In fact, in a recent experiment [51] it was seen that near the critical point significant deviations occur.

The preceding discussion shows that the Flory-Huggins picture has to be modified to understand the physics of polymer melts near the critical temperature. This modification can be performed by using a scaling approach. However, there is another serious problem coming up. Using Equ. (35) we can write for χ_{AB} on the coexistence curve ($\Delta\mu = 0$)

$$N\chi_{AB}(1-\Phi_V) = \frac{1}{m_{coex}} \ln\left[\frac{1-m_{coex}}{1-m_{coex}}\right], \tag{39}$$

whereas Equs. (31 and 34) give

$$(1-\Phi_V)\chi_{eff} = \chi_{AB}(1-\Phi_V) = \chi = \frac{z\varepsilon}{k_B T}. \tag{40}$$

Fig. 18.3 shows that the Flory-Huggins ansatz gives results which are far away from reality. It is especially striking that the longer the chains the stronger are the deviations. Not only that the absolute value of T_c is off by about a factor of two but there is also a significant Φ dependency.

These results were rather unexpected and we subsequently investigated this in detail. To do so we performed systematic studies of the symmetrical mixture of polymer chains with varying vacancy density Φ_V [21, 22, 24, 25]. In addition to the quantities already discussed above, we are particularly interested in the structural properties of the chains in the minority/majority phase [23].

Again one finds that the Flory-Huggins approximation does not describe the system. Also the more complicated modification by Guggenheim [20] improves the situation only marginally. Considering this, modifications are

a

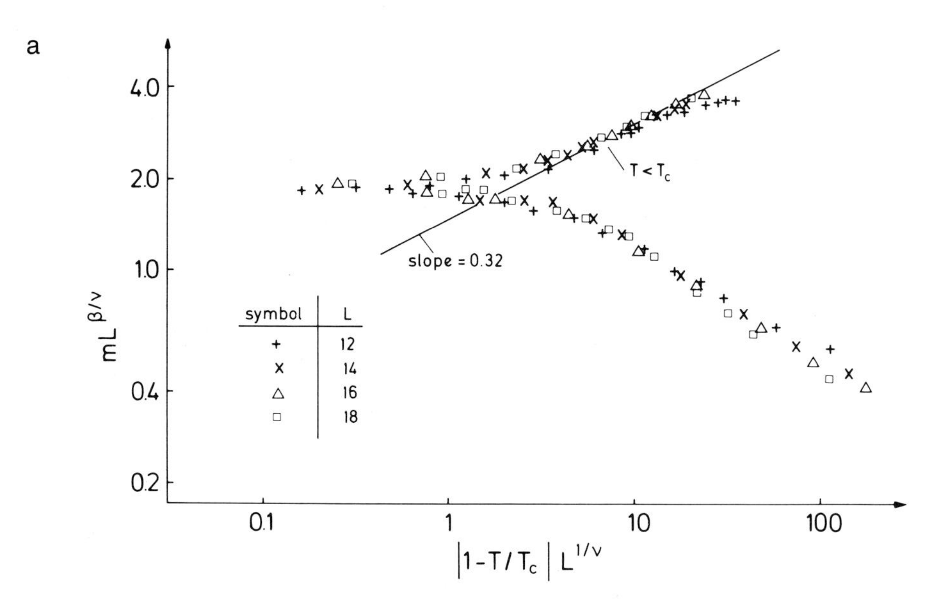

b

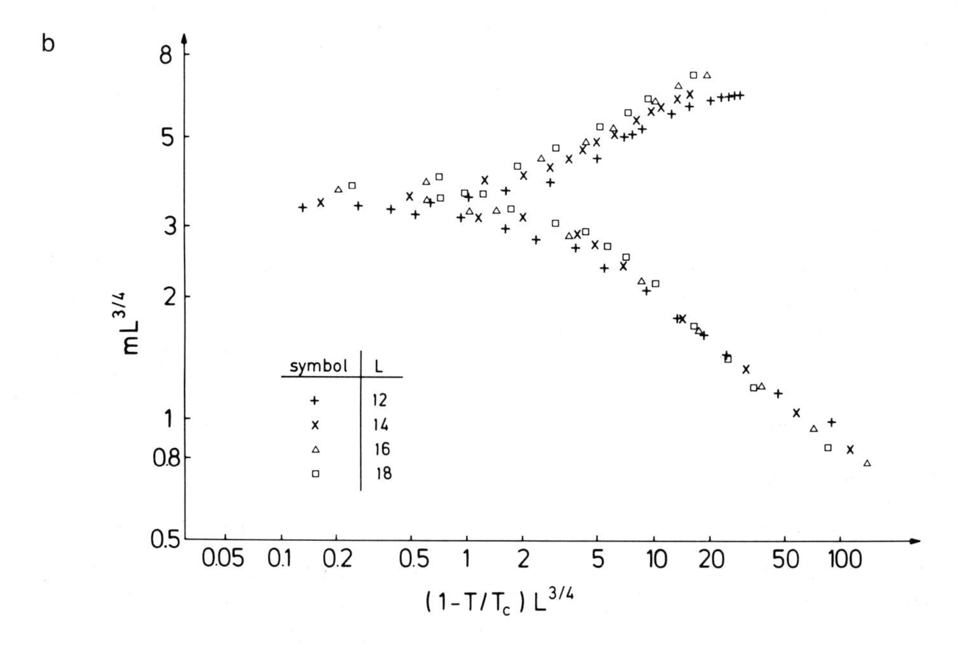

Fig. 18.2: Scaling plot of $mL^{\beta/\nu}$ vs $|1-T/T_c|L^{\beta/\nu}$ for N = 32 assuming Ising exponents (a) and mean field exponents (b). Taken from [20].

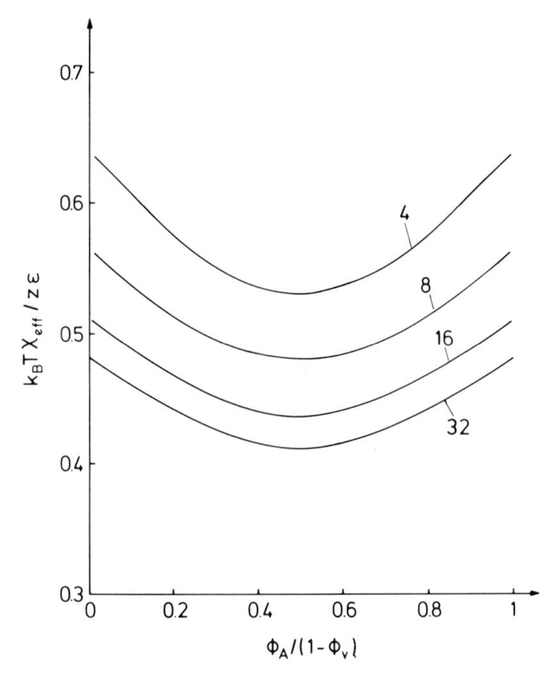

Fig. 18.3: Plot of the effective Flory-Huggins interaction parameter vs concentration for temperatures corresponding to the coexistence curve. Due to Equ. (40) these curves should be straigth lines at unity. From [20].

necessary which go beyond the mean field theories. To check this, first the number of contacts of the chains is investigated. It turns out that the vacancies lead to a strong decrease of the contacts between different species, much stronger than the mean field theories suggest. As one of the consequences the typical extension of the polymers changes as well. It is well known that the mean square end-to-end distance or the radius of gyration decreases with increasing density. However, what is important here, is that this decrease is not the same for the chains of the minority phase compared to the one of the majority phase. It already starts showing up in the one phase region. This change in the overall diameter of the chains has consequences for the theoretical description of the system. For wave vectors $\frac{2\pi}{k} \gtrsim \langle R_G^2 \rangle$, the radius of gyration, the scattering function of the individual chain is given by $S(k) = N(1-\frac{1}{3}\langle R_G^2 \rangle k^2)$ rather than by Equ. (5). Using the RPA approximation of de Gennes [6] (Equ. 4), we find that the mean extension of the chains directly modifies the collective structure function. S(0) is not changed by this, the change is of order k^2. However, since the extensions of the chains can be interpreted as an effective interaction range, this certainly is important for the Ginzburg criterion (Equ. 14) which

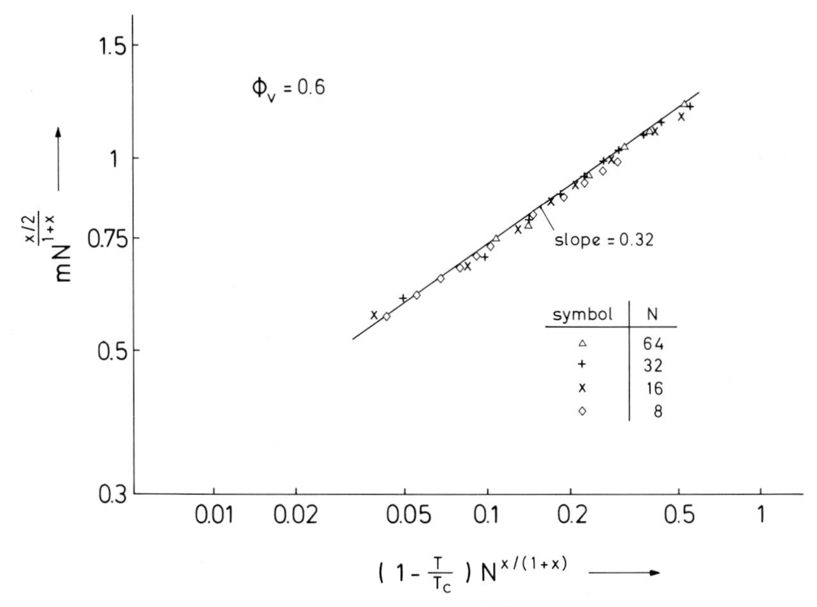

Fig. 18.4: Crossover scaling of m due to Equ. (42). From [22].

describes the crossover from the mean field to the Ising regime. This k dependence shows up in the coefficient of the gradient square term of the Ginzburg-Landau-functional. The Ginzburg criterion then yields for the crossover scaling function of the order parameter m (with $<R_G^2>$ the unperturbed radius of gyration)

$$m = (1-T/T_c)^{1/2} \tilde{f}_m \{(1-T/T_c) \cdot <R_G^2>^3/N^2\} \quad (41)$$

where $\tilde{f}$ is the crossover scaling function. However, this scaling does not work [21]. The reason is that this improved ansatz still does not take into account the contact distribution introduced by the vacancies of solvent molecules. Leibler et al. [52] use these blob effects to explain the effects of good solvents in polymer mixtures. They propose a scaling relation

$$m\, N^{x/[2(1+x)]} = \tilde{m}\, \{(1-T/T_c)\, N^{x/(1+x)}\},\ x \approx 0.22\ . \quad (42)$$

Using this modification the data all collapse onto one common curve as shown in Fig. 18.4.

Taking these results into account, it is clear that the Flory-Huggins theory is a useful tool for a first investigation. However, its limitations always should be kept in mind. Especially if one approaches the critical point or considers a system which includes solvents one has to be extremely careful.

18.3.2 Dynamics of a Polymer in a Tube: Test of the Reptation Concept

The reptation model for the dynamics of polymers in a melt assumes that a chain moves essentially along its own contour. The idea is that the interaction of the polymers leads to topological constraints which strongly favour the reptation motion [4]. This is essentially a one chain theory. This one chain then is supposed to move along a tube of diameter d_T, the "tube diameter". First we investigate how a chain moves along such a tube. In a real melt the tube itself is a random walk and not rigid, making it extemely difficult to check whether and how a chain moves along the tube. Thus we first consider a much simpler case, the statics and dynamics of a polymer in an infinite straight tube [26, 27]. For the simulations we used SAW's on a diamond lattice with chain lengths $100 \leq N \leq 800$. The chains were confined to a circular tube of diameter d_T with $16 \leq d_T \leq 32$. For the simulations we used the standard kink jump algorithm with moves involving 3 and 4 bonds. This algorithm is known to yield Rouse dynamics for non-reversal random walks [53] and thus allows for a dynamic interpretation on time and length scales less than the diffusion time / distance. A similar problem using the Zimm model was investigated by Brochard and de Gennes [54].

Fig. 18.5 gives typical snapshots of chains under consideration. Comparing persistence lengths one can estimate that the considered chain lengths and the considered tube diameters fall well into the regime of experimental interest [26]. For the static properties the data confirm our expectations such that

$$\langle R_{||}^2 (N,d_T) \rangle = N^{2\nu} f_{||}(N^{\nu}/d_T)$$
$$f_{||}(x) \propto \begin{cases} \text{const}, & x \to 0 \\ x^{-2+2/\nu}, & x \to \infty \end{cases} \tag{43a}$$

$$\langle R_{\perp}^2 (N,d_T) \rangle = N^{2\nu} f_{\perp}(N^{\nu}/d_T)$$
$$f_{\perp}(x) \propto \begin{cases} \text{const}, & x \to 0 \\ x^{-2}, & x \to \infty \end{cases} \tag{43b}$$

with $R_{||}$ ($R_{\perp}$) being the end-to-end distance along (perpendicular to) the tube axis. Unexpected was, however, that the mean square fluctuation of $R_{||}$ did not yet saturate to the expected $(\langle R^4 \rangle - \langle R^2 \rangle^2)_{||} / \langle R_{||}^2 \rangle^2 = \text{const}$ value. This is especially important since $\langle R_{||}^2 \rangle^{1/2}$ of our simulation can be interpreted as the "primitive path" length in the reptation regime theory. The knowledge of this fluctuation is important for the understanding of the crossover into the

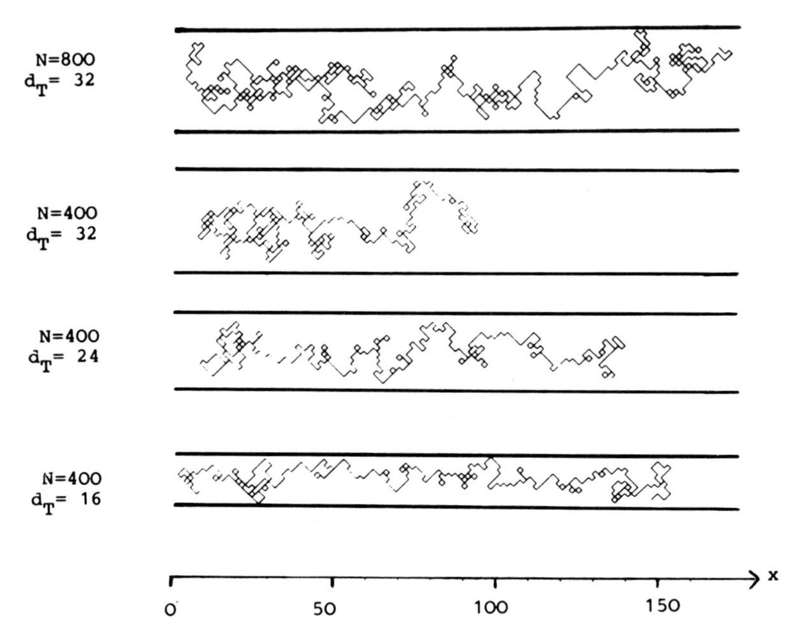

Fig. 18.5: Snapshot pictures of projections of the chains in the tube, d_T and N as indicated. From [26].

reptation regime [4]. Since we can also view our chains as a one dimensional lattice gas of hard spheres, it is clear that for $N \rightarrow \infty$ this saturation must occur [55]. Thus these data indicate that the enhanced mobility of the chain ends still dominates these fluctuation. Now, it is particularly interesting to check whether this has consequences for the dynamics of the chains as well. Following the Rouse dynamics in 1–d we expect for the mean square displacements g:

$$g_1(t) = \langle (\vec{r}_i(t) - \vec{r}_i(0))^2 \rangle ,$$

$$g_2(t) = \langle \{ (\vec{r}_i(t) - \vec{r}_{CM}(t)) - (\vec{r}_i(0) - \vec{r}_{CM}(0)) \}^2 \rangle \text{ and} \tag{44}$$

$$g_3(t) = \langle (\vec{r}_{CM}(t) - \vec{r}_{CM}(0))^2 \rangle ,$$

with $\vec{r}_{CM}$ being the position of the center of gravity of the chain and $\vec{r}_i$ the position of the i–th monomer, the following behaviour (σ being the unit of length and W denoting an elementary frequency unit)

$$g_{1,2}(t) \propto \sigma^2 (Wt)^{1/(1+1/2\nu)}; \; g_3(t) \propto \sigma Wt \tag{45a}$$

for times $1 \leq Wt \leq (d_T/\sigma)^{2(1+1/2\nu)}$. For larger times the monomers feel the constraints of the tube. There we have to distinguish between the parallel and perpendicular part

$$g_{1,2_{||}} \propto \sigma^2 (Wt)^{1/2} \; ; \; g_{3_{||}} \propto \sigma^2 Wt$$
$$g_{1,2\perp} \propto \text{const} \; ; \; g_{3\perp} \propto \text{const} \; . \tag{45b}$$

While for $Wt \gtrsim \langle R^2_{||} \rangle / \sigma^2$ we only see the overall diffusion of the chain:

$$g_{3_{||}} \propto g_{1_{||}} \propto \sigma^2 Wt \; ; \; g_{2_{||}} \propto \text{const.} \tag{45c}$$

For g_1 and g_3 the data nicely confirm the expected behaviour as quoted in Equ. 45a–c. Especially the fact, that the diffusion constant D is only a function of N ($D \propto 1/N$) but not a function of the tube diameter, shows that the considered diameters are big enough not to exhibit any blocked configurations. For $d_T = 16$ the data were not good enough to estimate dynamic properties to the accuracy needed. Fig. 18.6 shows a typical example for the motions of the chains. In this figure g_1, g_2 are calculated for the middle monomers. Note that there obviously is about a factor of 10 between the two times where g_2 approaches a constant and g_1 is governed by the overall diffusion. Asymptotically one expects this to be a constant factor independent of N. However, a more

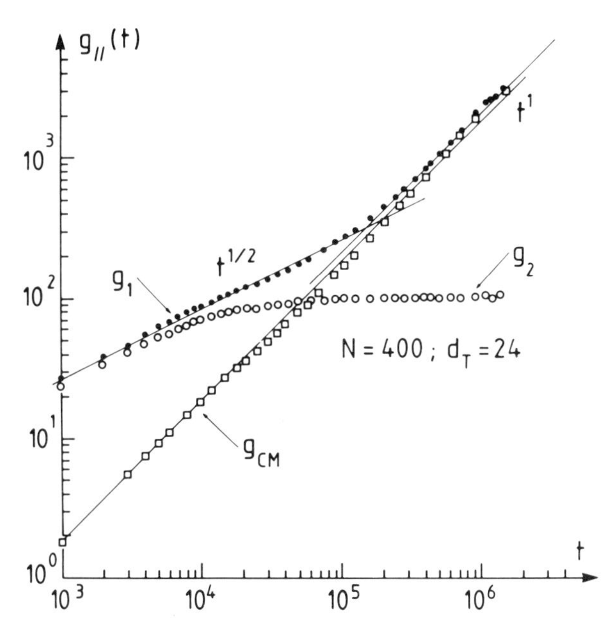

Fig. 18.6: Example of g_1,g_2,g_3 for a chain moving along the tube. From [26].

a

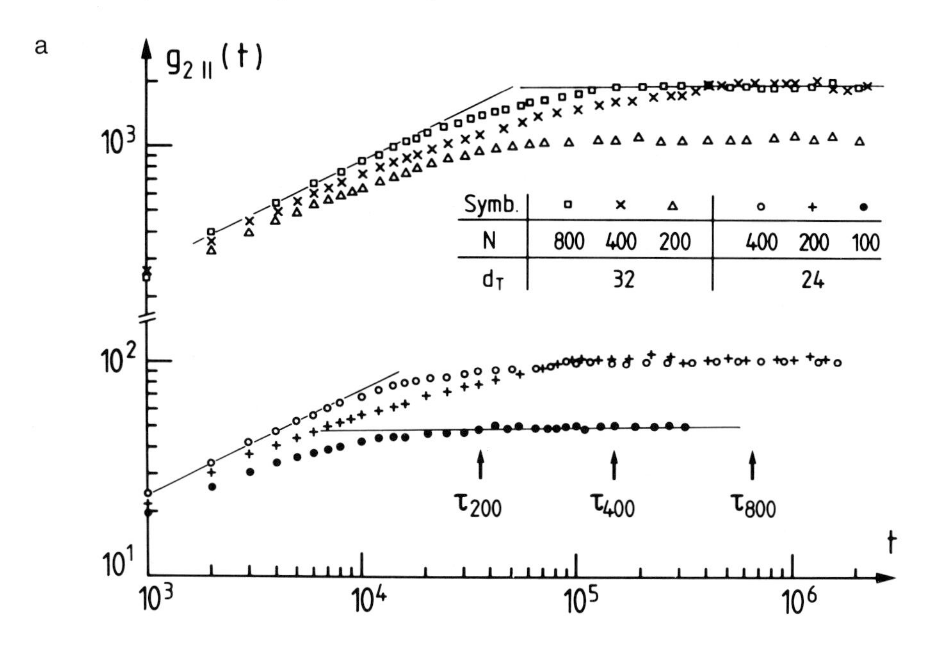

b

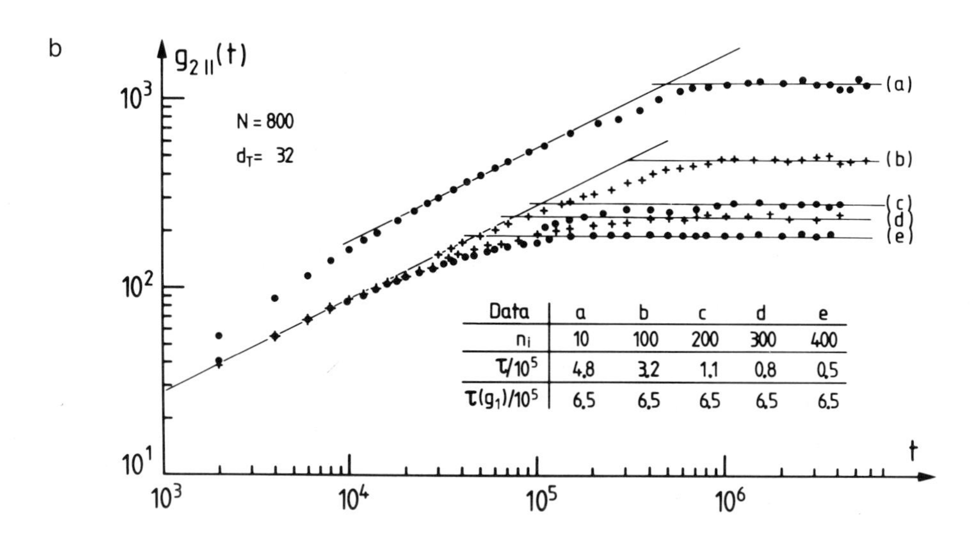

Data	a	b	c	d	e
n_i	10	100	200	300	400
$\tau/10^5$	4.8	3.2	1.1	0.8	0.5
$\tau(g_1)/10^5$	6.5	6.5	6.5	6.5	6.5

Fig.18.7: (a) $g_{2\parallel}(t)$ vt for various N and d_T. The arrows indicate the onset of the diffusion in $g_1(t)$. From [26]. (b) $g_{2\parallel}(t)$ for monomers of a chain of N = 800 monomers. τ gives the estimated relaxation times, whereas $\tau(g_1)$ shows that there is no significant effect on g_1. From [26].

detailed analysis of g_2 for different N and different positions along the chain reveals a rather unexpected result. Fig. 18.7 shows plots of g_2 for varying N (18.7a) and varying i (18.7b, compare Equ. 45). For long chains the plateau value of g_2, within the accuracy of the data, for the middle monomers is independent of N. This cannot be explained by the standard Rouse model. It especially means that inner monomers follow the diffusion of the center of mass at early times, independent of the length of the overall chain. It is not clear which consequences this has for the reptation of chains in a melt. Also the role of the excluded volume, used here to stretch the chains, is not completely understood in this context. Fig. 18.7b illustrates this for different monomers i along the chain for the case $N = 800$, $d_T = 32$. The outer monomers show a strongly increased plateau value compared to the middle ones. Especially the very last monomers display a dramatically increased mobility which certainly governs also the fluctuations in the mean square end-to-end distance as discussed above. It is interesting to note that g_1 is much less sensitive to the position of the monomer of the chain. A more detailed investigation of the time dependent structure function S(k,t) of the chain also reveals that depending on k the crossover to the "creep term" (chain diffusion along the tube) should occur at times t_{coh} strongly dependent on k, namely

$$t_{coh} \approx NW^{-1}(\sigma k)^{-2}. \tag{46}$$

This is especially important for the interpretation of experiments [56, 57] and for the discussion of memory function type theories [58] versus the reptation theory. However, to discuss these problems in detail would be beyond the scope of the present review.

18.3.3 Molecular Dynamics Simulation of a Polymer Melt

So far the simulations of polymers described here were confined to chains on a lattice. In order to allow for a dynamic interpretation of the data one has to have an algorithm which gives Rouse dynamics for NRRW's. One of the consequences is that this constrains the simulations to rather moderate densities. (Note that the case $\Phi_v = 0.2$ of Chapter 18.2.1 is an extremely high density, consequently the chains are confined to rather small $N \leq 32$.) For a general discussion see [48]. For a direct investigation of the dynamics of a polymer melt, however, one wishes to work at "realistic" densities. This can be achieved either by using extremely long chains on a lattice at a rather moderate density (e. g. $\varrho = 0.35$, $N \gtrsim 2000$ [53, 59]) or to work at very high density and to simulate chains of moderate length. For the latter case Monte Carlo simulations are not appropriate. A further disadvantage of Monte Carlo calculations for such systems is that the motion is given by random moves of single monomers or small subunits, such that cooperative effects are not directly taken into account. Here we describe a recent extensive molecular dynamics (MD) investigation of a melt of polymer chains [28]. In molecular dynamics one solves Newton's equations of motion for a collection of particles on the computer. This corresponds to a simulation of a microcanonical ensemble. However, the integration of such a typically highly non linear set of differential equations encounters problems with respect to the numerical stability of the calculations. Normally for MD times are simulated which are sufficient for particles to move a few nearest neighbour distances. For polymers the situation is much more different. Here we need simulations which cover such a time with respect to the typical chain diameter $<R^2(N)>^{1/2}$. That is why the computer time needed to run a microcanonical simulation of a model polymer melt is prohibitively large even on present day supercomputers. To overcome this problem we used a recently tested MD method where the monomers are weakly coupled to a heat bath and a frictional background [60]. The equation of motion is

$$\ddot{\vec{r}}_i = -\vec{\nabla} \sum_{i \neq j} U_{ij} - \Gamma \dot{\vec{r}}_i + \vec{W}_i(t) \ . \tag{47}$$

U_{ij} is the interaction potential between particles i and j of the system. The friction Γ and the random force $\vec{W}$ are correlated via the fluctuation dissipation theorem

$$<\vec{W}_i(t)\ \vec{W}_j(t')> = \delta_{ij}\delta(t-t')6k_BT\Gamma \ . \tag{48}$$

By choosing a weak friction the motion of the particles is still governed by the local interaction and only for times much larger than the local oscillation time the coupling to the background stabilizes the system. This method was tested for polymers in detail [61]. In the model we use, the particles interact with a purely repulsive Lennard Jones potential for distances of $r_{ij} \leq 2^{1/6}\sigma$ and for monomers which are neighbours along the chain we add an additional attractive term (Fene potential) which diverges at $|\vec{r}_i - \vec{r}_{i+1}| = 1.5\sigma$. The parameters of the potential were set to both allow a maximal time step and prevent bond cutting. The time step of integration then was $\Delta t = 0.006\,\tau$ and the temperature $kT = 1.0\varepsilon$ and $\Gamma = 0.5\tau^{-1}$. τ, t, ε are the typical Lennard Jones parameters (for more details see [28, 60]). The density used was $0.85\ \sigma^{-3}$. With these parameters chains up to N = 400 monomers were simulated. One key indication of the reptation concept is the behaviour of the mean square displacement

$$g_1(t) \propto \begin{cases} t^{1/2} & \tau < \tau_{N_e} \sim N_e^2 \\ t^{1/4} & \tau_{N_e} \lesssim t \lesssim \tau_N \sim N^2 \\ t^{1/2} & \tau_N < t < N^3 \ \tau_d \sim N^3/N_e \\ t & \tau_d < t \,. \end{cases} \tag{49}$$

τ_{N_e} is the relaxation time of an entanglement length containing N_e monomers. As already mentioned at the beginning of Chapter 18.3.2 this is the time when the monomer feels the topological constraint of the interaction of the monomers. After that time the Rouse relaxation has to occur along the tube, which itself is a random walk. This leads to the characteristic $t^{1/4}$ regime in the mean square displacement. The second $t^{1/2}$ regime then describes the diffusion along the original tube until the chain has lost all memory of its initial configuration. Fig. 18.8 gives the result of the MD simulation. The data clearly exhibit a $t^{1/4}$ regime in the mean square displacement. Especially important is also that the onset of the $t^{1/4}$ seems to be independent of the chain length. From this we get an entanglement length $N \cong 35$. Comparing this to real polymers and mapping the entanglement lengths yields a $\Delta x = 39$ Å (PDMS) and $\Delta x = 53$ Å (PS) for the displacements of the monomers at the onset of the $t^{1/4}$ regime. This was the first time a simulation based on collective motion was able to cover the regime from Rouse dynamics into the entangled regime.

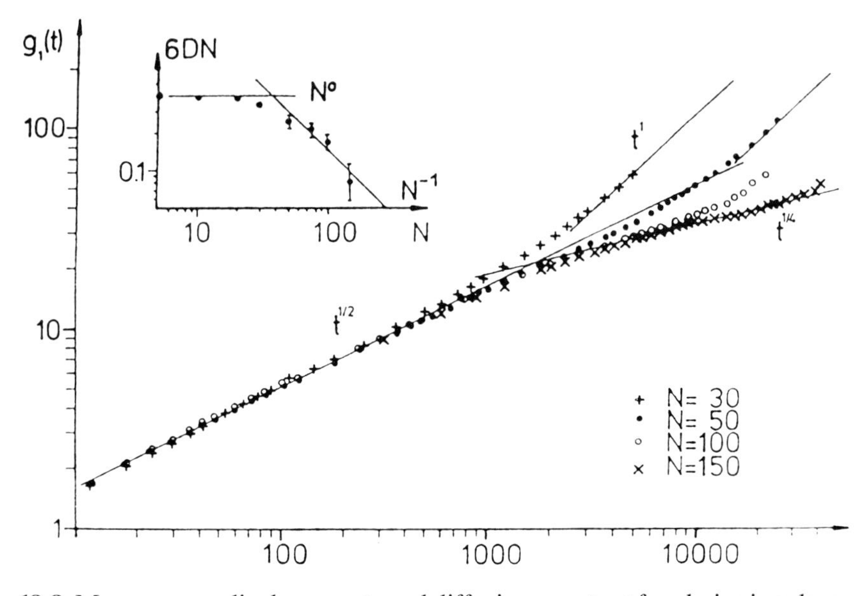

Fig. 18.8: Mean square displacements and diffusion constant for chains in a dense melt. From [28].

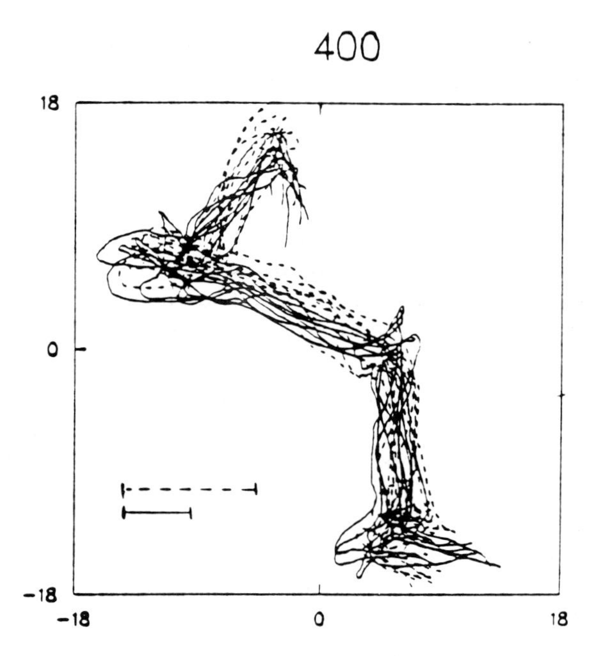

Fig. 18.9: Projection of the configuration of a primitive chain from a real chain of N = 400 monomers. The two bars give the width of the covered region for isotropic and reptation motion respectively. From [47].

However, do the chains really reptate or is the $t^{1/4}$ regime only the sign of an effective slowing down? To clear this we also analyzed the motion of the "primitive path" of chains of various lengths. The primitive path is constructed from the original chain as the chain built from the centers of gravity of all subchains of N_e monomers. Fig. 18.9 shows an example of configurations at different times up to $\Delta t = 24\,000\ \tau$. The data clearly show that the chain motion is governed by the motion along the contour. The width of the shaded region can directly be interpreted as the effective tube diameter for the primitive chain.

18.4 Conclusions

This review gave a short survey of a variety of problems connected with the understanding of melts of linear polymers. Within the Flory-Huggins mean field picture we were able to treat the spinodal decomposition of chains. A generalized Ginzburg Landau criterion yields a way to estimate the validity of this treatment. For the effect of polydispersity we found that there are no qualitative differences, however, problems arising from the variation of the diffusion constants with chain length are still not understood and need further investigation. The decomposition also may be affected by shear. Then the scattering intensities and consequently the density fluctuations are no longer isotropic. In a realistic situation there are always surfaces which might give rise to another effect. Assuming a sample in the two phase region the wall (surface) might be attracted for one of the species causing a wetting transition. For the growth of such a wetting layer a logarithmic behaviour is predicted. The surface enrichment of one species occurring in the one phase region predicted by the theory is consistent with recent experiments [43, 62, 63].

So far the considerations were confined to mean field calculations. Some of the mean field assumptions were tested numerically. The simulations yielded that the Flory-Huggins approach only provides a very rough description of the systems and probably should only be used for qualitative investigations. To gain more insight here extensive research both with numerical and analytical methods is needed.

Besides these global properties it is also important to understand the microscopic dynamics of the chains. The simulation of the chain in the straight tube showed that the enhanced mobility of the ends is very important for the understanding of the dynamics of melts. This also shows that we can only expect to see reptation for very long chains. Whether the chains move at all along their own contour was investigated by a molecular dynamics simulation. The analysis of the primitive chain, for the first time, clearly shows that with increasing chain length the dominant part of the motion occurs along the chain contour. This shows that the basic assumption of the reptation model is at least qualitatively correct.

Acknowledgements

One of us (K. Kremer) carried out initial stages of the research described here at the Institut für Festkörperforschung (IFF) of KFA Jülich in 1984, and spent 15 months in 1984/85 at the Corporate Research Science Laboratories, Exxon Research and Engineering, Annandale, USA. The molecular dynamics simulations of 18.3.3 were made possible by a computing time grant of the Höchstleistungsrechenzentrum (HLRZ) at Jülich.

18.5 References

[1] P. G. de Gennes: Scaling Concepts in Polymer Physics. Cornell University Press, Ithaca, N.Y. 1979.
[2] K. F. Freed: Renormalization Group Theory of Macromolecules. Wiley, New York 1987.
[3] J. des Cloizeaux, G. Jannink: Les Polymères en Solution: Leur Modélisation et leur Structure. Les Editions de Physique, Les Ulis (France) 1988.
[4] M. Doi, S. F. Edwards: The Theory of Polymer Dynamics. Clarendon Press, Oxford 1986.
[5] P. J. Flory: Principles of Polymer Chemistry. Cornell University Press, Ithaca, N.Y. 1953.
[6] P. G. de Gennes: Dynamics of fluctuations and spinodal decomposition in polymer blends I, J. Chem. Phys. 72 (1980) 4756–4763.
[7] P. A. Pincus: Dynamics of fluctuations and spinodal decomposition in polymer blends II, J. Chem. Phys. 75 (1981) 1996–2000.
[8] K. Binder: Collective diffusion, nucleation, and spinodal decomposition in polymer mixtures, J. Chem. Phys. 79 (1983) 6387–6409.
[9] G. R. Strobl: Structure evolution during spinodal decomposition of polymer blends, Macromolecules 18 (1985) 558–563.
[10] K. Binder: Dynamics of phase separation and critical phenomena in polymer mixtures, Colloid & Polymer Sci. 265 (1987) 273–288.
[11] H. Nakanishi, P. Pincus: Surface spinodals and extended wetting in fluids and polymer solutions, J. Chem. Phys. 79 (1983) 997–1003.
[12] I. Schmidt, K. Binder: Model calculations for wetting transitions in polymer mixtures, J. Physique 46 (1985) 1631–1644.
[13] T. E. Schichtel, K. Binder: Kinetics of phase separation in polydisperse polymer mixtures, Macromolecules 20 (1987) 1671–1681.
[14] N. Pistoor, K. Binder: Scattering functions and the dynamics of phase separation in polymer mixtures under shear flow, Colloid & Polymer Sci. 266 (1988) 132–140.
[15] N. Pistoor, K. Binder: Concentration fluctuation in polymer mixtures under shear flow: A phenomenological theory, in Proceedings of Workshop on Polymer Motion in Dense Systems, Grenoble, Sept. 1987. D. Richter and T. Springer (eds.), Springer, Berlin 1988, 285–289.
[16] I. Schmidt, K. Binder: Dynamics of wetting transitions: A time-dependent Ginzburg-Landau treatment, Z. Physik B67 (1987) 369–385.
[17] K. K. Mon, K. Binder, D. P. Landau: Monte Carlo simulation of the growth of wetting layers, Phys. Rev. B35 (1987) 3683–3685.
[18] A. Sariban, K. Binder, D.W. Heermann: A Monte Carlo test of the Flory-Huggins theory for polymer mixtures, Phys. Rev. B35 (1987) 6873–6876.
[19] A. Sariban, K. Binder, D.W. Heermann: Critical phenomena in polymer mixtures. Monte Carlo simulation of a lattice model, Colloid & Polymer Sci. 265 (1987) 424–431.
[20] A. Sariban, K. Binder: Critical properties of the Flory-Huggins lattice model of polymer mixtures, J. Chem. Phys. 86 (1987) 5859–5873.
[21] A. Sariban, K. Binder: Phase separation of polymer mixtures in the presence of solvent, Macromolecules 21 (1988) 711–726.

[22] A. Sariban, K. Binder: Monte Carlo Simulation of Thermodynamic and Structural Properties of Polymer Mixtures Including Solvent, Proceedings of Workshop on Polymer Motions in Dense Systems, Genoble, Sept. 1987. D. Richter and T. Springer (eds.), Springer, Berlin 1988, 301–306.
[23] A. Sariban, K. Binder: Interaction effects on polymer chain linear dimensions in polymer mixtures, Makromol. Chem. Rapid Commun. 189 (1988) 2357–2365.
[24] A. Sariban, K. Binder: Monte Carlo simulation of a lattice model for ternary polymer mixtures, Colloid & Polymer Sci. 266 (1988) 389–397.
[25] A. Sariban, K. Binder: Scattering from concentration fluctuations in polymer blends: A Monte Carlo investigation, Colloid & Polymer Sci. 267 (1989) 469–479.
[26] K. Kremer, K. Binder: Dynamics of polymer chains confined into tubes: Scaling theory and Monte Carlo simulations, J. Chem. Phys. 81 (1984) 6381–6394.
[27] K. Binder, K. Kremer: Simulation of polymers in confined geometries, in R. Pynn and A. Skjeltorp (eds.): Scaling Phenomena in Disordered Systems, Plenum Press, New York 1985, 525–536.
[28] K. Kremer, G. S. Grest, I. Carmesin: Crossover from Rouse to reptation, a molecular dynamics analysis, Phys. Rev. Lett. 61 (1988) 566–569.
[29] E. Helfand, Y. Tagami: Theory of the interface between immiscible polymers II, J. Chem. Phys. 56 (1971) 3592–3601.
[30] E. Helfand: Theory of inhomogeneous polymers, J. Chem. Phys. 62 (1975) 999–1005.
[31] M. G. Brereton, E.W. Fischer, Ch. Herkt-Maetzky, K. Mortensen: Neutron scattering from a series of compatible polymer blends: Significance of the Flory χ_F parameter, J. Chem. Phys. 87 (1987) 6144–6149.
[32] K. M. Hong, J. Noolandi: Theory of inhomogeneous multicomponent polymer systems, Macromolecules 14 (1981) 727–736.
[33] K. Binder: Nucleation barriers, spinodals, and the Ginzburg criterion, Phys. Rev. A29 (1984) 341–349.
[34] A. Baumgärtner, D.W. Heermann: Spinodal decomposition of polymer films, Polymer 27 (1986) 1777–1783.
[35] H.-O., Carmesin, D.W. Heermann, K. Binder: Influence of a continuous quenching procedure in the initial stages of spinodal decomposition, Z. Physik B65 (1986) 89–102.
[36] K. Binder, H. L. Frisch, J. Jäckle: Kinetics of phase separation in the presence of slowly releasing structural variables, J. Chem. Phys. 85 (1986) 1505–1512.
[37] T. Nose: Kinetics of phase separation in polymer mixtures, Phase Transitions 8 (1987) 245–260.
[38] T. Hashimoto: Structure formation in polymer mixtures by spinodal decomposition, in R. M. Ottenbride, L. A. Utracki, S. Inoue (eds.): Current Topics in Polymer Science, Volume II, Hanser Publishers, Munich-Vienna-New York 1987, pp. 199–242.
[39] J. F. Joanny: Effects of polydispersion in the separating of two molten polymers, C. R. Seances Acad. Sci. Ser. B286 (1978) 89–91.
[40] K.W. Kehr, K. Binder, S. M. Reulein: Mobility, interdiffusion, and tracer diffusion in lattice gas models of two-component alloys, Phys. Rev. B 39 (1989) 4891–4910.
[41] E. J. Kramer, P. Green, C. J. Palmstrom: Interdiffusion and marker movements in concentrated polymer-polymer diffusion couples, Polymer 25 (1984) 473–480.

[42] K. Binder, H. Sillescu: Diffusion, polymer-polymer, in J. L. Kroschwitz (ed.): Encyclopedia of Polymer Science and Engineering, 2nd ed., J. Wiley & Sons (1989) 297–315.
[43] R. J. Composto, R. S. Stein, E. J. Kramer, R. A. L. Jones, A. Mansour, A. Karim, G. P. Felcher: Surface enrichment in polymer blends. A neutron reflection test, Physica B 156 & 157 (1989) 434–436.
[44] K. Binder, H. L. Frisch: Interfacial profile between coexisting phases of a polymer mixture, Macromolecules 17 (1984) 2928–2930.
[45] I. Carmesin, J. Noolandi: First-order wetting transitions of polymer mixtures in contact with a wall, Macromolecules 22 (1989) 1689–1704.
[46] For a recent review on experimental questions see: W.W. Greassley, in M. Nagasawa (ed.): Molecular Conformation and Dynamics of Macromolecules in Condensed Systems, Elsevier, Amsterdam 1988.
[47] K. Kremer, G. S. Grest: Dynamics of entangled linear polymer melts: A molecular dynamics simulation, J. Chem. Phys. 92 (1990) 5057–5086.
[48] K. Kremer, K. Binder: Monte Carlo simulations of lattice models for macromolecules, Comp. Phys. Rep. 7 (1988) 259–312.
[49] K. Binder (ed.): Applications of the Monte Carlo Method in Statistical Physics, Springer, Berlin 1984.
[50] J. C. Le Guillou, J. Zinn-Justin: Critical exponents from field theory, Phys. Rev. B21 (1980) 3976–3998.
[51] D. Schwahn, K. Mortensen, H. Yee-Madeira: Mean field and Ising critical behavior of a polymer blend, Phys. Rev. Lett. 58 (1987) 1544–1546.
[52] D. Broseta, L. Leibler, J. F. Joanny: Critical properties of incompatible polymer blends dissolved in a good solvent, Macromolecules 20 (1987) 1935–1943.
[53] K. Kremer: Statics and dynamics of polymeric melts, Macromolecules 16 (1983) 1632–1638.
[54] F. Brochard, P. G. de Gennes: Dynamics of confined polymer chains, J. Chem. Phys. 67 (1977) 52–56.
[55] S. Alexander, P. A. Pincus: Diffusion of labelled particles on one-dimensional chains, Phys. Rev. B18 (1978) 2011–2012.
[56] D. Richter, A. Baumgärtner, K. Binder, B. Ewen, J. B. Hayter: Dynamics of collective fluctuations and Brownian motion in polymer melts, Phys. Rev. Lett. 47 (1981) 109–112.
[57] J. Higgins: Dynamics of polymermolecules, Physica 136 B (1986) 201–206.
[58] G. Ronca: Frequency spectrum and dynamic correlations in concentrated polymer liquids, J. Chem. Phys. 79 (1983) 1031–1043.
[59] Combining results of Ref. [28, 53] yields this estimate as minimal length.
[60] G. S. Grest, K. Kremer: Molecular dynamics simulation of polymers in the presence of a heat bath, Phys. Rev. A33 (1986) 3628–3631.
[61] G. S. Grest, K. Kremer, T. A. Witten: Structure of many arm star polymers: A molecular dynamcis simulation, Macromolecules 20 (1987) 1376–1383; G. S. Grest, K. Kremer, S. Milner, T. A. Witten: Relaxation of self-entangled many-arm star polymers, Macromolecules 22 (1989) 1904–1910.
[62] R. A. L. Jones, E. J. Kramer, M. H. Rafailovich, J. Sokolov, S. A. Schwarz: Surface enrichment in an isotopic polymer blend, Phys. Rev. Lett. 62 (1989) 280–283.
[63] J. Sokolov, M. H. Refailivich, R. A. L. Jones, E. J. Kramer: Enrichment depth profiles in polymer blends measured by forward recoil spectrometry (FRES). Appl. Phys. Lett. 54 (1989) 590–592.

Note added in Proof

Meanwhile there has been significant progress in the simulation of the dynamics of polymeric melts. In Ref. [47] a detailed comparison between the numerical results and experiments was performed. This enabled the authors to predict time and length scales for the onset of the slowing down in the motion for a variety of chemical species. The results are in excellent agreement with very recent neutron spin echo studies on PEP [A1]. Meanwhile we also performed a Monte Carlo Simulation using the bond fluctuation algorithm [A2], where the density was varied. The results compare very well to the data of Ref. [47]. Using density dependent monomeric mobility, we were able to scale the diffusion constants from the dilute free draining limit to the entangled dense melt.

[A1] D. Richter, B. Farago, L. J. Fetters, J. S. Huang, B. Ewen, C. Lartigue: Direct microscopic observation of the entanglements distance in a polymer melt, Phys. Rev. Lett. 64 (1990) 1389–1392.

[A2] W. Paul, K. Binder, D.W. Heermann, K. Kremer: Crossover scaling in semidilute polymer solutions: a Monte Carlo test, submitted to J. Phys. (Paris).

Appendix: Documentation of the Sonderforschungsbereich 41

Chairmen

W. Kern	1969–1971
W. Heitz	1971–1974
H. Höcker	1975–1977
H. Ringsdorf	1977–1980
E.W. Fischer	1980–1983
H. Sillescu	1983–1987

List of Participants

Name	Institute	Membership from–to
Section Mainz		
K. Berger	Physikalische Chemie	1981–1987
K. Binder	Physik	1984–1987
O. Bodmann	Physikalische Chemie	1969–1985
V. Böhmer	Physikalische Chemie	1976–1987
B. Ewen*	Physikalische Chemie	1978–1987
E.W. Fischer	Physikalische Chemie	1969–1987
	MPI-Polym.	from 1984
W. Heitz	Organische Chemie	1969–1975
H. Höcker**	Organische Chemie	1969–1977
V. Jaacks	Organische Chemie	1969–1971
H. Kämmerer	Organische Chemie	1969–1987
W. Kern	Organische Chemie	1969–1985
R. Kirste	Physikalische Chemie	1969–1987
G. Löhr	Physikalische Chemie	1972–1975
I. Lüderwald*/**	Organische Chemie	1977–1981
G. Meyerhoff	Physikalische Chemie	1969–1987

* "Promotion" supported by the Sonderforschungsbereich 41
** "Habilitation" supported by the Sonderforschungsbereich 41

M. Mutter	Organische Chemie	1978–1985
A. Müller	Physikalische Chemie	1981–1987
M. Przybylski*	Organische Chemie	1981–1987
H. Ringsdorf	Organische Chemie	1971–1987
M. Rothe	Organische Chemie	1969–1975
G. Schmidt	Physikalische Chemie	1971–1987
G.V. Schulz	Physikalische Chemie	1969–1987
H. Sillescu	Physikalische Chemie	1976–1987
H.W. Spiess**	Physikalische Chemie	1980–1987
	MPI-Polym.	from 1985
G. Strobl*/**	Physikalische Chemie	1976–1987
H. Stuhrmann**	Physikalische Chemie	1972–1978
I. Voigt-Martin	Physikalische Chemie	1971–1987
W. Vogt**	Organische Chemie	1969–1987
G. Wegner**	Physikalische Chemie	1969–1974
	MPI-Polym.	1984–1987
B. A. Wolf**	Physikalische Chemie	1972–1987
H. G. Zachmann	Physikalische Chemie	1969–1987

Section Darmstadt

Chairmen		
R. C. Schulz		1970–1975
D. Braun		1975–1987

D. Braun	Dt. Kunststoff-Inst.	1970–1987
E. Gruber*	TH Darmstadt	1970–1978
E. Heidemann	TH Darmstadt	1970–1987
U. Johnson	Dt. Kunststoff-Inst.	1970–1987
B. I. Jungnickel	Dt. Kunststoff-Inst.	1981–1987
W. Knappe	Dt. Kunststoff-Inst.	1970–1975
K. Lederer*	TH Darmstadt	1971–1975
R. C. Schulz	TH Darmstadt	1970–1974
	Mainz	1975–1987
I. Schurz	TH Darmstadt	1971–1975
I. H. Wendorff	Dt. Kunststoff-Inst.	1976–1987

Participating Institutes

from 1969	Inst. für Physialische Chemie, Universität Mainz
	Inst. für Organische Chemie, Universität Mainz
from 1970	Deutsches Kunststoff-Institut
1970	Institut für Makromolekulare Chemie TH Darmstadt

* "Promotion" supported by the Sonderforschungsbereich 41
** "Habilitation" supported by the Sonderforschungsbereich 41

1971–1975	Institut für Makromolekulare Chemie Abteilung Eiweiß und Leder, TH Darmstadt
from 1984	Institut für Physik, Universität Mainz
from 1984	Max-Planck-Institut für Polymerforschung

Heads of Projects

Name		Sub-project (No.)	from–to
K. C. Berger	(see G. Meyerhoff)		
K. C. Berger	Radiotracertechniken und Kinetik der nichtidealen Polymerisation	P28	1982–1984
K. Binder	Theoretische Untersuchungen zur Statik und Dynamik von Polymeren in konzentrierter Lösung und Schmelze	S38	1984–1987
O. Bodmann	Molekulare Konstanten und Wirkungs-Mechanismus des Enzyms Amylomaltase	41/69/1a	1969–1972
O. Bodmann	Struktur und biologische Aktivität von nativen und chemisch modifizierten Glucoseoxidasen	1b	1970–1972
O. Bodmann	Reversible Dissoziation und und Konformationsänderungen von Glucoseoxidasemolekülen	1b	1970–1972
O. Bodmann	Mechanismus der Glucose-oxidasen-Katalyse	1d	1970–1971
O. Bodmann	Konformationsumwandlungen von Enzymproteinen	III/1	1973–1975
V. Böhmer	(see H. Kämerer, W. Kern)		
V. Böhmer	Cyclopolymerisation als Hilfsmittel zur gezielten Synthese von Copolymeren	P24	1982–1985
D. Braun	Durchdringung der Makromoleküle	41/70/16a	1970–1972
D. Braun	Untersuchung an vernetzten Polymeren mit spaltbaren Netzbrücken	41/70/16b	1970–1972
D. Braun	Stabile Polyradikale	16c	1971–1975
from 1973:		I/16b	
D. Braun	Durchdringung und Vernetzung von Makromolekülen	I/16a	1973–1978
from 1976:		B10	
D. Braun	Mechanismus und Kinetik der Übertragungsstarts bei radikalischen Polymerisationen	P9	1979–1985

from 1982:	Geordnete Copolymere durch radikalische Polymerisation	P26	
D. Braun	Kinetik von Terpolymerisationen	P10	1979–1985
from 1982:	Darstellung und Eigenschaften von binären und ternären Copolymeren	P25	
B. Ewen	Untersuchung kollektiver Bewegungszustände in Kristallen aus Kettenmolekülen mit Hilfe der kohärenten inelastischen Neutronenstreuung	S12	1979
E.W. Fischer	(see G.V. Schulz, G. Strobl, J. H. Wendorff, H. G. Zachmann)		
E.W. Fischer	Fehlstellenbildung und Phasenumwandlung in Paraffinen und anderen Oligomeren	41/69/2a	1969–1972
E.W. Fischer	Schmelzverhalten von copolymeren Systemen	41/69/2b	1969–1972
E.W. Fischer	Struktur und Eigenschaften makromolekularer Einkristalle	41/69/2c	1969–1972
E.W. Fischer	Mechanische und dielektrische Relaxationserscheinungen in Hochpolymeren und Oligomeren	41/69/2d	1969–1972
E.W. Fischer	Röntgenstreuung bei extrem kleinen Streuwinkeln	2g	1970–1972
E.W. Fischer	Struktur der teilkristallinen Polymeren in Abhängigkeit vom Molekülbau und von der thermischen und mechanischen Vorgeschichte	V/2a	1973–1984
from 1976:		E2	
from 1979:	Strukturuntersuchungen an teilkristallinen Polymeren zur Aufklärung des Kristallisationsmechanismus	S7 (in cooperation with E13)	
E.W. Fischer	Mechanismus der Phasenumwandlung in Oligomeren und Polymeren	V/2b	1973–1978
from 1976:	Neutronenstreuuntersuchungen zur Struktur der Polymeren im amorphen und im kristallinen Zustand		
E.W. Fischer	Zusammenhang zwischen Struktur und molekularer Beweglichkeit in Polymeren und Oligomeren	V/2c	1973–1978
from 1976:		E6	

E.W. Fischer	Konformation der Fadenmoleküle in glasig erstarrten Polymeren und in Polymerschmelzen	V/2d	1973–1981
from 1976:	Strukturuntersuchungen an amorphen Polymeren mit Hilfe der Licht- und Röntgenstreuung sowie der optischen Aktivität	E11	
from 1979:	Struktur der amorphen Polymeren und der Einfluß physikalischer Vernetzung auf die Molekülbewegungen	S14	
E.W. Fischer	Kinetik der Phasentrennung in Multikomponenten-Systemen und Zusammenhang mit den mechanischen Eigenschaften	S24	1982–1986
E. Gruber	Lichtstreumessungen an intakten sowie teilweise und völlig denaturierten Helices	22	1970
E. Gruber	Untersuchung der Quartärstruktur von gelösten Hochpolymeren mit Hilfe der Lichtstreuungsmethode	22 (former 21b)	1971–1972
E. Gruber	Untersuchung der Quartärstruktur gelöster Hochpolymerer mit Lichtstreuung und rheologischen Methoden	III/22	1973–1981
from 1976:	Untersuchung von Multimerisaten und Assoziationsvorgängen in makromolekularen Lösungen mit Lichtstreuung und rheologischen Methoden	D5	
from 1979:		S4	
E. Gruber	(see J. Schurz) Untersuchungen der Viskoelastizität an Polymerlösungen mittels Messungen im Weissenberg-Rheogoniometer	III/21	1976–1978
E. Heidemann	Untersuchung der kurzkettigen Spaltprodukte, die bei der Spaltung von Kollagenketten mit Hydroxylamin auftreten	17a	1970–1971 changed 1972
E. Heidemann	Synthese und Strukturuntersuchungen an synthetischen Polypeptiden (sPP) mit bekannten Sequenzen	17b	1970–1972

E. Heidemann	Synthese und Strukturuntersuchungen an synthetischen Polypeptiden mit bekannten Sequenzen als Modelle für die Tripelhelix des Kollagens	I/17a	1973–1975
E. Heidemann	Ausarbeitung von Methoden zum Studium der Wechselwirkung von Proteinen untereinander (Bildung von Tertiär- und Quartärstrukturen bzw. strukturspezifische Zusammenlagerung bei Immunreaktionen enzymatischer Katalyse und Affinitätschromatographie)	I/17b	1973–1975
W. Heitz	(see M. Rothe)		
W. Heitz	Synthese und Strukturuntersuchungen an vernetzten Polymeren from 1972: Synthese von Oligomeren	41/69/3	1969–1972
W. Heitz G. Meyerhoff	Vernetzte Polymere als stationäre Phase bei chromatographischen Verfahren	3a	1970–1972
W. Heitz	Polymere Reagenzien	3b	1970–1971
W. Heitz	Synthese von Oligomeren	I/3a	1973–1975
W. Heitz	Vernetzte Polymere als stationäre Phase bei chromatographischen Verfahren	I/3b	1973–1975
H. Höcker	(see W. Kern, H. Ringsdorf, G.V. Schulz)		
H. Höcker	Synthese und Untersuchungen makrocyclischer Verbindungen, Darstellung und Umsetzung von lebenden Polymeren	41/69/4/7a	1969–1971
H. Höcker	Darstellung niedermolekularer und cyclischer Verbindungen mit Hilfe von metallorganischen Mischkatalysatoren	41/69/4/7b	1969–1971
H. Höcker	Die anionische "living" Polymerisation als Mittel zur Darstellung wohldefinierter Polymerer	I/4a	1973–1978
from 1976:	Die anionische Polymerisation als präparative Methode zur gezielten Synthese von Polymeren	B3	

H. Höcker	Herstellung niedermolekularer und cyclischer Kohlenwasserstoffe mit Hilfe von metallorganischen Mischkatalysatoren (Metathese)	I/4b	1973–1978
from 1976:	Die Metathese-Reaktionen als präparative Methode und als mechanistisches Problem	B4	
V. Jaacks	Untersuchungen über die Bildung und Eigenschaften von makrocyclischen Polymeren bei der kationischen Polymerisation oxacyclischer Monomerer	41/69/5	1969
V. Jaacks from 1972: W. Kern	Untersuchungen über die Entstehung von Makrozwitterionen bei ionischen Polymerisationen	5a	1970–1971
V. Jaacks from 1972: W. Kern	Untersuchungen über die Struktur der aktiven Zentren und der entstandenen Polymeren bei der ringöffnenden kationischen Polymerisation cyclischer Formale	5b	1971 changed 1972
U. Johnsen	Untersuchungen von Kettenbrüchen in mechanisch beanspruchten Polyamid-Fasern	V/18	1973–1978
from 1976:	Kettenbrüche in thermomechanisch und chemisch belasteten Polymeren. Methode der ESR	E15	
U. Johnson	Thermolumineszenzen in Polymeren	E8	1976–1981
from 1979:	Struktur und Relaxation amorpher Bereiche in teilkristallinen Polymeren	S13	
B. J. Jungnickel	Untersuchung des Kerreffektes in konzentrierten Lösungen und gequollenen Netzwerken	S40	1985–1987
H. Kämmerer	(see W. Kern)		
H. Kämmerer	Synthese und Untersuchung der Struktur (Endgruppen, Vernetzerbrücken, Verzweigungen) linearer, verzweigter und vernetzter Polymerer	41/69/6	1969

H. Kämmerer	Synthese und Untersuchung der der Struktur linearer, verzweigter und vernetzter Polymerer sowie oligomerer Modellverbindungen. Ausarbeitung quantitativer Bestimmungsmethoden für Endgruppen, Verzweigungen und Vernetzergruppen from 1971: 6a and 6b in cooperation Synthese und Untersuchung der Struktur linearer, verzweigter und vernetzter Polymerer sowie oligomerer Modellverbindungen Synthetische Matrizenreaktionen	6	1970–1972
H. Kämmerer	Herstellung und Untersuchung der Eigenschaften molekulareinheitlicher Phenol-Formaldehyd-Kondensate mit definierter Struktur und verwandter oligomerer Verbindungen. Synthetische Matrizenreaktionen	6a	1970 from 1971 6 and 6a in cooperation
H. Kämmerer from 1975 in cooperation with V. Böhmer	a) Untersuchung der Struktur linearer verzweigter und vernetzter Polymerer b) Herstellung und Untersuchung der Eigenschaften von molekulareinheitlichen phenolischen Mehrkernverbindungen und Phenol/Formaldehyd-Polykondensaten	I/6	1973–1978
from 1976:	Darstellung, Eigenschaften und Reaktionen von phenolischen Mehrkernverbindungen und Polykondensaten; Matrizenreaktion	B9	
W. Kern	(see V. Jaacks, G. Meyerhoff)		
W. Kern W. Vogt from 1972 only W. Vogt	Synthese und Untersuchung polymerer Ester von Phosphorsäuren	41/69/13/7	1969–1971
W. Kern from 1975 in cooperation with H. Höcker	Untersuchungen über die Entstehung von Makrozwitterionen bei der ionischen Polymerisation	I/7	1973–1981
from 1976:	Synthese Oligomerer und Polymerer zur Darstellung von definierten Netzwerken	B1	
from 1979:			

W. Kern H. Kämmerer V. Böhmer	Synthese reaktiver Oligomerer und Polymerer	P2	
W. Kern G. Wegner R. Kirste	Copolymerisation von oxacyclischen Monomeren (see B. A. Wolf)	II/7	1973–1974
R. Kirste	Kinetik der Umwandlungs- und Abbaureaktionen der DNS	41/69/8a	1969–1971
R. Kirste	Konformation von Polymeren in festem amorphen Zustand	41/69/8b	1969–1972
R. Kirste	Konformation von Copolymeren	41/69/8c	1969–1972
R. Kirste	Konformation von Copolymeren	III/8	1973–1987
from 1974:	Konformation von Fadenmolekülen in Lösung		
from 1976:	Konformation und Umgebung von verknäulten Fadenmolekülen in Lösung in Abhängigkeit von der Zusammensetzung der Systeme durch Streumethoden	D1	
from 1979:	Teilung: a) Konformation und Thermodynamik von Polymermischungen	 S1	
from 1982:	Konformation und Thermodynamik von Fadenmolekülen in Polymermischungen mit besonderer Struktur	S23	
from 1979:	b) Konformation und Wechselwirkung von Fadenmolekülen in wäßriger Lösung	S2	
R. Kirste	Konformation von Polymeren in festem amorphen Zustand in der Schmelze und in konzentrierten Lösungen	V/8	1973–1975
R. Kirste	Konformation und Umgebung von verknäulten Fadenmolekülen in Lösung in Abhängigkeit von der Zusammensetzung der Systeme durch Streumethoden	V/8	1975–1977
R. Kirste G. Strobl J. H. Wendorff H. Ringsdorf	Struktur und Dynamik flüssigkristalliner Polymerer / Fluktuation in flüssigkristallinen Polymeren	S26/27	1982–1985
W. Knappe	Struktur und rheologisches Verhalten hochpolymerer Schmelzen		1971–1972

W. Knappe	Spezifische Wärme von Polymeren in Abhängigkeit von Struktur und Vorgeschichte	V/19	1973–1975
K. Lederer	Röntgenkleinwinkelstreuung an gelösten Makromolekülen	III/23	1973–1975
G. Löhr	(see G.V. Schulz)		
I. Lüderwald	(see H. Ringsdorf)		
G. Meyerhoff	(see W. Heitz, G.V. Schulz)		
G. Meyerhoff G.V. Schulz	Molekulare Konstanten und Lösungseigenschaften von Oligomeren und Kondensationspolymeren	41/69/9a	1969–1972
from 1972: G. Meyerhoff	in cooperation with 9b: Thermodynamik makromolekularer Lösungen		
G. Meyerhoff G.V. Schulz from 1972 in cooperation with W. Kern K. C. Berger	Radikalpolymerisation – Molekülverzweigung von Polymeren	41/69/9c	1969–1972
G. Meyerhoff from 1972: K. C. Berger	Makromolekulare Instrumentation	9d	1970–1981
from 1976:		D10	
from 1979:		P15	
G. Meyerhoff	Diffusionsvorgänge in polymeren und oligomeren Lösungen from 1972: in cooperation with 9a (1969)	9b	1971
G. Meyerhoff	Radikalpolymerisation und Molekülverzweigung	II/9	1973–1981
from 1975:	Radikalpolymerisation		
from 1976:	Elementarreaktionen bei der radikalischen Polymerisation	C1 (in cooperation with II/11b)	
from 1979:		P12	
G. Meyerhoff	Charakterisierung von Copolymeren Makromolekulare Instrumentation	P20	1982–1987
G. Meyerhoff	Thermodynamik makromolekularer Lösungen	IV/9	1973–1984
from 1976:	Thermodynamik und Hydrodynamik gelöster Polymerer	D7	
from 1979:		S6	
from 1982:		S20	

G. Meyerhoff	Die Copolymerisation als kinetischer Prozeß	P21	1982–1984
A. Müller	(see G.V. Schulz)		
A. Müller	Kinetik und Mechanismus der Gruppentransferpolymerisation	P37	1985–1987
M. Mutter	Darstellung von Blockcopolymeren mit besonderen Eigenschaften	P16	1979–1984
from 1982:	Darstellung und Konformationsuntersuchung von Polypeptidblockcopolymeren mit besonderen Eigenschaften	P23	
M. Mutter	Synthese und Anwendung von hydrophilen Polymeren mit funktionellen Gruppen	P17	1979–1984
from 1982:	Synthese und Anwendung reaktiver Polymerer mit hydrophilen Eigenschaften	P34	
H. Ringsdorf	(see R. Kirste, J. H. Wendorf)		
H. Ringsdorf H. Höcker	Untersuchung der Abbaumechanismen und der Struktur von Monomeren und Polymeren mit Hilfe der Pyrolyse-Massenspektrometrie	12a	1972
H. Ringsdorf	Untersuchung von reaktiven Monomeren und Polymeren. Polyreaktionen von Vinylpyridiniumsalzen in wäßrigen Lösungen	12b, c	1972
H. Ringsdorf	Polyreaktionen in orientierten Medien	12c	1972
H. Ringsdorf	Untersuchungen der Abbaumechanismen und der Struktur von Monomeren, Oligomeren und Polymeren mit Hilfe der Pyrolyse-Massenspektrometrie	I/5a	1973–1976
from 1976:		B6	
from 1977: I. Lüderwald without H. Ringsdorf			1977–1981
from 1979:		P5	

H. Ringsdorf	Polyreaktionen in orientierten Medien und Untersuchung von reaktiven Monomeren und Polymeren	I/5b + 5c	1973–1981
from 1976:		B7	
from 1979:	Teilung:		
	Untersuchungen von Polyreaktionen in definierten Mono- und Multischichten	P6	
from 1982:	Synthese und Charakterisierung der strukturellen Eigenschaften membranbildender Monomerer und Polymerer	P32	1973–1987
	Mesogene Monomere und ihr Polymerisationsverhalten	P7	1973–1981
	Synthese von ionogenen und aggregatbildenden Monomeren und Untersuchung des Polymerisationsverhaltens	P8	1973–1981
H. Ringsdorf	Molekulare Struktur und Flüssigkristalleigenschaften	P31	1982–1987
H. Ringsdorf W. Vogt	Phospholipidmodelle mit Phosphonatstruktur	P33	1982–1987
M. Rothe W. Heitz from 1972 only M. Rothe	Oligomerensynthese an polymeren Trägern	41/69/10	1969–1971
M. Rothe	Konformation linearer und und cyclischer Peptide des L-Prolins	10a	1970–1975
M. Rothe	Molekulareinheitliche Kollagenmodelle	10b	1970–1975
M. Rothe	Oligoamidsynthese an polymeren Trägern	I/10c	1973–1974
M. Rothe	Untersuchungen über den Mechanismus der kationischen Lactam-Polymerisation	I/10d	1974–1975
G.F. Schmidt	Aufbau einer Meßanordnung für Röntgenbeugungsmessungen mit Hilfe eines energie-dispersiven bzw. ortsempfindlichen Detektors zur Untersuchung schnell ablaufender Änderungen der kristallinen Ordnung in makromolekularen Substanzen	V/25	1973–1978
from 1976:		E4	
G.V. Schulz	(see G. Meyerhoff, B. A. Wolf)		
G.V. Schulz	Bestimmung der Molekular-	41/69/11a	1969–1971

E.W. Fischer	gewichtsverteilung von nichtkristallisierenden Polymeren durch Elektronenmikroskopie		renaming 1972
from 1972: G.V. Schulz G. Meyerhoff	Bestimmung der Molekulargewichtsverteilung nach verschiedenen Methoden		
G.V. Schulz	Kinetik und Elementarprozesse der anionischen Polymerisation	41/69/11b	1969–1972
G.V. Schulz E.W. Fischer	Abbaukinetik und Struktur von Cellulose und pflanzlichen Zellwänden	41/69/11c	1969–1972
G.V. Schulz	Thermodynamische Untersuchung von Polymerlösungen in Abhängigkeit von der Temperatur und vom Druck. Messung der Intensität des Streulichts	11d	1970–1972
G.V. Schulz G. Meyerhoff from 1972: without G. Meyerhoff	Elementarprozesse der radikalischen Polymerisation	11e	1970–1972
G.V. Schulz	Abbaukinetik und Struktur von Cellulose	II/11a	1973–1975
G.V. Schulz	Elementarprozesse der radikalischen Polymerisation	II/11b	1973–1981
from 1976: G. Meyerhoff	C1 + P12		
G.V. Schulz	Bestimmung der Molekulargewichtsverteilung nach verschiedenen Methoden	III/11	1973–1975
G.V. Schulz G. Löhr from 1975: H. Höcker G.V. Schulz	Kinetik und Elementarprozesse der ionischen Polymerisation	II/27	1973–1987
from 1976:	Kinetik und Elementarprozesse der anionischen Polymerisation von polaren Monomeren	C3	
from 1979: G.V. Schulz H. Sillescu		P13	
from 1982: H. Sillescu A. Müller G.V. Schulz	Kinetik und anionische Aspekte der anionischen Polymerisation von polaren Monomeren	P29	1973–1987
G.V. Schulz	Kolonnenmethoden zur	D9	1976–1981

G. Meyerhoff	Bestimmung von Molekulargewichtsverteilung für Untersuchungen der Bildungsprozesse Polymerer		
from 1979 in cooperation with E.W. Fischer	Molekulargewichtsverteilung im Zusammenhang mit der Kinetik der Polymerisation	P14	
R. C. Schulz	Synthese und Eigenschaften optisch aktiver Polymerer	41/70/20a	1970–1972
R. C. Schulz	Photochemische Reaktionen zur Darstellung und Umwandlung von Makromolekülen	41/70/20b	1970–1987
from 1976:	Synthese und Umwandlung von Polymeren mit funktionellen Gruppen	B2	
from 1979:	Darstellung und Eigenschaften von reaktiven Polymeren	P1	
from 1982:	Funktionelle Polymere: Präparative Methoden und Eigenschaften	P30	
R. C. Schulz	Synthese und Eigenschaften optisch aktiver Polyester und Polysaccharid-Derivate	III/20	1973–1975
R. C. Schulz	Geordnete Copolymere	B12	1976–1985
from 1979:	Methoden zum Aufbau und zur Analyse von Copolymeren	P4	
from 1982:	Synthese und Eigenschaften von neuen Block- und Pfropfcopolymeren	P22	
R. C. Schulz	Synthese neuer Polyelektrolyte (mit P30)	P39	1985–1987
J. Schurz	Röntgenkleinwinkeluntersuchungen an gelösten Makromolekülen	21a	1970
J. Schurz	Untersuchung und Interpretation der Elastoviskosität an Lösungen von Polymeren	III/21	1973–1976
from 1976:	(see E. Gruber)		
H. Sillescu	(see G.V. Schulz)		

H. Sillescu	NMR-Untersuchungen zur Dynamik und Struktur von Polymerschmelzen	D3	1976–1987
from 1979:	NMR-Untersuchungen zur Dynamik und Struktur von Polymerschmelzen a) NMR-Relaxation in Polymerlösungen und -schmelzen b) Diffusion von Makromolekülen in Polymerschmelzen c) Struktur und molekulare Bewegung in teilkristallinen Polymeren d) Orientierungsverteilung in teilgeordneten Polymeren	S3	
from 1982: H. Sillescu H.W. Spiess	NMR-Untersuchungen zur Struktur und Dynamik von orientierten Polymeren	S32	
H. Sillescu	Diffusion von Makromolekülen in Polymeren	S31	1982–1987
H.W. Spiess	(see H. Sillescu)		
G. Strobl	(see R. Kirste)		
G. Strobl	Defektstrukturen in oligomeren Modellsystemen	E3	1976–1981
from 1979:	Molekulare Bewegung und Defektstrukturen in Kristallen aus Kettenmolekülen	S11	
G. Strobl	Ramanspektroskopische Untersuchungen zur Kettenanordnung und molekularen Beweglichkeit in oligomeren und polymeren Festkörpern	E7	1976–1984
from 1979: G. Strobl E.W. Fischer	Mechanismus des partiellen Schmelzens (mit E2)	S9	
from 1982:	Mechanismus der Kristallisation von statistischen Copolymeren	S34	
G. Strobl	Struktureigenschaften von Polyethylenoxid- Polypeptid-Blockcopolymeren	S25	1982–1984
H. B. Stuhrmann	Die Bestimmung des geometrischen Strukturfaktors von Makromolekülen durch Variation der Dichte des Lösungsmittels (Methode der Kontrastvariation)	III/28	1972–1978
from 1976:		D2	

W. Vogt	(see W. Kern, H. Ringsdorf)		
W. Vogt	Synthese und Untersuchung polymerer Ester von Phosphorsäuren	I/13	1973–1981
from 1976:		B5	
from 1979:	Polymere Ester von Säuren des Phosphors durch ringöffnende Polymerisation	P3	
W. Vogt	Polymere des 2,3-Diphenylbutadiens	P36	1985–1987
I. Voigt-Martin	Elektronenbeugungsuntersuchungen an Polymerschmelzen	E12	1976–1981
from 1979:		S15	
I. Voigt-Martin	Änderung der Morphologie und der Langperioden- und Kristalldickenverteilung beim Kristallisieren und Tempern von Polyäthylenterephthalat	S35	1982–1984
G. Wegner	(see W. Kern)		
G. Wegner	Topochemische Reaktionen von Monomeren mit konjugierten Dreifachbindungen und physikalische Eigenschaften der Polymerisate	41/69/14a	1969–1972
G. Wegner	Simultane Polymerisation und Kristallisation bei epitaktischen und topotaktischen Reaktionen	41/69/14b	1969–1972
G. Wegner	Polymerisation in fester Phase	V/14a	1973–1974
G. Wegner	Chemie der amorphen Bereiche teilkristalliner Polymerer	V/14b	1973–1974
G. Wegner	Struktur und Eigenschaften fester Polyelektrolyte am Beispiel der Ionene	P38	1985–1987
J. H. Wendorff	(see R. Kirste)		
J. H. Wendorff	Untersuchungen der Eigenschaften amorpher Polymerer mit Hilfe der elektrischen und magnetischen Doppelbrechung sowie der inelastischen Lichtstreuung	E14	1976–1986
from 1979:	Teilung:		
J. H. Wendorff E.W. Fischer	a) Dynamische Eigenschaften von Gläsern	S16	
from 1982: J. H. Wendorff	Untersuchung der dynamischen Eigenschaften von Polymeren	S28	

H. Ringsdorf	mittels der inelastischen Licht- und Neutronenstreuung		
	b) Untersuchungen zur Struktur und Dynamik von Polymeren und Modellsubstanzen mit Hilfe der elektrischen Doppelbrechung	S17	
from 1982:	Langzeitverhalten von Gläsern	S30	
B. A. Wolf G.V. Schulz from 1976: B. A. Wolf	Thermodynamische Untersuchung von Polymerlösungen	IV/26	1973–1987
	Thermodynamische Lösungseigenschaften von Kettenmolekülen, ihre Druckabhängigkeit und ihre kinetische Konsequenz	D6	
from 1979:		S5	
B. A. Wolf	Thermodynamische und rheologische Eigenschaften von Polymerlösungen; Druckeinflüsse und wechselseitige Abhängigkeit	S21	1982–1984
B. A. Wolf R. Kirste from 1985:	Lösungseigenschaften von Polyelektrolyten Thermoreversible Gelierung und Entmischung von Polymerlösungen	S22	1982–1987
H. G. Zachmann from 1972 in cooperation with E.W. Fischer	Statistische Thermodynamik des Kristallisierens und Schmelzens von Hochpolymeren	41/69/15	1969–1972
H. G. Zachmann	Experimentelle Untersuchung Kristallisation und Quellung von Hochpolymeren	15b	1970–1972
H. G. Zachmann	Untersuchung der magnetischen Kernresonanz von kristallinen Hochpolymeren	15d	1970–1973
H. G. Zachmann	Statistisch-thermodynamische Untersuchung der molekularen Ordnung in amorphen Hochpolymeren sowie der Kettenfaltung bei der Kristallisation	V/15a	1973–1978
from 1976:	Statistisch-thermodynamische Untersuchung der molekularen Ordnung in Schmelzen sowie der Kristallisation	E10	

H. G. Zachmann	Kristallisation, Schmelzen und Quellung von Hochpolymeren in Abhängigkeit vom Molekülbau und einer Molekülorientierung	V/15b	1973–1978
from 1976:	Kristallisation, Schmelzen und Quellen von Polymeren mit Störungen im regelmäßigen Aufbau der Moleküle	E1	
H. G. Zachmann	Untersuchung der Abhängigkeit des Dehnungs- und Bruchverhaltens von der Kristallinität, Verstreckung und Quellung	V/15c	1972–1978
from 1976:	Einfluß der Verstreckung auf die Orientierung und die Kristallisation	E5	
H. G. Zachmann	Anwendung der magnetischen Kernresonanz zur Untersuchung der Struktur der nicht kristallinen Bereiche im teilkristallinen Material sowie der Struktur der Schmelze	V/15d	1973–1978
from 1976:		E9	

Guests

Name	from	working in	from–to
Dr. A. Akimotoo		Uni Mainz Org. Chemie	1980/81
Dr. K. Aliev		Uni Mainz Org. Chemie	1976
B. Sc. P. Amareshwar		Uni Mainz Phys. Chemie	1980/81
Dr. J. E. Anderson		Uni Mainz Phys. Chemie	1976
Dipl.-Chem. J. Aramon-Stein	Barcelona	TH Darmstadt	1974
Dr. Steve Arnold		Uni Mainz Org. Chemie	1984
B. Sc. M. Avram-Waganoff	Costa-Rica	TH Darmstadt	1974
Dr. H. Balard		TH Darmstadt	1976
Dr. A. Banihashemi	Teheran	TH Darmstadt II	1974/87
Dipl.-Chem. E. Bezdadea		TH Darmstadt	1976
Dr. Madhusudan M. Bhagwat		Uni Mainz Org. Chemie	1980/81
Dr. A. Bledzki		TH Darmstadt	1976
Dr. W. Borsig		TH Darmstadt	1971
Dr. C. Boudevska	Sofia	Dt. Kunststoff-Institut	1974/75
Dr. M. G. Brereton	Univ. of Leeds	Uni Mainz Phys. Chemie	1979,82,84
Prof. C. D. Callihan	Baton-Rouge/ USA	TH Darmstadt	1973

Prof. Dr. L. A. Carpino	Univ. of Massachusetts, Amherst	Uni Mainz Org. Chemie	1975
Prof. Dr. I. C.W. Chien	Univ. of Massachusetts, Amherst	Uni Mainz Org. Chemie	1983
Dozent Dr. J. Coupek	Prag	Uni Mainz Org. Chemie	1969/70
Dr. J. Cosp		TH Darmstadt	1973
Dipl.-Chem. D. Day		Uni Mainz Org. Chemie	1976
Dr. J. Dayantis		Uni Mainz Phys. Chemie	1976
Dr. Praunil C. Deb		Uni Mainz Phys. Chemie	1982
Dipl.-Ing. M. Deselnicu	Romania	TH Darmstadt	1972
F. Drislane		Uni Mainz Phys. Chemie	1974
Dr. M. Doherty	Univ. of Florida	Uni Mainz Phys. Chemie	1984
Dr. S. Fakirov	Sofia	Uni Mainz Phys. Chemie	1971–1974/78
Prof. Dr. J. Furukawa	Science Univ. Tokyo	Uni Mainz Org. Chemie	1976
Prof. Dr. G. Glöckner	TU Dresden	Uni Mainz Phys. Chemie	1985
Prof. Dr. J. E. Guillet	Univ. of Toronto	Uni Mainz Phys. Chemie	1981
Prof. Dr. Hogen-Esch	Gainsville/ Florida	Uni Mainz Phys. Chemie	1982
Dr. S. Hayashi	Ueda/Japan	Dt. Kunststoff-Institut	1976
Dr. T. Hirano		Uni Mainz Org. Chemie	1976
Dr. Y. Hiragi	Japan	Uni Mainz Phys. Chemie	1975/76
Dr. D. Holden		Uni Mainz Org. Chemie	1982
R. A. Tianbao Huang		Uni Mainz Phys. Chemie	1982
Prof. Dr. Norio Ise	Kyoto Univ.	Uni Mainz Org. Chemie	1980/81
Dr. Bing-Zheng Jiang		Uni Mainz Phys. Chemie	1980–1982
Dr. F. Jones	Keele/England	Uni Mainz Org. Chemie	1970/71
Dr. K. Kaji		Uni Mainz Phys. Chemie	1976
Dipl.-Phys. A. Kalepki		Uni Mainz Phys. Chemie	1976
Dr. M. Kato		Uni Mainz Org. Chemie	1974
Prof. Dr. T. Kawai	Tokyo Institute of Technology	Uni Mainz Org. Chemie	1978
Dr. T. Kelen	Budapest	Uni Mainz Org. Chemie	1969/70
Dr. Jedrzej Kielkiewicz	Univ. Warschau	Uni Mainz Org. Chemie	1981/82
Dr. J. Kiji	Osaka	TH Darmstadt	1971/72
Prof. Dr. E. J. Klabunovski	Moskau	TH Darmstadt	1972
Prof. Dr. W. Klein	Boston Univ.	Uni Mainz Physik	1984
Dr. M. Kloczewiak		TH Darmstadt	1974/75
Dr. Fevzi Köksal		Uni Mainz Phys. Chemie	1980/81
Prof. Dr. R. Koningsveld	Netherlands	Uni Mainz Phys. Chemie	1984
B. Sc. K. Krishna		Uni Mainz Phys. Chemie	1980/81
Prof. Dr. S. Krozer		TH Darmstadt	1972–1974

Prof. Dr. R.W. Lenz	Univ. of Massachusetts, Amherst	Uni Mainz Org. Chemie	1972/73
Dr. G. R. Lescano		Uni Mainz Org. Chemie	1983
Dr. E. Lillie	Belfast	TH Darmstadt	1972/73
Dr. E. Macchi		Uni Mainz Phys. Chemie	1976
Dr. Eugen Maderek	Warschau	Uni Mainz Phys. Chemie	1980/81
Dr. K. H. Mahabadi		Uni Mainz Phys. Chemie	1976
Dr. J. Majnusz		Uni Mainz Org. Chemie	1971
Dr. D. Marr-Leisy		Uni Mainz Org. Chemie	1983
Dr. R. Mateva		Uni Mainz Org. Chemie	1971
Dr. R. Matewa		Uni Mainz Phys. Chemie	1972/73
Prof. Dr. M. Miura		Uni Mainz Org. Chemie	1980/81
Prof. Dr. G. Montaudo	Univ. di Catania/Italy	Uni Mainz Phys. Chemie	1980
Prof. Dr. P. Munk	Univ. of Texas/Austin	Uni Mainz Phys. Chemie	1979/80
Dr. Jagannath Nath		Uni Mainz Org. Chemie	1980/81
Prof. Dr. E.W. Neusse		Uni Mainz Org. Chemie	1976
Dr. J. Nobbs		Uni Mainz Phys. Chemie	1976
Dr. S. P. Nunes	Brazil	Uni Mainz Phys. Chemie	1984
Dr. Hiroyuki Ohmo	Kyoto	Uni Mainz Org. Chemie	1982
Prof. Dr. K. F. O'Driscoll	Univ. of Waterloo/ Canada	Uni Mainz Org. Chemie	1976/77
Dr. S. Orban	Budapest	Dt. Kunststoff-Institut	1976
Dr. T. Pakula		Uni Mainz Phys. Chemie	1976
Prof. Dr. St. Penczek	Polish Academy of Science, Lodz	Uni Mainz Org. Chemie	1980/81
Dr. B. Pion	La Plata	Uni Mainz Org. Chemie	1982/83
Prof. Dr. S. Potnis	Bombay	Uni Mainz Org. Chemie	1969/70
Dr. Monty Reichert		Uni Mainz Org. Chemie	1984
Dr. C. Rentsch		Uni Mainz Org. Chemie	1975/76
Dr. R. Richards	Univ. of Salford	Dt. Kunststoff-Institut	1974/75
Dipl.-Ing. A. Rodriguez	Gualdajara/ Mexico	Eiweiß und Leder	1974
Prof. Dr. S. Samsonija		Uni Mainz Org. Chemie	1972
Dr. T. Sato		Uni Mainz Org. Chemie	1976
Dr. E. Schofield		Uni Mainz Org. Chemie	1976
Prof. Dr. J. M. Schultz	Newark/ Delaware	Uni Mainz Phys. Chemie	1975/82/83
Dr. H. Schuttenberg	La Plata	TH Darmstadt	1970/71
Dr. I. Seganov		Uni Mainz Phys. Chemie	1973
Prof. Dr. V. Shibeav		Uni Mainz Org. Chemie	
Dr. M. Shimomura		Uni Mainz Org. Chemie	1982
Dr. S. Slomkowski		Uni Mainz Org. Chemie	1976
Dr. J. Speakmann	Glasgow	Uni Mainz Org. Chemie	1970

Prof. Dr. J. F. Sproviero	Buenos Aires	TH Darmstadt	1972
Dip.-Ing. Stolarzewicz	Zabrze/Poland	TH Darmstadt	1971
Dr. S. Subramanian		Uni Mainz Phys. Chemie	1980/81
Dr. P. Tørmälä		Dt. Kunststoff-Institut	1976
Prof. Dr. N.W. Tschoegl		Uni Mainz Phys. Chemie	1976
Prof. Ch. B. Tsvetanov		Uni Mainz Phys. Chemie	1982
Dr. B. Turcsanyi	Budapest	TH Darmstadt	1970/71
Dr. R. Turner	Gainesville/ Florida	TH Darmstadt	1971/72
Dr. Sc. V. Váskowá	Academy of Sciences Prag	Uni Mainz Phys. Chemie	1983
Dr. L. Wloinski		Uni Mainz Org. Chemie	1976
Dr. H. Yamaguchi	Osaka	TH Darmstadt	1973/74
Prof. G. S.Y. Yeh	Univ. of Michigan	Uni Mainz Phys. Chemie	1984
Prof. Dr. Hyuk Yu	Univ. of Wisconsin	Uni Mainz Phys. Chemie	1984
T. Zowade		Uni Mainz Org. Chemie	1971

Promotion of Young Scientists

Year	Habilitationen	Dissertationen	Diplom-Arbeiten
1969		17	4
1970		20	15
1971	1	24	21
1972	2	24	23
1973	2	26	25
1974		29	18
1975	1	32	26
1976	1	26	12
1977		48	33
1978/79		16	21
1980/81	1	24	35
1982		13	18
1983	1	15	16
1984/85		35	30
1986/87		24	16

Funding

The Sonderforschungsbereich 41 was supported by grants of the Deutsche Forschungsgemeinschaft totalling DM 54 069 400 in the period 1969–1987.

Erschienene und in Vorbereitung befindliche Abschlußbücher der Sonderforschungsbereiche

Sonderforschungsbereiche 1969–1984
Bericht über ein Förderprogramm der Deutschen Forschungsgemeinschaft
Herausgegeben von Karl Stackmann und Axel Streiter
1985. XII, 552 Seiten mit 124 Abbildungen und 16 Tabellen.
Gebunden. ISBN 3-527-27701-3

Ökonomische Prognose-, Entscheidungs- und Gleichgewichtsmodelle
Ergebnisse aus dem gleichnamigen Sonderforschungsbereich
der Universität Bonn
Herausgegeben von Wilhelm Krelle
1986. XII, 447 Seiten mit 40 Abbildungen und 12 Tabellen.
Gebunden. ISBN 3-527-27702-1

Atmosphärische Spurenstoffe
Ergebnisse aus dem gleichnamigen Sonderforschungsbereich.
Johann-Wolfgang-Goethe-Universität Frankfurt, Johannes-Gutenberg-Universität Mainz, Max-Planck-Institut für Chemie Mainz 1970–1985.
Herausgegeben von Ruprecht Jaenicke
1987. VI, 443 Seiten mit 207 Abbildungen, davon 44 in Farbe, und 32 Tabellen.
Gebunden. ISBN 3-527-27703-X

Technik der Wärmekraftwerke
Beiträge zur Kernkraftwerksforschung
Ergebnisse aus dem Sonderforschungsbereich „Wärmekraftwerk"
der Universität Stuttgart
Herausgegeben von Rudolf Quack und Jakob Wachter (†)
1987. X, 276 Seiten mit 112 Abbildungen und 10 Tabellen.
Gebunden. ISBN 3-527-27704-8

Flexibles Fertigungssystem
Beiträge zur Entwicklung des Produktionsprinzips
Ergebnisse aus dem Sonderforschungsbereich „Fertigungstechnik" der Universität Stuttgart
Herausgegeben von Karl Tuffentsammer (†), Alfred Storr, Kurt Lange, Günter Pritschow, Hans-Jürgen Warnecke
Bearbeitet von Manfred Berger
1988. XXXIV, 362 Seiten mit 164 Abbildungen, davon 5 in Farbe, und 8 Tabellen.
Gebunden. ISBN 3-527-27706-4

Processing Structures for Perception and Action
Final Report of the Sonderforschungsbereich „Kybernetik" 1969–1983
Herausgegeben von Hans Marko, Gert Hauske und Albrecht Struppler
1988. XVII, 278 Seiten mit 124 Abbildungen und 3 Tabellen.
Gebunden. ISBN 3-527-27705-6

Fortschritte in der Semantik
Ergebnisse aus dem Sonderforschungsbereich 99 „Grammatik und sprachliche Prozesse" der Universität Konstanz
Herausgegeben von Arnim von Stechow und Marie-Theres Schepping
1988. X, 233 Seiten mit 8 Abbildungen.
Gebunden. ISBN 3-527-17023-5 (Acta humaniora)

Mensch, Arbeit und Betrieb
Beiträge zur Berufs- und Arbeitskräfteforschung. Ergebnisse aus dem Sonderforschungsbereich „Theoretische Grundlagen sozialwissenschaftlicher Berufs- und Arbeitskräfteforschung" der Universität München
Herausgegeben von Karl Martin Bolte
1988. VI, 328 Seiten mit 7 Abbildungen und 1 Tabelle.
Gebunden. ISBN 3-527-17025-1 (Acta humaniora)

Pathomechanismen entzündlicher rheumatischer Erkrankungen bei Mensch und Tier
Ergebnisse aus dem Sonderforschungsbereich „Pathomechanismen der rheumatoiden Entzündung bei Mensch und Tier" der Medizinischen und der Tierärztlichen Hochschule Hannover
Herausgegeben von Helmuth Deicher
1989. VII, 379 Seiten mit 111 Abbildungen und 52 Tabellen.
Gebunden. ISBN 3-527-27708-0

Sicherheit und Wirtschaftlichkeit großer und schneller Handelsschiffe
Ergebnisse aus dem Sonderforschungsbereich „Schiffstechnik und Schiffbau“
Herausgegeben von Otto Geisler und Harald Keil
1989. XII, 453 Seiten mit 231 Abbildungen und 12 Tabellen.
Gebunden. ISBN 3-527-27710-2

Biological Signal Processing
Final Report of the Sonderforschungsbereich „Biologische Nachrichtenaufnahme und -verarbeitung“. Grundlagen und Anwendung 1972–1986
Herausgegeben von Hans Christoph Lüttgau und Reinhold Necker
1989. XII, 200 Seiten mit 75 Abbildungen und 1 Tabelle.
Gebunden. ISBN 3-527-27709-9

20 Jahre Sonderforschungsbereiche
Herausgegeben von Axel Streiter
1989. VIII, 141 Seiten mit 7 Abbildungen und 12 Tabellen.
Broschur. ISBN 3-527-27713-7

Satellitengeodäsie
Ergebnisse aus dem gleichnamigen Sonderforschungsbereich der Technischen Universität München
Herausgegeben von Manfred Schneider
1990. XIV, 458 Seiten mit 120 Abbildungen und 59 Tabellen.
Gebunden. ISBN 3-527-27712-9

Fortschritte in der Psychiatrischen Epidemiologie
Ergebnisse aus dem Sonderforschungsbereich „Psychiatrische Epidemiologie“
Herausgegeben von Martin H. Schmidt
1990. X, 265 Seiten mit 29 Abbildungen und 36 Tabellen.
Gebunden. ISBN 3-527-27714-5

Leicht und Weit. Zur Konstruktion weitgespannter Flächentragwerke
Ergebnisse aus dem Sonderforschungsbereich 64 „Weitgespannte Flächentragwerke“ der Universität Stuttgart
Herausgegeben von Günther Brinkmann
1990. XII, 335 Seiten mit 214 Abbildungen.
Gebunden. ISBN 3-527-27711-0

Wachstum und Differenzierung von Zellen
Molekularbiologische, zellbiologische und energetische Aspekte
Ergebnisse aus dem Sonderforschungsbereich „Zellenergetik und Zelldifferenzierung" der Universität Marburg (1973–1988)
Herausgegeben von Friedhelm Schneider und Wolfgang Wesemann
1991. XV, 330 Seiten, 91 Abbildungen und 17 Tabellen.
Gebunden. ISBN 3-527-27716-1

Von der Quelle zur Karte
Abschlußbuch des Sonderforschungsbereichs
„Tübinger Atlas des Vorderen Orients"
Herausgegeben von Wolfgang Röllig
1991. XIII, 286 Seiten, 46 Abbildungen und 6 Tabellen.
Gebunden. ISBN 3-527-17027-8 (Acta humaniora)

Qualitätskriterien der Versuchstierforschung
Ergebnisse aus dem Sonderforschungsbereich „Versuchstierforschung" der Tierärztlichen Hochschule Hannover
Herausgegeben von Klaus Gärtner
1991. Ca. 400 Seiten, ca. 94 Abbildungen und ca. 44 Tabellen.
Gebunden. ISBN 3-527-27717-X